国家鲆鲽类产业技术体系
年度报告
（2016）

国家鲆鲽类产业技术研发中心　编著

中国海洋大学出版社
·青岛·

图书在版编目(CIP)数据

国家鲆鲽类产业技术体系年度报告. 2016/国家鲆鲽类产业技术研发中心编著. —青岛: 中国海洋大学出版社, 2017. 9

ISBN 978-7-5670-1589-0

Ⅰ. ①国… Ⅱ. ①国… Ⅲ. ①鲆科—海水养殖—研究报告—中国—2016 ②鲽科—海水养殖—研究报告—中国—2016 Ⅳ. ① S965. 399

中国版本图书馆 CIP 数据核字(2017)第 232037 号

出版发行　中国海洋大学出版社
出 版 人　杨立敏
社　　址　青岛市香港东路 23 号
邮政编码　266071
网　　址　http://www.ouc-press.com
电子信箱　dengzhike@sohu.com
订购电话　0532-82032573(传真)
责任编辑　姜佳君　邓志科
电　　话　0532-88334466
印　　制　日照报业印刷有限公司
版　　次　2017 年 12 月第 1 版
印　　次　2017 年 12 月第 1 次印刷
成品尺寸　185 mm × 260 mm
印　　张　28. 25
字　　数　550 千
印　　数　1—800
定　　价　60. 00 元

国家鲆鲽类产业技术体系2016年工作亮点

图1 鲆鲽类产业发展研讨会暨中国水产流通与加工协会大菱鲆分会成立

图2 “鲆鲽类工业化循环水养殖标准体系建设”研讨会

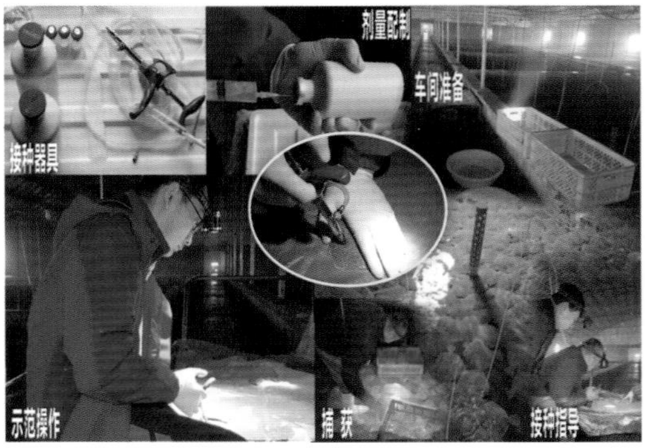

图3 疫苗生产性免疫接种操作规程培训及应用示范推广

图4 央视等媒体记者团现场调研

图5 体系共举办产业技术调研、技术咨询、培训会76次

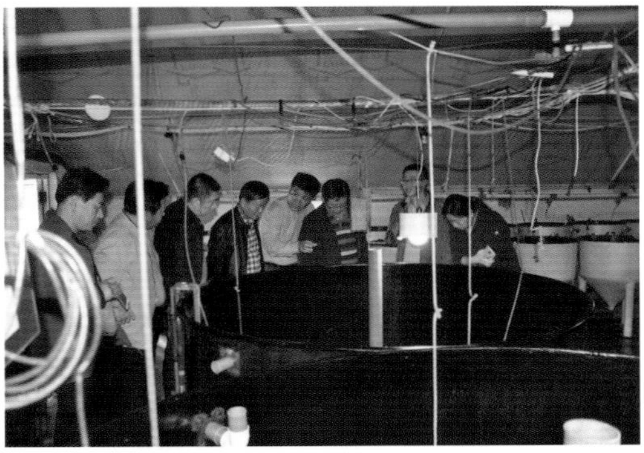

图6 体系共完成阶段性成果现场验收5项

图7 产业经济调研跟踪

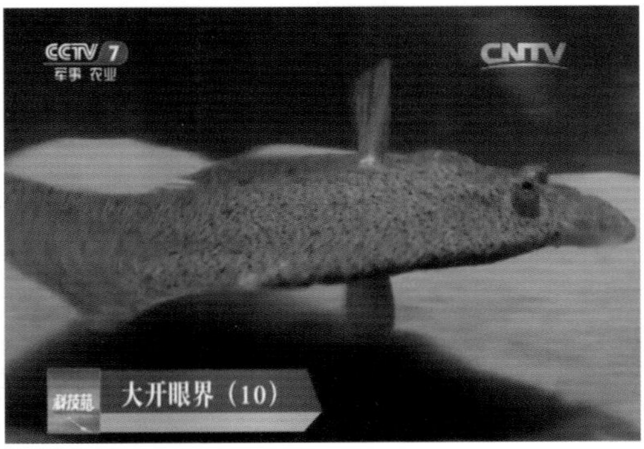

图8 注重宣传鲆鲽类产业技术体系特色与文化

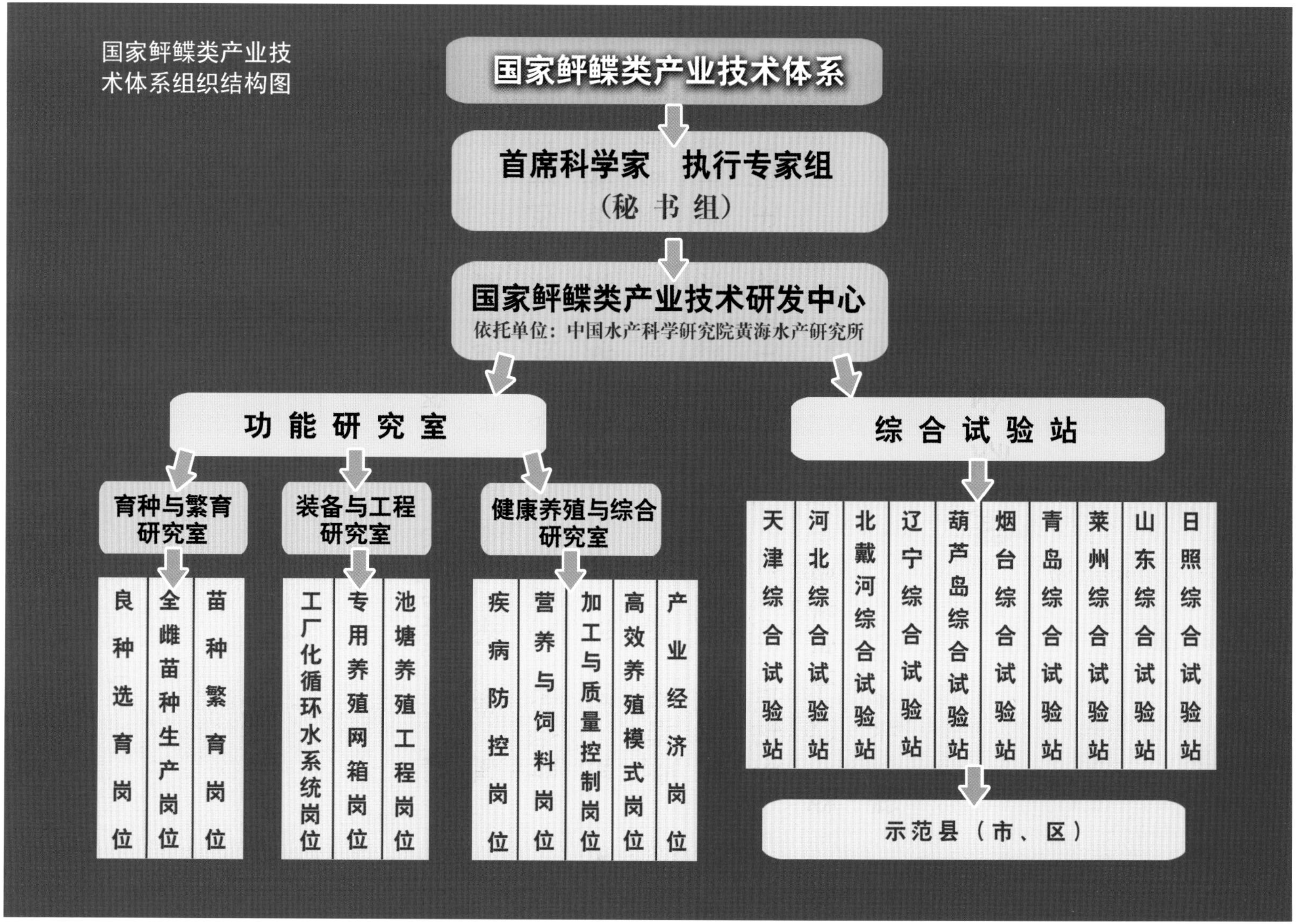
国家鲆鲽类产业技术体系组织结构图
国家鲆鲽类产业技术体系
首席科学家　执行专家组
（秘 书 组）
国家鲆鲽类产业技术研发中心
依托单位：中国水产科学研究院黄海水产研究所
功能研究室
育种与繁育研究室
良种选育岗位
全雌苗种生产岗位
苗种繁育岗位
装备与工程研究室
工厂化循环水系统岗位
专用养殖网箱岗位
池塘养殖工程岗位
健康养殖与综合研究室
疾病防控岗位
营养与饲料岗位
加工与质量控制岗位
高效养殖模式岗位
产业经济岗位
综合试验站
天津综合试验站
河北综合试验站
北戴河综合试验站
辽宁综合试验站
葫芦岛综合试验站
烟台综合试验站
青岛综合试验站
莱州综合试验站
山东综合试验站
日照综合试验站
示范县（市、区）

前　言

鲆鲽类，包括鲆科、鲽科、鳎科等鱼类，属于海洋底层鱼类中经济价值较高的一个大类。因其分类地位特殊、生物多样性丰富、生态分布广泛、口感良好，以及营养价值较高等特点，一直是国际上重要的捕捞和养殖对象，深受国内外研究者与消费者的喜爱。我国鲆鲽类的繁育与养殖研究始于20世纪50年代末，而鲆鲽类养殖业则兴起于20世纪90年代初，现已形成了以大菱鲆（又名多宝鱼）、褐牙鲆、半滑舌鳎等为代表的海水鱼类养殖支柱产业。

自2008年现代农业产业技术体系全面启动建设以来，国家鲆鲽类产业技术体系面向产业发展需求，依托优势科研与产业资源，以“四化养殖”（装备工程化、技术精准化、生产集约化、管理智能化）为核心，集成“六大板块”（育种与繁育、养殖模式与工艺、装备与工程、营养饲料、疾病防控、加工与产品质量安全）系统工程，努力实践着我国鲆鲽类工业化养殖的创新理念，研创出累累硕果。到2016年年底，鲆鲽类体系累计研发新技术25项、研制新设备及新产品32项、培育并通过审定新品种3个、授权专利146件、发表论文550余篇、出版专著12部、制定各类标准与技术规范15项，研制出世界首个大菱鲆迟钝爱德华氏菌活疫苗，共获得包括2项国家级二等奖在内的各级科技奖励20余项。相关技术与成果的推广应用，有力地支撑了我国鲆鲽类养殖产业的稳步发展，鲆鲽类养殖最高年产量达到14.05万吨，与体系建设前的2007年产量相比，增长了95.4%，实现新增产值300多亿元，辐射带动稳定就业50 730人，新增从业人员6 930人，在开拓我国全新的海洋产业、保障水产品有效供给、改善国民膳食结构、提供沿海渔民就业机会和繁荣“三农”经济等方面作出了重要贡献。

《国家鲆鲽类产业技术体系年度报告（2016）》由国家鲆鲽类产业技术研发中心编著，“现代农业产业技术体系专项资金（CARS-50）”资助。本书概括了国家鲆鲽类产业技术体系2016年度的研究内容与成果，主要包括鲆鲽类产业技术研发进展、鲆鲽类主产区调研报告、2016年度研究论文选编、轻简化实用技术、获奖或鉴定成果汇编及专利技术简介等。国家鲆鲽类产业技术体系全体岗位科学家、综合试验站团队参与了编写工作，体系秘书组对书稿进行了整合、审阅和补充。

2017年，依据我国现代化农业发展和农业供给侧结构性改革对科技的新需求，“国家鲆鲽类产业技术体系”更名为“国家海水鱼产业技术体系”，由单一的鲆鲽类扩容为整个海水鱼类，并在产业链各环节配齐了岗位设置。新体系将面向海水鱼类养殖产业发展需求，以“生态友好、生产发展、设施先进、产品优质”为目标，继往开来，再创佳绩。

由于编写时间仓促、学科交叉内容较多，书中错误和疏漏之处在所难免，敬请广大读者批评指正并给予谅解。

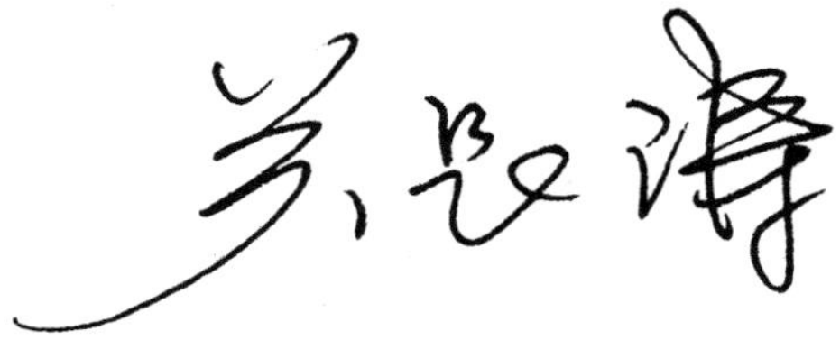

2017 年 7 月 28 日

目　　次

第四篇　轻简化实用技术

第五篇　获奖、鉴定及验收成果汇编

第六篇　专利技术简介

附　录

Contents

Chapter 3 Selected publications in full-text

Chapter 4 Concise practical technology

Chapter 5 Awards and achievements appraised

Chapter 6 Summaries of patents

Appendix

第一篇

鲆鲽类产业技术研发进展

2016年度鲆鲽类产业技术发展报告

国家鲆鲽类产业技术体系

1　国际鲆鲽类生产与贸易概况

1.1　世界鲆鲽类捕捞及养殖情况

据联合国粮农组织(FAO) 2016年数据，2014年世界鲆鲽类主产区总产量1 237 351 t，同比2013年增加0.8%，其中捕捞产量1 042 230 t，同比减少0.05%，养殖产量195 121 t，同比增加8.8%。黄盖鲽、格陵兰庸鲽和欧鲽占主要份额，箭齿鲽产量大增。生产格局基本稳定，主产国为美国、中国、俄罗斯、韩国、日本和印度，各主产国产量增减互现。2016年，欧盟成员国在大西洋北部水域的鲆鲽类捕捞配额总量为264 114 t。西班牙是欧洲地区鲆鲽类养殖规模最大的国家，2014年大菱鲆养殖量7 766.6 t，较2013年增长12.6%，达历年来最高水平。其中，鳎的养殖量大幅增加。2016年，太平洋北部海域太平洋庸鲽捕捞配额约为13 558 t，比2015年增加2.3%，北太平洋阿留申群岛附近水域纳入配额制体系的鲆鲽类总配额为360 602 t，与2015年持平，大西洋西北部海域格陵兰庸鲽和黄尾鲽配额为27 966 t，较2015年减少577 t。2015年，日本鲆鲽类产量5.13万吨，较2014年减少6.6%。其中捕捞量4.88万吨，同比减少6.7%，包括牙鲆0.8万吨，鲽鱼(黄盖鲽、格陵兰庸鲽、欧鲽等)4.08万吨；养殖量0.25万吨，同比减少3.8%，主要为牙鲆。2000年以来日本牙鲆养殖量持续下滑，占其海水鱼类养殖总量的比重也逐渐下降。据韩国统计网站数据，2015年上半年，韩国养殖牙鲆产量同比增加，产值同比减少，平均价格同比下跌21.1%，养殖面积233.7万平方米，同比增加15.7万平方米，增幅7.2%，饲料使用量107 923 t，同比增加22.0%。

1.2　世界鲆鲽类贸易情况

2016年，主要出口国的鲆鲽类出口贸易变化各不同。出口排名第二、三位的美国和丹麦的鲆鲽类出口规模扩大，出口额同比分别增加了7.0%和13.5%(丹麦为上半年的数据)，韩国鲆鲽类出口额增加5.0%。出口排名前十位的冰岛、加拿大和西班牙的鲆鲽类出口规模出现不同程度萎缩，出口额同比分别下降4.6%、8.2%和11.2%(加拿大和西班牙为上半年的数据)。除了冰岛，其他国家鲆鲽类出口均价变化不大。冰岛的鲆鲽类出口均价普遍回落，其主要出口品种格陵兰庸鲽和欧鲽的出口均价同比下跌约10%。

2016年,主要进口国的鲆鲽类进口贸易规模略有缩小。进口排名第二的日本进口规模萎缩,美国和韩国的进口量同比分别减少1.6%和2.9%。上半年,欧盟国家的鲆鲽类贸易活跃度增加,加拿大的鲆鲽类产品消费旺盛,进口排名第四、五、六位的意大利、加拿大和西班牙的鲆鲽类进口规模扩大,进口额同比分别增加了13.5%、52.5%和8.7%。

2016年,世界主要市场鲆鲽类产品的集散情况表现各异。在产地国(如西班牙)销量基本稳定,在其他国家销量与2015年相比增幅较大。价格方面,2016年总体高于2015年。鲜舌鳎、鲜大菱鲆、牙鲆和大西洋比目鱼的价格呈震荡上升态势,鲜养殖大菱鲆、鲜野生大菱鲆、黄盖鲽、帆鳞鲆、欧鲽、柠檬鲽、大菱鲆和川鲽的价格呈震荡下降态势,冷冻舌鳎、冷冻舌鳎鱼片、冷冻大菱鲆和其他鲆鲽类产品的价格平稳。亚洲市场的销量与去年同期相比显著下降,价格波动较大。特别是鲽鱼,销量波动大,总体呈下降态势。

受欧元区危机影响,挪威部分出口至西班牙、意大利和葡萄牙的海产品价格走低。贸易商表示在欧元区销售海产品,尤其鱼类产品,变得越来越难,其价格也处于下滑趋势。帆鳞鲆对西班牙的出口价格一直波动。在欧洲,西班牙、意大利和葡萄牙是三大海产品消费国,但货币贬值导致苏格兰海产品出口受阻,在西班牙、意大利、希腊和葡萄牙的出口销售严重减少,欧债危机改变了欧元区消费者的消费习惯,尤其在西班牙,民众在减少外出就餐的同时加大了在超市购买海产品的频率和数量,捕捞量难以满足消费者的需求。在西班牙,人们对鱼类消费传统非常看重,消费者更喜欢鲜海产品。法国的水产品消费生鲜品占33%,加工冰藏品占31%。在英国,加工冷冻水产品消费约占水产品总消费量的50%。

2 国内鲆鲽类生产与贸易概况

2.1 国内鲆鲽类养殖生产情况

根据国家鲆鲽类产业技术体系各综合试验站调查数据,2016年跟踪示范区县鲆鲽类工厂化养殖面积为721.9万平方米,与2015年同比下降3.8%,与2014年同比基本持平。网箱养殖面积为24.0万平方米,与2015年同比下降26.2%,与2014年同比降幅为33.4%。池塘养殖面积为4 490亩[①],与2015年同比下降58.4%,与2014年同比降幅为47.43%。三大主要养殖品种中,大菱鲆、牙鲆、半滑舌鳎的工厂化养殖面积分别为601.5万平方米、31.6万平方米、87.5万平方米,占工厂化养殖总面积的比重分别为83.3%、4.4%及12.1%,合计占鲆鲽类工厂化养殖总面积的比重为99.8%。2016年大菱鲆工厂化养殖面积与2015年同比下降7.5%,与2014年同比降幅为2.4%。牙鲆养殖面积与2015年同比下降11.4%,与2014年同比下降30.8%。半滑舌鳎养殖面积与2015年同比增长45.2%,与2014年同比基本持平。2016年体系跟踪示范区县鲆鲽类产品总产量为5.5万吨,与2015年同比下降8.4%,与2014年同比降幅为26.5%。其中,大菱鲆产量为4.5万吨,占年度鲆鲽类产

① 亩为非法定单位,考虑到生产实际,本书继续保留。1亩 ≈ 666.7 m^2。

品总产量的比为 81. 8%，与 2015 年同比下降 5. 9%，与 2014 年同比降幅为 24. 8%；牙鲆产量为 4 036. 3 t，占总产量的比为 7. 3%，与 2015 年同比下降 54. 2%，与 2014 年同比降幅为 59. 2%；半滑舌鳎产量为 5 945. 7 t，占总产量的比为 10. 8%，与 2015 年同比增长 82. 4%，与 2014 年同比增长 24. 3%。

2.2　国内鲆鲽类贸易情况

2016 年，我国鲆鲽类进出口额增量减少。2016 年，鲆鲽类进出口总额 79 923. 03 万美元（同比增加 2. 93%），进出口总量 261 949. 60 t（同比减少 1. 06%）。其中，出口额 47 145. 88 万美元（与 2015 年持平），进口额 32 777. 15 万美元（同比增加 7. 49%），出口量 100 315. 48 t（同比减少 5. 08%），进口量 161 634. 12 t（同比增加 1. 60%），进出口总额和进出口总量分别占我国同期水产品进出口总体的 2. 65%和 3. 16%。出口额占该项进出口总额的 58. 99%，进口量占该项进出口总量的 61. 70%。综合冻比目鱼类的进出口规模占统计最大项。加工冷冻比目鱼片的出口规模最大，占鲆鲽类出口总额的 74. 92%，出口总量的 73. 51%，出口额和出口量同比分别减少 8. 10%和 8. 15%。作为单种类，冻格陵兰庸鲽进出口规模锐减：进口额（6 519. 02 万美元）和进口量（13 580. 70 t）同比分别减少 18. 12%和 30. 12%；出口额（2 402. 43 万美元）和出口量（3 462. 13 t）同比分别增加了 6. 34%和减少了 14. 22%。

我国鲆鲽类进出口贸易涉及 62 个国家和地区，其中，出口 49 个国家和地区，进口来自 29 个国家和地区。出口额规模前 10 位的国家和地区依次是日本、美国、加拿大、中国台湾省、韩国、荷兰、法国、波兰、西班牙和德国，出口该 10 个方向的金额合计占该项出口总额的 86. 99%，出口量占该项总量的 84. 82%。同比 2015 年，对美国、中国台湾省和西班牙的出口额大增，特别是对中国台湾省和西班牙的出口规模，金额和数量同步大增。对日本、韩国、法国、波兰和荷兰的出口额减少，其中，对日本和韩国的出口量同比分别减少了 79. 9%和 74. 8%。

同期，全国有 16 个省市有鲆鲽类产品进出口记录。出口有辽宁、山东、福建、吉林、浙江、广东、上海和河北等 8 个省市口岸，与 2015 年比较，除上海口岸出口锐减外，其他均呈现增加态势，而且增幅巨大。出口仍集中在辽宁和山东。辽宁的鲆鲽类产品出口规模占总体的 50%以上，出口额和出口量分别占总体的 53. 48%和 60. 92%，山东的鲆鲽类产品的出口额和出口量分别占总体的 30. 54%和 25. 03%。从进口来看，主要集中在辽宁、山东和吉林。同比 2015 年，除了上海、北京、福建和江苏进口额锐减外，其他多数省市的进口额均增加，特别是辽宁、山东、吉林、浙江、湖北和黑龙江的进口额同比 2015 年增幅巨大，进口量也普遍同步增长。其中，辽宁的鲆鲽类产品进口规模仍很大，进口额和进口量分别占总体的 51. 58%和 67. 24%。细分贸易方式，一般贸易进出口规模显著增长，来料加工和进料加工形式金额增加明显。

2016 年，我国的养殖大菱鲆出口规模显著扩大，海外市场的开拓力度明显增强。我国养殖大菱鲆(冷冻)出口额 81.60 万美元,出口量 77.68 t,主要出口马来西亚、俄罗斯和加拿大,第三季度开拓了乌克兰市场。平均价格 10.50 美元/千克。其中,福建出口 36.85 t,辽宁出口 33.43 t,广东出口 7.40 t。对马来西亚的出口量比 2015 年全年多 3 倍,平均单价翻倍。

3 国际鲆鲽类产业技术研发进展

3.1 育种与繁育技术

3.1.1 遗传育种

2016 年度,国外对鲆鲽鱼类主要品种的遗传改良,仅对大菱鲆的研究取得了一定进展,对其他鲆鲽鱼类选育的研究尚未见报道。西班牙维戈高等科学研究委员会海洋研究所、西班牙圣地亚哥大学和纽约基因组中心等 12 所大学及研究所联合开展了大菱鲆全基因组的测序研究,为大菱鲆分子标记辅助育种提供有价值的信息。西班牙圣地亚哥•德•孔波斯特拉大学、西班牙农业和食品技术研究所动物育种系、西班牙国家研究委员会海洋研究所和法国国家农业研究所等 8 所大学及研究所为促进育种计划,共同完成了大菱鲆基因组资源研究。

3.1.2 繁育技术

目前在国际上针对重要养殖鲆鲽类苗种人工繁育技术的工艺流程已经建立,为了提升优质苗种生产技术工艺,实现鲆鲽类苗种产业可持续发展,通过加强基础研究来实现对优质苗种生产技术精准调控,成为当前国际鲆鲽类苗种繁育产业技术研发的一个重要趋势。2016 年国际鲆鲽类苗种繁育基础研究主要集中在大菱鲆、塞内加尔鳎、牙鲆、星斑川鲽、大西洋庸鲽、半滑舌鳎、欧鳎等重要鲆鲽类养殖品种。相关学者就大菱鲆苗种的早期发育、性别决定、营养强化、应急胁迫等方面展开了研究,取得了一系列的研究成果。这些研究为大菱鲆、塞内加尔鳎、牙鲆、星斑川鲽、大西洋庸鲽、半滑舌鳎、欧鳎等养殖品种苗种培育过程中的营养精准强化调整、早期发育规律、苗种培育设施改进等研究提供了理论依据。

3.2 养殖模式与工程技术

3.2.1 工厂化养殖

重点关注活鱼起捕分级技术研究。20 世纪五六十年代国外开始对吸鱼泵技术进行研究,荷兰 KUBBE 公司研制了真空吸气卸鱼装置,美国马可公司研制了离心式潜水吸鱼泵,1988 年美国 ETI 公司研发了 SILKSTREAM 射流吸鱼泵等。至今,这些设备仍在水产养殖发达国家被广泛使用。

3.2.2 活鱼分级

活鱼分级主要有箱式分级装置、板式分级机、分级槽、柔性分级网和池内水平杆分级机等。最近的研究报道是关于美国研发的气提起捕分级装备技术,该技术利用气力提升装置将活鱼抽吸出养殖池，再通过格栅进行分级处理。

3.2.3　海水网箱技术

挪威继续引领深海养殖技术发展，2016年先后报道了其研制的“巨蛋”养殖设施，中国承建的养殖水体达25万立方米、7 000多万美元的半潜式深海养鱼场，以及正在设计的高度自动化的深海养殖工船等。养殖水域深海化、养殖设施大型化、养殖操作自动化、生产管理信息化是深远海养殖工程与科技发展的主要趋势。

3.2.4　池塘养殖技术研发

孟加拉国科研人员利用池塘养殖代谢副产物(底泥、粪便、残饵等)养殖一种可用作奶牛饲料的牧草，有效解决了池塘养殖底质容易恶化的问题，还获得了牧草收益。法国科学家提出在池塘养殖系统中增加水草养殖来调控水环境和小生境生态平衡。还有学者研究了接力养殖过程中鱼类生理适应机制与高效生长调控技术。另外，池塘“鱼菜共生”养殖模式研究方面，研究人员在水生植物/蔬菜选择、养殖生长调控、水质净化效果等方面开展研究，但在海水池塘“鱼菜共生”研究方面进展不多。

3.3　疾病防控技术

2016年国际上公开报道的鲆鲽类疾病防控技术主要集中于黏着杆菌、美人鱼发光杆菌、黏孢子虫和病毒性出血性败血症病毒(VHSV)等细菌、病毒和寄生虫病原。葡萄牙和埃及联合报道了鳎鱼重要病原鱼黏着杆菌致病机制方面的研究进展。西班牙报道了大菱鲆病毒性出血性败血症病毒的研究现状和塞内加尔鳎对美人鱼发光杆菌免疫机制的研究成果。日本和韩国报道了牙鲆和大菱鲆黏孢子虫病病原。西班牙圣地亚哥大学开展了大菱鲆抗黏孢子虫及抗纤毛虫药物筛选研究。韩国和挪威研究机构开发了一种牙鲆VHSV活疫苗并进行了实验室评价，免疫保护率可达80%以上。在商业化研制方面，国外除早已广泛应用的商业化疫苗，未见新疫苗产品上市的报道。在鱼病防控方面，多价载体疫苗、活疫苗、亚单位疫苗及以环境友好技术为基础的微生态制剂、免疫增强剂等新型水产药物依然是国际渔药界的研制热点。

3.4　营养与饲料技术

国外有关鲆鲽类营养研究的重点主要涉及3个方面：鱼粉替代蛋白源开发、鱼油替代脂肪源开发和饲料添加剂开发。这3方面的研究体现出两个特点：实用性更强，为实际生产中鱼油和鱼粉的替代提供技术指导；对营养代谢机理研究的深入，从分子机制上阐述鱼粉和鱼油替代引起的代谢变化。研究对象包括大菱鲆、牙鲆、半滑舌鳎、塞内加尔鳎、大西洋庸鲽等。在替代蛋白源方面，由于复合蛋白源的氨基酸平衡性更好，实用性更佳，因此，研究聚焦在复合蛋白源替代上。添加剂的研究聚焦在降脂、维护鱼体健康和产品安全等方面，研究对象有植酸酶、益生菌等。此外，2016年，许多研究者将更多的研究聚焦于仔稚鱼和亲鱼研究上，这反映出国内外对鲆鲽类营养研究的重视和深入。

3.5 产品质量安全控制与加工技术

研究发现,大菱鲆在常温加冰条件下的优势腐败菌为气单胞菌、柠檬酸杆菌和哈夫尼氏菌。对于保持水产品鲜度以及延长贮藏时间的研究也有了新的进展:水产品的腐败变质多是微生物活动引起的,微生物在合适的环境下生长繁殖并产生胞外酶等物质,分解食品基质,导致水产品的腐败。最新研究发现,微生物的群体感应现象(QS)会参与到食品的腐败进程当中。已有研究发现,波罗的海希瓦氏菌分泌的一种信号分子——环肽可以加速冷藏大黄鱼的腐败变质。目前最新研究表明,希瓦氏菌可以利用其他细菌分泌的QS信号分子或胞外产物来调节自身菌群优势,因此检测大菱鲆冷藏过程中的微生物信号分子类型并探究其对大菱鲆腐败的影响,可以为大菱鲆保鲜提供新思路。

4 国内鲆鲽类产业技术研发进展

4.1 育种与繁育技术

4.1.1 大菱鲆良种选育

研究了大菱鲆选育新品种的生长特征、大菱鲆微卫星分子标记分析以及大菱鲆转录组解析——繁育、生长及免疫相关基因的发掘和遗传标记的鉴定研究。构建了大菱鲆成鱼多组织混合样本的转录组数据库,全面了解了大菱鲆雌雄基因表达情况,筛选了与生殖、生长和免疫响应相关的候选基因。

4.1.2 牙鲆育种新技术的应用

开展了牙鲆全基因组序列研究、牙鲆淋巴囊肿抗病家系的构建、生长和抗病性能的分析、牙鲆抗迟缓爱德华氏菌性状的遗传评估、雌核发育牙鲆联合快速发育遗传统一性研究、相关性状标记的开发等,同时,开展了全基因组重测序技术研究、牙鲆双单倍体、克隆系的制备,为牙鲆良种克隆化奠定了材料基础。对半滑舌鳎的遗传改良,开展了半滑舌鳎生长相关的微卫星标记及优势基因型研究、生长相关性状的表型和遗传参数研究,在半滑舌鳎微卫星连锁图谱上进行鳗弧菌抗病相关的性状QTL定位研究等。

4.1.3 亲鱼培育等环节的相关技术工艺

目前,国内在鲆鲽类重要养殖品种亲鱼培育、受精卵生产和苗种培育等环节的相关技术工艺体系已基本成熟,能够满足生产需求。近年来相关的应用基础和技术研发主要针对亲鱼培育过程中的光温环境因子调控、营养强化调控、促熟、催产和授精等技术,针对苗种培育过程中的病害预防,降低白化及黑化率、提高生长率和成活率的营养强化技术,以及微藻、光合细菌和中草药制剂使用的环境调控等技术。2016年国内报道了有关大菱鲆、半滑舌鳎、牙鲆、星斑川鲽、圆斑星鲽、钝吻黄盖鲽等鲆鲽类人工繁育的最新研究进展。

4.2　养殖模式与工程技术

4.2.1　养殖模式

开展了大菱鲆工厂化健康养殖技术研究，分析了养殖密度对大菱鲆生长、固有免疫与氧化应激水平的影响，完成了大菱鲆应对关键水质因子胁迫的生理相应研究。

4.2.2　工厂化养殖

开展了综合标准化研究，建立了“鲆鲽类循环水养殖标准体系表”，研发了一种滴淋式臭氧混合吸收塔，进行了初步试用。开展了生物流化床技术化研究，完成了滤器挂膜规律和水处理性能研究以及滤料微生物群落结构分析。

4.2.3　鲆鲽类网箱养殖

国内首次开展了鲆鲽类耐流性能和升降式网箱水动力特性的基础性研究，采用升降式网箱养殖牙鲆，养殖成活率高达 98.9%，同时本年度还设计建造了养殖水体达 6 万立方米的大型浮绳式围网。

4.2.4　池塘养殖

开展了岩礁池塘工程化养殖模式研究，设计了增氧环流系统和集污控制减排系统，并完成了牙鲆高效养殖试验与示范，提高了池塘养殖效率和生态化水平。在大连地区进行了工程化池塘循环水养殖模式推广应用，并对养殖池塘进行了内循环的创新设计。开展了鱼类体色调控机制、生长生理健康评价等研究，支撑了池塘养殖鲆鲽类体色异常和健康生长调控技术构建。

4.3　疾病防控技术

采用基因工程手段获得多株免疫效果显著的海水鱼用减毒活疫苗，免疫保护率达到 80% 以上，主要针对鲆鲽类细菌性病害腹水病（迟钝爱德华氏菌）和弧菌病（鳗弧菌）两种重要病害防治对象。于 2016 年向农业部提交了大菱鲆腹水病弱毒活疫苗 I 类新兽药生产文号申报，即将获批。鳗弧菌减毒活疫苗正在接受农业部中监所新兽药注册证申报复核检验。在抗病毒和寄生虫的多价载体疫苗、菌蜕疫苗以及其他创新疫苗等多种新型鲆鲽疫苗上也展开了大量的临床前研究。在应用基础研究领域，进行了大菱鲆黏膜免疫机制方面的研究和大菱鲆养殖密度对免疫应答影响的研究。

4.4　营养与饲料技术

国内有关鲆鲽类营养研究的重点主要涉及替代蛋白源开发、脂肪代谢机制、添加剂开发以及亲鱼营养研究 3 个方面。研究对象包括大菱鲆、半滑舌鳎和星斑川鲽等。

4.4.1 替代蛋白源研究

2016年的研究聚焦在复合蛋白源替代上，因为复合蛋白源较单一的蛋白源氨基酸平衡性更好，养殖效果明显，实际生产中应用更广泛。

4.4.2 脂肪代谢机制研究

研究了亚麻油替代鱼油对大菱鲆脂肪代谢、免疫以及肝脏中转铁蛋白基因表达的机理。

4.4.3 添加剂研究

聚焦在降脂、保证鱼体健康等方面，研究对象有茶多酚、益生菌和姜黄素等。在鲆鲽类亲鱼营养学研究方面，研究了花生四烯酸大菱鲆亲鱼性类固醇激素的影响以及南极磷虾粉对半滑舌鳎雄性亲鱼繁殖性能的影响，填补了国内在亲鱼营养领域的研究空白。

4.5 产品质量安全控制与加工技术

4.5.1 鲆鲽类加工产品研制

研发并推广了大菱鲆一鱼多吃产品，将整鱼开发成鱼头、鱼块、鱼糜、鱼片、鱼皮等冰鲜产品，并在酒店进行推广应用。

4.5.2 食品安全性评价

通过确立反相高效液相色谱测定嘌呤含量的方法，并对大菱鲆不同部位进行嘌呤含量的测定，建立了持续稳定的高尿酸血症鹌鹑动物模型，研究了尿酸前体物质嘌呤的长期摄入与个体血尿酸水平之间的关系，明确了造成痛风的原因不仅仅是食物中所含嘌呤。

4.5.3 检测样品前处理

制备氧化石墨烯，探究了其对水中氟喹诺酮类药物的吸附和解析特性，并采用高效液相色谱进行了药物浓度的检测研究。

4.5.4 方便食品的研发

分别研究了腊制多宝鱼、半干多宝鱼和即食清蒸多宝鱼方便食品的加工工艺，并对加工过程中的质量安全进行控制，建立了标准化、科学化的菜肴检测技术。

2016年主产区鲆鲽类产业运行分析

产业经济岗位

1　引言

便于业界、管理部门、科研单位等有关部门及人员掌握2016年鲆鲽类产业经济运行情况，以国家鲆鲽类产业技术体系各综合试验站跟踪示范区县调查数据为基础，以产业经济岗位团队的调研数据为补充，撰写出本报告。报告中所指的跟踪调查区域包括辽宁、河北、天津、山东、江苏、福建等省市40多个区县。具体是：辽宁的东港市、庄河市、大连旅顺口区、甘井子区、大洼县、瓦房店市、葫芦岛龙港区、绥中县、兴城市，河北的秦皇岛山海关区、昌黎县、乐亭县、滦南县、黄骅市、唐山丰南区、曹妃甸区，天津的汉沽区、塘沽区、大港区，山东的日照岚山区、日照开发区、日照东港区、荣成市、文登市、乳山市、威海环翠区、昌邑市、龙口市、莱州市、招远市、莱阳市、海阳市、烟台牟平区、烟台开发区、烟台芝罘区、蓬莱市、青岛黄岛区、潍坊滨海区，江苏的赣榆县、如东县，福建的东山县等区县。其中，辽宁、河北、天津、江苏、福建的数据完全采用了各综合试验站提供的数据；山东进行鲆鲽类养殖的区县共有20多个，其中青岛黄岛区、潍坊滨海区、蓬莱市、烟台开发区、烟台芝罘区、海阳市、烟台牟平区、荣成市、日照开发区等9个区县采用了产业经济岗位团队与烟台市水产研究所联合采集的数据，其余区县采用了各综合试验站提供的数据。除特别说明，各指标的数据均包括上述地区。微观层面的数据来自对养殖者的跟踪调查。各综合试验站、烟台水产研究所及有关各方在数据采集中给予了大力帮助和支持，在此一并致以诚挚的谢意！

2　2016年跟踪示范区县鲆鲽类养殖面积变动情况

2.1　不同养殖模式养殖面积变动情况

2.1.1　工厂化养殖面积年内小幅增长，长期看呈下降趋势

跟踪调查数据显示，2016年跟踪示范区县鲆鲽类工厂化养殖面积在637.9万平方米～656.7万平方米之间变动。从季度变动看，年内环比先增后持平。与2015年同比的变动趋势则先下降后持平再增长，即第一季度与2015年同比下降8.4%，第二季度下降5.5%，

第三季度同比基本持平,第四季度同比增长 2. 3%。与 2014 年同比呈下降趋势,降幅均在 10. 0%以下,分别为 1. 2%、0. 5%、3. 4%及 3. 5%。

在此,需要提醒注意的是,虽然 2016 年鲆鲽类工厂化养殖面积长期看呈下降趋势,但鲆鲽类三大主要养殖品种的工厂化养殖面积变动情况有所不同。

2.1.2 山东和辽宁仍为鲆鲽类养殖产业的主要集聚地,与其他地区相比,始终具有明显的比较优势

2016 年山东示范区县鲆鲽类工厂化养殖面积与 2015 年同比呈下降趋势,降幅分别为 13. 5%、12. 5%、4. 1%及 6. 8%。2016 年第四季度,山东鲆鲽类工厂化养殖面积约为 290. 0 万平方米,占本季度鲆鲽类工厂化养殖总面积的比为 45. 0%。

2016 年辽宁示范区县鲆鲽类工厂化养殖面积前三季度总体保持在 200 万平方米左右,同比 2015 年同期(210. 0 万平方米左右)小幅下降。2016 年辽宁第四季度鲆鲽类养殖面积为 220. 0 万平方米,与 2015 年同比增长 8. 7%,占本季度鲆鲽类工厂化养殖总面积的比为 33. 8%。

河北鲆鲽类工厂化养殖面积与 2015 年同比均呈增长趋势。其中第一季度养殖面积为 79. 1 万平方米,同比增长 0. 9%;第二季度养殖面积为 99. 9 万平方米,同比上涨 16. 1%;第三季度养殖面积为 102. 0 万平方米,同比涨幅为 23. 2%;第四季度的养殖面积与第三季度基本持平,与 2015 年同比增长 27. 4%。

2016 年天津鲆鲽类工厂化养殖面积与 2015 年同比呈先下降后增长之态势。第一季度养殖面积为 4. 1 万平方米,与 2015 年同比下降 3. 7%;第二季度养殖面积为 4. 0 万平方米,同比增长 29. 4%;第三季度养殖面积增长至 6. 9 万平方米,环比增长 72. 5%,同比大幅增长 167. 1%;第四季度的养殖面积与第三季度环比基本持平,与 2015 年同比涨幅达 89. 3%。根据跟踪调查的数据可以看出,2016 年第三季度天津鲆鲽类工厂化养殖面积同比大幅增长的主要原因是汉沽区的部分养殖生产者扩增了大菱鲆和半滑舌鳎的养殖生产计划,加大了养殖产品的投苗量。

江苏 2016 年鲆鲽类工厂化养殖面积年内各季度环比小幅下降,与 2015 年同比降幅分别为 3. 5%、8. 1%、12. 9%及 20. 5%。受气候因素影响,与往年一样,福建仅第一季度和第四季度有工厂化养殖,养殖面积分别为 1. 5 万平方米及 3 000 m^2,与 2015 年同比第一季度呈增长趋势,涨幅为 75%,第四季度则下降 53. 3%。

2016 年跟踪示范区县鲆鲽类工厂化养殖面积与 2014 年同比均呈下降趋势,降幅均在 5. 0%以内。从各地区的变动情况看,辽宁鲆鲽类工厂化养殖面积与 2014 年同比先降后增,即前三季度同比下降,第四季度同比增长,但其变动幅度均在 10%以内。山东及江苏的鲆鲽类工厂化养殖面积与 2014 年同比均呈下降趋势。山东前三季度的降幅在 10%以内,第四季度下降 16. 7%。江苏第一季度下降 11. 5%,第二季度降幅为 15. 3%,第三季度下降 19. 4%,第四季度降幅为 26. 1%。河北省的养殖面积与 2014 年同比则均呈增长趋势,涨幅分别为 17. 7%、23. 8%、26. 9%及 36. 6%。天津鲆鲽类工厂化养殖面积与 2014 年同比,第

一季度增长 4. 4%，第二季度下降 19. 2%，第三、第四季度分别增长 65. 1%及 61. 4%。福建的养殖面积与 2014 年第一季度同比增长 73. 0%，第四季度同比下降 81. 6%。

2.1.3　网箱养殖面积年内先增后减再增，与 2015 年同比先下降后增长再减少

2016 年跟踪示范区县鲆鲽类网箱养殖分布于辽宁省、山东省及福建省。第一季度，辽宁省无鲆鲽类网箱养殖生产情况记录，山东省和福建省的养殖面积为 19. 2 万平方米，与 2015 年同比下降 26. 73%。其中，山东省的养殖面积为 12. 0 万平方米，占总养殖面积的比为 62. 6%。第二季度鲆鲽类网箱养殖面积为 27. 1 万平方米，与 2015 年同比增长 37. 4%。辽宁、山东和福建均有养殖，其养殖面积占总养殖面积的比例分别为 16. 6%、73. 8%及 9. 6%。2016 年第三季度，鲆鲽类网箱养殖面积为 22. 0 万平方米，与 2015 年同比增长 29. 4%。其中，辽宁的养殖面积为 2. 0 万平方米，山东的养殖面积为 20. 0 万平方米。第四季度仅山东和福建有网箱养殖，养殖总面积与 2015 年同比下降 13. 0%。

2016 年跟踪示范区县鲆鲽类网箱养殖面积与 2014 年同比前三季度呈增长趋势，其中第一季度大幅增长 148. 6%，第二、第三季度分别增长 34. 6%及 22. 2%，第四季度同比微幅下降 3. 0%。

2.1.4　池塘养殖面积长期看呈下降趋势

2016 年跟踪示范区县仅辽宁省和山东省有鲆鲽类池塘养殖生产记录。第一季度鲆鲽类池塘养殖面积为 2 600 亩，与 2014 年及 2015 年相比均大幅增长，涨幅分别为 225. 0%及 348. 3%。其中，辽宁的养殖面积为 2 000 亩，山东的养殖面积为 600 亩。第二、第三季度，鲆鲽类池塘养殖面积分别为 4 600 亩，与 2015 年同比均呈下降趋势，降幅为 43. 2%及 57. 4%，与 2014 年同比分别下降 42. 4%及 33. 3%。第四季度池塘养殖面积为 1 600 亩，与 2015 年同比下降 32. 8%，与 2014 年同比降幅为 74. 9%。

2. 2　各品种工厂化养殖面积的变动情况及其地区分布

根据体系各综合实验站调查数据显示，2016 年我国鲆鲽类示范区县有工厂化养殖的品种有大菱鲆、牙鲆、半滑舌鳎、星突江鲽及漠斑牙鲆，在往年的跟踪数据中，圆斑星鲽也有少量的工厂化养殖面积。不同品种的工厂化养殖面积构成、区域分布及其变动情况见表 1。

表 1　跟踪示范区县鲆鲽类各养殖品种工厂化养殖面积变动情况　（单位：m^2）

地区	时间	大菱鲆	牙鲆	半滑舌鳎	星突江鲽	圆斑星鲽	漠斑牙鲆	总计
辽宁	2014 年第四季度	2 063 000	8 500	0	0	0	0	2 071 500
	2015 年第四季度	2 038 000	5 500	0	0	0	0	2 043 500
	2016 年第四季度	2 191 000	30 000	0	0	0	0	2 221 000
河北	2014 年第四季度	497 000	177 700	99 500	0	0	0	774 200
	2015 年第四季度	568 500	176 800	85 000	0	0	0	830 300
	2016 年第四季度	575 000	201 500	281 000	0	0	0	1 057 500

(续表)

地区	时间	大菱鲆	牙鲆	半滑舌鳎	星突江鲽	圆斑星鲽	漠斑牙鲆	总计
天津	2014 年第四季度	20 520	11 600	10 250	0	0	0	42 370
	2015 年第四季度	10 000	2 800	23 340	0	0	0	36 140
	2016 年第四季度	26 300	0	42 100	0	0	0	68 400
山东	2014 年第四季度	2 741 600	61 000	722 500	17 000	0	0	3 542 100
	2015 年第四季度	2 717 500	56 000	387 000	10 000	0	0	3 170 500
	2016 年第四季度	2 472 600	59 000	418 000	5 000	0	0	2 954 600
江苏	2014 年第四季度	330 100	0	21 000	0	0	2 000	353 100
	2015 年第四季度	325 000	0	2 500	0	0	1 000	328 500
	2016 年第四季度	260 000	0	0	0	0	1 000	261 000
福建	2014 年第四季度	0	16 285	0	0	0	0	16 285
	2015 年第四季度	0	6 430	0	0	0	0	6 430
	2016 年第四季度	0	3 000	0	0	0	0	3 000
2014 年第四季度各省总计		5 652 220	275 085	853 250	17 000	0	2 000	6 799 555
2015 年第四季度各省总计		5 659 000	247 530	497 840	10 000	0	1 000	6 415 370
2016 年第四季度各省总计		5 524 900	293 500	741 100	5 000	0	1 000	6 565 500

与往年一样，结合示范区县跟踪调查数据可以看出，工厂化养殖一直是我国鲆鲽类养殖的主要养殖模式，而大菱鲆、牙鲆及半滑舌鳎始终是我国鲆鲽类养殖的三大主要品种。在鲆鲽类三大主要养殖品种中，大菱鲆养殖占主导地位的格局也从未发生变动。

2016 年第四季度，跟踪示范区县鲆鲽类总的工厂化养殖面积为 656. 6 万平方米左右，与 2015 年同比微幅增长 2. 3%，与 2014 年同比下降 3. 4%。2016 年第四季度，鲆鲽类工厂化养殖总面积与 2015 年同比虽微幅增长，但主要养殖品种中，各养殖品种的变动情况有所不同，即大菱鲆的工厂化养殖面积同比呈下降趋势，降幅为 2. 4%，而牙鲆及半滑舌鳎的养殖面积则分别增长 18. 6%及 48. 9%。出现 2016 年第三季度大菱鲆工厂化养殖面积同比减少，牙鲆及半滑舌鳎养殖面积同比增长的主要原因：一是大菱鲆养殖生产受 2015 年日照案件负面影响，再度打击养殖户的信心，继而导致市场萎缩；二是牙鲆市场价格较大菱鲆维持在相对较高价位，而半滑舌鳎市场价格 2016 年开始处于上行通道，在此利好因素的影响下，部分养殖生产者加大了对牙鲆及半滑舌鳎养殖投资。

在 2016 年第四季度 656. 6 万平方米的鲆鲽类工厂化养殖面积中，大菱鲆的养殖面积为 552. 5 万平方米，牙鲆养殖面积为 29. 4 万平方米，半滑舌鳎养殖面积为 74. 1 万平方米，占总养殖面积的比分别为 84. 2%、4. 5%及 11. 3%。2016 年第四季度，大菱鲆 552. 5 万平方米的工厂化养殖面积与 2015 年和 2014 年同比呈下降趋势，降幅在 5. 0%以内。牙鲆养殖面积与 2015 年同比增长 18. 6%，与 2014 年同比涨幅为 6. 7%。半滑舌鳎养殖面积与 2015 年同比也呈增长趋势，涨幅为 48. 9%，与 2014 年同比下降 13. 1%。

2.2.1 大菱鲆养殖面积年内基本平稳、同比均呈下降趋势

跟踪示范区县调查数据显示，2016 年大菱鲆工厂化养殖面积基本平稳，与 2015 年同比均呈下降趋势，即第一季度下降 5.2%，第二季度降幅为 7.7%，第三、第四季度分别下降 4.4%及 2.4%。与 2014 年同比第一、第二季度分别增长 5.8%及 3.6%，第三、第四季度小幅下降 1.3%及 2.3%。从各地区分布情况看，在 2016 年第四季度大菱鲆的总养殖面积中，山东大菱鲆的工厂化养殖面积占总养殖面积的比重为 44.8%，辽宁占 39.7%，河北占 10.4%，江苏占 4.7%，天津占 0.5%。

2.2.2 牙鲆养殖面积年内小幅波动，同比先下降后增长

2016 年跟踪示范区县牙鲆的工厂化养殖面积在 23.0 万平方米～29.4 万平方米之间变动，呈先增后减再增态势，但其变动幅度较小。从同比的变动趋势看，则先减少后增长，即 2016 年第一季度，牙鲆的工厂化养殖面积同比下降 22.3%，第二、第三季度分别增长 16.0%及 15.8%，第四季度涨幅为 18.6%。与 2014 年同比则先减少后增长，即前三季度分别下降 8.9%、5.5%及 3.3%，第四季度增长 6.7%。从各地分布情况看，在 2016 年第四季度各地牙鲆的工厂化养殖面积中，河北养殖占比最大，为 68.7%，其次是山东，为 20.1%，再次是辽宁，为 10.2%，最后为福建，占比仅为 1.0%。

2.2.3 半滑舌鳎养殖面积年内持续增长，同比先下降后增长

2016 年跟踪示范区县半滑舌鳎养殖面积在 47.8 万平方米～74.1 万平方米之间变动，年内呈持续增长态势。与 2015 年同比，第一季度下降 29.7%，第二至第四季度分别增长 10.2%、44.7%及 48.9%。与 2014 年同比，第一至第四季度分别下降 41.5%、23.6%、16.1%及 13.1%。从区域分布情况看，2016 年第一季度，辽宁、河北、天津、山东及江苏均有养殖，第二至第四季度仅山东、河北及天津有半滑舌鳎养殖生产记录。在 2016 年第四季度 74.1 万平方米的半滑舌鳎工厂化养殖面积中，山东的养殖面积最大，占比为 56.4%，其次是河北，占比为 37.9%，最后为天津，占比为 5.7%。

2.2.4 鲆鲽类小品种养殖面积变动情况

2016 年除大菱鲆、牙鲆、半滑舌鳎鲆鲽类三大主要养殖品种外，山东和江苏还有星突江鲽及漠斑牙鲆的工厂化养殖生产记录，养殖面积分别为 5 000 m^2 和 1 000 m^2。2016 年山东 5 000 m^2 的星突江鲽工厂化养殖面积同比呈下降趋势；江苏 1 000 m^2 的漠斑牙鲆工厂化养殖面积同比持平。而与 2014 年同比星突江鲽和漠斑牙鲆的养殖面积均呈下降趋势。

2.3 各品种网箱和池塘养殖面积变动情况

2.3.1 各品种网箱养殖面积变动情况

跟踪调查数据显示，2016 年有网箱养殖的品种为大菱鲆和牙鲆，这种养殖模式的品种分布与前两年相比均未发生变动。就跟踪示范区县大菱鲆和牙鲆的网箱养殖面积来看，2016 年第一季度，大菱鲆和牙鲆的网箱养殖面积为 19.2 万平方米。其中大菱鲆的养殖面积

为4.6万平方米,分布于福建省的东山县。牙鲆的养殖面积为14.6万平方米,分布于山东省的荣成市和福建省的东山县,其中山东省牙鲆的网箱养殖面积为12.0万平方米,福建的养殖面积为2.6万平方米。第二季度,大菱鲆的网箱养殖面积为12.5万平方米,分布于山东省的荣成市和辽宁省的旅顺口区,牙鲆的养殖面积与一季度相比未发生变动。第三季度,大菱鲆的养殖面积为10.0万平方米,与第二季度相比,山东的大菱鲆网箱养殖未发生变动,辽宁的养殖面积有所减少。第三季度牙鲆网箱养殖面积仅山东省有统计,养殖面积为12.0万平方米。第四季度大菱鲆和牙鲆养殖面积分别为12.6万平方米和12.8万平方米。

2.3.2 各品种池塘养殖面积变动情况

2016年鲆鲽类跟踪示范区县辽宁和山东有鲆鲽类池塘养殖生产记录,养殖品种均为牙鲆。2016年山东牙鲆池塘养殖面积为600亩,同比先下降后增长,即前三季度分别下降14.3%、40.0%及83.8%,第四季度增长114.3%。辽宁一季度的养殖面积为2 000亩,第二、第三季度养殖面积分别为4 000亩,第四季度则无养殖生产记录。2015年第一季度辽宁无池塘养殖生产记录,第二、第三季的养殖面积为7 000亩。

3 2016年跟踪示范区县鲆鲽类成鱼养殖存量变动情况①

与往年一样,为了叙述方便,下文中统一将1两以下(平均25克/条)、1～3两(平均2两/条)、3～7两(平均0.5斤/条)、7两～1斤(平均0.85斤/条)、1～1.5斤(平均1.25斤/条)、1.5～2.0斤(平均1.75斤/条)、2斤以上(平均2.5斤/条)②规格的鱼分别称为鱼苗、小鱼、中鱼、大鱼、标准商品鱼、大商品鱼及超大商品鱼。

3.1 鲆鲽类产品总存量分布状况

鲆鲽类产品季末总存量年内先增后减,同比呈先减后增再减的态势。2016年跟踪示范区县鲆鲽类养殖产品季末存量变动情况见表2。结合表2可以看出,2016年第一季度,鲆鲽类产品的季末存量为4.1万吨,第二季度末鲆鲽类产品的季末存量增长为4.5万吨,环比涨幅为8.8%。第三季度末存量与第一季度基本持平,环比下降8.4%,第四季度末较第三季度继续下降,为3.7万吨,环比降幅为10.1%。从长期发展来看,2016年第一季度,鲆鲽类产品的季末存量与2015年同比下降2.7%,第二季度同比增长6.0%,第三季度同比降幅为3.9%,第四季度同比降幅为13.3%。

① 本报告因赣榆县2014年第一季度的季末存量没有填写,考虑到数据的可比性,故在本报告的存量分析中,不包含赣榆县的数据。望有关各方在与前面的报告相比时,请注意调查区域变动带来的影响。谢谢!

② 斤、两为非法定单位,考虑到生产实际,本书继续保留。1斤 = 500克,1两 = 50克。

表 2　2016 年跟踪调查区域鲆鲽类主要产品存量变动情况　（单位：t）

品种	2015 年年末	2016 年第一季度	2016 年第二季度	2016 年第三季度	2016 年第四季度
大菱鲆	34 533. 5	34 272. 5	36 400. 4	31 185. 5	27 902. 45
牙鲆	4 401. 4	3 912. 4	4 389. 1	5 145. 2	4 512. 05
半滑舌鳎	3 397. 2	2 751. 4	3 762. 3	4 499. 4	4 301. 094
三大品种合计	42 332. 1	40 936. 2	44 551. 9	40 830. 1	36 715. 59
所有产品合计	42 403. 4	40 994	44 608. 7	40 876. 9	36 762. 89

结合表 2 可以看出，与往年一样，跟踪调查区域鲆鲽类产品季末总存量变动情况完全取决于三大主要养殖品种存量的变动，而在三大主要养殖品种中，大菱鲆是主要贡献者。

3. 2　大菱鲆季末存量分布状况

大菱鲆季末存量年内环比及同比均为先增长后减少。2016 年第一季度，跟踪示范区县大菱鲆季末存量为 3. 4 万吨左右，环比基本持平，与 2015 年同比呈增长趋势，涨幅为 1. 7%，与 2014 年同比下降 6. 7%。第二季度大菱鲆季末存量约为 3. 6 万吨，环比增长 6. 2%，与 2015 年同比增长 12. 4%，与 2014 年同比下降 4. 1%。第三季度大菱鲆季末存量在 3. 1 万吨左右，环比、同比均呈下降趋势，环比下降 14. 3%，与 2015 年同比下降 7. 3%，与 2014 年同比下降 11. 8%。第四季度大菱鲆的季末总存量为 2. 79 万吨，环比下降 10. 5%，与 2015 年同比下降 19. 2%，与 2014 年同比降幅为 21. 0%。

从历年数据看，2016 年第四季度大菱鲆 2. 79 万吨的季末存量是近三年来的最小值。结合产品市场分析，出现这一变动趋势的主要因素：首先是在政府部门、中国水产流通与加

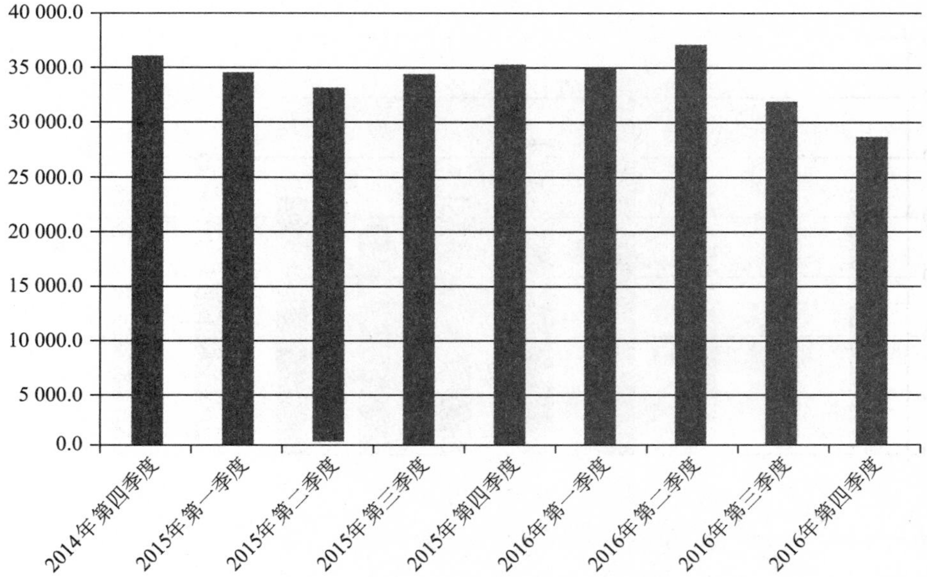

图 1　跟踪示范区县大菱鲆产品季末存量变动情况（单位：t）

工协会大菱鲆分会、产业技术体系、养殖生产者等相关各方的共同正面宣传及纠正错误信息的努力下,消费者信心逐渐恢复;其次是连续两年大菱鲆行情低迷,导致很多大菱鲆养殖从业者退出,养殖生产量减少;再次是自 2016 年 9 月中旬开始,受中秋和国庆双节拉动消费市场需求,产品价格开始回升,养殖生产者出鱼量增加。

根据不同规格的大菱鲆季末存量变动情况分析,在大菱鲆季末存量中,主要构成分别以标准商品鱼、大鱼及中鱼为主。从不同规格产品存量看,2016 年第四季度末大菱鲆小苗存量为 520. 3 t,与 2015 年同比下降 39. 4%,与 2014 年同比降幅为 41. 1%。小鱼季末存量为 3 114. 5 t,与 2015 年和 2014 年同比均呈下降趋势,降幅为 5. 3%及 6. 9%。中鱼季末存量为 6 207. 2 t,与 2015 年同比下降 5. 2%,与 2014 年同比降幅为 13. 0%。大鱼季末存量为 8 641. 0 t,与 2015 年同比下降 15. 8%,与 2014 年同比降幅为 15. 7%。达到商品鱼规格产品(标准商品鱼、大商品鱼及超大商品鱼)存量为 9 421. 6 t,与 2015 年同比降幅为 30. 6%,与 2014 年同比下降 31. 3%。

3. 3 牙鲆季末存量分布状况

牙鲆季末存量年内环比先增长后减少,同比先下降后增长。从不同规格的产品存量看:2016 年牙鲆鱼苗、小鱼的季末存量先增后减;中鱼及大鱼的季末存量前三季度持续增长,第四季度开始下降;达到商品鱼规格的产品季末存量变动趋势为先减少后增长。在各种规格产品季末存量的综合作用下,2016年跟踪示范区县牙鲆季末总存量年内环比先增长后减少:2016 年第一季度末牙鲆总存量为 3 912. 4 t;第二季度末存量为 4 389. 1 t,环比增长 12. 2%;第三季度牙鲆的季末存量较第二季度持续增长,为 5 145. 2 t,环比涨幅为 17. 2%;第四季度,牙鲆季末总存量为 4 512. 1 t,环比下降 12. 3%。从长期发展变动趋势看:2016 年牙鲆

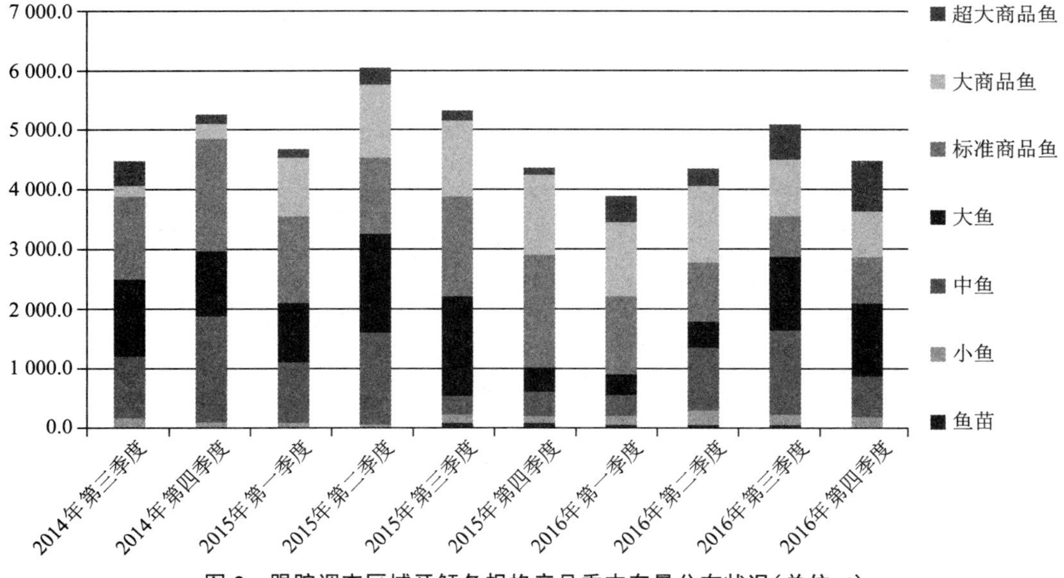

图 2 跟踪调查区域牙鲆各规格产品季末存量分布状况(单位:t)

季末总存量与2015年同比前三季度均呈下降趋势，降幅分别为17.1%、28.1%及4.4%，第四季度同比增长2.5%；与2014年同比则前三季度均呈增长趋势，第一季度涨幅较大，为56.2%，第二、第三季度分别增长14.7%及14.5%，而第四季度则下降15.1%。

2016年牙鲆达到商品鱼规格的产品季末总存量环比先下降后增长。即第一季度末的存量为2 986.7 t，第二季度下降为2 599.3 t，环比降幅为13.0%，到第三季度末存量为2 264.3 t，环比降幅为12.9%，第四季度末存量为2 400.8 t，环比增长6.0%。2016年第一至第四季度，达到商品鱼规格牙鲆季末存量与2015年同比第一季度增长14.4%，第二至第四季度分别下降8.3%、28.7%及29.3%。

3.4　半滑舌鳎季末存量分布状况

半滑舌鳎季末存量年内先增后减、同比先下降后增长。2016年半滑舌鳎季末存量年内环比先增后减。2016年第一季度半滑舌鳎的季末存量为2 751.4 t，第二季度末的存量为3 762.3 t，环比增长36.7%，第三季度末半滑舌鳎存量继续增长为4 499.4 t，环比涨幅为19.6%，第四季度末存量为4 301.1 t，环比下降4.4%。与2015年同比，第一季度下降21.4%，第二至第四季度分别增长7.5%、32.1%及26.6%。与2014年同相比，第一、第二季度分别下降23.0%及9.7%，第三季度呈增长趋势，涨幅为13.8%，第四季度同比下降5.5%。

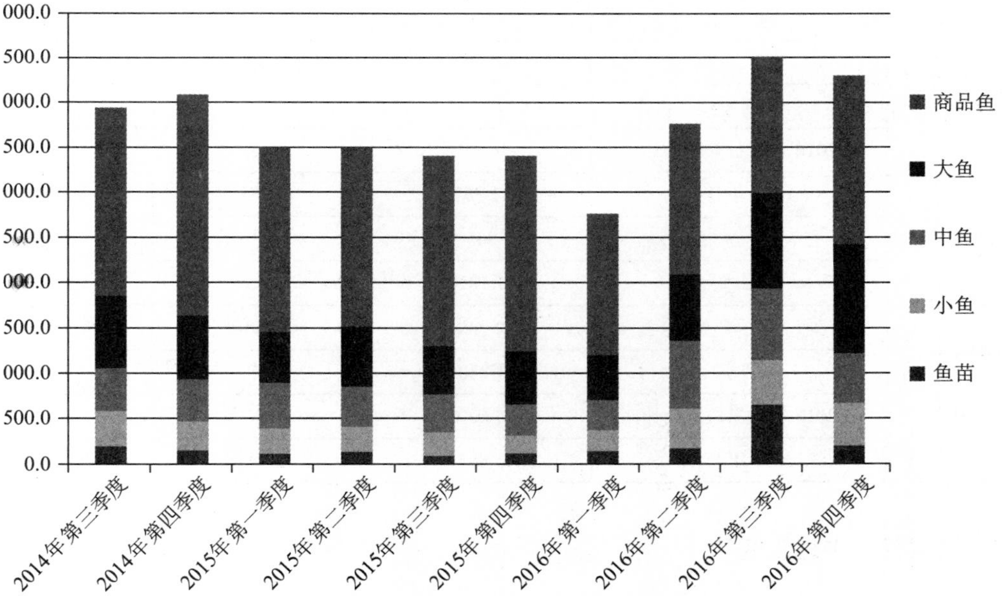

图3　跟踪调查区域半滑舌鳎各规格产品季末存量分布状况（单位：t）

从不同规格的产品存量看，在2016年第四季度末4 301.1 t的半滑舌鳎存量中，鱼苗、小鱼、中鱼、大鱼、标准商品鱼、大商品鱼及超大商品鱼的存量占总存量的比例分别为4.4%、

10.9%、13.2%、27.7%、27.2%、6.7%及9.9%，其中达到上市规格的商品鱼总存量为1 883.2 t，占季末总存量的比例为43.8%，环比呈增长趋势，涨幅为24.6%，与2015年同比下降12.9%，与2014年同比降幅为23.2%。

4 2016年跟踪示范区县鲆鲽类苗种销量变动情况

2016年跟踪调查区域鲆鲽类三大主要养殖品种苗种销量与2015年同比的变动趋势为：大菱鲆、牙鲆呈下降趋势，半滑舌鳎增长。

表3 跟踪示范区县鲆鲽类主要养殖品种苗种销量变动情况 （单位：万尾）

年份	季度	大菱鲆	牙鲆	半滑舌鳎
2014年	第一季度	3 922	679	480
	第二季度	2 899	2 515	898
	第三季度	3 332	1 845	210
	第四季度	1 925	415	232
	2014年前三季度合计	10 153	5 039	1 588
	2014年合计	12 078	5 454	1 820
2015年	第一季度	3 172	326	47
	第二季度	3 830	2 451	485
	第三季度	2 542	1105	125
	第四季度	1 813	405	170
	2016年前三季度合计	9 544	3 882	657
	2015年合计	11 357	4 287	827
2016年	第一季度	805	93	148
	第二季度	1 722	1 500	423
	第三季度	2 080	1 090	812
	第四季度	1 280	212	470
	2016年前三季度合计	4 607	2 683	1 383
	2016年合计	5 887	2 895	1 853

4.1 大菱鲆苗种销量变动

跟踪调查数据显示，2016年第一季度，跟踪示范区县大菱鲆苗种销量为805万尾，第二季度为1 722万尾，第三季度2 080万尾，第四季度1 280万尾。年内大菱鲆苗种销量先增后减。2016年大菱鲆苗种销量总计为5 887万尾，与2015年和2014年同比均呈下降趋势，降幅为48.2%及51.3%。2016年大菱鲆苗种同比持续减少的原因主要还是因为产业受2015

年日照“多宝鱼”事件的影响，产品价格行情持续低迷，养殖生产者亏损，导致很多抗风险能力小、自身调节能力差的养殖从业者退出，进而养殖量减少。

4.2　牙鲆苗种销量变动

2016年牙鲆苗种年内销量呈现出先增后减的变动趋势，其中第一季度销量为93万尾，第二季度大幅增长为1 500万尾，第三季度牙鲆苗种销量较第二季度减少410万尾，第四季度销量为212万尾。2016年牙鲆苗种销量累计为2 895万尾，与2015年同比下降32.5%，与2014年同比降幅为46.9%。

4.3　半滑舌鳎苗种销量变动

2016年第一至第四季度，半滑舌鳎共计销售苗种1 853万尾，年内环比持续增长。从同比情况看，2016年半滑舌鳎苗种总销量与2015年同比大幅增长124.1%，与2014年同比基本持平。

5　主要养殖品种病害情况

5.1　大菱鲆病害情况

跟踪调查数据显示，在2016年第四季度12月份所跟踪调查的鲆鲽类养殖生产者中，有大菱鲆养殖的有43户，其中报告有病害的有16户，占大菱鲆总养殖户数的比例为35.8%。在16户的大菱鲆病害报告中，发病率最高的是腹水病，占总发病率的比例为50.8%，其次是肠道白浊，占比为39.2%，再次是纤毛虫病，其发病率占总发病率的比例为22.5%，再次为红嘴病，占比16.6%。最后分别为烂尾病和其他寄生虫病，在16户的大菱鲆养殖病害报告中各有一户发生，发病率均为4.9%。

5.2　牙鲆病害情况

在2016年第四季度12月份所跟踪调查的鲆鲽类养殖生产者中，有牙鲆养殖的有14户，其中报告有病害发生的有5户，占牙鲆养殖户数的比例为42.8%。而值得注意的是在所有的病害报告数中，腹水病的发病率仍然最高，占总发病率的比例为51.5%，其次是红嘴病，其在总发病率中的占比为28.8%，然后是烂尾病，其发病率在总发病率中的占比为22.1%。最后是纤毛虫病、红底病及淋巴囊肿病，其发病率在总的病害报告中均有一户发生，占总发病率的比重均为6.8%。

5.3　半滑舌鳎病害情况

在2016年第四季度12月份所跟踪调查的鲆鲽类养殖生产者中，有半滑舌鳎养殖的有

15户,其中报告有病害发生的有8户,占半滑舌鳎总养殖户数的比例为40.8%。而在半滑舌鳎的病害报告中,烂尾病的发病率与2015年同期一样,仍为最高,达55.1%,其次是腹水病,占总发病率的比例为40.9%,最后是肠道白浊病、寄生虫及红底病,在2016年第三季度9月跟踪调查区域半滑舌鳎养殖中均有一户养殖生产者报告有该病害发生,占总发病率的比重均为12.3%。

6 鲆鲽类养殖生产投入要素价格变动情况

6.1 饲料价格

6.1.1 主要品牌配合饲料价格

根据鲆鲽类产业经济岗位调查数据显示,2016年鲆鲽类养殖生产所使用的配合饲料升索、常兴、海康、七好的价格变动情况如下:

首先看升索。根据调查的数据来看,2016年升索饲料的价格平均为16元/千克,年内价格基本平稳,与2015年同比价格基本持平,与2014年同比,每千克涨幅为14.3%。

其次看常兴。2016年常兴饲料的平均价格与2015年同比持平,为17元/千克,2014年同期,其均价为14元/千克,2016年常兴饲料的销售价格与2014年同比每千克涨幅为21.4%。

再次看海康。2016年海康饲料的平均价格为16元/千克,与2015年同比增长7.1%,与2014年同比涨幅为14.2%。

最后看七好。2016年度鲆鲽类专用饲料七好的均价为15元/千克,与2015年同比微幅增长7.1%,与2014年同比每千克降幅为15.4%。

6.1.2 冰鲜饲料鱼价格

从长期价格变动情况看,2016年度山东玉筋鱼的平均价格为7元/千克,辽宁皮条鱼、天津皮条鱼、河北皮条鱼的平均价格则分别为6元/千克。2016年度山东玉筋鱼、辽宁皮条鱼、天津皮条鱼、河北皮条鱼平均价格与2015年同比基本持平。

6.2 鱼苗价格

大菱鲆苗种价格:跟踪调查的数据表明,2016年跟踪调查区域大菱鲆苗种价格在0.8元/尾~1.8元/尾之间波动,其销售均价为1.2元/尾。2016年跟踪调查区域大菱鲆苗种销售均价高于2015年同期价格,同比涨幅为18.9%,与2014年同比呈下降趋势,降幅为7.6%。

牙鲆苗种价格:2016年第一至第四季度,跟踪调查区域牙鲆苗种销售价格在0.8元/尾~1.5元/尾之间波动,即最高价为1.5元/尾,出现在2016年的9、10月份,最低价0.8元/尾发生在3月份。2016年第一至第三季度,跟踪调查区域牙鲆苗种销售均价为

1 元／尾。从长期价格波动情况看，2016 年牙鲆苗种销售均价与 2015 年同比基本持平，与 2014 年同比呈下降趋势，降幅分别为 26. 6%。

半滑舌鳎苗种价格：2016 年跟踪示范区县半滑舌鳎仅 7 月份和 9 月份有苗种销售，销售均价为 2. 1 元／尾。2016 年半滑舌鳎苗种销售均价与 2015 年同比呈增长趋势，涨幅为 8. 4%，与 2014 年同比下降 6. 5%。

综上所述，2016 年大菱鲆、半滑舌鳎苗种销售均价与 2015 年同比呈上涨趋势，牙鲆基本持平。而与 2014 年同比，无论是大菱鲆、牙鲆还是半滑舌鳎，其销售均价均呈下降趋势。

6. 3　临时工工资

从零工工资看，2016 年与 2015 年相比略有增长。跟踪调查数据表明，跟踪示范区县 2016 年鲆鲽类养殖零工平均工资为 88. 9 元／（人•天），同比微幅上涨 3. 2%。而与往年一样，除了看到跟踪调查区域鲆鲽类养殖零工工资平均水平同比有所上涨以外，还需注意到不同地区在零工工资方面的巨大差异。其中，辽宁葫芦岛龙港区，河北的曹妃甸区、丰南区，江苏如东等地较高，最高时达到 150 元／（人•天）；而山东文登、辽宁绥中等地较低，为 75 元／（人•天）～ 80 元／（人•天）。

6. 4　电费价格

跟踪调查数据表明，跟踪示范区县鲆鲽类养殖用电价格变动不是很大。从年度平均值来看，2016 年平均电价为 0. 61 元／千瓦时，与 2015 年同比基本持平。虽然 2016 年鲆鲽类养殖用电均价同比基本持平，但值得注意的是，与往年一样，有部分地区差异较大。绝大多数地区养殖生产者的电价在 0. 5 元／千瓦时～ 0. 65 元／千瓦时之间，但少数几个地区因借用别人电压器等原因，电费出现过高现象，比如辽宁的大洼为 1. 1 元／千瓦时，天津的大港为 1 元／千瓦时。

7　鲆鲽类主要品种价格变动趋势

7. 1　大菱鲆价格变动趋势

跟踪调查的数据显示，2016 年跟踪示范区县大菱鲆的出池价格变动情况数据基本一致，故本报告继续选择我国大菱鲆工厂化养殖最集中、规模最大的主产地之一辽宁兴城的大菱鲆出池价格进行分析。

7.1.1　长期价格变动趋势

从长期价格变动情况看，2016 年兴城大菱鲆的出池均价为 34 元／千克，2015 年同期的出池均价为 40. 8 元／千克，同比呈下降趋势，降幅为 17. 1%。2014 年同期兴城大菱鲆的出池均价与 2015 年同期基本持平。

7.1.2 年内价格变动趋势

2016年1至12月份，兴城大菱鲆出池价格在24元/千克至49元/千克之间波动，年内的变动趋势为先增长后下降回升。即1月份大菱鲆的出池价格为24元/千克，2月份价格上升为32元/千克，2至5月份的出池价格在30元/千克～32元/千克的价位变动，6、7月份价格再次下降，并在28元/千克的价位运行，8月份开始，价格在7月份的基础上小幅回升，为30元/千克，且之后的几个月内，价格陆续小幅上涨，直至11月份，兴城大菱鲆的出池价格为49元/千克。这是2015年7月济南多宝鱼“药残”风波后的最高价。而12月份的价格较11月份则又小幅下降，为47元/千克。

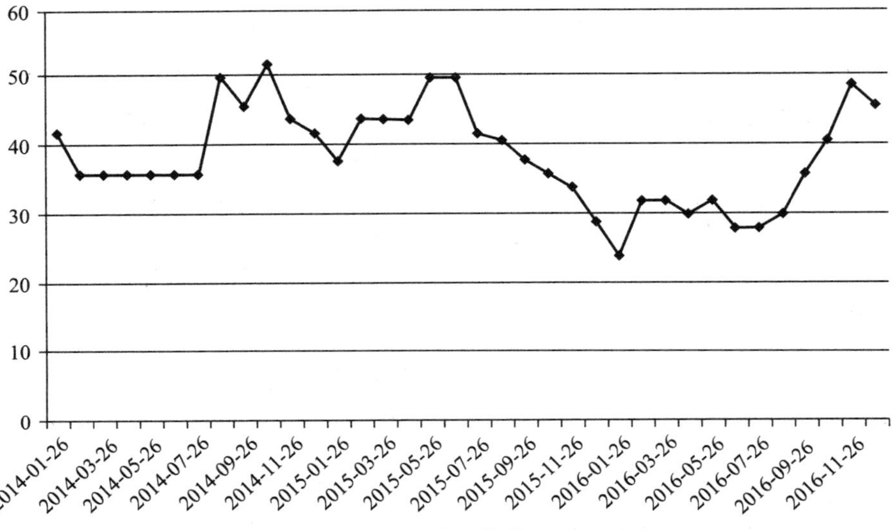

图4 兴城大菱鲆价格变动趋势图(元/千克)

7.1.3 后市价格预测

以前期数据为基础，以计量模型预测，结果表明，如果不发生特殊冲击(比如药残、重大政策变动等)，短期内，大菱鲆价格将不仅能够弥补正常成本，而且可使生产者获取正常利润及一定的风险利润。从长期看，根据上述跟踪示范区县大菱鲆季末存量数据分析，2016年第四季度，大菱鲆的季末总存量为2.79万吨，环比、同比均呈下降趋势，环比降幅为10.5%，同比下降19.2%，与2014年同比降幅为21.0%。据此可以看出，产品季末存量减少，后市产品的市场供给也将有所减少。

在前面的分析中已指出，此次产品季末存量下降从需求角度看，主要原因是在政府部门、中国水产流通与加工协会大菱鲆分会、产业技术体系、养殖生产者等相关各方的共同正面宣传及纠正错误信息的努力下，消费者信心逐渐恢复。从供给角度看，受前期低价的影响，很多大菱鲆养殖从业者退出，养殖生产量减少，其次是养殖生产者在负面预期的影响下集中出鱼，导致存量减少。总之，结合各方面来看，如果不发生重大事故及宏观经济冲击，后期大菱鲆价格将持续一段时间维持在比较理想的价位。

7.2　牙鲆价格变动趋势

7.2.1　长期价格变动趋势

从长期价格走势来看，2016年河北昌黎牙鲆的出池均价为40元/千克，与2015年同比增长11.1%，与2014年同比下降4.8%。

7.2.2　年内价格变动趋势

从年内价格变动情况看，2016年1至12月份，河北昌黎牙鲆的出池价格在34元/千克～48元/千克的价位变动，其变动趋势总体看呈上涨趋势。2016年牙鲆38元/千克的最低价与2015年同比增长5.6%，而48元/千克的价格则是2014年9月份以后首次出现的最高价。

7.2.3　后市价格预测

2016年牙鲆的市场价格利好上扬。但值得注意的是，根据存量数据分析来看，在2016年第四季度牙鲆季末存量中，虽然达到商品鱼规格的产品存量与2015年同比呈下降趋势，降幅为29.3%，但其他规格的产品存量同比则大幅增长，增幅达109.5%。据此推断，即在其他条件不变的情况下，中长期内，牙鲆达到有效供给的产品存量较当前将比较充裕，因此其产品市场价格将会面临较大压力。

7.3　半滑舌鳎价格变动趋势

7.3.1　长期价格变动趋势

跟踪调查数据显示，从长期价格变动情况来看，2016年山东烟台的半滑舌鳎出池价格在将近2年之久的历史低谷价位运行后，于2016年开始反弹回暖。

7.3.2　年内价格变动趋势

2016年半滑舌鳎出池均价为186元/千克，与2014年和2015年同比分别增长44.7%及38.0%。2016年1至7月份，半滑舌鳎价格持续增长，即从1月份的145元/千克上涨为210元/千克，8月份价格与7月份持平，9至12月份，半滑舌鳎的出池价格维持在200元/千克的价位。而2016年半滑舌鳎价格出现上涨的原因在《2016年鲆鲽类产业发展及市场动态报告分析》中均有阐述，主要是受产品有效供给减少的影响。

7.3.3　后市价格预测

根据跟踪示范区县半滑舌鳎季末存量分析，2016年第四季度，半滑舌鳎虽然达到商品鱼规格的产品季末存量与2015年同比减少，但其他规格的产品季末存量则增长66.4%。据此提醒相关各方注意：如果消费偏好、宏观经济政策、养成率等因素未出现大的变动，一个生长周期以后，半滑舌鳎的供给较目前将会大幅增长，因此长期来看半滑舌鳎价格存在较大价格走低的风险。

8 成本收益分析

以2016年度鲆鲽养殖企业(或个体)连续跟踪调研数据为基础,以产业经济岗位团队的调研数据为补充,对鲆鲽类三大主要养殖品种成本构成及收益变化进行探讨,分析其变化趋势,结果表明:2016年大菱鲆的成本利润率为-19.7%,牙鲆的成本利润率为6.6%,半滑舌鳎的成本利润率为69.3%。2012年至2016年,鲆鲽类三大主要养殖品种大菱鲆、牙鲆及半滑舌鳎的成本利润变动情况见图5。

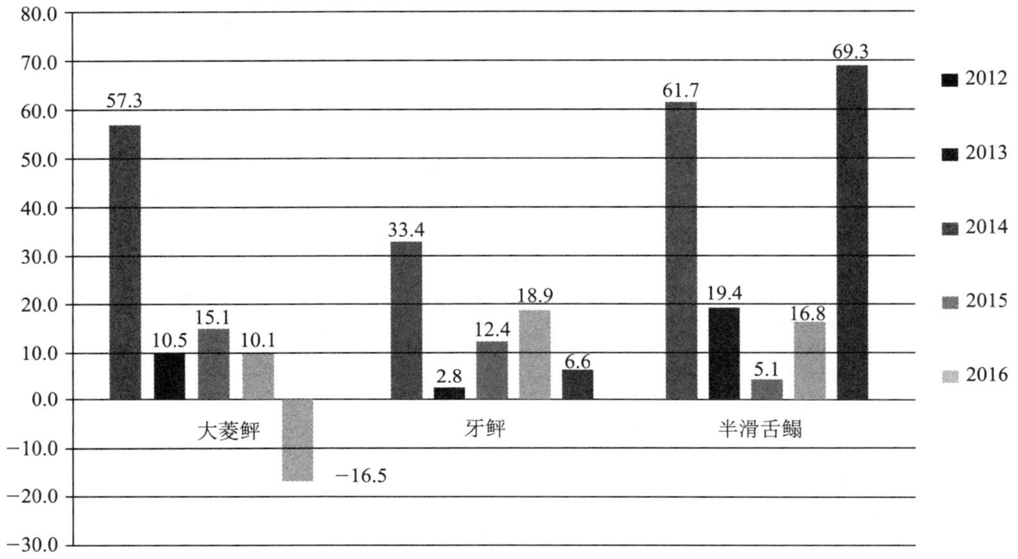

图5 2012～2016年鲆鲽类三大养殖品种成本利润率(%)

关于2016年度鲆鲽类养殖经济效益分析的详细信息,请参阅本团队跟踪研究的《2016年鲆鲽类产业经济效益分析及技术经济跟踪研究》报告。

9 存在的问题

9.1 大菱鲆价格长期低迷,产业警情严重

2015年10月以来大菱鲆出池价格持续走低,警情风险一直持续到2016年年底,尤其是2016年1月至9月大菱鲆出池价格均处巨警区,9月之后价格有所回升,警情方从负向巨警转变成负向重警,再到轻警和无警状态。

9.2 大菱鲆、牙鲆养殖经济效益下降

我国大菱鲆养殖占鲆鲽类养殖量80%以上。然而2016年其销售价格持续低迷,与往

年同比均呈下降趋势。受产品价格持续低迷因素的挤压，大菱鲆养殖经济效益下降。2016年大菱鲆的养殖成本利润率为 −16.53%，2015 年和 2014 年分别为 10.07% 及 15.13%。2016 年牙鲆成本利润率为 6.6%，2015 和 2014 年分别为 18.92% 及 12.44%。

9.3　产品消费方式单一，加工业发展缓慢

目前，鲆鲽类产品国内消费仍然以鲜活产品消费为主。加工产品消费占比很小，且加工产品的原料主要来自于国外进口，国内养殖产品进入加工消费的占比不到 5%。加工产业不发达不仅增加了产品运销成本，同时也在一定程度上增加了产品质量安全隐患，不利于产品品牌的打造及产业转型提升。事实上，加工消费方式单一、加工产业发展滞后已经成为了我国鲆鲽类产业发展瓶颈。

9.4　大菱鲆消费者信心脆弱，市场维系曾举步维艰

受 2015 年日照－济南多宝鱼风波的影响，消费者对大菱鲆的产品质量安全曾严重质疑，产业体系对百度搜索关键词持续跟踪结果表明，“多宝鱼有毒吗”这一信息日搜索量曾于 2016 年 2 月下旬上升至“多宝鱼”相关关键词的第三位，其搜索量高达 14.5 万次，“多宝鱼致癌”则于 2016 年 7 月份取代“多宝鱼有毒吗”成为相关关键词的第三位，日搜索量曾一度达 36.1 万多次。至 11 月下旬，“多宝鱼致癌”已退为相关关键词第四位，但日关注量仍有 2 万多次。受这些负面信息的影响，市场销售曾一度困难重重。

表 4　“多宝鱼”信息百度大数据搜索跟踪

时间	关键词					
	第一位	搜索数量	第三位	搜索数量	第五位	搜索数量
2015 年 4 月	多宝鱼的做法	815 500	多宝鱼多少钱一斤	200 000	多宝鱼怎么杀	712 000
2015 年 7 月	多宝鱼的做法	649 000	多宝鱼做法大全	13 400	多宝鱼图片	947 000
2015 年 9 月	多宝鱼的做法	19 000	多宝鱼多少钱一斤	333 000	多宝鱼做法大全	17 800
2015 年 12 月	多宝鱼的做法	82 300	多宝鱼有毒吗	1 460 000	多宝鱼做法大全	70 000
2016 年 3 月	多宝鱼做法	491 000	多宝鱼有毒吗	145 000	多宝鱼图片	1 720 000
2016 年 6 月	多宝鱼的做法	346 000	多宝鱼多少钱一斤	73 800	多宝鱼致癌	315 000
2016 年 7 月	多宝鱼的做法	268 000	多宝鱼致癌	361 000	多宝鱼做法大全	177 000
2016 年 8 月	多宝鱼的做法	245 000	多宝鱼致癌	374 000	多宝鱼做法大全	183 000
2016 年 9 月	多宝鱼的做法	240 000	多宝鱼致癌	279 000	多宝鱼做法大全	190 000
2016 年 10 月	多宝鱼的做法	251 000	多宝鱼致癌	201 000	多宝鱼做法大全	196 000
2016 年 11 月	多宝鱼怎么收拾	47 200	多宝鱼吧	1 180 000	多宝鱼饲料	396 000
2016 年 12 月	多宝鱼吧	1 200 000	多宝鱼做法	265 000	多宝鱼饲料	454 000

10 对策建议

10.1 转变财政支渔理念，在兼顾生产的同时投入更大力度进行市场拓展

就水产养殖业而言，转方式、调结构迫在眉睫。产业调控政策需要对此加以引导。随着国内消费者收入水平提高、城市化加剧，水产品消费层次差异化已比较明显。作为中档（大菱鲆、牙鲆）及中高档（半滑舌鳎）水产品，鲆鲽类产品是满足消费者需求的重要组成部分，而其产品质量始终并将继续被社会高度关注。为满足消费者需求，同时拉动产业向质量型、效益型、环境友好型转变，建议转变过去以燃油补贴等形式支持渔业的方式，转而支持：① 循环水养殖等环境友好型产业，包括鲆鲽类工业化循环水养殖；② 产品质量安全追踪技术研发与应用，以便于保障包括鲆鲽类在内的水产品质量安全，从而为提振消费者信心和福利、促进加工业发展、拓展产品市场提供支持；③ 养殖生产者协会、专业合作社及龙头企业的品牌打造及市场开发，以配合国家“一带一路”倡议，打造国际知名的中国鲆鲽类品牌，引领国际鲆鲽类产业发展。

10.2 继续推进产业整合，提升从业者的宣传、沟通能力

从多次的食品安全事件及其危机应对、风险管理经验与教训中可以看出，消费者的消费行为、消费心理需要引导、维护与巩固。如何方能引导消费者理性选择？最关键的可能就是业界、管理部门与主流媒体加强沟通，加大对消费者的正面宣传，而且要持之以恒地引导与宣传。然而，这种沟通与宣传仅靠单个的生产者很难取得规模效应。因此，加大产业合作，尽快建立和完善区域性生产者协会，并在此基础上完善全国生产者协会就显得十分必要。目前大菱鲆产业已成立了全国性的加工流通分会，但仍有诸多工作需要做，其产业影响需要提升，建议以业界会员的会费为主，政府以服务购买等形式适当予以补贴，科研院所和高校适当予以相关辅助，进一步强化其功能，以发挥其提供“准公共产品”的功能。与此同时，建议牙鲆、半滑舌鳎以及其他主要水产品（不仅是鲆鲽类）产业尽快建立类似的产业组织，以提升行业自律能力，适度控制产业规模，真正适应市场经济的发展。

10.3 以水产养殖保险制度为支撑，通过产品质量信用担保机制的建立，提升消费者信心，为鲆鲽类产品加工业提供支撑

包括鲆鲽类养殖业在内，水产养殖业是高投入、高风险的行业，现代水产养殖业需要水域、鱼苗、饲料、设施、技术等要素投入，而这一切都依赖于资金投入。本产业技术体系调研表明，养殖者信用贷款需求强度高，这种需求在出现种种“风波”“事件”而导致产品滞销时尤其强烈，但往往难以得到有效满足。而贷不到款的原因，主要是风险高、额度小，金融业难以承受。若建立养殖保险制度，则可以在一定程度上化解产业风险，平抑产业波动。

与此同时，保险与产品质量信用担保制度的缺失还在一定程度上阻碍了鲆鲽类加工业

的发展。原因在于加工企业因担心鲆鲽类养殖产品质量而不敢加工产品，因为即便加工出来，也可能通过不了质量检验，尤其是出口质量检验。若建立产品质量担保与水产养殖保险制度，则在出现产品质量问题时加工企业可以向担保者追讨，从而消除后顾之忧。

2016 年 5 月 17 日，李克强总理已经批示：“要有针对性地加大政策扶持力度，大力发展政府支持的融资担保和再担保机构。”建议各级政府出台具体政策措施，支持设立为水产养殖业服务的融资性担保机构，完善税收优惠举措，为有效化解水产养殖业“担保难”起到托底作用，并促进水产品加工业的发展。

10.4 转变重生产、轻市场，重技术、轻管理的技术推广方式，推进标准化生产及现代化生产经营管理

推进生产技术的标准化及经营管理的现代化是提升产品质量，推动产业转方式、调结构的重要内容。在过去相当长的一段时间内，水产技术推广系统将主要精力放在了新品种的引进与推广方面，建议以后将其工作重点转移到既有重要品种，包括鲆鲽类产品养殖生产技术的标准化及经营管理方式现代化的推广方面。

10.5 充分重视信息传播方式变化对消费者心理与行为的影响

互联网是把“双刃剑”，在推动社会进步的同时，也成为一些人凭空捏造、传播虚假信息的工具，成为造谣、传谣的载体，严重影响了包括鲆鲽类在内的水产养殖业的发展。近年来随着微信、微博等自媒体的广泛运用，信息传播方式正在深刻变化。主要体现在：信息传播速度大大加快，信息的滞留时间大大延展，消除负面信息的代价昂贵。因此，建议业界有关各方充分关注信息传播方式的变化对产品消费者行为的影响，建立专门机构，配备专门人员，建立网络谣言监测与防谣预警机制，重视日常无事发生时的防谣工作，加大正面宣传，并在谣言发生时，第一时间辟谣。

10.6 引导理性投资，防范产业风险

2017 年，半滑舌鳎存量较 2016 年将增大，产业风险高。大菱鲆产业中仍然存在食品质量安全、网络谣言影响等产业风险，市场价格警情虽不如 2015 年下半年及 2016 年严重，但仍然存在一定风险。建议加强宣传与引导，防止业界非理性投资。

（岗位科学家　杨正勇）

鲆鲽类良种选育技术研发进展

良种选育岗位

1 工作内容

重点开展了大菱鲆新品种“多宝1号”生长特征分析、生产性能评估以及高密度遗传连锁图谱的构建和耐温性状的遗传评估，并利用分子标记辅助育种技术构建了选育四代耐高温品系。

在2016年度，开展了“多宝1号”生长特征分析，开展了“多宝1号”和普通养殖群体进行养殖生产对比实验，完成“多宝1号”生产性能评估以及“多宝1号”和普通养殖群体的生长特征比较，完成了大菱鲆耐温性状(UTT)的遗传评估，完成了大菱鲆耐温选育群体的耐温性能比较，完成了大菱鲆高密度遗传连锁图谱的构建，利用分子标记辅助育种技术构建选育四代耐高温品系，完成了大菱鲆“多宝1号”大规模苗种的生产和推广工作。

1.1 开展了“多宝1号”生长特征分析

基于Logistic、Gompertz、von Bertalanffy和Chapman-Richards 4个典型的非线性S型生长曲线，对大菱鲆选育新品种“多宝1号”的生长性能进行了研究。首先对所采用的4种典型非线性模型进行研究。基于“一般系统论”推导出基于4种模型的缓慢生长区间、快速生长区间、渐进生长区间以及快速生长区间长度公式(表1)。

表1 4种非线性模型的生长特征(时间)

模型	Chapman-Richards	Logistic
缓慢生长期	$\left(0,\ \frac{1}{k}\ln\left[-\frac{B\left(m+2-\sqrt{m^2+4m}\right)}{2(m-1)}\right]\right)$	$\left(0,\ -\frac{1}{k}\ln\frac{2+\sqrt{3}}{B}\right)$
快速生长期	$\left(\frac{1}{k}\ln\left[-\frac{B\left(m+2-\sqrt{m^2+4m}\right)}{2(m-1)}\right],\ \frac{1}{k}\ln\left[-\frac{B\left(m+2+\sqrt{m^2+4m}\right)}{2(m-1)}\right]\right)$	$\left(-\frac{1}{k}\ln\frac{2+\sqrt{3}}{B},\ -\frac{1}{k}\ln\frac{2-\sqrt{3}}{B}\right)$
渐进生长期	$\left(\frac{1}{k}\ln\left[-\frac{B\left(m+2+\sqrt{m^2+4m}\right)}{2(m-1)}\right],\ +\infty\right)$	$\left(-\frac{1}{k}\ln\frac{2-\sqrt{3}}{B},\ +\infty\right)$
快速生长区间长度	$\ln[(m+2+\sqrt{m^2+4m})/(m+2-\sqrt{m^2+4m})]/k$	$\ln[(2+\sqrt{3})/(2-\sqrt{3})]/k$

（续表）

模型	Gompertz	von Bertalanffy
缓慢生长期	$\left(0, -\frac{1}{k}\ln\frac{3+\sqrt{5}}{2B}\right)$	$\left(0, -\frac{1}{k}\ln\frac{4+\sqrt{7}}{9B}\right)$
快速生长期	$\left(-\frac{1}{k}\ln\frac{3+\sqrt{5}}{2B}, -\frac{1}{k}\ln\frac{3-\sqrt{5}}{2B}\right)$	$\left(-\frac{1}{k}\ln\frac{4+\sqrt{7}}{9B}, -\frac{1}{k}\ln\frac{4-\sqrt{7}}{9B}\right)$
渐进生长期	$\left(0, -\frac{1}{k}\ln\frac{3-\sqrt{5}}{2B}, +\infty\right)$	$\left(0, -\frac{1}{k}\ln\frac{4-\sqrt{7}}{9B}, +\infty\right)$
快速生长区间长度	$\ln[(3+\sqrt{5})/(3-\sqrt{5})]/k$	$\ln[(4+\sqrt{7})/(4-\sqrt{7})]/k$

利用所采集的“多宝 1 号”体重数据对 4 种模型进行拟合，以拟合度 R^2（决定系数）、均方误差 MSE 和赤池信息量准则 AIC 三项指标为准则，筛选最适模型。结果发现（表 2），Gompertz 模型有最低的 MSE（6 421. 870 6）和 AIC（65. 132 2），且有第二高的拟合度（0. 990 8）（和第一高的拟合度几乎相等），这表明 Gompertz 模型是拟合“多宝 1 号”的最适模型。利用 Gompertz 模型分析“多宝 1 号”的生长特征，“多宝 1 号”的缓慢生长期、快速生长期和渐进生长期分别为（0，10. 23）、（10. 23，26. 78）和（26. 78，0）（单位：月龄），快速生长区间长度为 16. 55 月龄（表 3）。

表 2　“多宝 1 号”基于 4 种非线性模型的模拟参数

模型	Logistic	Gompertz	von Bertalanffy	Chapman-Richards
决定系数（R^2）	0. 990 6	0. 990 8	0. 989 8	0. 990 9
均方误（MSE）	6 574. 177 3	6 421. 870 6	7 127. 175 4	7 436. 316 1
A	2 156. 69	2 788. 11	3 584. 70	2 788. 24
B	112. 61	8. 60	1. 28	0. 001 3
k	0. 25	0. 12	0. 07	0. 12
m	—	—	—	0. 999 8
赤池信息准则（AIC）	65. 343 1	65. 132 2	66. 070 0	68. 452 2

表 3　大菱鲆“多宝 1 号” 4 种非线性模型的生长特征　（单位：月龄）

模型	缓慢生长期	快速生长期	渐进生长期	时长
Chapman-Richards	（0，7. 58）	（7. 58，23. 62）	（23. 62，0）	16. 04
Logistic	（0，13. 63）	（13. 63，24. 16）	（24. 16，0）	10. 53
Gompertz	（0，10. 23）	（10. 23，26. 78）	（26. 78，0）	16. 55
von Bertalanffy	（0，7. 86）	（7. 86，30. 58）	（30. 58，0）	22. 72

1.2 基于点评估完成了完成“多宝1号”生产性能评估

在2016年度,完成了河北昌黎公司和烟台龙口公司于2015年6月购买大菱鲆“多宝1号”苗种养殖10个月(表4),山东潍坊公司和山东威海公司于2015年8月购买大菱鲆“多宝1号”苗种养殖10个月(表5),山东蓬莱公司、山东日照公司和青岛胶南公司分别于2016年2月、2016年2月和2016年1月购买大菱鲆“多宝1号”苗种养殖9个月(表6),共三批次“多宝1号”苗种与同期非选育苗种的生长性能对比试验。每一批次的“多宝1号”苗种生长速度和成活率均高于比非选育苗种,显示出良好的育种成效。

表4 第一批推广苗种养殖10个月的生产性能

推广公司	购苗时间	购苗数量	苗种规格	体重/g			成活率		
				“多宝1号”均重	非选育苗均重	提高百分比	“多宝1号”成活率	非选育苗种成活率	提高的百分比
河北昌黎公司	2015年6月	15 000尾	5 cm	419.23	265.43	57.94%	96.65%	78.91%	22.48%
烟台龙口公司	2015年6月	5 000尾	5 cm	401.11	270.41	48.33%	95.61%	77.87%	22.78%

表5 第二批推广苗种养殖10个月的生产性能

推广公司	购苗时间	购苗数量	苗种规格	体重/g			成活率		
				“多宝1号”均重	非选育苗均重	提高百分比	“多宝1号”成活率	非选育苗种成活率	提高的百分比
山东潍坊公司	2015年8月	18 000尾	5 cm	396.45	261.77	51.45%	96.15%	76.34%	25.95%
山东威海公司	2015年8月	10 000尾	5 cm	387.18	265.18	46.01%	95.84%	76.32%	25.58%

表6 第二批推广苗种养殖9个月的生产性能

推广公司	购苗时间	购苗数量	苗种规格	体重/g			成活率		
				“多宝1号”均重	非选育苗均重	提高百分比	“多宝1号”成活率	非选育苗种成活率	提高的百分比
山东蓬莱公司	2016年2月	15 000尾	5 cm	318.46	222.58	43.08%	96.11%	77.43%	24.13%
山东日照公司	2016年2月	12 000尾	5 cm	322.58	219.31	47.09%	95.97%	76.56%	25.35%
青岛胶南公司	2016年1月	20 000尾	5 cm	328.12	217.27	51.01%	95.96%	76.65%	25.19%

1.3 完成了大菱鲆耐高温及相关重要生长性状的遗传评估

以山东烟台天源水产有限公司在2007和2010年所构建的大菱鲆G_1和G_2代选育家

系为材料，进行耐高温实验。以大菱鲆耐高温选择指数(UTT)作为耐高温性状指标。对构建的选育家系进行耐高温胁迫实验，对每代UTT进行数据统计同时统计实验鱼的体重。基于两性状动物模型，进行UTT及其与重要生长性状的遗传评估。耐高温和生长性状遗传参数表明(表7)，耐热性的遗传力是0.1106 ± 0.0799，属低等遗传力，而体质量的遗传力为0.2394 ± 0.1410，属中等遗传力。耐高温性状和体质量性状间表型相关和遗传相关结果表明(表8)，两性状间的表型相关是较低的正相关，为0.075 ± 0.026，而两性状间的遗传相关是负相关，为 -0.019 ± 0.011。

表7　耐高温和体重性状的方差组分和遗传力

性状	方差组分				遗传力
	σ_p^2	σ_a^2	σ_f^2	σ_e^2	h^2
耐热性(UTT)	88 849.79	9 829.338	76.92	78 943.53	0.110 6 ± 0.079 9
体质量	270.575 2	64.766 9	28.858 3	176.95	0.239 4 ± 0.141 0

σ_p^2为表型方差，σ_a^2为加性遗传方差，σ_f^2为全同胞家系方差，σ_e^2为残差，h^2为遗传力

表8　耐高温和体重性状的遗传相关和表型相关

性状	耐热性(UTT)	体质量
耐热性(UTT)		-0.019 ± 0.011
体质量	0.075 ± 0.026	

对角线右侧为遗传相关，对角线左侧为表型相关

1.4　完成了大菱鲆“多宝1号”全周年大规模苗种的生产实验

2016年，烟台天源水产有限公司从2015年构建的快速生长核心育种群选留1 500尾雌性亲鱼，从高成活率核心育种群选留雄性亲鱼600尾。目前，烟台天源水产有限公司保存育种材料共计6 950尾。其中，来源于快速生长核心育种群的雌性亲鱼5 100尾，来源于高成活率核心育种群的雄性亲鱼1 850尾。本年度，通过动物模型BLUP法遗传评估，从2015年构建的快速生长核心育种群选留1 500尾雌性亲鱼，从高成活率核心育种群选留雄性亲鱼600尾。通过快速生长(♀) × 高成活率(♂)交配繁育“多宝1号”优质苗种，经养殖对比，比普通大菱鲆平均体重和成活率分别提高37%和25.6%，“多宝1号”新品种选育性状的优势比选育初期更为优良。

2016年，基于制订的“‘多宝1号’优质苗种全周年供应培育计划”，利用2014、2015年度选留的大菱鲆亲鱼，在3～4月、5～6月和10～12月，分三批进行苗种培育，超额完成体系年度任务书规定的苗种的生产考核指标(表9)。

表 9 2016 年“多宝 1 号”苗种生产

批次	布卵时间	生产 5 cm 苗种时间	生产苗种数量	推广数量	推广地
第一批	3～4 月	6～7 月	75 万尾	60 万尾	日照养殖场、河北养殖场、葫芦岛地区养殖场、蓬莱宗哲养殖场和烟台开发区养殖场等地
第二批	5～6 月	8～9 月	35 万尾	35 万尾	大连天正有限公司、天津立达公司、蓬莱京鲁有限公司、牟平泰华有限公司、莱州养殖场、文登养殖场、乳山养殖场、威海养殖场等地
第三批	10～11 月	12 月和 2017 年 1～2 月	30 万尾	17 万尾	福建省，山东海阳、乳山、威海、龙口等地

经过制订完善、合理的苗种培育计划，本年度按照体系任务实现了“多宝 1 号”优质苗种的全周年供应，保证养殖产品均衡上市，实现产业效益的最大化。

1.5 完成了大菱鲆耐温选育群体的耐温性能比较

实验室分别于 2008、2011 和 2014 年对构建的耐温选育群体进行了耐温性能评估，对群体间的耐温性能进行了比较。2008 年对耐温选育一代进行耐温实验，27 ℃条件下，选育一代各家系的死亡率是(78.5 ± 2.31)%，平均死亡率为 49%，对照组的平均死亡率为 72.7%；2011 年对耐温选育二代进行了耐温性能评估实验，在 27 ℃条件下，死亡率平均值仅为 10.26%，对照组的死亡率平均值为 50.95%。将最高实验温度上调为 28 ℃，耐温选育二代品系的平均死亡率为 32.83%，而对照组的平均死亡率为 88.9%，耐温家系在耐温性方面表现出显著的提高。2014 年耐温性能评估实验的最高实验温度定为 28 ℃，耐温选育三代的平均死亡率 20.53%，对照组的平均死亡率为 86.52%，与耐温选育一、二代相比，耐温选育三代的耐温性能有了明显提升(表 10)。

表 10 28 ℃高温培育下各实验组大菱鲆幼鱼死亡历时时域及死亡率

	最早死亡时间 / h	死亡历时时域 / h	平均死亡率 / %
选育一代	39	39～1 121	62.14
选育二代	48	48～1 142	32.83
选育三代	56	56～1 201	20.53

1.6 完成了大菱鲆高密度遗传连锁图谱的构建

在实验室原有的 SSR 遗传连锁图谱基础上，利用高通量测序技术，以大菱鲆 F_2 群体为材料，构建高密度遗传连锁图谱。

1.6.1 SLAF-seq 数据分析

利用 Illumina HiSeq 2500 测序平台，亲本和 119 个子代共获得 418.26 MB read 数据，测序平均 Q30 为 86.8%，平均 GC 含量为 42.63%，经过生物信息学分析，共开发 600 329

个 SLAF 标签，其中多态性 SLAF 标签为 194 513 个，可以用于遗传连锁图谱构建的标签有 134 126 个，构建遗传图谱时的有效多态性为 22. 34%。上图标记亲本平均测序深度为 207. 34×，子代平均测序深度为 44. 30×。

1.6.2　遗传连锁图谱构建

SLAF 标签和 SSR 标记的上图标记数为 5 003 个，包括 SLAF 标签数为 4 823 个，其中 SNP 位点标记为 5 773 个，SSR 标记数为 180 个，分析得到 22 个连锁群。其中雄性图谱包含 2 894 个标记，总图距为 2 444. 19 cM，平均图距为 0. 84 cM；雌性图谱含有 3 092 个标记，总图距为 3 402. 65 cM，平均图距 1. 10 cM；合并图谱含有 5 003 个标记，总图距 3 132. 88 cM，平均图距 0. 63 cM（图 1）。共注释基因 357 个，其中，GO 注释 43 个，KEGG 注释 210 个，包括 170 个代谢通路，蛋白库注释 88 个。

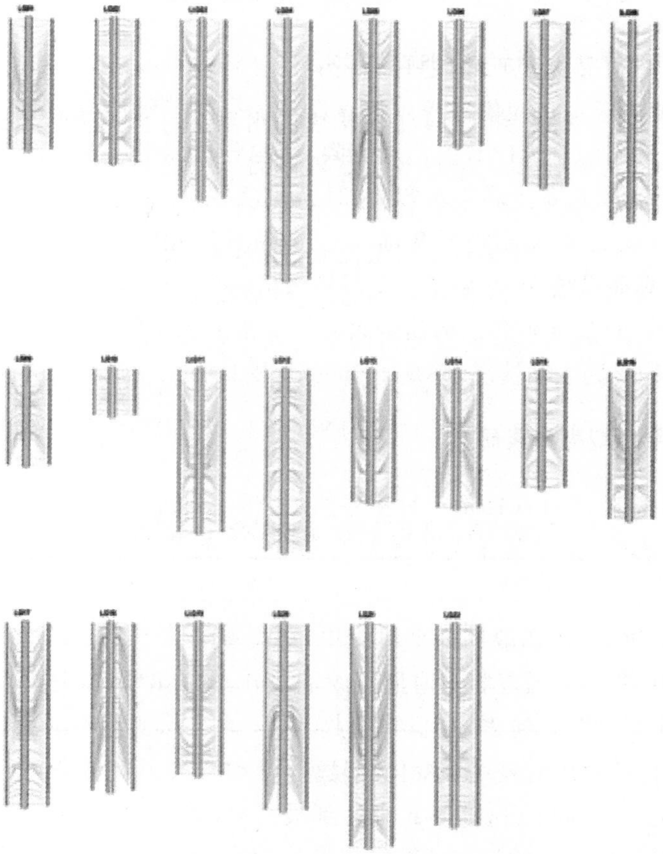

图 1　大菱鲆高密度遗传联锁图谱

1. 7　利用分子标记辅助育种技术构建选育四代耐高温品系

2016 年春季从耐高温纯系中挑选亲本，构建选育四代耐高温品系，其中亲本需要符合的标准为:① 亲本个体间遗传距离较远,② 含有耐高温性状紧密连锁的 QTL 标记。根据课题组前期耐高温相关分子标记筛选工作的反复验证,共确定 3 个 SSR 标记为与耐高温性状显著相关的标记。共构建 34 个家系。

1.8 完成了大菱鲆“多宝 1 号”优质苗种及耐高温品系苗种的生产及其推广

完成了大菱鲆“多宝 1 号”推广:培育大菱鲆“多宝 1 号”苗种 150 万尾,推广大菱鲆“多宝 1 号”苗种 112 万尾,推广到山东胶南、日照、招远、海阳、乳山、威海、龙口、蓬莱、烟台开发区养殖场,福建养殖场,葫芦岛养殖场等。175 kg 受精卵推广到乳山、威海、天津等养殖场。完成了培育推广大菱鲆耐高温性状苗种 42 万尾,推广至江苏、福建等地。

2 前瞻性研究

开展了大菱鲆 C 型凝集素研究,通过分析其活性特点以及环境胁迫对其表达调控的影响,探讨其在大菱鲆先天免疫应答过程中的潜在功能以及养殖环境中的应用。开展了大菱鲆 Lily 型凝集素研究,通过 *Sm*LTL 的基因克隆、结构分析和组织表达,对其特性进行分析,并通过环境胁迫实验、蛋白重组和凝集实验,对 *Sm*LTL 的功能展开深层次的研究,为大菱鲆 Lily 型凝集素的进一步研究奠定理论基础,为大菱鲆疾病预防和控制新途径打下基础。利用 Illumina-Solexa 高通量技术,开展了大菱鲆转录组解析——繁育、生长及免疫响应相关基因的发掘和遗传标记的鉴定研究。构建了大菱鲆成鱼多组织混合样本的转录组数据库,全面了解了大菱鲆雌雄基因表达情况,筛选了与生殖、生长和免疫响应相关的候选基因。

2.1 开展了大菱鲆 C 型凝集素研究

大菱鲆 C 型凝集素(*Sm*Lec1)是一种重要的免疫因子,通过分析其活性特点以及环境胁迫对其表达调控的影响,探讨其在大菱鲆先天免疫应答过程中的潜在功能以及养殖环境中的应用。

通过荧光定量 PCR,在选取的 8 个组织中,C 型凝集素在肝脏中的表达量显著,在其他组织中均不显著(图 2)。温度胁迫时肝组织的 *Sm*Lec1 mRNA 表达量受到温度的影响显著,在 19 ℃开始升高,22 ℃到 27 ℃呈降低趋势,其他组织变化不显著(图 3)。盐度胁迫时 *Sm*Lec1 mRNA 在各盐度的表达量变化不显著(图 4)。鳗弧菌感染实验显示,感染 2 h 起 *Sm*Lec1 mRNA 表达量开始上升,在 6 h 时达到最高值,随后有所下降(图 5)。实验利用原核表达系统进行重组表达,并通过纯化获得重组目的蛋白(图 6、图 7)。活性分析结果显示:*Sm*Lec1 重组蛋白对所选 6 种细菌的凝集作用存在差异,对鳗弧菌和爱德华氏菌具有明显的凝集作用(图 8)。在选取的 8 种糖中,木糖、果糖、蔗糖、乳糖和甘露糖中能够抑制 *Sm*Lec1 重组蛋白的凝集活性(表 11)。在理化因子(pH、温度)中,低 pH 和高温对 *Sm*Lec1 重组蛋白的凝集活性产生抑制(表 12、表 13)。

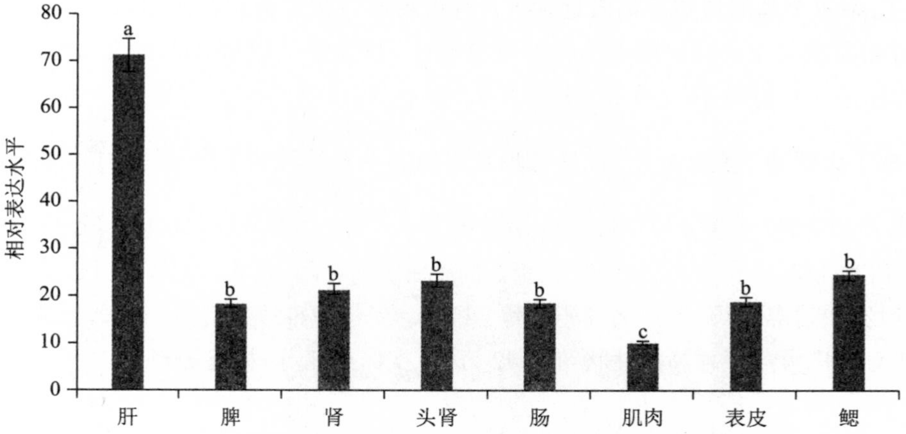

图 2　大菱鲆 *Sm*Lec1 mRNA 在不同组织中的表达情况

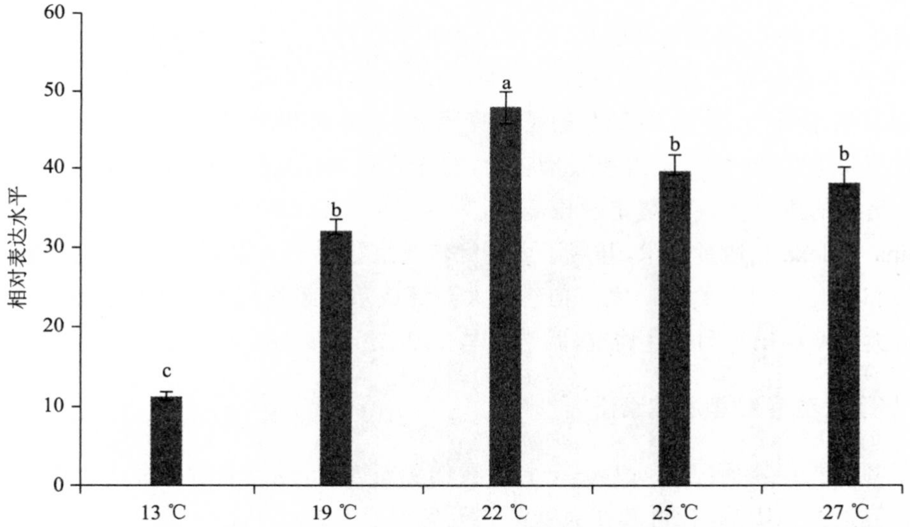

图 3　大菱鲆肝脏 *Sm*Lec1 mRNA 在不同温度下的表达情况

2.2　开展了大菱鲆 Lily 型凝集素研究

通过大菱鲆 LTL（*Sm*LTL）的基因克隆、结构分析和组织表达，对其特性进行分析，并通过环境胁迫实验、蛋白重组和凝集实验，对 *Sm*LTL 的功能展开深层次的研究。通过 RACE 技术克隆得到大菱鲆 Lily 型凝集素的 cDNA 全长为 569 bp（图 9），包括编码 112 个氨基酸的 339 bp 开放阅读框。通过生物信息软件对其进行分析，结果显示：*Sm*LTL 在第 30～99 个氨基酸处有两个甘露糖结合位点（Q*x*D*x*N*x*V*x*Y）（*x* 是任何氨基酸残基）和一个特殊位点（T*x*-T*x*G*x*R*x*V）形成一个 β-棱镜架构（图 10）。

根据组织的 qPCR 分析显示，*Sm*LTL mRNA 的于表皮、鳃、肠的表达量显著，在其他组织的表达量不显著（图 11）。环境胁迫（温度和盐度）对 LTL 表达水平的调控实验结果显

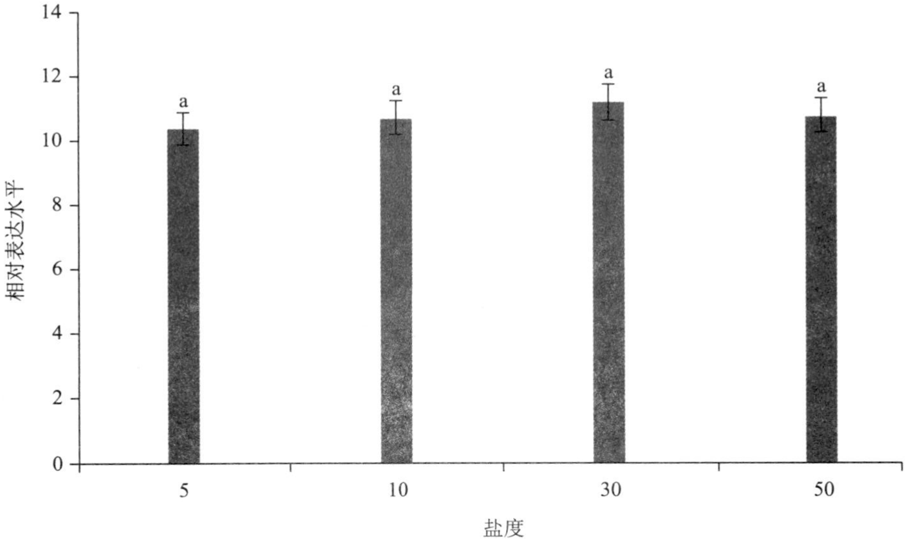

图 4　大菱鲆肝脏 *Sm*Lec1 mRNA 在不同盐度下的表达情况

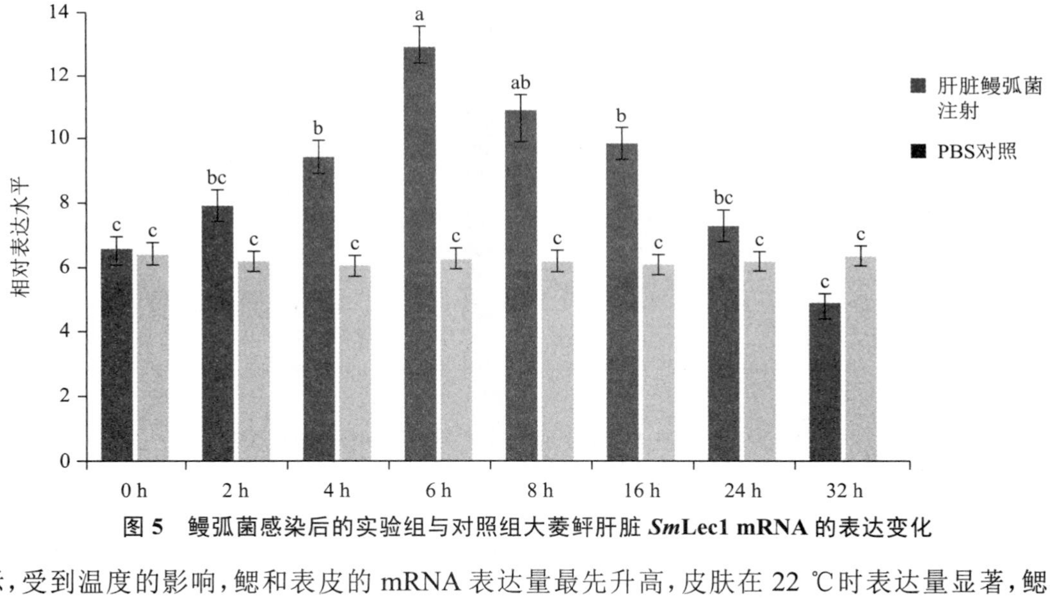

图 5　鳗弧菌感染后的实验组与对照组大菱鲆肝脏 *Sm*Lec1 mRNA 的表达变化

示，受到温度的影响，鳃和表皮的 mRNA 表达量最先升高，皮肤在 22 ℃时表达量显著，鳃在 25 ℃时表达量显著，随后呈降低趋势，肠的表达量变化不明显(图 12)。盐度胁迫条件下，鳃表达量变化并不显著，表皮在盐度 5 时表达量最显著，肠在盐度 30 时表达量最显著(图 13)。

通过蛋白重组技术，获得 *Sm*LTL 蛋白(图 14、图 15)。不同浓度重组 *Sm*LTL 蛋白在大肠杆菌系统中能够凝集小鼠血细胞且没有钙依赖性，凝集的效果与蛋白浓度正相关，在 *D*-甘露糖存在的情况下被抑制(图 16)。

另外，*Sm*LTL 表现出对爱德华氏菌和鳗弧菌的凝集活性(图 17)。同时 *Sm*LTL 对纤毛虫具有杀伤性作用，浓度为 100 μg/mL 在 24 h 后能致死 60%的纤毛虫(表 14)。

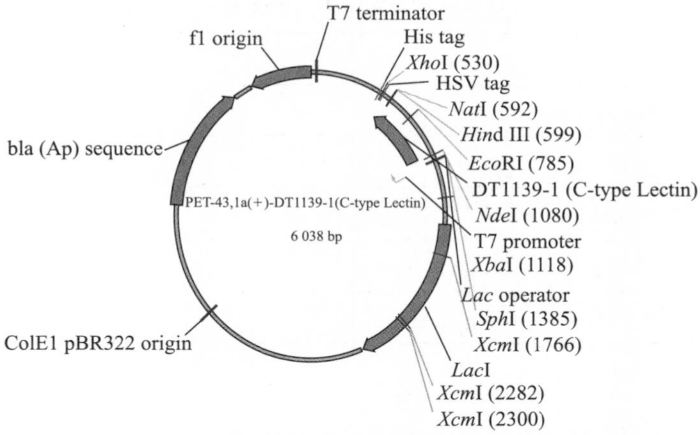

图 6　大菱鲆 Lec1 重组蛋白表达载体 pET43. 1a-DT1139-1 图谱

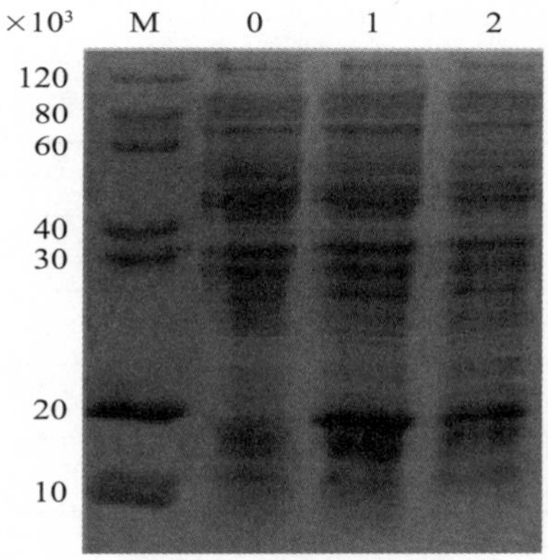

图 7　SDS-PAGE 分析大菱鲆 Lec1 重组蛋白于 BL21(DE3) 表达情况

M: marker; 0: 对照; 1: 15 ℃诱导过夜; 2: 37 ℃诱导 4 h

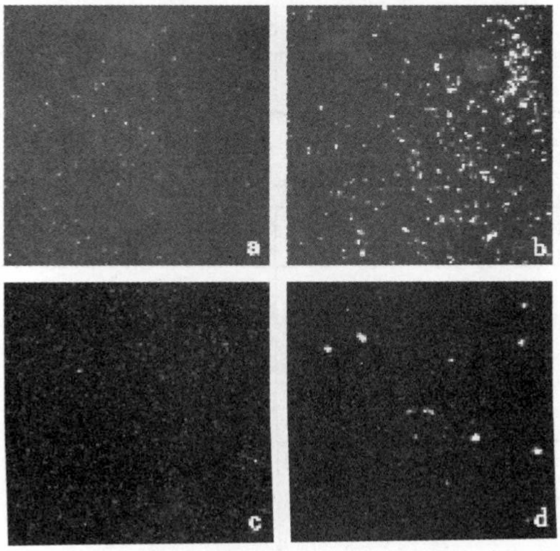

图 8　*Sm*Lec1 对 *Edwardsiella*（爱德华氏菌）和 *V. anguillarum*（鳗弧菌）具有凝集作用

a 为爱德华氏菌在无 *Sm*Lec1 蛋白下的正常对照，b 为爱德华氏菌与 *Sm*Lec1 蛋白发生凝集，c 为鳗弧菌在无 *Sm*Lec1 蛋白下的正常对照，d 为鳗弧菌与 *Sm*Lec1 蛋白发生凝集

表 11 糖特异性结合重组蛋白凝集鳗弧菌的活性

糖类	最小抑菌浓度
葡萄糖	NI[a]
D-木糖	NI[b]
D-果糖	NI[b]
蔗糖	NI[b]
麦芽糖	NI[a]
乳糖	NI[b]
半乳糖	NI[a]
甘露糖	NI[b]

NI[a] 表示在 >400 mmol 浓度仍不抑制凝集，NI[b] 表示在 50 mmol 浓度时即可抑制凝集

表 12 pH 对 C 型凝集素稳定性影响

pH	凝集活性
2	NI[b]
3	NI[b]
4	NI[b]
5	NI[b]
6	NI[b]
7	NI[a]
8	NI[a]
9	NI[a]

表 13 温度对 C 型凝集素稳定性影响

温度 / ℃	凝集活性
4	NI[a]
10	NI[a]
25	NI[a]
37	NI[b]
45	NI[b]
60	NI[b]

NI[a] 表示在 >400 mmol 浓度仍不抑制凝集，NI[b] 表示在 50 mmol 浓度时即可抑制凝集

```
ACATGGGGGCCCCCTCTTCAGCTGCTTCTCAGACACAAGGACCCGCCACAACTGAATC    58

ATGAACAGAAACTCCATCAGCACGGACCAGGAGCTGCGTAAGGGAGAGTTCCTCATGAGT  118
M  N  R  N  S  I  S  T  D  Q  E  L  R  K  G  E  F  L  M  S

GTGAATGGCGAGTTCAAAGCCATCTTCCAGGATGACGGCAACTTTGTCATCTACAAATGG  178
V  N  G  E  F  K  A  I  F  Q  D  D  G  N  F  V  I  Y  K  W

TCTCCCATTTGGGATACTAAGACATGTGGGAAAAACCCATTTCGAGTCCTTCTGCAACCG  238
S  P  I  W  D  T  K  T  C  G  K  N  P  F  R  V  L  L  Q  P

GATAACAACCTGGTTATGTACGACAAATTGTCTAAACCAGTTTGGGCCACTGGCACCCAC  298
D  N  N  L  V  M  Y  D  K  L  S  K  P  V  W  A  T  G  T  H

TCAAACCAAGCCAATCAAAGGATGCGCCTGACCTTGACTGATGGAGGTCGGCTGGTTCTT  358
S  N  Q  A  N  Q  R  M  R  L  T  L  T  D  G  G  R  L  V  L

GATAAAGATGGAGGTGAAATTTGGGGTGCTGGAGGATAATCTTAGCAATGAAAAGTGCTC  418
D  K  D  G  G  E  I  W  G  A  G  G  *

TATAAATGAAATTTATTATCATTATTATTACTAAATAGATTCCAGAATTCAAACAGAATT  478

CATGATGTGGCAACACACTGTACACGGTTTGCTCTGAGATTGAAATAAAGAGCTCTGATG  538

TTTCGAAAAAAAAAAAAAAAAAAAAAAAAAA                               569
```

图 9　*Sm*LTL 完整的 cDNA 和推导蛋白序列

三个特定基序以灰色突出显示

2.3　利用 Illumina-Solexa 高通量技术，开展了大菱鲆转录组解析——繁育、生长及免疫相关基因的发掘和遗传标记的鉴定研究

利用 Illumina-Solexa 高通量技术，构建了大菱鲆成鱼多组织混合样本的转录组数据库，全面了解了大菱鲆雌雄基因表达情况，筛选了与生殖、生长和免疫响应相关的候选基因，为大菱鲆性腺发育、生长发育及免疫响应等重要生物过程的阐释提供基础数据，并检测到大量的分子遗传标记，为标记辅助育种研究提供宝贵的标记资源。

主要结果如下：

（1）基于 Illumina 平台测定了大菱鲆成年雌鱼和雄鱼各 1 尾主要组织（肝、脾、肾、脑、精巢或卵巢、肌肉）的转录组，共得到 4.42 GB 的平均长度 100 bp 的 read，通过严格的拼接组装，获得 106 643 个 contig 序列，经过去冗余处理后，得到 71 107 个 unigene 序列（平均长度 892 bp）（表 15），其中注释的有 24 052 个，有 12 087 个的 E 值范围为 9E-10～1E-110，有 7 162 个注释基因的 E 值为 0（图 18）。

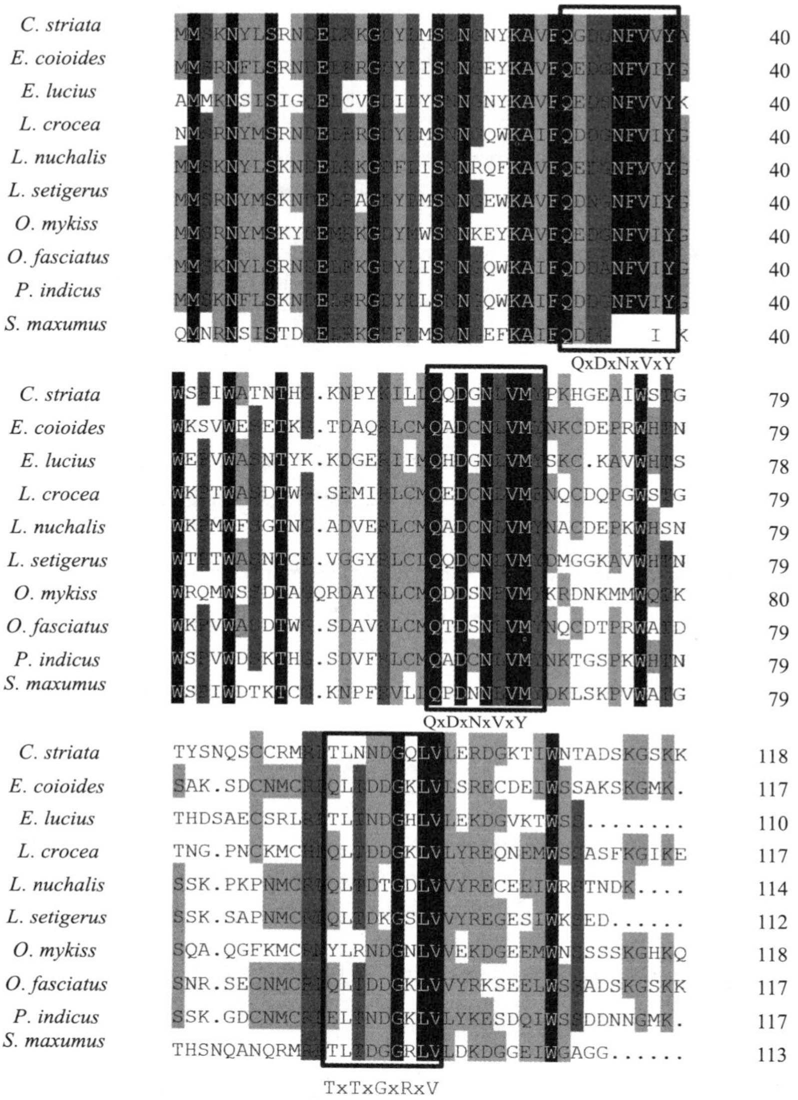

图 10 *Sm*LTL 与其他同源蛋白多序列比对

依次为线鳢、斜带石斑鱼、白斑狗鱼、大黄鱼、颈带鲾、黑鮟鱇、虹鳟、条石鲷、鲬鱼和大菱鲆的 Lily 型凝集素。3 个 *D*-甘露糖结合位点被突出显示

(2) 筛选出 40 个与性别决定和性腺分化相关的基因,既有已知在精巢发育发挥作用的基因(*ar*、*mis*、*sox9*、*sox6*),也有首次在大菱鲆中报道的新基因(如 *hsp90a*、*dnali1*、*ropn1l*),其中与精巢发育相关的最重要新基因是 *dmrt1*。鉴定出的卵巢发育相关基因较多,既包括有明确调节作用的 *cyp19a*、*zpc5*、*zar1*、*gtc5*、*gdf9*,也发现了 *sox17*、*sox19*、*foxl2*、*vtgr* 等在大菱鲆性腺发育中尚未研究的基因。此外,还挖掘出一些功能有待明确的新基因(*dmrt2*、*dmrt3*、*sox8a*)(表 16)。

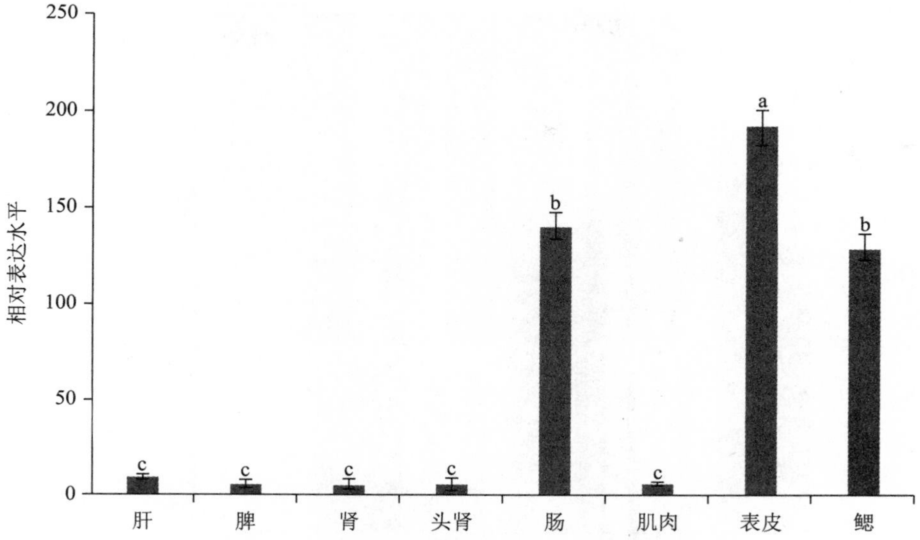

图 11 荧光定量 PCR 检测 *Sm*LTL 在各组织的表达量

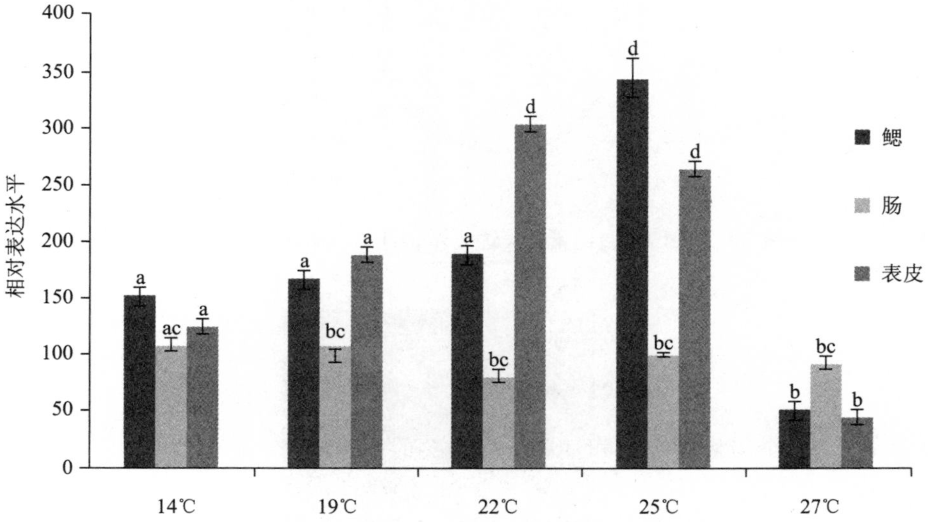

图 12 不同温度下 *Sm*LTL mRNA 在鳃、肠和表皮的表达量

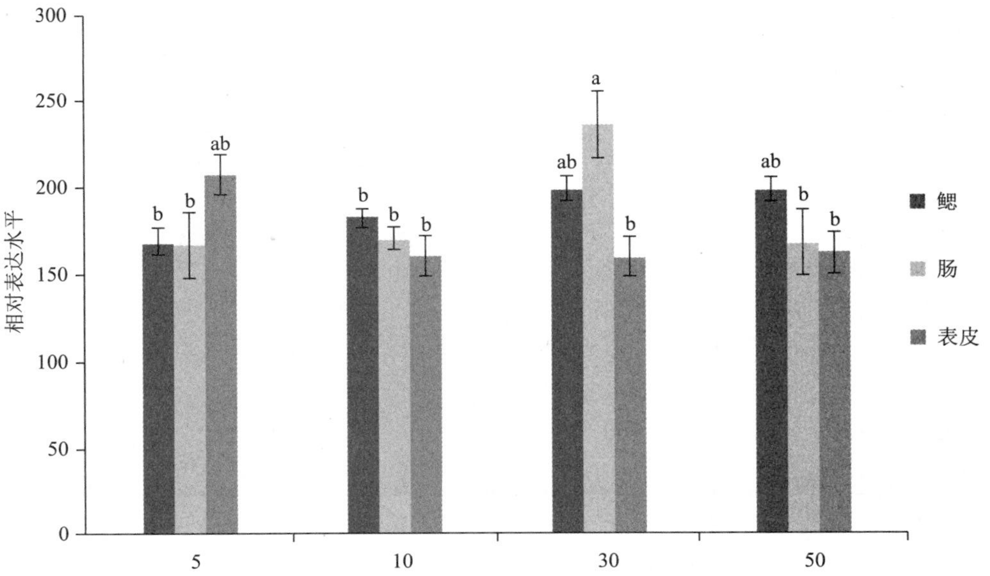

图 13　不同盐度下 ***Sm*LTL mRNA** 在鳃、肠和表皮的表达量

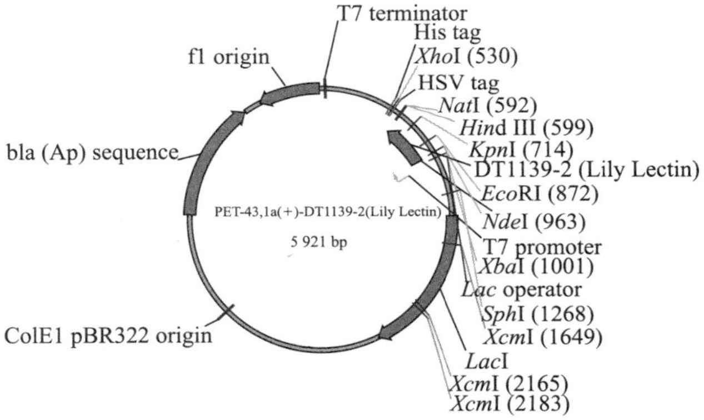

图 14　***Sm*LTL** 重组蛋白表达载体 **pET43. 1a-DT1139-2** 图谱

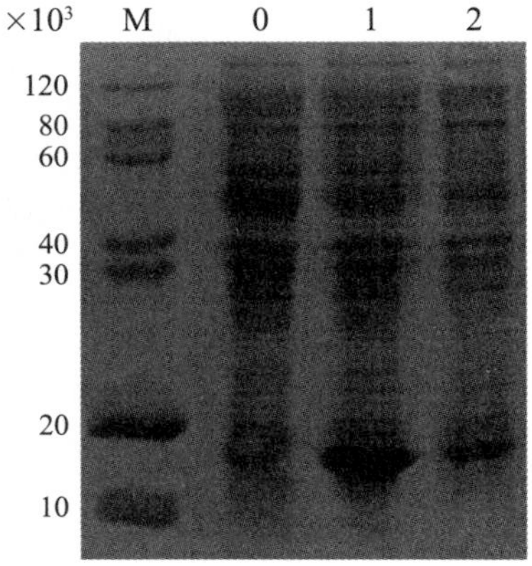

图 15　**SDS-PAGE** 分析 ***Sm*LTL** 重组蛋白于 **BL21（DE3）** 表达情况

M：marker；0：对照；1：15 ℃诱导过夜；2：37 ℃诱导 4 h

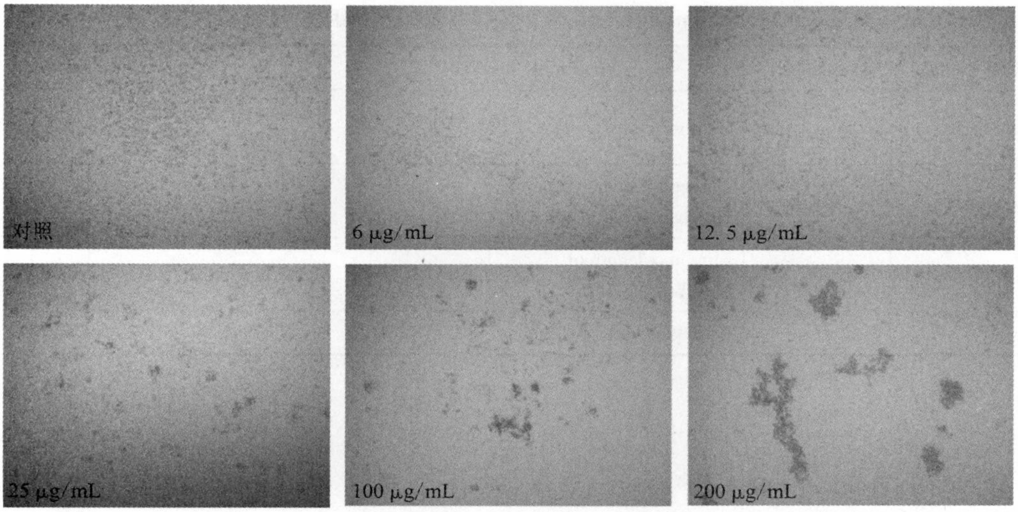

图 16　不同浓度 *Sm*LTL 蛋白与小鼠红细胞发生凝集

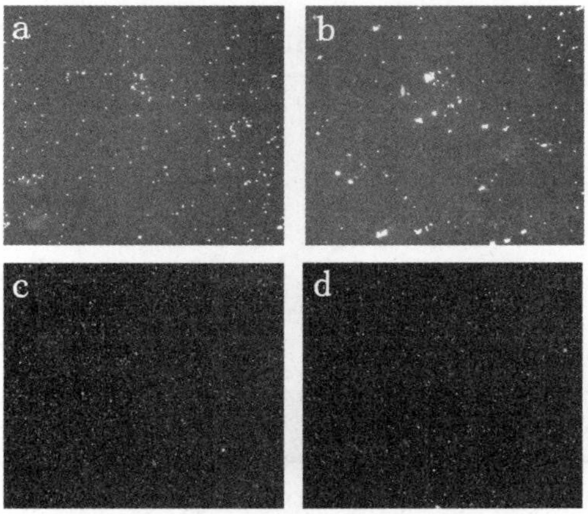

图 17　重组 *Sm*LTL 蛋白对 *Edwardsiella*（爱德华氏菌）和 *V. anguillarum*（鳗弧菌）具有凝集作用

a 为爱德华氏菌的正常对照，b 为爱德华氏菌与重组 *Sm*LTL 蛋白发生凝集

c 为鳗弧菌的正常对照，d 为鳗弧菌与重组 *Sm*LTL 蛋白发生凝集

表 14　*Sm*LTL 对纤毛虫活性的影响

时间 / h	不同浓度下的死亡率 / %					
	200 μg/mL	100 μg/mL	50 μg/mL	25 μg/mL	12. 5 μg/mL	对照
3	0	0	0	0	0	0
6	3	1	0	0	0	0
12	45	28	20	14	6	0
24	80	60	52	30	24	10
48	100	100	100	100	100	100

表 15　Illumina 平台测出大菱鲆转录组的序列组装、注释信息

Raw results（after trimming）		Assembly results		Annotation results	
Clean bases/G	8. 844	Contigs	106 643	Nr and Swiss-Prot annotations	24 052
Read pairs	44 219 773	Unigenes（after eliminating redundancy）	71 107	COG hits	37 058
Read length/bp	100	Min-max length of unigenes	201～17 407	GO mapped	16 540
		Average length of unigenes	892	KEGG hits	11 938

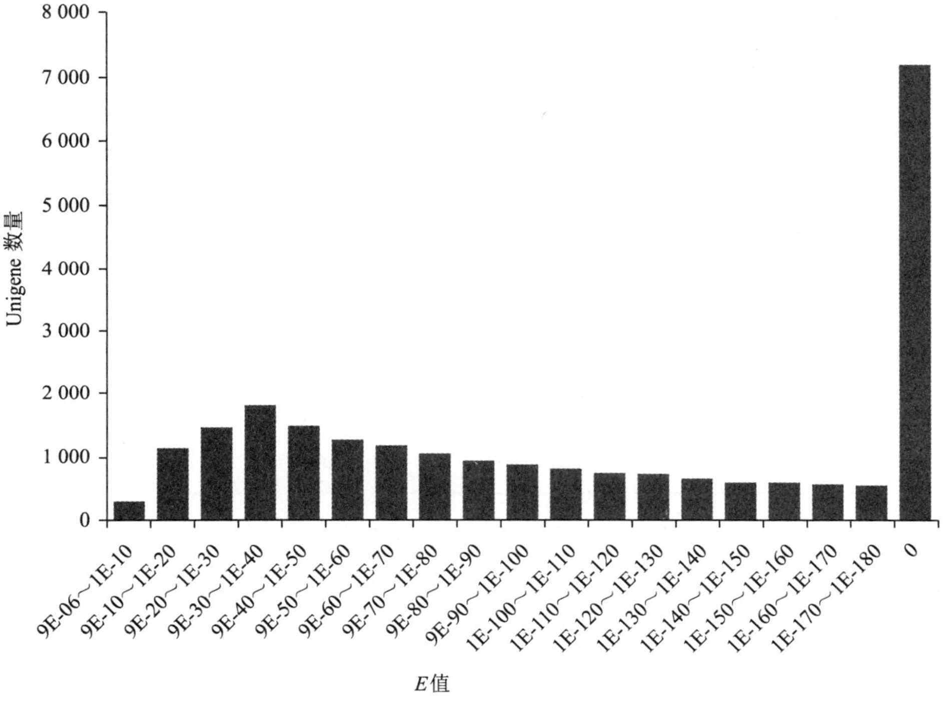

图 18　大菱鲆转录组中注释到 NR 数据库的基因的 *E* 值分布

表 16　大菱鲆转录组中鉴定出的与性别决定和性腺发育相关的基因

	Gene name	Annotation	Query name	Length / bp	Expression quantity（♂/♀）	*Q* value	Trend
Male-biased	***dmtr1***	Doublesex and mab-3-related transcription factor 1	comp30078_c2_seq1	566	6. 43	1. 36E-10	Up
	fshr	Follicle stimulating hormone receptor II	comp34552_c0_seq1	3 389	2. 39	2. 73E-70	Up

（续表）

	Gene name	Annotation	Query name	Length / bp	Expression quantity（♂/♀）	*Q* value	Trend
Male-biased	*sox9*	SRY-box containing protein 9	comp36634_c7_seq1	1 879	−0. 20	0. 47	
	sox6	SRY-box containing protein 6-like	comp21968_c1_seq1	560	2. 53	0. 023	
	ar	Androgen receptor	comp34574_c1_seq1	411	−0. 51	0. 69	
	gsf	gonadal soma derived factor 1	comp38118_c0_seq1	2 047	1. 33	1. 4E-160	Up
	mnd1	meiotic nuclear division protein 1 homolog	comp34969_c0_seq1	1 598	−0. 75	2. 7E-07	
	mis	Müllerian inihibiting substance	comp15802_c0_seq1	3 092	0. 22	0. 477	
	hsp90a	Cytosolic heat shock protein 90 alpha	comp18173_c0_seq1	294	0. 07	1. 00	
	dnali1	Axonemal dynein light intermediate polypeptide 1	comp21140_c0_seq1	1 007	8. 30	5. 61E-32	Up
	ropn1l	Ropporin-1-like protein-like isoform X2	comp28799_c0_seq1	1 720	3. 26	3. 45E-19	Up
Female-biased	*cyp19a*	P450 aromatase	comp20558_c0_seq1	2 026	−7. 64	5. 1E-45	Down
	zpc5	Zona pellucida glycoprotein 5	comp20390_c0_seq1	1 801	−12. 65	0	Down
	zar1	Zygote arrest protein 1-like	comp20043_c0_seq1	1 446	−12. 54	0	Down
	gtc	Gonadotropin alpha	comp28377_c1_seq1	830	−11. 45	0	Down
	gdf9	Growth differentiation factor 9	comp30868_c0_seq1	3 680	−5. 76	0	Down
	star	Steroidogenic acute regulatory protein	comp23962_c0_seq1	967	0. 32	1	
	start5	StAR-related lipid transfer protein 5-like	comp35171_c2_seq1	5 165	−1. 50	0	Down
	start7	StAR-related lipid transfer protein 7-like	comp37065_c0_seq1	3 481	−2. 00	0	Down
	sox17	SRY-box containing protein 17	comp26556_c1_seq1	266	1. 07	0. 48	

（续表）

	Gene name	Annotation	Query name	Length / bp	Expression quantity（♂/♀）	*Q* value	Trend
Female -biased	*sox19*	SRY-box containing protein 19	comp34381_c0_seq1	2 961	−2. 04	0	Down
	foxl2	Forkhead transcription factor L2	comp34087_c3_seq1	1 401	−5. 00	2. 68E−46	Down
	vtgr	Vitellogenin receptor isoform 1	comp36664_c17_seq1	3 808	−3. 35	0	Down
	zp4	Zona pellucida sperm-binding protein 4-like	comp30721_c0_seq1	2 238	−9. 38	3. 1E−122	Down
	zp3	Zona pellucida sperm-binding protein 3-like	comp29565_c0_seq1	1 247	−13. 66	0	Down
	zp	Zona pellucida sperm-binding protein	comp28111_c0_seq1	1 390	−12. 61	0	Down
	fem1c	Protein fem-1 homolog C-like	comp35674_c0_seq1	4 180	−2. 02	1. 2E−151	Down
	fzd3	Frizzled-3-like	comp36698_c1_seq1	3 514	−1. 84	5. 9E−209	Down
	42sp43	P43 5*S* RNA-binding protein-like	comp20349_c2_seq1	1 515	−8. 56	0	Down
	gtf3a	Transcription factor IIIA-like	comp17728_c0_seq1	1 282	−14. 12	0	Down
	wee2	Wee1-like protein kinase 2-like isoform X1	comp36348_c0_seq1	1 904	−7. 57	0	Down
	ovca2	Ovarian cancer-associated gene 2 protein homolog	comp31026_c0_seq1	1 363	−5. 28	2. 2E−289	Down
	ctss	Cathepsin S-like	comp38095_c0_seq1	2 038	−8. 92	0	Down
Bisexual	*vasa*	Vasa	comp36885_c2_seq1	2 658	−0. 21	0	
Uncon-firmed	*dmrt2*	Doublesex and mab-3-related transcription factor 2	comp30078_c0_seq1	1 113	−4. 68	9. 19E−07	Down
	dmrt3	Doublesex and mab-3-related transcription factor 3	comp29037_c2_seq1	1 392	−1. 55	0. 00073	Down

(续表)

	Gene name	Annotation	Query name	Length / bp	Expression quantity (♂/♀)	*Q* value	Trend
Uncon-firmed	***dmrt5***	Doublesex and mab-3-related transcription factor 5	comp556746_c0_seq1	291	−1. 51	0. 519288	
	sox8a	SRY-box containing protein 8a	comp36634_c2_seq1	2 437	−1. 54	2. 35E-13	Down
	shbg	Sex hormone binding globulin	comp17530_c0_seq1	2 212	−1. 33	4. 1E-62	Down
	cyp11b	Steroid 11-beta-hydroxylase	comp248709_c0_seq1	422	1. 49	0. 23	

加粗的基因是大菱鲆转录组中新发现基因

(3) 注释的生长相关基因有 GH/IGF-I 内分泌调控轴的主要组分，如生长激素及其受体基因(*gh*/*ghr*)、类胰岛素生长因子及其受体基因(*igf*/*igfr*)、生长激素释放/抑制激素基因(*ghrh*/*ghih*)，它们是鱼类骨骼肌生成的主要调控基因。肌生调节因子 5 基因(*myf5*)是重要的肌肉生长调节因子，其内含子单碱基突变与牛体重增加相关，表明该基因可以作为经济品种标记辅助育种的候选基因。还有在大菱鲆首次发现的促生长因子 ESTs，如脂肪酸结合蛋白基因(*fabp*)、催乳激素及其受体基因(*prl*/*prlr*)(表 17)。

表 17 大菱鲆转录组中鉴定出的生长和肌肉发育相关的基因

	Gene name	Annotation	Query name	Length/bp	Expression quantity (♂/♀)	*Q* value	Trend
Components of GH/IGF-I axis	*gh*	Growth hormone precursor	comp17564_c0_seq1	775	−10. 66	4. 01E-123	Down
	ghr	Growth hormone receptor	comp34224_c0_seq1	1 177	−0. 78	0. 04	
	ghrh	Growth hormone releasing hormone	comp374142_c0_seq1	325	2. 66	0. 10	
	ghih	Growth hormone inhibiting hormone-like	comp23996_c0_seq1	1 004	0. 32	0. 59	
	sstr	Somatostatin receptor	comp24799_c1_seq1	1 202	−1. 13	0. 0039	
	igf	Insulin-like growth factors	comp38041_c8_seq1	3 754	−0. 88	1. 85E-45	
	igfr	IGF receptor	comp33325_c0_seq1	3 008	−0. 44	0. 040	
	igfbp	IGF binding protein	comp32674_c4_seq1	1 454	−1. 43	4. 03E-51	Down

(续表)

	Gene name	Annotation	Query name	Length/bp	Expression quantity (♂/♀)	Q value	Trend
Transforming growth factors	*mstn*	Myostatin	comp20346_c1_seq1	2 996	0. 16	0. 60	
	myf5	Myogenic regulatory factor 5	comp37586_c17_seq1	1 393	−2. 65	1. 21E−34	Down
Others	*fabp*	Fatty acid-binding protein	comp38300_c0_seq1	606	−4. 13	0	Down
	htr	5-Hydroxytryptamine receptor	comp26484_c1_seq1	1 741	1. 66	0. 48	
	prl	Prolactin	comp38254_c0_seq1	747	−15. 12	0	Down
	prlr	Prolactin receptor	comp37125_c1_seq1	6 331	−2. 36	2. 99E−188	Down

(4)首次在大菱鲆中报道了参与免疫响应的基因:① Toll 样受体基因 *tlr1*、*tlr21*、*tlr22*,白介素基因 *il15*、*il34*,白介素受体基因 *il1r1*、*il18r1*,这些基因参与 Toll 样受体信号通路;② 参与细胞凋亡的 *TRAF2* 基因;③ 新的补体 *C9*;④ 其他的免疫分子基因,如胶原凝集素基因(*colec*)、选择素基因(*slep*)(表 18)。

表 18 大菱鲆转录组中新发现的免疫响应相关基因

	Gene name	Annotation	Query name
Toll-like receptor signaling pathway	*tlr1*	Toll-like receptor 1	comp32484_c3_seq1
	tollip	Toll-interacting protein	comp21400_c0_seq1; comp23286_c0_seq1; comp23286_c1_seq1
	traf3	PREDICTED: TNF receptor-associated factor 3-like	comp22885_c1_seq1; comp23509_c0_seq1
	tlr21	Toll-like receptor 21	comp32484_c3_seq1
	tlr22	Toll-like receptor 22	comp36502_c11_seq1
	il34	Interleukin-34	comp30033_c0_seq1
	il18r1	Interleukin-18 receptor 1	comp3046_c0_seq1; comp22532_c0_seq1
	il1r1	Interleukin-1 receptor 1	comp28977_c1_seq1
Apoptosis or programmed cell death	*traf2*	PREDICTED: TNF receptor-associated factor 2-like	comp26413_c0_seq1; comp26467_c3_seq1; comp26467_c5_seq1
Complement pathway	*c9*	PREDICTED: Complement component C9-like	comp37672_c0_seq1

（续表）

	Gene name	Annotation	Query name
Other immune molecules	*slep*	Selectin P	comp35827_c0_seq1
	colec	Collectin 10 / 11 / 12	comp31521_c1_seq1; comp31124_c0_seq1; comp30265_c0_seq1; comp31642_c0_seq1

（5）从转录组鉴定出大量的遗传标记，包括 21 192 个 SSR 和 8 642 个 SNP（图 19）。随机选取了 100 个位点进行引物合成和鉴定，其中有 70 对引物扩增出目标片段。用 1 个大菱鲆家系的 30 个个体进行多态性评估，发现 17 个 EST-SSR 位点至少是中等多态性(图 20)。

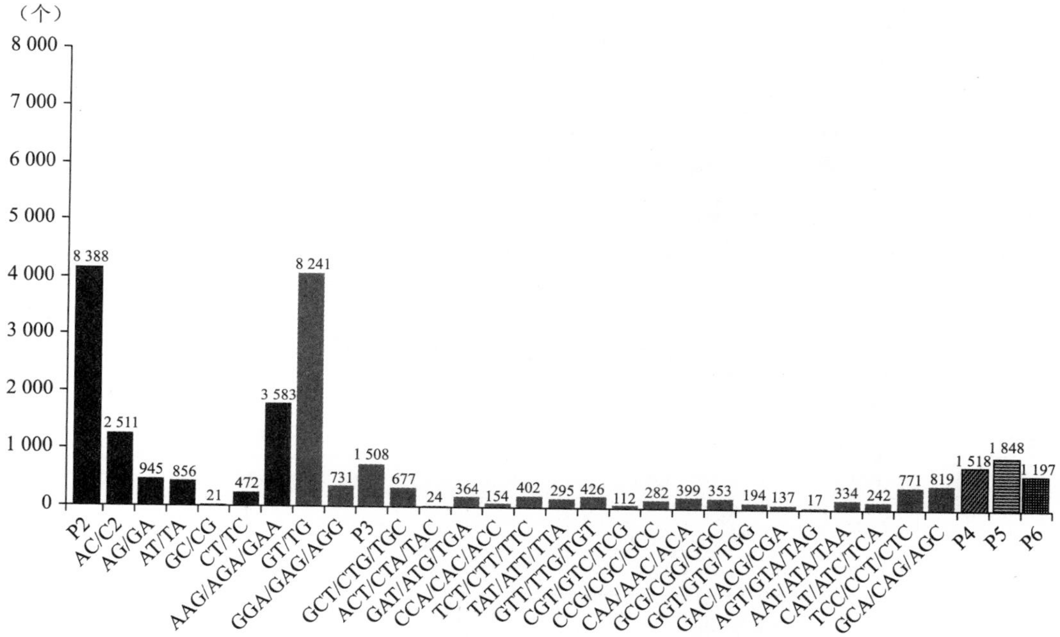

图 19　大菱鲆转录组中基于基序长度的 SSR 分布

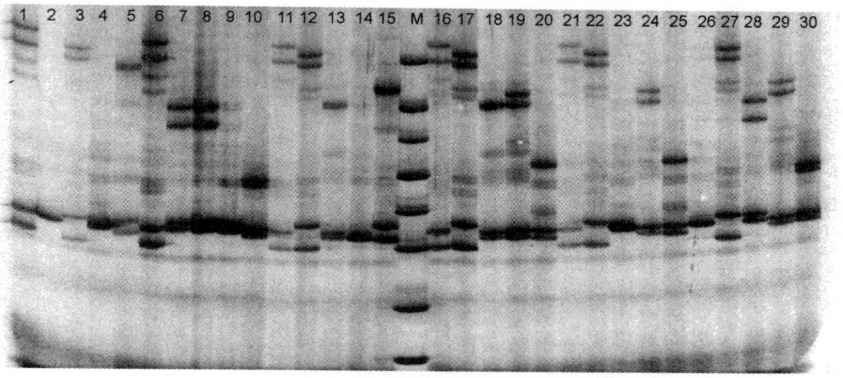

图 20　引物(**comp15993_c0_seq1**)扩增大菱鲆 **30** 个个体的聚丙烯酰胺凝胶电泳图

本研究共获得 8 642 个 SNP，其中有 4 894 个转换（Transition，Ts）、3 748 个颠换（Transversion，Tv），Ts：Tv＝1.306：1。四种类型的转换是最常见的 SNP 类型，而 GC/CG 是数量最少的颠换（图 21）。多态信息含量分析和连锁不平衡检验证明，得到的大多数 SNP 位点能够用于大菱鲆的遗传学分析（表 19）。21 个 SNP 位点所在的 20 个 contig 均有明确、有效的功能注释，涉及细胞周期调控、细胞骨架、能量转换及 RNA 的加工修饰等（表 20）。

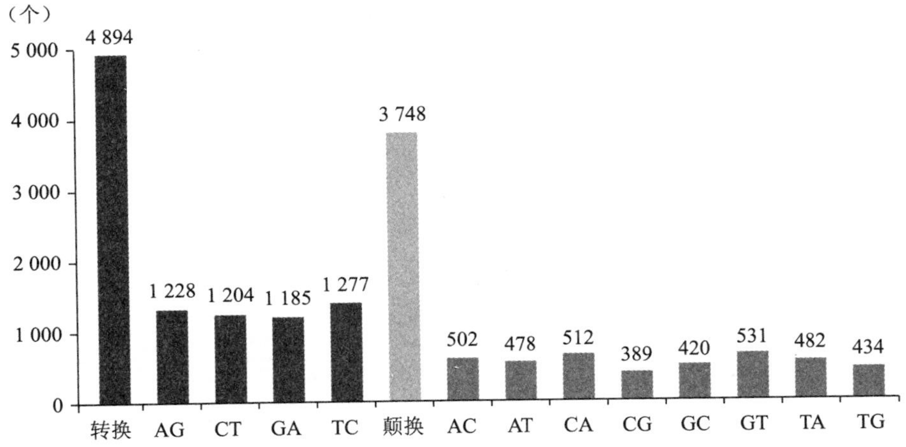

图 21　大菱鲆转录组检测到的 SNP 类型分布

表 19　大菱鲆 21 个 SNP 位点的遗传多态性

SNP 位点	多态信息含量	有效等位基因数	平均观测杂合度	平均期望杂合度	最小等位基因频率
8S82	0.370	1.958	0.854	0.495	0.427 1
CS1	0.373	1.986	0.915	0.502	0.457 4
CS10	0.373	1.983	0.907	0.502	0.453 5
CS3	0.374	1.958	0.505	0.495	0.427 1
CS4	0.374	1.992	0.723	0.503	0.468 1
CS5	0.235	1.374	0.325	0.276	0.162 5
CS6	0.259	1.441	0.333	0.310	0.188 9
CS9	0.374	1.991	0.841	0.503	0.465 9
S10	0.371	1.968	0.447	0.497	0.436 2
S11	0.446	2.051	0.638	0.518	0.042 6
S12	0.375	2.000	1.000	0.505	0.500 0
S14	0.308	1.614	0.511	0.384	0.255 3
S15	0.358	1.878	0.575	0.472	0.372 3
S16	0.374	1.990	0.605	0.503	0.465 1
S22	0.344	1.792	0.447	0.447	0.329 8

（续表）

SNP 位点	多态信息含量	有效等位基因数	平均观测杂合度	平均期望杂合度	最小等位基因频率
S3	0. 305	1. 600	0. 417	0. 379	0. 250 0
S42	0. 332	1. 724	0. 333	0. 425	0. 300 0
S62	0. 366	1. 932	0. 479	0. 488	0. 4062
S63	0. 375	1. 999	0. 896	0. 505	0. 489 6
S7	0. 226	1. 352	0. 256	0. 264	0. 153 8
S9	0. 374	1. 992	0. 648	0. 503	0. 468 8

表 20　20 个 SNP 位点所在基因的注释信息

SNP locus	Query name	Annotation	Length of contig /bp
8S82	comp30956	carbonic anhydrase 4-like	1 645
CS1	comp38245	ribosomal protein L7a	1 089
CS10	comp31070	uridine-cytidine kinase 2-B-like	1 108
CS3	comp28103	cyclin-A2-like	2 668
CS4	comp35299	ribonucleoside-diphosphate reductase subunit M2-like isoform X2	1 638
CS5	comp38118	tubulin alpha-1C chain isoform X1	2 207
CS6	comp17868	mid1-interacting protein 1-B-like	702
CS9	comp18331	gem-associated protein 6-like	1 064
S10	comp29672	RNA-binding protein PNO1	1 298
S11	comp32333	makorin 4 protein	2 310
S12	comp32257	microfibril-associated glycoprotein 4-like	1 547
S14	comp33063	eautophagy-related protein 101-like	1 444
S15	comp31115	ATPase，Na^+/K^+ transporting，beta 2a polypeptide	1 149
S16	comp28490	tumor suppressor protein p53	3 138
S22	comp28329	ATP synthase subunit alpha，mitochondrial-like	2 306
S3	comp17656	caprin-2-like	3 546
S42	comp31217	ubiquitin carboxyl-terminal hydrolase isozyme L5-like isoform 1	1 527
S62/63	comp28258	putative deoxyribonuclease TATDN2-like	2 760
S7	comp29640	calphotin-like	2 824
S9	comp27994	nucleolar RNA helicase 2-like	3 633

（岗位科学家　马爱军）

鲆鲽类全雌苗种生产技术研发进展

全雌苗种生产岗位

1 牙鲆

1.1 全雌苗种生产

亲鱼数量:依托北戴河综合试验站,现保有牙鲆各种育种材料共计 12 540 尾。其中性成熟的野生亲鱼、抗病亲鱼、克隆亲鱼、优良家系及其伪雄鱼计 1 384 尾,后备亲鱼 2 156 尾,共计 3 540 尾。2016 年新制备各种优良家系后备亲鱼 9 000 余尾,其中伪雄鱼 1 000 尾。

牙鲆优质受精卵生产和推广:2016 年 1～2 月,对牙鲆养殖主产区全雌卵需求时间和需求量进行调研,并根据市场需要制定全雌牙鲆亲鱼培育促熟方案,开始亲鱼促熟培育。3 月中旬开始生产全雌牙鲆优质受精卵。2016 年度共生产全雌牙鲆优质受精卵约 2 亿粒,共推广约 6 360 余万粒(53 kg)。推广地区有辽宁、河北、天津以及山东等地。目前,北戴河综合试验站生产的全雌牙鲆优质受精卵,可以满足全国养殖牙鲆的需求。

1.2 非照射法诱导牙鲆雄核发育双单倍体及雄核发育克隆系的建立

雄核发育是一种胚胎只具有父本核遗传信息的特殊生殖方式。在雄核发育中,母本卵子只提供胚胎发育所需的营养物质以及线粒体 DNA。传统的雄核发育诱导技术,需要对未受精的卵子进行射线照射灭活处理。通常这种灭活处理采用伽马射线、X 射线或紫外线作为照射源。对这些特定的照射源,另需特定的安全防护装备以确保安全性。同时,在照射过程中,需要特殊的保护液以确保卵子的受精能力。对照射源以及保护液的要求,让传统的雄核发育方法在日常诱导中操作甚为不便。

2011 年,日本学者 Morishima 等在泥鳅(*Misgurnus anguillicaudatus*)中发明了一种新的雄核发育单倍体诱导方法。他们将刚受精的卵子置于 0 ℃～3 ℃的水中处理 60 min,在所孵化的仔鱼中,获得了超过 30%的雄核发育单倍体。随后,利用四倍体泥鳅所产的二倍体精子激活卵子,并进行冷休克处理获得了可存活的二倍体雄核发育泥鳅。利用单倍体精子激活卵子,并先进行冷休克处理诱导雄核发育,随后热休克处理抑制卵裂的方法,制备了泥鳅的雄核发育双单倍体。在斑马鱼(*Danio rerio*)中,利用冷休克加热休克的方法,同样成功地获得了雄核发育双单倍体，并利用雄性雄核发育双单倍体所产的精子激活卵子，再进行一轮雄核发育诱导，制备了斑马鱼的雄核发育克隆系。

牙鲆雌性个体大于雄性，但在一些雄性个体中，会有特殊的优良性状。通过雄核发育的方法，可以快速的固定这些优良性状。但是基于传统照射法诱导牙鲆雄核发育的方法尚未见报道。因此，我们通过参数优化，在牙鲆上首次建立了基于冷休克的非照射法雄核发育单倍体诱导方法，并利用静水压对单倍体胚胎加倍，获得了牙鲆雄核发育双单倍体。

首先是进行冷休克持续时间的优化。将刚受精的牙鲆卵放入 0 ℃的水浴中冷休克处理 15 min、30 min、45 min 和 60 min，统计各处理组的受精率、孵化率、畸形率以及单倍体率。结果显示，对照组的受精率（65. 88% ± 6. 77%）和 30 min、45 min 以及 60 min 组的差异不显著（$P > 0.05$），但和 15 min 组（46. 27% ± 14. 92%）差异显著（图 1）。各冷休克处理组之间受精率的差异不显著。对照组的孵化率最高，为 60. 82% ± 0. 48%。在各冷休克处理组中，15 min 组的孵化率（32. 64% ± 1. 11%）和 30 min 组（46. 76% ± 5. 79%）差异显著。在畸形率指标上，15 min 组（46. 78% ± 2. 40%）最高，和其他处理组差异显著（$P < 0.05$）。

流式细胞仪倍性检测结果表明，在所有对照组的畸形仔鱼中，除了 1 尾为 2. 6 倍体外，其余都为二倍体，且未检测到单倍体（表 1）。在冷休克处理组中，单倍体、三倍体以及超二倍体被检测到。在 15 min 和 30 min 组中，检测到部分二倍体（图 2）。15 min 组的单倍体率（9. 12% ± 1. 38%）显著高于其他组，且差异显著（$P < 0.05$）（图 1）。

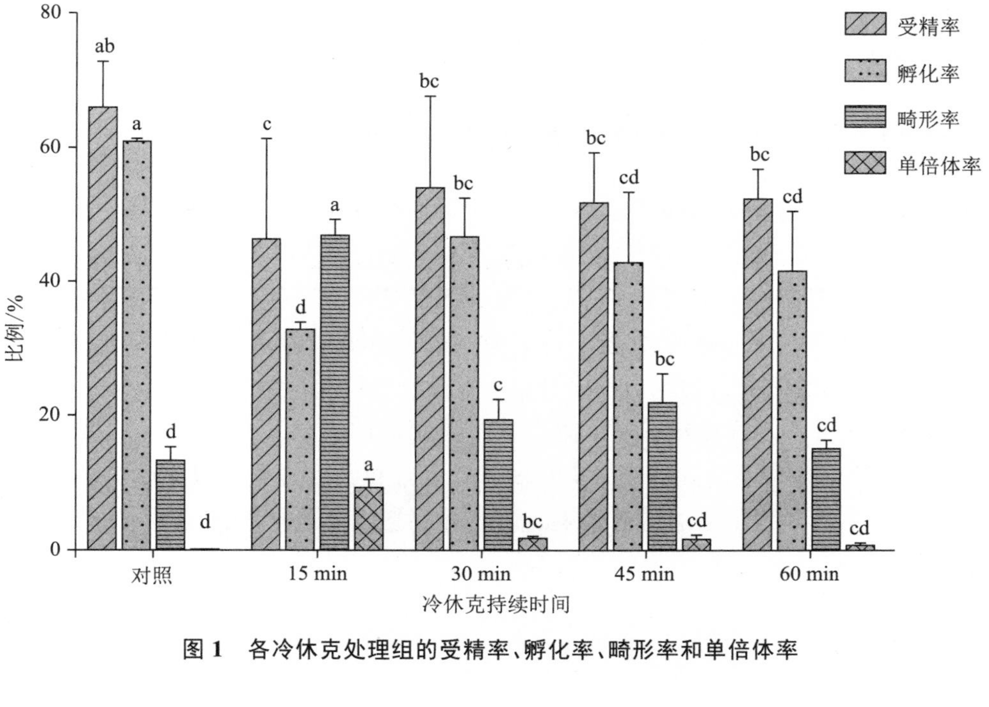

图 1　各冷休克处理组的受精率、孵化率、畸形率和单倍体率

表 1　各处理组畸形仔鱼倍性检测结果

处理组	仔鱼数 / 尾	倍性				
		1N	2N	3N	Hyper-2N	嵌合体
对照组	32	0	31	0	1[a]	0
0 ℃，15 min	70	43	2	12	13[b]	0
0 ℃，30 min	47	9	3	18	15[c]	2[d]
0 ℃，45 min	46	7	0	28	10[e]	1[f]
0 ℃，60 min	29	2	0	20	7[g]	0

a: 2. 6N; b: 2. 2N 4 尾、2. 4N 2 尾、2. 6N 3 尾、2. 8N 4 尾; c: 2. 2N 3 尾、2. 6N 9 尾、2. 8N 3 尾; d: 1N～4. 6N, 1. 5N～2N; e: 2. 2N 3 尾、2. 4N 2 尾、2. 6N 3 尾、2. 8N 2 尾; f: 2. 8N～3N; g: 2. 2N 1 尾、2. 4N 1 尾、2. 6N 1 尾、2. 8N 3 尾

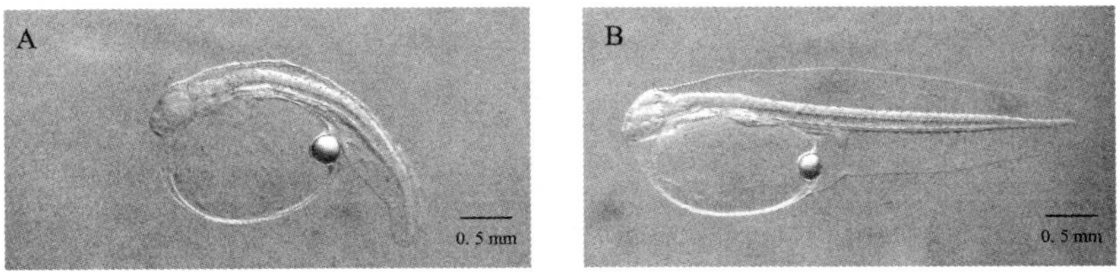

图 2　牙鲆单倍体仔鱼（A）和二倍体仔鱼（B）

利用两个微卫星标记 *Poli1359TUF* 和 *HLJYP38* 对单倍体仔鱼的全父本遗传进行鉴定。在对照组的二倍体仔鱼中，每个个体都有一对等位基因，一个来自于母本，另外一个来自于父本；而在每个单倍体仔鱼中，只检测到等位基因中的一个基因，且来自于父本（表 2）。

表 2　单倍体和二倍体仔鱼的微卫星检测结果

位点	母本	父本	冷休克单倍体	对照组二倍体
Poli1359TUF	194 bp / 194 bp	210 bp / 210 bp	210 bp（8 尾）	194 bp / 210 bp（5 尾）
HLJYP38	190 bp / 190 bp	184 bp / 184 bp	184 bp（8 尾）	190 bp / 184 bp（5 尾）

上述结果表明，对刚受精的牙鲆卵在 0 ℃进行冷休克处理可以诱导雄核发育单倍体，15 min 是最佳的冷休克持续时间。

单倍体因发育畸形而不能存活。为了获得可存活的雄核发育双单倍体，我们将 0 ℃、15 min 冷休克处理后的胚胎转移至 17 ℃水浴中培育 60 min，随后再用 63. 74 MPa 的静水压处理 6 min，从而达到倍性加倍的效果。在所有 5 个批次的诱导中，各诱导组的受精率在 32. 30% 到 73. 99%间，表明了组间的高差异性（表 3）。在出膜期，各组均有正常仔鱼出膜，但正常仔鱼比例的组间差异同样很大。流式细胞倍性检测结果表明各组均有二倍体存活，但二倍体比例很低，在 0. 81% 到 1. 79%之间。

表 3　各冷休克加静水压处理组的受精率、孵化率、二倍体率等

处理组	所用卵个数	受精卵个数（受精率 / %）	孵化仔鱼尾数（孵化率 / %）	正常仔鱼尾数（正常率 / %）	开口期存活仔鱼尾数（存活率 / %）	二倍体仔鱼尾数（二倍体率 / %）
1	359	116（32. 30）	86（23. 96）	49（56. 98）	32（37. 21）	4（1. 11）
2	735	289（39. 32）	190（25. 85）	75（39. 47）	43（22. 63）	8（1. 08）
3	1107	649（58. 63）	257（23. 22）	110（42. 80）	85（33. 07）	10（0. 81）
4	223	165（73. 99）	113（50. 67）	32（28. 31）	28（24. 78）	4（1. 79）
5	1232	695（56. 41）	377（30. 60）	127（33. 69）	71（18. 83）	15（1. 22）

这些检测到的二倍体可能是雄核发育双单倍体。我们利用覆盖牙鲆 24 个连锁群的 36 个微卫星标记对来自于处理组 2 的 8 尾二倍体的纯合度进行了检测。结果表明，所有 8 尾二倍体在 36 个位点都为纯合，未检测到杂合位点，可以确认为雄核发育双单倍体（表 4）。因此，通过冷休克加静水压的方法，可以成功诱导牙鲆的雄核发育双单倍体。

表 4　微卫星纯合度检测结果

位点（连锁群）	父本	二倍体子代	位点（连锁群）	父本	二倍体子代
BDHYP39（1）	266 bp / 266 bp	266 bp / 266 bp（8 尾）	*BDHYP468*（12）	264 bp / 282 bp	264 bp / 264 bp（3 尾） 282 bp / 282 bp（5 尾）
BDHYP239（2）	156 bp / 156 bp	156 bp / 156 bp（8 尾）	*Poli609TUF*（13）	192 bp / 202 bp	192 bp / 192 bp（6 尾） 202 bp / 202 bp（2 尾）
Poli1167TUF（3）	174 bp / 190 bp	174 bp / 174 bp（4 尾） 190 bp / 190 bp（4 尾）	*Poli1446TUF*（13）	204 bp / 228 bp	204 bp / 204 bp（6 尾） 228 bp / 228 bp（2 尾）
Poli1831TUF（3）	98 bp / 132 bp	98 bp / 98 bp（5 尾） 132 bp / 132 bp（3 尾）	*BDHYP216*（14）	180 bp / 192 bp	180 bp / 180 bp（7 尾） 192 bp / 192 bp（1 尾）
Poli1361TUF（4）	180 bp / 200 bp	180 bp / 180 bp（3 尾） 200 bp / 200 bp（5 尾）	*BDHYP27*（14）	208 bp / 222 bp	208 bp / 208 bp（4 尾） 222 bp / 222 bp（4 尾）
Poli222TUF（4）	171 bp / 177 bp	171 bp / 171 bp（5 尾） 177 bp / 177 bp（3 尾）	BDHYP116（15）	236 bp / 236 bp	236 bp / 236 bp（8 尾）
Poli1072TUF（4）	218 bp / 236 bp	218 bp / 218 bp（3 尾） 236 bp / 236 bp（5 尾）	*Poli1937TUF*（16）	198 bp / 216 bp	198 bp / 198 bp（3 尾） 216 bp / 216 bp（5 尾）
Poli641TUF（5）	192 bp / 228 bp	192 bp / 192 bp（6 尾） 228 bp / 228 bp（2 尾）	*Poli954TUF*（16）	112 bp / 178 bp	112 bp / 112 bp（3 尾） 178 bp / 178 bp（5 尾）
Poli2022TUF（5）	154 bp / 164 bp	154 bp / 154 bp（6 尾） 164 bp / 164 bp（2 尾）	*Poli1235TUF*（17）	116 bp / 128 bp	116 bp / 116 bp（3 尾） 128 bp / 128 bp（5 尾）
Poli445TUF（6）	182 bp / 206 bp	182 bp / 182 bp（4 尾） 206 bp / 206 bp（4 尾）	*Poli230TUF*（18）	114 bp / 140 bp	114 bp / 114 bp（5 尾） 140 bp / 140 bp（3 尾）
Poli2046TUF（7）	154 bp / 186 bp	154 bp / 154 bp（3 尾） 186 bp / 186 bp（5 尾）	*BDHYP160*（18）	186 bp / 204 bp	186 bp / 186 bp（2 尾） 204 bp / 204 bp（6 尾）
Poli803TUF（7）	154 bp / 192 bp	154 bp / 154 bp（4 尾） 192 bp / 192 bp（4 尾）	*BDHYP170*（19）	208 bp / 226 bp	208 bp / 208 bp（4 尾） 226 bp / 226 bp（4 尾）

(续表)

位点 (连锁群)	父本	二倍体子代	位点 (连锁群)	父本	二倍体子代
Poli194TUF (8)	156 bp / 164 bp	156 bp / 156 bp (4 尾) 164 bp / 164 bp (4 尾)	*BDHYP427* (20)	242 bp / 242 bp	242 bp / 242 bp (8 尾)
Poli106TUF (8)	118 bp / 128 bp	118 bp / 118 bp (5 尾) 128 bp / 128 bp (3 尾)	*BDHYP271* (21)	150 bp / 172 bp	150 bp / 150 bp (3 尾) 172 bp / 172 bp (5 尾)
Poli1450TUF (8)	132 bp / 156 bp	132 bp / 132 bp (5 尾) 156 bp / 156 bp (3 尾)	*Poli2TUF* (22)	136 bp / 136 bp	136 bp / 136 bp (8 尾)
BDHYP455 (9)	182 bp / 182 bp	182 bp / 182 bp (8 尾)	*Poli1817TUF* (23)	168 bp / 182 bp	168 bp / 168 bp (4 尾) 182 bp / 182 bp (4 尾)
BDHYP345 (10)	248 bp / 248 bp	248 bp / 248 bp (8 尾)	*Poli1836TUF* (24)	246 bp / 258 bp	246 bp / 246 bp (6 尾) 258 bp / 258 bp (2 尾)
BDHYP58 (11)	224 bp / 224 bp	224 bp / 224 bp (8 尾)	*Poli1953TUF* (24)	171 bp / 177 bp	172 bp / 172 bp (5 尾) 200 bp / 200 bp (3 尾)

在成功制备牙鲆雄核发育双单倍体的基础上,我们利用一尾雄性双单倍体所产的精子激活卵子,随后用优化的参数进行冷休克加静水压处理制备雄核发育克隆系。基于三个批次诱导所获得的受精率为 43. 14% ± 8. 32% (所用卵数:579～914),孵化率为 29. 54% ± 9. 65%,二倍体率为 0. 52% ± 0. 33%。因为诱导效率较低,我们用了 400 mL 卵来进行大规模的诱导,在 6 月龄时,共有 45 尾二倍体存活。在这 45 尾存活的二倍体中,随机选取 21 尾进行纯合性和克隆性的验证。经过 65 个微卫星标记检测发现,在 21 尾二倍体中,只有 1 尾在 5 个位点为杂合,60 个位点为纯合,剩余 20 尾在所有 65 个位点都为纯合,且各尾鱼在每个位点的基因型相同。微卫星的结果表明,这 20 尾鱼为克隆。

牙鲆非照射法雄核发育诱导技术的建立以及雄核发育克隆系的成功制备,为充分利用杂种优势,实现牙鲆良种的克隆化奠定了坚实的基础。

1.3 牙鲆抗淋巴囊肿病家系选育及生长和抗病性能分析

选取 2012 年制备的经淋巴囊肿病毒染毒后获得的未患病的 8 尾雌性和 5 尾雄性牙鲆,制备 5 个普通家系(1 尾雌鱼配 1 尾雄鱼) C1～C5 和 3 个雌核发育家系 G1～G3。另外,以未经病毒感染的 1 尾野生雌鱼配 1 尾野生雄鱼所获得的家系作为对照组。分别在 120 d、180 d、240 d 和 300 d 时,从每个家系随机挑选 60 尾进行生长性状的测量。并使用 Motic Images Plus 2. 0 软件测量全长、体长、头长、体高、尾柄长和尾柄高。

染毒试验自 2015 年 10 月 12 日开始,至 2016 年 6 月 3 日结束。将荧光标记完的各家系鱼平均分成 4 个池子(每个池子 42 m^2)进行混合饲养。试验期间,养殖水温维持在 13 ℃～23 ℃,盐度 10～27。在染毒试验过程中,定期观察染病情况,并进行记录。试验结束后,统计各家系健康个体和患病个体的比例,并测量各家系抗病和患病个体体重和体长。判定是否抗病的标准为:如果用肉眼观察到鱼体表或鳃长有颗粒状囊肿,则认定该鱼为患病个体;

如果体表或鳃未观察到颗粒状囊肿，则判定该鱼为抗病个体。

各家系生长的比较结果表明，在体重性状上，120 d 和 240 d 时，家系 G2 的体重显著高于其他家系，差异显著($P < 0.05$)。180 d 时，家系 G1、G2 和 C5 之间的体重差异不显著，但体重均高于其他家系($P < 0.05$)。而 300 d 时，家系 G1 的体重值最高，其次为家系 G2 和 G3。在各个日龄的体重变异系数(CV)上，家系 G2 均在 20%以内，且日龄间变动较小，而家系 C2 在 180 d 时的变异系数达到了 81.63%，可以看出此阶段家系内个体间在体重上存在着较大的差异(表 5)。不同日龄各家系的绝对增重率分析结果显示，除家系 G3 之外，其余家系在 121 d～180 d 期间的生长速度最快，家系 C1、C2、C5 和 G2 在 181 d～240 d 期间的生长速度快于 241 d～300 d 期间，而 C3、C4、G1 和对照组则相反。各家系在 181 d～240 d 和 241 d～300 d 期间的生长速度相近，且快于 121 d～180 d 期间。所有家系的生长速度在 1 d～120 d 期间最慢。家系的生长速度比较表明：1 d～120 d 和 121 d～180 d 期间，家系 G2 的生长速度最快，其次为家系 G1；181 d～240 d 期间，家系 G3 的生长速度最快，其次为 G2；而 241 d～300 d 期间，家系 G1 的生长速度最快，其次为 G3(表 6)。

表 5　9 个牙鲆选育家系不同日龄体重的比较

家系	体重/g				变异系数 CV/%			
	120 d	180 d	240 d	300 d	120 d	180 d	240 d	300 d
对照组	23.787 ± 8.227^{a}	110.500 ± 24.321^{a}	156.926 ± 45.47^{ac}	216.339 ± 59.759^{a}	34.59	22.01	28.97	27.62
C1	32.178 ± 8.045^{b}	101.963 ± 32.863^{a}	161.632 ± 47.407^{ab}	209.638 ± 50.929^{a}	25.02	32.19	29.33	24.29
C2	32.330 ± 5.562^{b}	102.942 ± 84.034^{a}	155.408 ± 40.424^{a}	206.357 ± 48.757^{a}	17.20	81.63	26.01	23.63
C3	33.067 ± 10.713^{b}	112.330 ± 35.359^{a}	156.204 ± 51.747^{ab}	216.708 ± 65.400^{a}	32.39	31.48	33.13	30.18
C4	39.715 ± 8.968^{cd}	111.115 ± 30.699^{a}	166.243 ± 48.416^{abd}	224.757 ± 58.509^{a}	22.58	27.63	29.13	26.03
C5	36.122 ± 9.328^{bc}	139.223 ± 28.315^{b}	191.065 ± 35.599^{c}	234.363 ± 45.219^{ab}	25.83	20.34	18.63	19.30
G1	41.070 ± 9.881^{de}	144.496 ± 36.098^{b}	195.509 ± 52.468^{cde}	287.753 ± 82.231^{c}	24.06	24.98	26.84	28.58
G2	44.928 ± 7.750^{e}	160.750 ± 28.550^{b}	225.488 ± 37.020^{e}	262.490 ± 47.827^{bc}	17.25	17.76	16.42	18.22
G3	25.558 ± 3.630^{a}	103.923 ± 20.141^{a}	182.920 ± 32.078^{bc}	261.965 ± 53.673^{bc}	14.20	19.38	17.54	20.49

同列中不同上标字母表示不同家系间体重差异显著($P < 0.05$)

表 6　9 个牙鲆选育家系不同日龄的绝对增重率的比较

家系	绝对增重率/(g/d)			
	1 d～120 d	121 d～180 d	181 d～240 d	241 d～300 d
对照组	0.198^{9}	1.445^{4}	0.774^{8}	0.990^{4}
C1	0.268^{7}	1.163^{9}	0.995^{3}	0.800^{7}
C2	0.269^{6}	1.177^{8}	0.875^{5}	0.849^{6}
C3	0.276^{5}	1.321^{5}	0.731^{9}	1.009^{3}
C4	0.331^{3}	1.190^{7}	0.919^{4}	0.975^{5}

（续表）

家系	绝对增重率 /（g/d）			
	1 d～120 d	121 d～180 d	181 d～240 d	241 d～300 d
C5	0.301 ^{4}	1.718 ^{3}	0.864 ^{6}	0.722 ^{8}
G1	0.342 ^{2}	1.724 ^{2}	0.850 ^{7}	1.537 ^{1}
G2	0.374 ^{1}	1.930 ^{1}	1.079 ^{2}	0.617 ^{9}
G3	0.213 ^{8}	1.306 ^{6}	1.317 ^{1}	1.318 ^{2}

上角标数字表示增重率排序

表 7　9 个牙鲆选育家系不同日龄体长的比较

家系	体长 / cm				变异系数 CV / %			
	120 d	180 d	240 d	300 d	120 d	180 d	240 d	300 d
对照组	12.641 ± 2.116^{bc}	20.505 ± 1.823^{cd}	20.967 ± 2.420^{ab}	24.552 ± 2.408^{a}	16.77	8.87	11.54	9.82
C1	11.921 ± 1.233^{b}	19.981 ± 2.439^{cd}	22.529 ± 2.334^{b}	25.453 ± 2.315^{ab}	10.32	12.21	10.34	9.12
C2	12.552 ± 1.018^{bc}	18.782 ± 1.920^{ab}	22.457 ± 2.144^{b}	25.399 ± 2.173^{ab}	8.13	10.22	9.53	8.54
C3	11.980 ± 1.437^{b}	19.583 ± 2.285^{bc}	20.392 ± 2.361^{a}	25.302 ± 3.147^{ab}	12.02	11.64	11.57	12.45
C4	13.578 ± 1.330^{d}	19.849 ± 1.865^{bd}	21.141 ± 2.043^{ac}	26.078 ± 2.546^{b}	9.79	9.55	9.65	9.78
C5	12.787 ± 2.404^{bd}	19.718 ± 1.449^{bd}	22.414 ± 1.944^{b}	25.375 ± 2.146^{ab}	18.76	7.35	8.66	8.47
G1	13.413 ± 1.99^{cd}	21.042 ± 1.718^{de}	22.481 ± 2.364b^{cd}	25.714 ± 2.838^{ab}	8.95	8.17	10.50	11.05
G2	13.575 ± 1.583^{d}	21.821 ± 1.749^{e}	23.947 ± 1.709^{d}	26.067 ± 1.489^{b}	11.63	8.02	7.14	5.72
G3	10.902 ± 0.705^{a}	18.035 ± 1.408^{a}	20.997 ± 1.535^{ac}	25.033 ± 2.099^{ab}	6.51	7.82	12.77	8.39

同列中不同上标字母表示不同家系间体长差异显著（$P < 0.05$）

从表 7 可见，在所有家系中，C4 和 G2 在 120 d 和 300 d 时的体长最长，和其他家系差异显著（$P < 0.05$），而 180 d 和 240 d 时，家系 G2 的体长要大于其他家系（$P < 0.05$）。在绝对增长率指标上，除家系 C4 之外，其他家系在 121 d～180 d 期间的体长生长速度最快，其次是 1 d～120 d 期间，家系 C4 在 1 d～120 d 期间的体长生长速度要快于 121 d～180 d 期间。181 d～240 d 期间，除了家系 C2 和 G2 的体长生长速度要快于 241 d～300 d 期间，其余家系此期间的体长生长速度均要慢于 241 d～300 d 期间（表 8）。

表 8　9 个牙鲆选育家系不同日龄的绝对增长率

家系	绝对增长率 /（cm/d）			
	1 d～120 d	121 d～180 d	181 d～240 d	241 d～300 d
对照组	0.105 ^{4}	0.131 ^{3}	0.007 ^{9}	0.060 ^{3}
C1	0.099 ^{6}	0.134 ^{2}	0.040 ^{4}	0.043 ^{6}
C2	0.105 ^{4}	0.104 ^{8}	0.061 ^{1}	0.049 ^{5}

（续表）

家系	绝对增长率 /（cm/d）			
	1 d～120 d	121 d～180 d	181 d～240 d	241 d～300 d
C3	0. 100 [5]	0. 127 [4]	0. 014 [8]	0. 082 [1]
C4	0. 113 [1]	0. 105 [7]	0. 022 [7]	0. 082 [1]
C5	0. 107 [3]	0. 116 [6]	0. 045 [3]	0. 049 [5]
G1	0. 112 [2]	0. 127 [4]	0. 024 [6]	0. 054 [4]
G2	0. 113 [1]	0. 138 [1]	0. 036 [5]	0. 035 [7]
G3	0. 091 [7]	0. 119 [5]	0. 050 [2]	0. 067 [2]

上角标数字表示增长率排序

不同家系在不同日龄体重和体长生长的比较表明，虽然各家系牙鲆在不同时期有各自不同的生长规律，呈现出家系间生长规律的不一致性。但在所有家系中，家系 G2 各个时期的生长表现始终是排名靠前的。

自 2015 年 10 月染毒试验开始至 2016 年 2 月底，各家系均未发现有患淋巴囊肿的个体。自 2016 年 3 月开始，逐渐有个别鱼开始患淋巴囊肿，其典型症状为患病鱼体表、鳃或吻部等增生白色或血红色淋巴颗粒，增生的淋巴颗粒成团或成簇，形如菜花。而抗病个体在整个染毒试验养殖过程中，身体的各个部位均未有淋巴囊肿物出现（图 3）。随着养殖时间的增加，患病鱼所增生的囊肿物越来越多，而且患病鱼比例开始增加。至 5 月中旬开始至 6 月 3 日，患病个体的比例保持稳定，未有新的患病个体出现。因此，在 6 月 3 日统计各家系抗病和患病个体的比例。统计结果显示，对照组的抗病保护率最低，只有 59. 57%，而实验组中，G2 的抗病保护率最高（97. 20%），其次为 C5 （84. 49%），C1、C2、G1 和 G3 家系抗病保护率在 70%～80%之间，C4 最低，为 66. 09%（图 4）。在所有家系中，C5 和 G2 的抗病保护率超过了我们设定的 80%的基准线，为抗病家系。家系 C3 的抗病保护率为 79. 10%，表现也较为优秀。

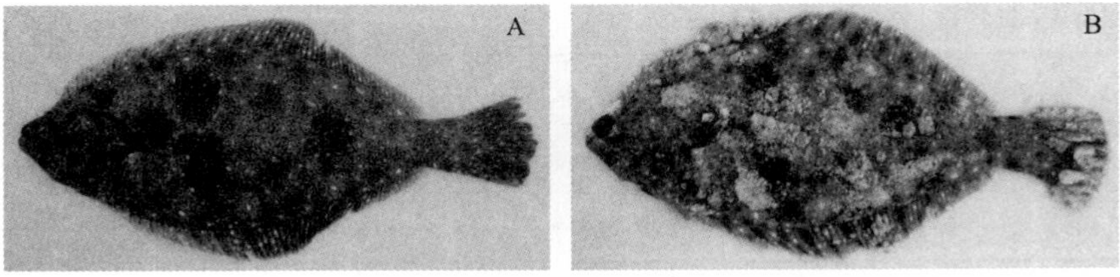

图 3　牙鲆淋巴囊肿抗病个体（A）和患病个体（B）

对各家系抗病个体和患病个体的体重和体长分别进行配对 t 检验。表 9 中的结果显示，在体重性状上，家系 C1 和 G2 抗病个体的平均体重显著高于患病个体（$P < 0.05$），其余家系抗病和患病个体间的体重差异不显著（$P > 0.05$）。家系 G2 抗病个体的平均体重要比对照组抗病个体的平均体重高 24. 47%。在体长性状上，对照组抗病个体的体长要长于患病个

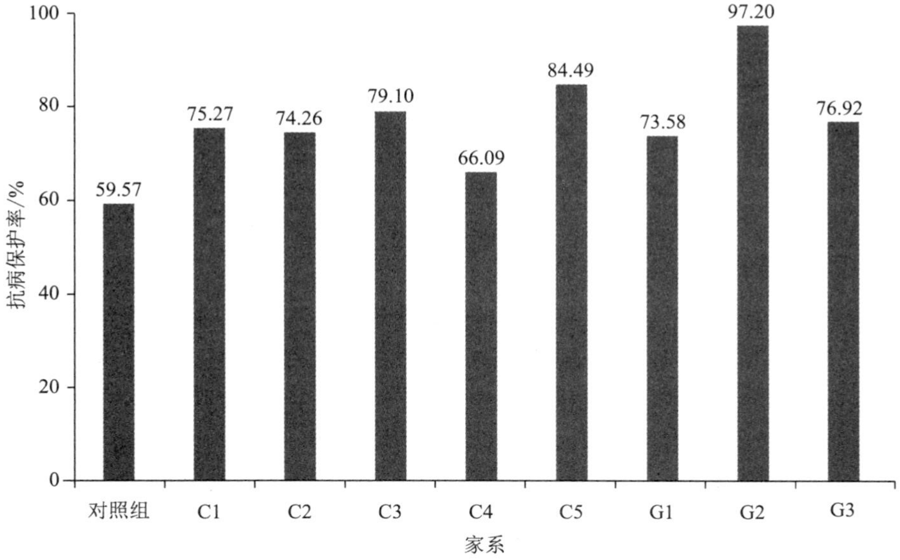

图 4　牙鲆各家系淋巴囊肿抗病保护率

表 9　390 d 牙鲆各家系淋巴囊肿抗病个体和患病个体间体重、体长的配对 *t* 检验

家系	体重 / g		*P* 值	体长 / cm		*P* 值
	抗病	患病		抗病	患病	
对照组	290. 211 ± 91. 338	266. 274 ± 81. 349	0. 302	28. 598 ± 2. 169	26. 314 ± 2. 357	0. 001
C1	304. 700 ± 68. 881	245. 100 ± 79. 109	0. 027	28. 005 ± 2. 392	26. 581 ± 3. 326	0. 162
C2	254. 600 ± 83. 603	216. 700 ± 73. 184	0. 083	26. 485 ± 2. 512	25. 740 ± 2. 672	0. 296
C3	231. 571 ± 76. 191	293. 429 ± 105. 718	0. 072	25. 348± 3. 013	28. 060 ± 3. 053	0. 028
C4	278. 421 ± 99. 694	280. 474 ± 84. 053	0. 948	27. 359 ± 2. 754	27. 719 ± 2. 434	0. 689
C5	277. 330 ± 72. 092	399. 445 ± 453. 801	0. 274	27. 589 ± 2. 213	28. 255 ± 3. 245	0. 257
G1	390. 464 ± 99. 800	408. 721 ± 81. 290	0. 621	30. 087 ± 2. 484	30. 345 ± 2. 428	0. 715
G2	361. 164 ± 60. 706	277. 282 ± 90. 150	0. 012	29. 581 ± 2. 073	27. 268 ± 2. 797	0. 073
G3	377. 505 ± 119. 298	376. 235 ± 105. 442	0. 962	28. 216 ± 2. 720	27. 959 ± 3. 282	0. 735
平均值	304. 255 ± 100. 133	305. 847 ± 189. 665	0. 923	27. 844 ± 2. 742	27. 484 ± 3. 072	0. 217

体，差异显著（$P < 0.05$），而家系 C3 患病个体的体长要显著长于抗病个体（$P < 0.05$），其余家系抗病和患病个体间的差异不显著（$P > 0.05$）。从所有个体总体上分析，无论体重还是体长，抗病和患病个体间的差异均不显著。

390 d 各家系体重、体长和淋巴囊肿抗病性的相关性分析结果显示，体重、体长与抗病性为正相关，但相关性均不显著（$P > 0.05$）（表 10）。上述结果表明，从总体上而言，牙鲆个体是否患淋巴囊肿病对生长不会造成明显的影响，而且个体的患病与否和生长快慢无直接关联。

表 10 390 d 牙鲆体重、体长与淋巴囊肿病抗性的相关性

	体重	体长
Pearson 相关系数	0. 009	0. 087
P	0. 863	0. 110

综上，本研究通过对牙鲆淋巴囊肿抗病个体间的杂交以及人工诱导减数分裂雌核发育的方式，建立了实验家系，并在淋巴囊肿病高发区进行了染毒养殖试验。同时，对各家系的生长情况进行了动态跟踪测量和分析，获得了一个既抗病又生长快的优良家系（抗病保护率：97. 20%；390 d 抗病个体平均体重比对照组抗病个体平均体重高 24. 47%）。以此家系为基础继续选育，有望获得牙鲆抗淋巴囊肿速生新品种。

1. 4 牙鲆抗淋巴囊肿病的全基因组关联分析

全基因组关联分析（Genome-wide association study，GWAS）是一种对全基因组范围内的常见遗传变异（单核苷酸多态性和拷贝数）基因总体关联分析的方法，该方法以自然群体为研究对象，以长期重组后保留下来的基因（位点）间连锁不平衡（Linkage disequilibrium，LD）为基础，将目标性状表型的多样性与基因（或标记位点）的多态性结合起来分析，可直接鉴定出与表型变异密切相关且具有特定功能的基因位点或标记位点。采用 GWAS 技术在全基因组范围内进行研究，能够一次性对多个性状进行定位，适用于定位性状关联区间、功能基因研究、开发性状选育标记等方面的研究。

GWAS 技术最早应用于人类遗传疾病的研究，其基本原理简述如下：在一定群体中选择病例组和对照组（对于数量性状可以是连续分布的群体），比较全基因组范围内所有 SNP 位点的等位基因或者基因型频率在病例组和对照组之间的差异。若某个 SNP 位点的等位基因或基因型在病例组中出现的频率明显高于或低于对照组，则认为该位点与性状存在关联性。之后，根据该位点在基因组中的位置和连锁不平衡关系推测可能的性状相关基因。近年来，GWAS 技术也被推广应用于水产领域。

为了在全基因组水平解析牙鲆淋巴囊肿抗病和患病个体的遗传差异，筛选抗病分子标记和功能基因，在 2016 年，我们对 182 尾牙鲆（抗病和患病个体各 91 尾）进行了全基因组重测序和 GWAS 分析。样品经过 DNA 片段末端修复、加 ploy A 尾、加测序接头、纯化、PCR 扩增等步骤完成整个文库制备。制备好的文库在经过文库质量检测，确保文库质量之后，进行 Illumina HiSeq PE150 高通量测序。

经过对测序数据的严格过滤，得到高质量的clean data。对 182个牙鲆产出数据进行统计，测序数据量为 446. 4 GB，平均每个样品 2. 36 GB，高质量的 clean data 数据量为 430. 08 GB，平均每个样品 2. 43 GB，表明建库测序成功（表 11）。

表 11　部分牙鲆测序数据产出统计

样品	原始碱基数 / bp	过滤后碱基数 / bp	有效率 / %	错误率 / %	Q20 / %	Q30 / %	GC 含量 / %
LV1	2 266 495 800	2 244 171 900	99. 02	0. 03	95. 82	90. 32	42. 61
LV10	2 248 458 000	2 224 363 500	98. 93	0. 03	95. 7	90. 15	42. 69
LV100	2 082 913 500	2 074 054 800	99. 57	0. 04	94. 77	88. 09	42. 26
LV101	3 611 918 700	3 598 482 000	99. 63	0. 03	96. 81	92. 29	41. 92
LV102	3 362 380 200	3 336 565 500	99. 23	0. 03	96. 47	91. 61	42. 32
LV103	3 997 270 200	3 975 734 400	99. 46	0. 03	96. 71	92. 09	42. 23

有效的高质量测序数据通过 BWA 软件(参数:mem -t 4 -k 32 - M)比对到参考基因组,比对结果经 SAMTOOLS 去除重复(参数:rmdup)。

牙鲆参考基因组大小为 534 300 861 bp,群体样本平均比对率为 94. 29%,对基因组的平均测序深度为 4. 04,平均覆盖度为 94. 04%(至少有一个碱基的覆盖)。

样本比对率反映了样本测序数据与参考基因组的相似性,覆盖深度和覆盖度能够直接反应测序数据的均一性及与参考序列的同源性。各个样本的比对结果显示,它们与参考基因组的相似度达到重测序分析的要求,同时又有非常不错的覆盖深度和覆盖度。

表 12　部分样品测序深度及覆盖度统计

样品	比对到基因组 read 数	Read 总条数	比对率 / %	平均测序深度	1 × 覆盖率 / %	4 × 覆盖率 / %
LV1	14 149 440	14 961 146	94. 57	3. 86	93. 93	42. 39
LV10	14 014 341	14 829 090	94. 51	3. 81	93. 68	41. 55
LV100	12 964 299	13 827 032	93. 76	3. 48	92. 62	34. 89
LV101	22 432 488	23 989 880	93. 51	5. 57	98. 36	72. 02
LV102	21 044 471	22 243 770	94. 61	5. 24	97. 97	67. 15

利用 SAMTOOLS 等软件进行群体 SNP 的检测。利用贝叶斯模型检测群体中的多态性位点,对获得的 SNP 进行过滤,共获得 733 301 个高质量的 SNP 用于后续分析。在经过群体进化树分析、主成分分析以及连锁不平衡分析之后,利用 GEMMA 软件的混合线性模型(Mixed linear model, MLM)对 SNP 标记和抗病性状进行关联分析,并通过关联的显著度(*P* value),筛选出潜在的候选 SNP。

分析结果在图 5 中表示。该图为曼哈顿图,为遗传标记效应值即经 *F* 检验的全基因组 *P* 值按染色体上物理位置排序图,横坐标为基因组坐标,纵坐标 $-\log_{10}P$, *P* 值越小关联性越强,表现为纵坐标越大。曼哈顿图中水平的虚线表示显著性水平,当 $-\log_{10}P > 6$ 时,认为该 SNP 与该性状显著关联。

由图 5 可以看到,本次分析得到与淋巴囊肿抗病性状显著相关的 SNP 标记 356 个,其

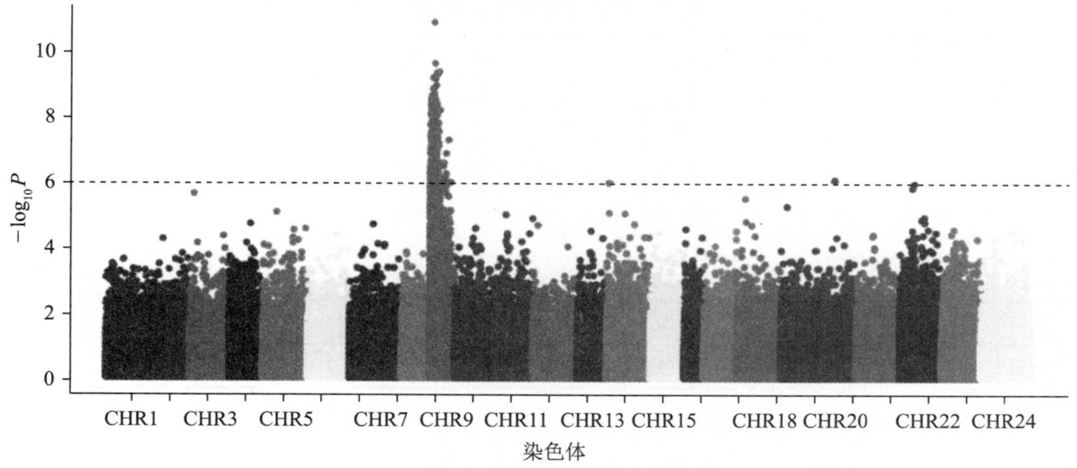

图 5　表型全基因组关联分析曼哈顿图

中只有 1 个位于牙鲆第 20 号染色体上，剩余的 355 个位于牙鲆第 9 号染色体上。和抗病性状显著相关 SNP 在第 9 号染色体上高度富集，表明该染色体上存在潜在的抗病基因。目前，进一步的分析验证工作正在进行中。

1.5　牙鲆生长性状相关 SNP 的发掘

为了深入的挖掘牙鲆种子资源中的优势基因或等位基因，利用北戴河综合试验站培育的各种牙鲆优良资源和丰富的育种材料，对与牙鲆生长相关基因的等位基因进行发掘。通过候选基因关联分析法，对牙鲆 GH、MSTN 基因的 SNP 进行开发，并与牙鲆生长性状进行关联分析，获得与生长性状相关的 SNP 标记，这些 SNP 标记可用于牙鲆标记辅助选育，以缩短育种周期。

首先对数量性状差异明显的 20 个野生个体进行 GH、MSTN 基因重测序，进行序列比对，确定 SNP 位点。重测序和序列分析结果显示。GH 基因中确定一个多态的位点 SNP2-2，多态性为 G-A，位置在第 4 内含子 33 bp（Intron 4-33），属于同义突变。MSTN 基因中获得 7 个多态性 SNP：SNP1-2，G-A，Exon 1-87，密码子 CAG-CAA，氨基酸 Gln29Gln，氨基酸位置 29 为同义突变；SNP1-4，A-T，Intron 1-86；SNP2-4，T-C，Intron 2-725；SNP2-6，C-T，Intron 2-741；SNP2-8，C-T，Exon 3-30，密码子 GAC-GAT，氨基酸 Asp252Asp，氨基酸位置 252 为同义突变；SNP2-9，A-C，Exon 3-156，密码子 CCC-CCA，Pro303Pro，氨基酸位置 30 为同义突变；SNP3-7，C-A，3′UTR-330。

利用制备的杂合克隆 3 个家系、雌核发育 3 个家系，标记后同池混养。培育至 15 月龄时，每个家系随机选取 20 尾，共计 120 尾牙鲆进行数量性状测量和 SNP 分型。

牙鲆生长性状的描述性统计检验结果显示性状符合正态分布，可用于 ANOVA 分析（表 13）。

表 13　牙鲆生长性状描述性统计分析

性状	数量	最小值	最大值	平均数	CV/%	Skewness	Kurtosis
体重	120	18.900	950.300	420.61 ± 19.63	4.67	0.360	−1.072
全长	120	22.800	48.340	34.39 ± 0.53	1.55	0.292	−0.801
体长	120	19.450	43.690	30.00 ± 0.51	1.68	0.312	−0.842
体高	120	7.280	17.980	11.91 ± 0.21	1.43	0.225	−0.913
头长	120	5.660	11.810	8.26 ± 0.12	2.53	0.371	−0.692
尾柄长	120	1.523	4.929	2.91 ± 0.07	1.79	0.434	−0.666
尾柄高	120	1.630	3.930	2.79 ± 0.05	1.77	0.213	−0.940

利用多重 SNaPshot SNP 分型方法对 120 尾牙鲆进行 SNP 位点分型。SNaPshot SNP 分型是一种以单碱基延伸原理为基础,同时利用多重 PCR 对多个已知 SNP 位点进行遗传分型的方法。在一个含有测序酶,四种荧光标记的 ddNTP,紧挨多态位点 5′ 端的不同长度延伸引物和 PCR 产物模板的反应体系中,引物延伸一个碱基即终止,经 ABI3730 测序仪跑胶后,根据峰的颜色可知掺入的碱基种类,从而确定该样本的基因型,根据峰移动的胶位置确定该延伸产物对应的 SNP 位点。

运用 ANOVA 方法将 SNP 位点的基因型与体重、体长、全长、体高、头长、尾柄长、尾柄高等性状进行关联分析,结果显示 SNP2-2、SNP2-6、SNP2-8、SNP3-7 与生长性状显著相关($P<0.05$),GH 基因中 SNP2-2 与生长性状显著相关($P<0.05$)(表 14)。个体验证结果表明 AA 为优势基因型,与 AG 基因型个体生长性状均值差异显著($P<0.05$),与 GG 的均值差异极显著($P<0.01$)。

表 14　GH 基因 SNP2-2 多重比较结果

性状	SNP2-2 基因型	均值	标准差	最小值	最大值
体重/g	AA	455.45[a*]	105.45	246.39	664.51
	AG	429.41[a]	103.76	223.71	635.12
	GG	389.54[b]	96.07	199.07	580.00
全长/cm	AA	35.73[a*]	3.00	29.78	41.67
	AG	34.14[a]	2.95	28.29	39.98
	GG	33.35[b]	2.73	27.93	38.76
体长/cm	AA	30.94[a*]	2.81	25.37	36.51
	AG	29.67[a]	2.77	24.19	35.15
	GG	29.02[b]	2.56	23.94	34.09
体高/cm	AA	12.38[a*]	1.16	10.07	14.68
	AG	11.78[b]	1.14	9.51	14.05
	GG	11.42[b]	1.06	9.33	13.52

同行不同字母表示差异显著($P<0.05$),* 表示差异极显著($P<0.01$)

MSTN 基因分析结果显示(表 15)：SNP2-6 位置为 Intron 2 的 741 bp，突变型为 C-A，基因型 AA 为优势基因型，该基因型的个体数量性状均值显著高于 AC 和 CC 型个体；SNP2-8 位于 Exon 3 的 30 bp，突变型为 C-T，密码子 GAC-GAT，氨基酸 Asp252Asp，氨基酸位置 252 为同义突变，CT 为优势基因型，数量性状均值均极显著高于其他个体；SNP3-7 位置为 3′UTR 的 330 bp，突变型为 C-A，AC 为优势基因型。

对在 MSTN 基因所检测到的 SNP 标记不同生长速度(F 组：生长快速家系 60 尾，S 组：生长缓慢家系 60 尾)的家系中的频率进行 *G* 检验。验证结果显示：优势基因型(SNP2-6：AA，SNP2-8：CT，SNP3-7：AC)与劣势基因型(SNP2-6：CC，SNP2-8：AA、SNP3-7：AA、AC)出现频率在 F 组和 S 组中存在极显著差异(表 16)，并且优势型的个体数量性状均值显著高于劣势基因型个体。MSTN 基因中与生长性状相关的三个 SNP 可用于牙鲆分子辅助育种，快速选育出生长快的牙鲆，以缩短育种周期。

2　大菱鲆

2.1　大菱鲆全雌苗种生产技术研究

2.1.1　减数分裂型雌核发育后备亲鱼培育情况

从本岗位设立以来，课题组累计诱导培育减数分裂型雌核发育大菱鲆苗种 4 批，每批苗种都选留了一定数量生长速度快、体型正常的个体作为后备亲鱼培育，用于性别遗传决定机制和全雌苗种生产研究。截至 2016 年年末，课题组的试验现场青岛综合试验站保有 8 龄、7 龄和 6 龄后备亲鱼合计 56 尾(表 17)，分为春季和秋季两个繁殖群体进行培育管理。2016 年春季繁殖群体 22 尾雌鱼中，21 尾卵巢顺利成熟、排卵，雄鱼 8 尾，均有精液产生。冬季繁殖群体雌鱼 12 尾，雄鱼 14 尾，目前性腺发育状态良好，预期均可作为繁育亲鱼使用。大菱鲆减数分裂型雌核发育后备亲鱼的培育为后续超雌鱼的甄别和制备工作奠定了基础。

2.1.2　有丝分裂型雌核发育亲鱼培育情况

培育大菱鲆有丝分裂型雌核发育诱导群体 1 批次，现保有 3 龄后备亲鱼 31 尾，其中雌鱼 15 尾，雄鱼 16 尾，分春季和冬季两个繁殖群体培育和调控(表 18)。2016 年春季成熟雌鱼 3 尾，产精液雄鱼 5 尾，另有雌雄鱼合计 13 尾调控至冬季产卵。大菱鲆有丝分裂型雌核发育亲鱼的制备和培育，明确了大菱鲆性别遗传决定类型，所产生的超雌鱼一方面可直接用于大菱鲆全雌苗种的生产，另一方面还可以对所生产的全雌卵进行减数分裂雌核发育诱导，快速扩繁超雌亲鱼群体，满足全雌苗种规模化生产的需要。

2.1.3　全雌苗种生产中试试验

2.1.3.1　全雌苗种中试试验结果

2016 年 6 月有丝分裂型雌核发育子代雌鱼成熟 3 尾，PIT 编号为 9344、9324 和 9359，

表 15　MSTN 基因中各 SNP 位点与表型的多重比较结果

位点	基因型	样本数	体重 / g	全长 / cm	体长 / cm	体高 / cm	头长 / cm	尾柄长 / cm	尾柄高 / cm
SNP2-6	AA	60	452. 71 ± 92. 50[a*]	35. 17 ± 2. 65[a*]	30. 79 ± 2. 47[a*]	12. 20 ± 1. 03[a*]	8. 04 ± 0. 61[a*]	2. 93 ± 0. 39[a*]	2. 73 ± 0. 25[a*]
	AC	31	418. 00 ± 101. 29[a]	34. 82 ± 2. 90[a]	30. 26 ± 2. 71[a]	11. 91 ± 1. 12[a]	7. 81 ± 0. 67[b]	2. 69 ± 0. 43[b]	2. 68 ± 0. 27[b]
	CC	29	354. 93 ± 101. 16[b]	32. 25 ± 2. 89[b]	27. 78 ± 2. 70[b]	11. 03 ± 1. 12[b]	7. 22 ± 0. 66[b]	2. 34 ± 0. 43[b]	2. 45 ± 0. 27[b]
SNP2-8	AA	20	166. 74 ± 135. 31[a*]	28. 22 ± 3. 87[a*]	24. 10 ± 3. 62[a*]	9. 23 ± 1. 50[a*]	6. 68 ± 0. 89[a*]	2. 07 ± 0. 57[a]	2. 14 ± 0. 37[a]
	AC	96	372. 88 ± 133. 64[b*]	34. 02 ± 3. 82[b*]	29. 28 ± 3. 57[b*]	11. 77 ± 1. 48[b*]	7. 83 ± 0. 88[b*]	2. 59 ± 0. 57[b*]	2. 66 ± 0. 36[b*]
	CT	4	686. 0 ± 91. 34[c*]	39. 99 ± 2. 61[c*]	35. 45 ± 2. 44[c*]	14. 14 ± 1. 01[c*]	8. 55 ± 0. 60[c*]	3. 30 ± 0. 39[c*]	3. 06 ± 0. 25[c*]
SNP3-7	AA	3	329. 60 ± 125. 95[a]	31. 36 ± 3. 60[a]	27. 14 ± 3. 37[a]	10. 87 ± 1. 40[a]	7. 05 ± 0. 83[a]	2. 41 ± 0. 53[a]	2. 41 ± 0. 34[a]
	AC	39	550. 50 ± 90. 16[b*]	37. 90 ± 2. 58[b*]	33. 12 ± 3. 37[b*]	13. 22 ± 1. 0[b*]	8. 58 ± 0. 59[b*]	3. 07 ± 0. 38[b*]	2. 98 ± 0. 24[b*]
	CC	78	345. 50 ± 89. 47[a]	32. 99 ± 2. 56[a]	28. 56 ± 2. 39[a]	11. 05 ± 0. 99[a]	7. 43 ± 0. 59[a]	2. 48 ± 0. 38[a]	2. 48 ± 0. 24[a]

同行不同字母表示差异显著（$P < 0.05$），* 表示差异极显著（$P < 0.01$）

表 16　SNP 标记基因型验证结果

SNP 标记	基因型	F 组	S 组	*G* 检验	*P* 值
SNP2-6	AA	43	17	40. 557	$P < 0.001$
	AC	17	14		
	CC	0	29		
SNP2-8	AA	0	20	26. 667	$P < 0.001$
	CC	56	40		
	CT	4	0		
SNP3-7	AA	0	3	42. 462	$P < 0.001$
	AC	56	22		
	CC	4	35		

表 17　大菱鲆减数分裂型雌核发育后备亲鱼

序号	标记	体重 / kg	性别	布卵时间	序号	标记	体重 / kg	性别	布卵时间
1	9262	4. 18	♀	2008 年 11 月	31	8872	3. 46	♀	2010 年 4 月
2	8891	3. 78	♀	2010 年 4 月	32	9246	4. 74	♀	2008 年 11 月
3	9026	4. 28	♀	2010 年 4 月	33	9202	4. 70	♀	2008 年 11 月
4	8822	3. 28	♀	2010 年 4 月	34	8827	3. 62	♀	2010 年 4 月
5	9249	4. 24	♀	2008 年 11 月	35	8802	4. 14	♀	2010 年 4 月
6	9271	3. 06	♀	2009 年 4 月	36	9265	4. 08	♀	2008 年 11 月
7	8867	3. 28	♀	2010 年 4 月	37	8857	3. 22	♂	2010 年 4 月
8	8868	2. 32	♀	2010 年 4 月	38	8855	2. 74	♂	2010 年 4 月
9	8883	3. 16	♀	2010 年 4 月	39	8880	2. 34	♂	2010 年 4 月
10	8881	2. 70	♀	2010 年 4 月	40	0076	2. 66	♂	2010 年 4 月
11	8847	4. 24	♀	2010 年 4 月	41	8884	2. 70	♂	2010 年 4 月
12	8873	2. 86	♂	2010 年 4 月	42	9016	2. 86	♂	2010 年 4 月
13	8805	2. 58	♂	2010 年 4 月	43	0938	2. 90	♂	2010 年 4 月
14	8860	1. 90	♂	2010 年 4 月	44	8858	3. 68	♀	2010 年 4 月
15	8850	2. 86	♂	2010 年 4 月	45	9049	3. 58	♀	2010 年 4 月
16	9290	5. 04	♀	2008 年 11 月	46	8854	4. 58	♀	2010 年 4 月
17	8886	3. 46	♀	2010 年 4 月	47	9251	3. 94	♀	2008 年 11 月
18	8840	3. 54	♀	2010 年 4 月	48	9220	3. 72	♀	2008 年 11 月
19	8852	3. 32	♀	2010 年 4 月	49	9226	4. 56	♀	2008 年 11 月
20	8896	3. 92	♀	2010 年 4 月	50	9039	2. 20	♂	2010 年 4 月
21	9237	3. 38	♀	2009 年 4 月	51	8842	3. 02	♂	2010 年 4 月

(续表)

序号	标记	体重/kg	性别	布卵时间	序号	标记	体重/kg	性别	布卵时间
22	9057	4.30	♀	2010年4月	52	8844	2.88	♂	2010年4月
23	9041	4.46	♀	2010年4月	53	9566	2.90	♂	2010年4月
24	9089	4.16	♀	2010年4月	54	8890	2.72	♂	2010年4月
25	8871	3.80	♀	2010年4月	55	9001	3.02	♂	2010年4月
26	9244	3.98	♀	2009年4月	56	9096	3.10	♂	2010年4月
27	8834	2.56	♂	2010年4月					
28	0031	2.32	♂	2010年4月					
29	8863	2.20	♂	2010年4月					
30	8829	2.56	♂	2010年4月					

表18　大菱鲆有丝分裂型雌核发育亲鱼

序号	电子标记	性别	体重/g	
			2015年11月3日	2016年5月26日
1	9369	♂	980	1 240
2	9325	♀	1 100	981
3	9379	♀	960	1 202
4	9309	♂	600	776
5	9333	♀	620	563
6	9388	♂	980	998
7	9310	♂	640	663
8	9366	♀	1 440	1 834
9	9392	♂	740	902
10	9356	♀	580	877
11	9335	♀	700	817
12	9344	♀	1 560	1 864
13	9341	♂	940	944
14	9358	♀	980	925
15	9363	♂	1 360	1 326
16	9318	♀	900	1 146
17	9375	♂	1020	867
18	9326	♂	1 140	1 244
19	9328	♂	1 260	1 203
20	9359	♀	1 500	1 771

（续表）

序号	电子标记	性别	体重 / g	
			2015 年 11 月 3 日	2016 年 5 月 26 日
21	9329	♂	1 000	1 134
22	9324	♀	1 460	1 600
23	9354	♂	1 780	1 855
24	9377	♂	1 600	1 956
25	9311	♂	940	930
26	9374	♂	1 400	1 445
27	9387	♀	500	611
28	9384	♀	980	1 283
29	9350	♀	1 036	1 268
30	9317	♂	987	1 113
31	9347	♀	1 145	1 317

利用成熟超雌鱼与正常雄鱼交配，培育 3 个子代家系，苗种 4 月龄左右，统计各家系得的苗种数量和成活率，3 个全雌苗种家系苗种数量总计 21 090 尾（图 6），各家系苗种成活率分别为 5. 64%、25. 20%和 12. 71%，平均成活率 14. 52%，对照组平均成活率 15. 40%，两者无显著性差异（表 19）。

图 6　大菱鲆全雌苗种育苗车间（左）和全雌苗种（右）

表 19　2016 年大菱鲆全雌苗种生产结果

序号	亲鱼	受精卵 / g	布卵日期	苗种数量	成活率 / %
1	9344	3	6 月 19 日	203	5. 64
2	9324	63	6 月 18 日	19 057	25. 20
3	9359	12	6 月 18 日	1 830	12. 71
4	对照 1	30	6 月 18 日	6 425	17. 85
5	对照 2	20	6 月 23 日	3 108	12. 95

2.1.3.2 3个全雌苗种家系的性别比例

苗种4月龄时,从3个全雌苗种家系和2个对照样分别随机抽取20尾苗种,解剖,肉眼鉴定子代性别。此时幼鱼性腺位于体腔下缘,大部分已经分化成型,左右两侧卵巢均为囊状结构,呈倒戟状且中部向后突出,精巢位于膀胱表面,左右两侧各有1条白色细带状结构(图7),肉眼很容易区分卵巢和精巢。结果如图8所示,两个对照组雌性比例分别为65%和40%,3个超雌鱼家系中,9344后代全部为雌性,9324和9359家系中有1尾性腺呈薄膜状覆盖于膀胱表面,形态类似卵巢,但缺乏典型的囊状结构(图9),无法准确鉴别是否是卵巢。

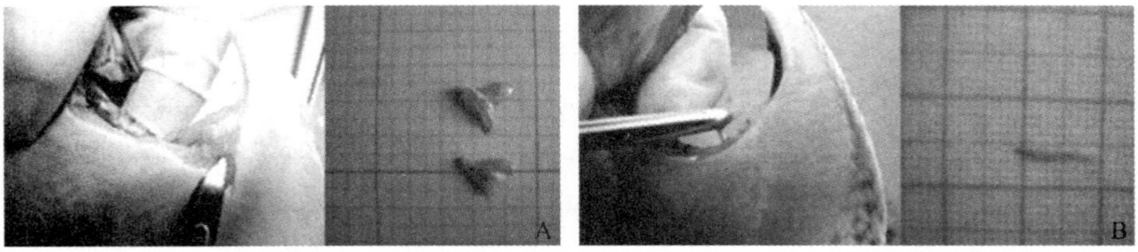

图7 4月龄大菱鲆幼鱼卵巢(A)与精巢(B)解剖学特征

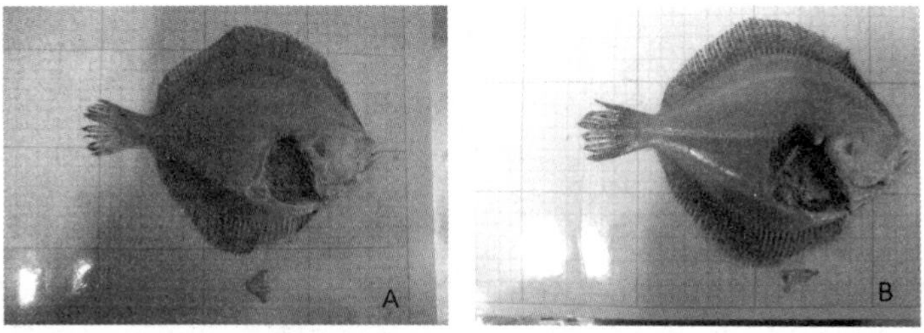

图8 全雌苗种正常发育(A)与发育迟滞(B)卵巢的解剖学特征(波恩染色)

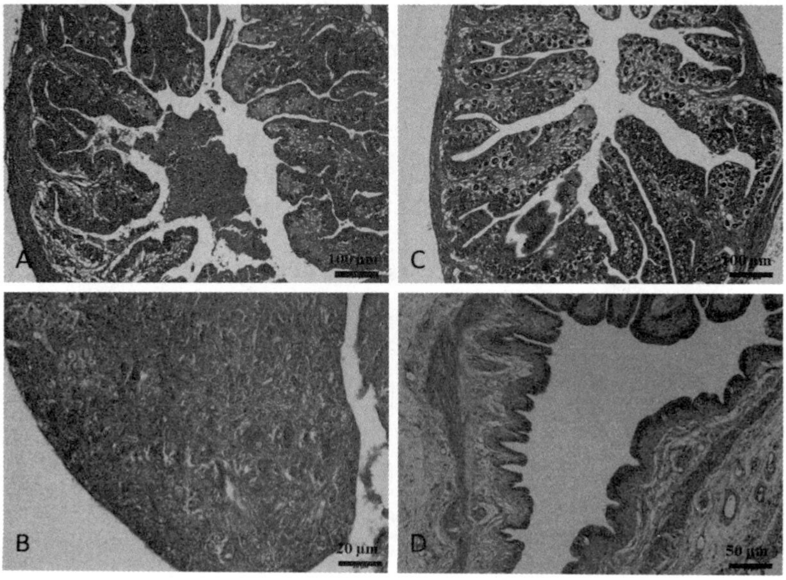

图9 大菱鲆幼鱼对照组雌鱼卵巢(A)、精巢(B)和全雌苗种正常卵巢(C)、发育迟滞卵巢(D)的组织学特征

对上述生殖腺样品采用组织学切片的方法进行性腺复核，肉眼可以确认性别的超雌子代家系和对照组幼鱼雌性卵巢腔结构明显，卵原细胞数目较多，有初级卵母细胞，细胞呈卵圆形，体积大于卵原细胞。雄性精巢内输精管和精小叶结构明显，精原细胞数目较多，出现初级精母细胞。肉眼不能鉴定性别的 2 尾超雌鱼子代，生殖腺具卵巢腔的基本结构，但未见有卵原细胞和卵母细胞，可以认定为发育迟滞的卵巢(图 8、图 9)。综合解剖学和组织学检测结果，可以认定 3 个超雌鱼子代家系个体的雌性比例达到 100%。

另外，从 70 日龄至 150 日龄，对雌鱼 9324 子代家系和对照组的幼鱼性别比例进行了长期跟踪检测。每次取样 10 尾～20 尾，解剖判断其性别，结果如图 10 所示。110 日龄前，对照组和全雌苗种因苗种规格较小，都有少量个体难以通过解剖学特征区分性别；110 日龄以后，对照组所有个体都可以通过解剖特征区分雌雄，雌雄比例接近 1∶1；全雌苗种幼鱼中，始终有少量迟滞发育的卵巢存在，性别比例为 100%雌性。

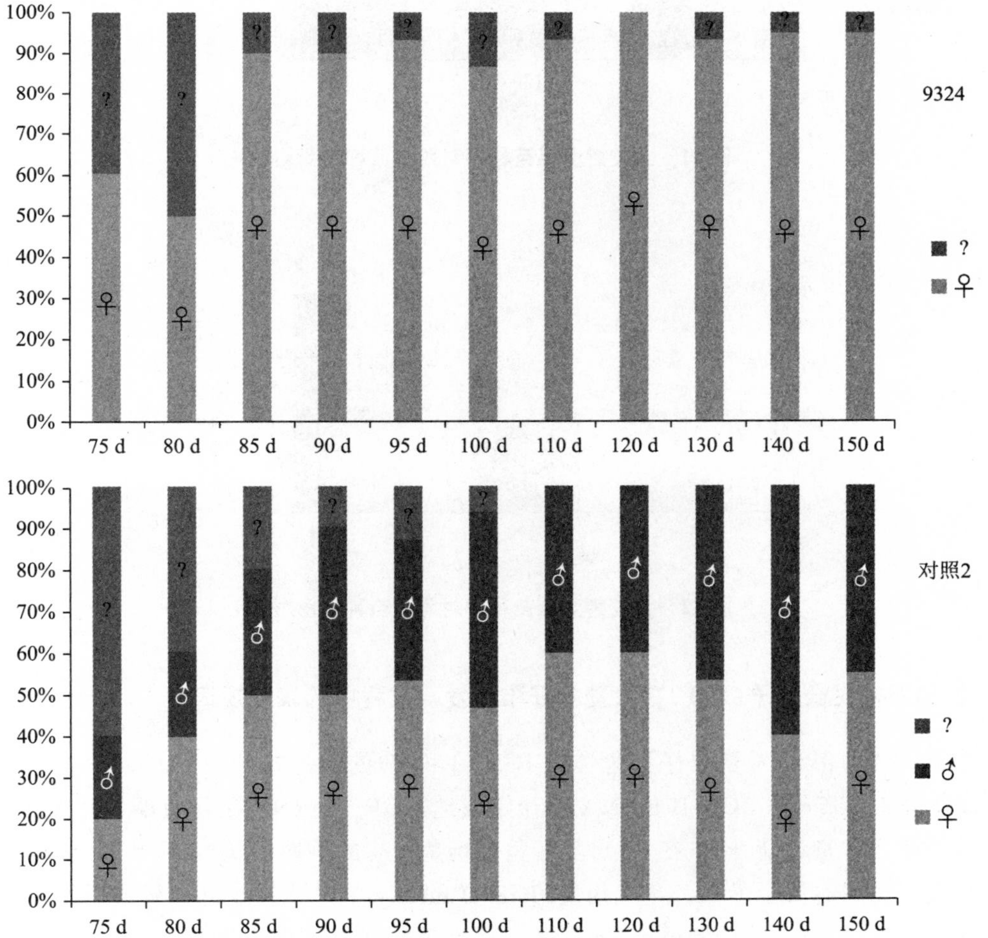

图 10　9324 超雌鱼家系和对照组 2 的幼鱼跟踪检测的性别比例

2.1.3.3 超雌鱼各家系子代生长情况

为评估全雌苗种子代生长性状,对构建的各全雌苗种家系幼鱼的生长情况进行了跟踪和观测。结果表明,在孵化后 75 d、120 d 和 140 d,9324 家系的全雌苗种与对照组之间的全长和体重无显著差异,而 9344 和 9359 家系苗种 120 d 和 140 d,全长和体重均显著高于对照组(图 11、图 12),由于超雌鱼家系的养殖密度有所差异,这种差异可能不能用全雌苗种的遗传来解释,但从目前检测结果看,各家系苗种的平均规格比对照组整齐。

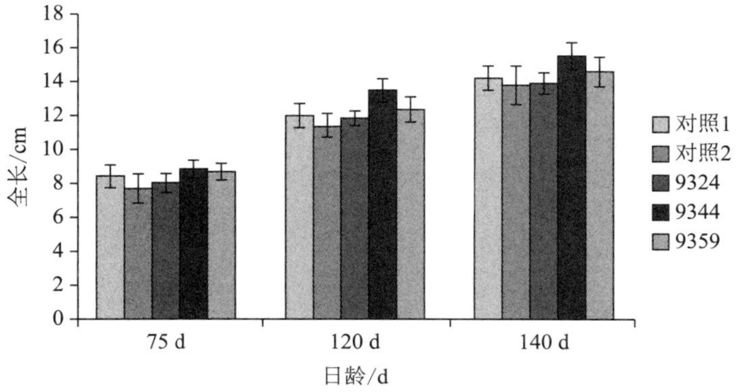

图 11 大菱鲆全雌苗种和对照组苗种全长对比

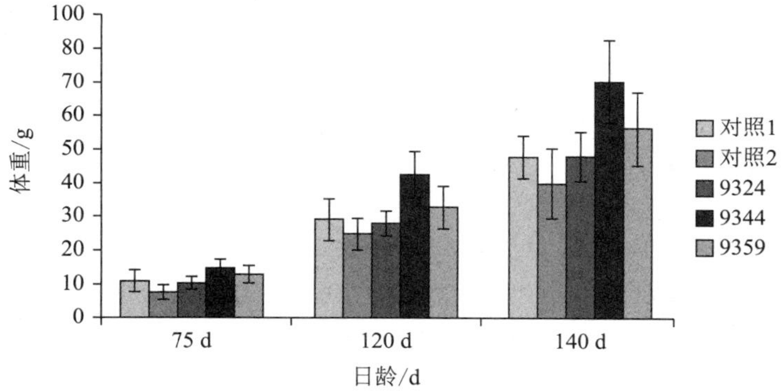

图 12 大菱鲆全雌苗种和对照组苗种体重对比

2.2 胰岛素样生长因子(IGF)在大菱鲆仔稚鱼发育过程中的表达规律

对 3～34 日龄的大菱鲆仔稚鱼体内 IGF (I 和 II)的表达规律进行了研究。孵化后 3 日龄仔鱼体内 IGF-I 和 IGF-II mRNA 都开始表达,IGF-I mRNA 在仔鱼期表达量平稳,从 18 日龄开始表达量逐渐升高,在 22 日龄仔鱼向稚鱼变态的早期达到高峰,此后逐渐降低,在 34 日龄变态完成时降至最低。与此相反,IGF-II mRNA 表达水平在仔鱼孵化后即呈逐渐升高趋势,在孵化后第 8 天达到最高,随后其表达水平呈逐渐下降趋势,在变态中期下降到最低,此后虽有略有上升,但直至变态后期其表水平与变态中期无显著差异(图 13)。

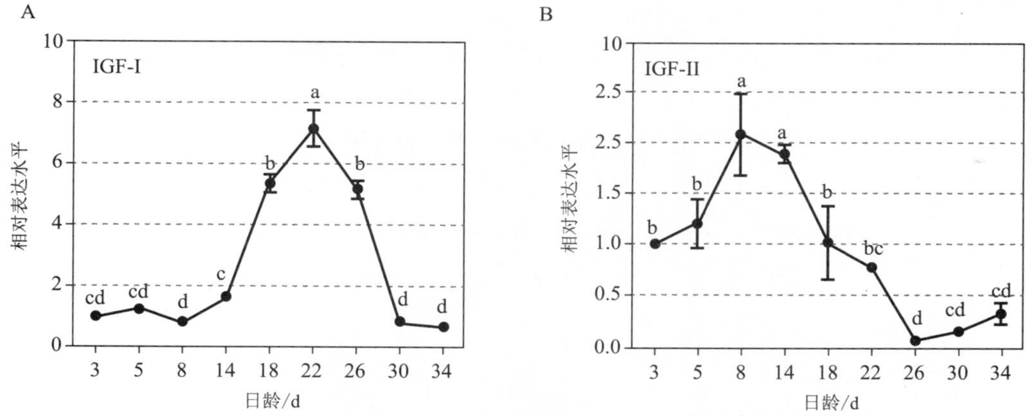

图 13　IGF-I 和 IGF-II mRNA 在大菱鲆仔幼鱼变态反应阶段表达变化

2.3　大菱鲆维甲酸合成酶和降解酶编码基因的克隆

通过 RT-PCR 方法，设计简并引物扩增得到大菱鲆维甲酸合成酶基因 *aldh1a2* 和降解酶 *cyp26a1*、*cyp26b1*，分别获得长度为 1 305 bp、1 064 bp 和 1 026 bp 的保守片段，经 Blast 分析验证为目的基因。后续拟通过 RACE 技术获得 3 个基因的全长序列，并对所得基因全长进行生物信息学分析和时空表达模式的研究（图 14）。

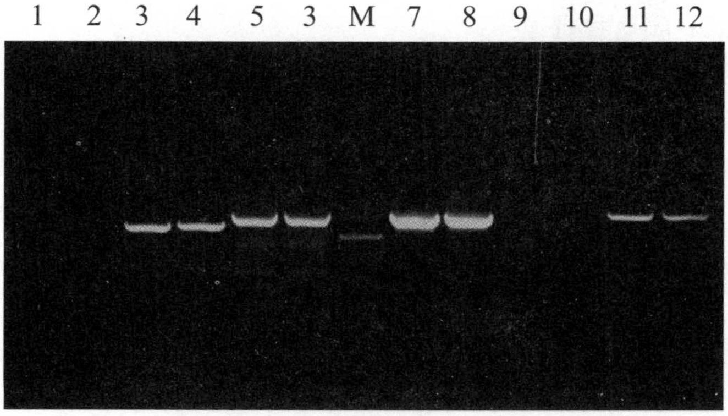

图 14　大菱鲆维甲酸合成酶 *aldh1a2* 和降解酶 *cyp26a1*、*cyp26b1* 基因克隆

M: Marker DL2000; 1、2: *aldh1a21*; 3、4: *aldh1a22*; 5、6: *cyp26b11*; 7、8: *cyp26b12*; 9、10: *cyp26a11*; 11、12: *cyp26a12*

（岗位科学家　王玉芬）

鲆鲽类苗种繁育技术研发进展

苗种繁育岗位

1 鲆鲽类繁育技术综述

目前针对重要养殖鲆鲽类苗种人工繁育技术工艺流程已经建立,为了提升优质苗种生产技术工艺,实现鲆鲽类苗种产业可持续发展,通过加强基础研究来实现对优质苗种生产技术精准调控,成为当前鲆鲽类苗种繁育产业技术研发的一个的重要趋势。2016年国际上鲆鲽类苗种繁育基础研究主要集中在大菱鲆、塞内加尔鳎、牙鲆、星斑川鲽、大西洋庸鲽、半滑舌鳎、欧鳎等重要鲆鲽类养殖品种。相关学者就大菱鲆苗种的早期发育、性别决定、营养强化、应急胁迫等方面展开了研究,取得了一系列的研究成果。研究发现大菱鲆仔鱼在变态期类胰岛素生长因子I表达量呈现先升高后降低趋势,而类胰岛素生长因子II在变态前表达量升高,进入变态期后表达量迅速下降,证实类胰岛素生长因子参与了大菱鲆幼鱼早期生长发育调控。促滤泡激素受体(*fshr*)基因主要在大菱鲆雌性亲鱼卵巢中表达,表达量随着卵黄原前体进入细胞初期逐渐升高,在卵黄蛋白形成期达到最高值,证明*fshr*参与卵巢中卵黄蛋白合成和促进卵细胞发育。通过对大菱鲆肝脏、脾、肾、大脑、性腺和肌肉转录组分析,发现了44个基因(如*cd98*、*gpd1*和*cry2*)参与卵巢分化及形成,12个基因(如*ace*、*capn8*和*nxph1*)参与精巢分化及形成,同时获得了与生长相关基因(*ghrh*、*myf5*、*prl*/*prlr*)以及首次获得了免疫相关基因(*tlr1*、*tlr21*、*tlr22*、*il15*、*il34*),丰富了大菱鲆基因资源。在营养强化研究方面,研究发现在大菱鲆幼鱼饵料中肉骨粉替代鱼粉达45%时,会导致幼鱼的特定生长率显著降低,营养积累变慢,同时还会影响幼鱼雷帕霉素靶蛋白(TOR)免疫调控和类胰岛素生长信号基因调控通路,从而抑制葡萄糖和脂质合成代谢,加快幼鱼肌肉蛋白分解和肝脏脂质分解代谢过程。大菱鲆变态前仔鱼对鱼肝油强化的卤虫饵料捕食更具主动性,能加强仔鱼促食欲基因(如*gal*、*npy*、*agrp2*)的高表达,从而促进进食。随着亚麻籽油替代鱼油比例的提高,幼鱼鱼体中的蛋白含量会显著下降,100%替代组中幼鱼的脂肪酸去饱和酶基因和甾醇酰基转移酶基因表达显著升高,肉碱棕榈酰基转换酶基因却显著下降,当亚麻籽油替代量低于66.7%时,不会对大菱鲆幼鱼的生长和免疫造成显著影响。大菱鲆幼鱼可以利用饵料中的碳水化合物,当饲料中蛋白水平为39%,碳水化合物和脂质比为18∶12时,可以显著促进大菱鲆幼鱼的生长。采用植物乳杆菌P8发酵的大豆粉作为大菱鲆幼鱼饵料蛋白替代量达45%时，会显著抑制幼鱼肠道中丙氨酸氨基转移酶和天冬氨酸氨基转移酶的活性，破坏幼鱼肠道结构，抑制营养吸收，从而影响幼鱼生长发育。随着饲料中鱼水解蛋白物(UF)添加量的增加，大菱鲆幼鱼肠道中的肽转运蛋白基因(*pept1*)表达量显著降低，影响了幼鱼的正常发育，建议饲料中添加低水平的UF。在应急胁迫研究方面，研究发现在热胁迫条件下，牙鲆幼鱼的乳酸、丙二酮、过氧化氢酶等指标存在显著的应急变化，大菱鲆幼鱼的乳酸、丙二酮、超氧化物歧化酶等指标存在显著应急变化，研究为评估牙鲆和大菱鲆是否处于应激反应状态提供了生物指示标记。在低温胁迫条件下，随着温度的降低，大菱鲆幼鱼血液中红细胞、白细胞及血红蛋白浓度持续下降，同时血浆中丙氨酸氨基转移酶和肌酸激酶的酶活力显著上升，乳酸脱氢酶活力呈现先升高后降低的趋势，热激蛋白 *hsp90* 基因表达水平在各个组织无显著变化。采用不同浓度和梯度的亚硝酸盐处理大菱鲆幼鱼发现，随着亚硝酸盐浓度(0.4 mmol/L～0.8 mmol/L)的提高，幼鱼血液中的谷丙转氨酶(GPT)、谷氨酸草氨酸转移酶(GOT)、碱性磷酸酶(ALP)、补体C3和C4表达量显著升高，而免疫球蛋白(IgM)、溶酶体(LYS)、类胰岛素生长因子Ⅰ(IGF-Ⅰ)的表达量显著降低，高浓度亚硝酸盐能够导致大菱鲆幼鱼出现生理和免疫缺陷，影响生长和存活。在大菱鲆培育池中添加高浓度的银离子化合物以及乙炔雌二醇会分别抑制雌性幼鱼雌激素酮和雄性幼鱼中雄烯二酮激素的合成水平。在苗种培育设备研发及光照、水温调控等研究方面，研究发现养殖设施背景色能够影响大菱鲆幼鱼的生长、代谢，白色、灰色、粉色和黑色背景可以作为大菱鲆培育的驯养颜色。大菱鲆幼鱼在短波长光谱(绿光和蓝光)条件下的生长速率显著高于长波长光谱(红光)，绿光培育的幼鱼呈现较高死亡率。在变态前期17 ℃～18 ℃时仔鱼体长最长，温度为21 ℃～22 ℃条件下仔鱼最先完成变态并且体长最长，低温条件下13 ℃～14 ℃仔鱼的肌肉细胞结构发育缓慢。研究发现封闭循环水养殖条件下增加换水流量至800 L/h可以显著促进大菱鲆幼鱼的健康生长。在塞内加尔鳎营养强化研究方面也取得了一些研究进展，研究发现植物来源的脂肪酸饲料能够显著促进塞内加尔鳎幼鱼肝脏中肉碱棕榈酰基转移酶(*cpt1*)基因的高表达，从而加快肝脏脂肪积累。高水平脂肪酸(鱼油100%和植物油70%)添加饵料皆能降低肝脏和肠道中脂肪酸合成酶的表达。采用不同来源脂肪酸饲料投喂塞内加尔鳎幼鱼发现，鱼肝油配制的饲料能够显著提高塞内加尔鳎幼鱼的生长率、成活率，缩短变态时间以及加快肠道组织发育成熟，研究还揭示鱼肝油饲料能够降低肝脏脂肪的积累，并且促进脂肪转运基因(*apob*、*apoc2*、*apoe*)和磷脂代谢调控基因的高表达，促进幼鱼的健康生长。塞内加尔鳎孵化后幼鱼的前体细胞可以合成自身所需的DHA，在后期低水平含量的DHA饵料虽然不会影响仔稚鱼的成活率，但会显著影响塞内加尔鳎仔稚鱼的生长。饲料中植物蛋白组成比例和益生菌添加不会影响塞内加尔鳎幼鱼的生长，但却能显著影响幼鱼肠道菌群的组成以及免疫应激蛋白基因的表达变化，益生菌的添加还能够促进肠道抗氧化应激蛋白基因(*gpx*和 *cat*)的上调。在欧鳎营养强化研究方面，通过人工投喂沙蚕和配合饲料对比发现，投喂沙蚕组的欧鳎幼鱼生长速率、摄食以及营养转化效率显著高于配合饲料组，配合饲料中高脂质有可能影响了欧鳎幼鱼的生长。在牙鲆、星斑川鲽和大西洋庸鲽早期生长发育研究方面，研究发现采用甲状腺激素(T3和T4)合成抑制剂2-巯基-1-甲基咪唑处理的牙鲆和星斑川鲽仔稚鱼的体高显著低于正常组,研究表明甲状腺素的表达能显著促进细胞增殖,相似的研究结果也在斑马鱼中得到证实。鲆鲽鱼类眼睛移位机理一直不清楚,通过对牙鲆和星斑川鲽的神经轴胚进行转录组分析获得了与9个眼睛发育相关的基因,尤其是带有Asp77Glu的Tgfβr2仅在眼睛右移的星斑川鲽中表达。以往的研究认为甲状腺激素(TH)参与了硬骨鱼变态发育调控,对变态期大西洋庸鲽仔鱼头部、皮肤和胃肠道组织进行转录组分析,结果显示98%的差异表达基因在变态期并未与甲状腺激素相关基因进行应答,并且获得的甲状腺激素相关应答基因中有145个差异基因在变态期是下调的,甲状腺激素相关基因是否决定鲆鲽类变态过程有待进一步验证。在半滑舌鳎常染色体上克隆得到精巢特异蛋白激酶(*altesk 1*)基因,研究发现该基因转录表达仅发生在卵巢中,在半滑舌鳎雌鱼中*altesk 1*启动子低水平的甲基化导致其在卵巢中大量表达。上述研究为大菱鲆、塞内加尔鳎、牙鲆、星斑川鲽、大西洋庸鲽、半滑舌鳎、欧鳎等养殖品种苗种培育过程中的营养精准强化调整、早期发育规律、苗种培育设施改进等研究提供了理论依据。

近年来国内对鲆鲽类重要养殖品种的应用基础和技术研发主要针对亲鱼培育过程中的光温环境因子调控、营养强化调控、促熟、催产和授精等技术,针对苗种培育过程中的病害预防、白化及黑化率降低、提高生长率和成活率的营养强化技术,及微藻、光合细菌和中草药制剂使用的环境调控等技术。2016年国内报道了有关大菱鲆、半滑舌鳎、牙鲆、星斑川鲽、圆斑星鲽、钝吻黄盖鲽等鲆鲽类人工繁育的最新研究进展。在鲆鲽类苗种生产技术研究方面,李仲明(2016)阐述了半滑舌鳎亲鱼培育、苗种培育及后期养殖管理环节的关键技术流程。在大菱鲆苗种培育过程中适时投喂开口活饵料,延长鲜活饵料投喂时间,适时进行饵料转换和水质调控,可以有效提高苗种成活率和降低苗种白化率。总结了大菱鲆种业发展历程及相关前沿技术应用,提出当前由于缺乏有效的亲鱼管理,导致出现亲鱼近交,结果造成严重的种质衰退,急需种质遗传改良。在鲆鲽类性别决定机制研究方面,*sox*(SRY-related HMG-box)基因、分泌家族基因*shh*(Sonic Hedgehog)和E3泛素连接酶基因*trim36*(Tripartite motif-containing 36)在半滑舌鳎雌雄性别分化过程中表达量呈现显著的性别二态性,与其雌雄性腺分化及性腺发育密切相关。从数量性状基因座位互作角度研究了半滑舌鳎性别决定的遗传学机制,发现常染色体也参与到了半滑舌鳎性别决定过程。孕酮受体膜组分1(*pgrmc1*)基因主要参与半滑舌鳎卵巢的发育和成熟过程,并在繁殖期介导孕酮调控卵母细胞的最终成熟。*dazl*基因在牙鲆精巢中的表达显著高于卵巢,母源性*dazl* mRNA参与了胚胎发育中原始生殖细胞的形成。空通气孔同源框1(*emx1*)基因表达存在组织特异性,在牙鲆卵巢中的表达量丰富,精巢中次之。在骨骼早期发育过程中的研究发现,骨形态发生蛋白*bmp4*基因在半滑舌鳎胚胎发育过程中呈先高后低表达模式,仔鱼期表达量再次升高。发生性逆转的遗传半滑舌鳎雌鱼在性腺发育过程中存在明显的雌雄兼性性状。半滑舌鳎乙酰辅酶A羧化酶(*acc1*)基因主要在肝脏和全脑等生脂组织中表达,基因表达量与鱼油添加量呈负相关。大菱鲆苗种在变态前视网膜视敏度高、光敏性低,变态后视敏度降低、光敏性增强,变态后的大菱鲆光感受系统不发达。RNA/DNA 和蛋白质/DNA 比率结果表明大菱鲆胚体的生长是细胞增殖和细胞增大周期性交替的过程。在鲆鲽类数量性状研究方面，通过微卫星标记获得了 29 个与半滑舌鳎雌性个体生长相关联的优势基因型。牙鲆在不同生长时期，影响体重的主要形态性状不同，不同阶段适用的最优拟合模型也不同。正常体色牙鲆群体中全长是影响体重的最重要的生长性状，其次是体高，白化牙鲆群体中体长是影响体重的最主要的生长性状。在营养强化及饵料替代研究方面，在大菱鲆饲料中添加不同种类的营养物质如浒苔（<5%）、瑞士乳杆菌、低水平水解鱼蛋白（UF-5）、海洋脂肪酶、中叔丁基氢醌（<150 mg/kg）等可以有效提高大菱鲆幼鱼的生长速率和成活率。研究还发现，在大菱鲆幼鱼饲料中添加核苷酸对幼鱼生长无明显的促进或抑制作用，但对其血清中的溶菌酶活性具有促进作用。在圆斑星鲽幼鱼饲料中添加 10% 的发酵豆粕对其生长、饲料利用和形体指标无显著性影响，过量添加会抑制鱼体的生长。在半滑舌鳎稚鱼饲料中添加谷氨酰胺能够显著增强其鱼体溶菌酶活力，增强其非特异性免疫力。在大菱鲆幼鱼饵料替代研究方面，通过采用不同来源的蛋白源如黄粉虫（20% ～ 25%）、鸡肉粉（<20%）、肉骨粉（<45%）替代鱼粉发现，不会影响幼鱼的生长发育。在大菱鲆幼鱼饲料中用糊精替代鱼油既影响幼鱼糖和脂类代谢又影响其免疫力。星斑川鲽幼鱼能够有效地利用酶解大豆蛋白，酶解大豆蛋白替代 69. 03% 以下的鱼粉对幼鱼生长和饲料利用不会产生不利的影响，且对体组织脂肪蓄积有一定的改善作用。南极磷虾粉替代 10% ～ 30% 的鱼粉可以提高圆斑星鲽幼鱼的生长性能和非特异性免疫力。在精子低温冷冻保存技术研究方面，研究发现激活剂的渗透压决定了夏牙鲆精子能否激活并显著影响了激活后精子的运动率。通过筛选获得了钝吻黄盖鲽低温冷冻保存的最适稀释液和抗冻剂，为钝吻黄盖鲽种质保存提供了基础。在杂交制种技术研究方面，星突江鲽与石鲽的正反杂交种在眼睛位置、鳞片和鳍的形态特征上更倾向于星突江鲽，而侧线、可数、可量性状则更多地遗传了母本的形态特征。在应急胁迫研究方面，低盐胁迫条件下，半滑舌鳎幼鱼的代谢水平始终维持在较高状态，渗透压调节器官中糖代谢作用增强，并且抑制生长。牙鲆幼鱼对短期低溶解氧含量有较强的适应能力，能够通过提高摄食率，在较短的恢复期获得完全补偿生长。上述研究为大菱鲆、半滑舌鳎、牙鲆、星斑川鲽、圆斑星鲽、钝吻黄盖鲽等亲鱼性腺诱导发育成熟技术，苗种培育过程中环境因子调控、营养强化技术、精子冷冻保存以及杂交制种技术的优化提供了理论参考。

2　苗种繁育岗位年度工作综述

2. 1　鲆鲽鱼类优质亲鱼培育和优质受精卵生产技术完善研究

大菱鲆与牙鲆优质亲鱼培育和优质受精卵生产示范：留种和保育大菱鲆亲鱼 1 500 余尾、牙鲆亲鱼 700 余尾，通过采用细化和提升温、光调控及激素诱导亲鱼性腺发育成熟的方法，分别实现大菱鲆、牙鲆性腺同步成熟率达 90% 以上。筛选优质的雄性大菱鲆、牙鲆种鱼并诱导获得优质精子，采用新筛选的稀释液和抗冻剂进行冷冻保存，共冻存优质精子 300 余毫升，解冻复活率皆大于 85%以上。采用盐度、温度、流水、光照等环境因子调控相结合的手段完善了大菱鲆、牙鲆优质受精卵规模化培育技术工艺，实现大菱鲆受精率、孵化率、苗种成活率分别达到 85%、80%、75%，示范生产大菱鲆、牙鲆优质受精卵 250 余千克，示范生产大菱鲆优质苗种 360 余万尾。

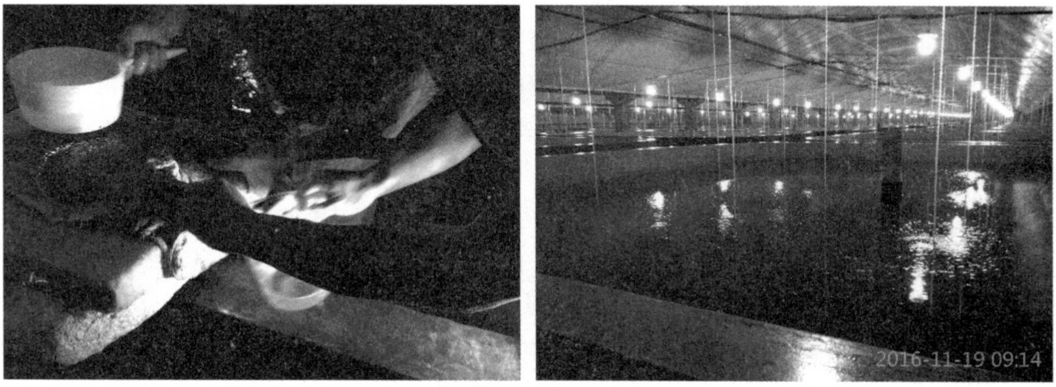

图 1　鲆鲽类优质受精卵生产

牙鲆与夏鲆种间杂交优质亲鱼培育和优质杂交受精卵生产示范：留种保育优质夏鲆种鱼 530 余尾、牙鲆种鱼 700 尾。筛选优质的雄性夏鲆种鱼并诱导获得大量优质精子，并进行冷冻保存，共冻存优质精子 100 余毫升。通过采用性腺发育同步成熟诱导技术、延长雄性亲鱼产精期及增精技术、提高受精率人工授精技术，完善了优质杂交受精卵规模化生产技术工艺，实现杂交受精率达 72%，受精卵孵化率 70%以上，示范生产优质杂交受精卵 50 余千克。苗种培育过程中的通过优化温、光调控、营养转换与强化，实现受精卵孵化率 70%以上，苗种成活率 50%以上，共示范生产杂交苗种 250 余万尾。

鲆鲽鱼类正常体色苗种培育技术改进的生产性实验研究：调整牙鲆、夏鲆亲鱼培育温度调控策略，补充静息期低温处理措施，营养强化时机与剂量的调整。通过对杂交苗种进行光照、温度调控，营养强化及培育密度控制，提高了正常体色苗种比例，实现正常体色率达 95%以上。

2.2　鲆鲽鱼类繁育样本采集及其种质资源评价

大菱鲆种质资源评价样本采集：采集大菱鲆主产区威海、烟台、日照、青岛等地大菱鲆繁育群体和子代群体 32 个群体，共计 1 100 余尾，通过采用酶切、连接、预扩增和选择性扩增筛选，从 20 对引物组合中筛选获得了 5 对高效扩增的 AFLP 引物组合。其中，威海圣航水产科技公司储备了来自智利北、智利南、法国、挪威、冰岛等地大菱鲆种质群体，本研究收集了上述繁育群体和子代群体 8 个，共计 400 余尾，用于研究典型苗种生产模式下大菱鲆种质资源的遗传变化趋势。

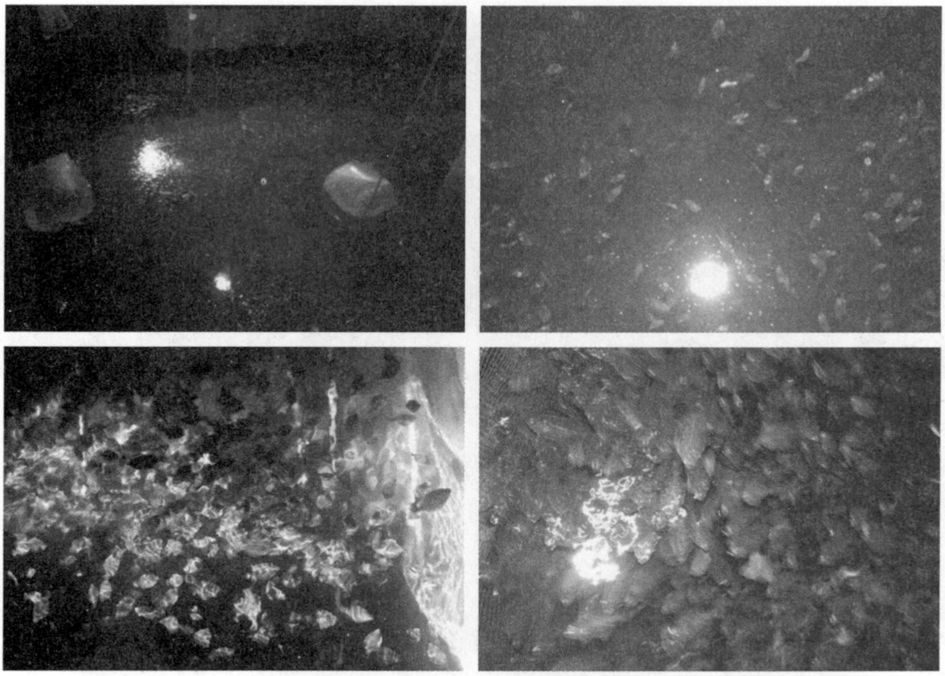

图 2　杂交鲆受精卵生产及苗种培育

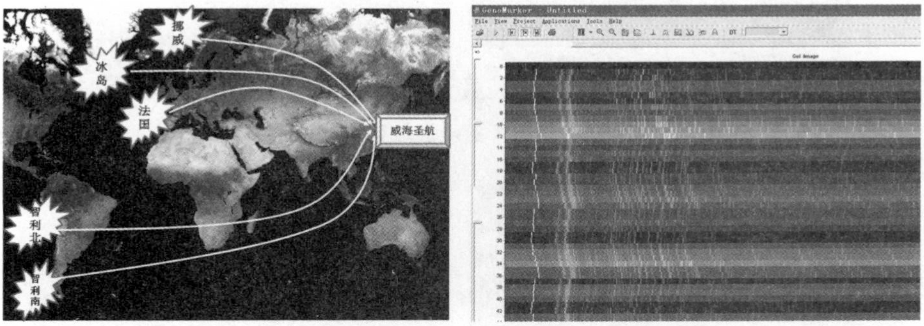

图 3　大菱鲆繁育核心种质资源来源及样本采集

半滑舌鳎种质资源评价样本采集：采集半滑舌鳎主产区莱州、日照等地半滑舌鳎繁育群体和子代群体 8 个群体，共计 230 余尾。通过筛选获得了 4 对高效扩增的 AFLP 引物组合。

2.3　雌核发育牙鲆、大菱鲆规模化诱导及伪雄鱼诱导培育研究

本年度继续通过雌核发育家系群体诱导，构建牙鲆、大菱鲆育种材料，并对大菱鲆同质雌核发育诱导技术进行了进一步优化，再次尝试了规模化诱导，目前正在培育牙鲆、大菱鲆雌核发育家系 / 群体分别为 45 个、19 个。还对 2014 年诱导的生长快和耐高温家系 / 群体进行了筛选。进一步，从转录组水平对牙鲆雌核发育单倍体、同质二倍体和异质二倍体等胚胎进行了基因差异表达分析，还对牙鲆性别表型形成机制进行了研究。以上工作和研究为

下一步育种工作提供基础和依据。

雌核发育牙鲆规模化诱导和培育——异质雌核发育:2016 年度春、夏季新诱导获得了牙鲆异质雌核发育家系/群体 8 个,正在培育 7 月龄鱼苗约 8 000 尾。

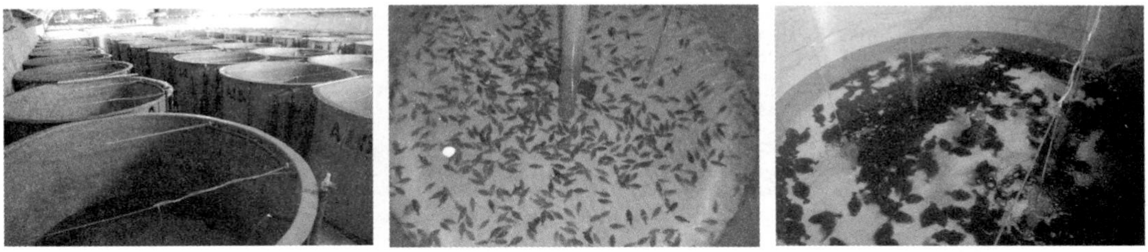

图 4　2016 年批量诱导牙鲆异质雌核发育群体

本年度继续培育 2008～2014 年诱导获得的牙鲆异质雌核发育家系/群体 29 个,正在培育 2 龄以上成鱼约 2 500 尾(全长 30 cm～40 cm)(图 5、图 6)。各家系/群体均进行了生长跟踪观察。

图 5　2014 年批量诱导牙鲆异质雌核发育家系培育情况

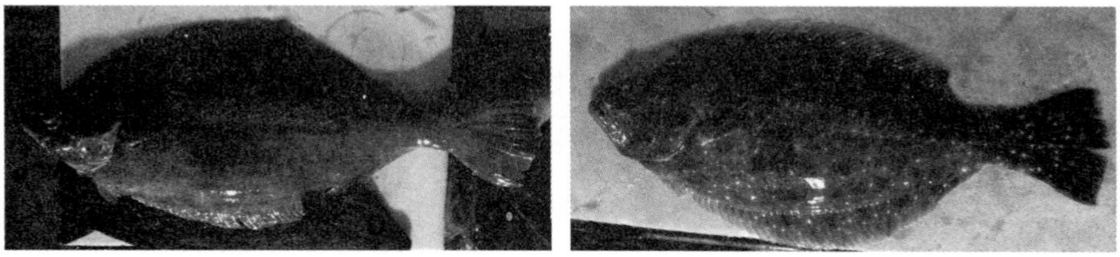

图 6　2007～2013 年诱导异质雌核发育牙鲆

2016 年春季,利用 2010 年诱导的已经性成熟的雌核发育家系,进行了雌核发育子二代的诱导,获得子二代家系 1 个,目前有 7 月龄鱼苗约 800 尾(图 7)。

雌核发育牙鲆规模化诱导和培育——同质雌核发育:2016 年诱导获得了牙鲆同质雌核发育群体 6 个,经过混养目前正在培育同质雌核发育群体 4 个,培育 7 月龄鱼苗约 400 尾。正在培育 2014 年春季、秋季及 2015 年春季诱导的同质雌核发育群体苗种共约 200 尾。

2.4　生长快、耐高温牙鲆家系/群体筛选

生长情况跟踪观察及生长优势家系/群体筛选:对培育的牙鲆雌核发育家系/群体生长

图 7　牙鲆异质雌核发育子二代培育情况

情况进行跟踪测量，初步筛选得到具有生长优势的家系 / 群体 7 个。

耐高温家系 / 群体筛选：对 2014 年春季诱导的部分雌核发育牙鲆家系及群体进行耐温实验，筛选出耐高温及不耐高温家系各 2 个，并筛选得到乳酸、过氧化氢酶和丙二醛等可作为耐温性状相关的生理标志物。

2. 5　雌核发育大菱鲆规模化诱导及培育研究

大菱鲆异质雌核发育：2016 年诱导获得了大菱鲆异质雌核发育家系 / 群体 3 个，正在培育 7 月龄苗种约 1 900 余尾。

本年度继续培育 2014 年秋、冬季及 2015 年诱导获得的大菱鲆异质雌核发育家系 / 群体 14 个，正在培育 21 月龄以上成鱼约 250 尾。

大菱鲆同质雌核发育：2016 年进一步优化了大菱鲆同质雌核发育的诱导条件，正在培育 7 月龄鱼苗 40 尾。

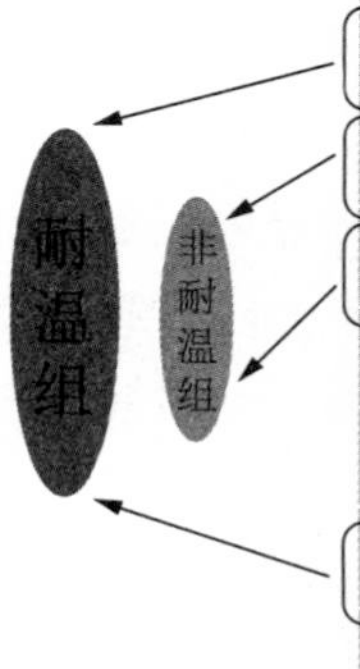

牙鲆家系	31℃存活时间 /h
OF2014SE-F1	7.78±0.33
OF2014SE-F2	9.28±0.82
OF2014SE-F3	10.71±1.12
OF2014SE-F4	7.92±0.00
OF2014SE-F5	7.65±0.41
OF2014SE-F6	9.83±0.47
OF2014SE-F7	9.88±0.56
OF2014SE-F8	10.42±1.72
OF2014SE-F9	17.64±7.03
OF2014SE-G	10.04±7.12

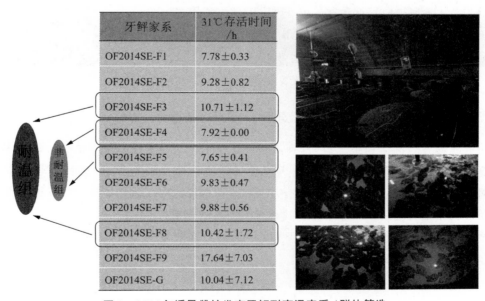

图 8　2014 年诱导雌核发育牙鲆耐高温家系 / 群体筛选

图 9　2014～2015 年诱导大菱鲆异质雌核发育家系/群体

2.6　生长快、耐高温大菱鲆家系筛选

生长情况跟踪观察及生长优势家系/群体筛选:对培育的大菱鲆雌核发育家系/群体生长情况进行跟踪测量,筛选得到具有生长优势的家系/群体 3 个。

耐高温家系/群体筛选:对 2014 年及 2015 年诱导的部分雌核发育大菱鲆家系及群体进行耐温实验,筛选出耐高温及不耐高温家系各 2 个,并筛选得到乳酸、超氧化物歧化酶和羰基等可作为耐温性状相关的生理标志物。

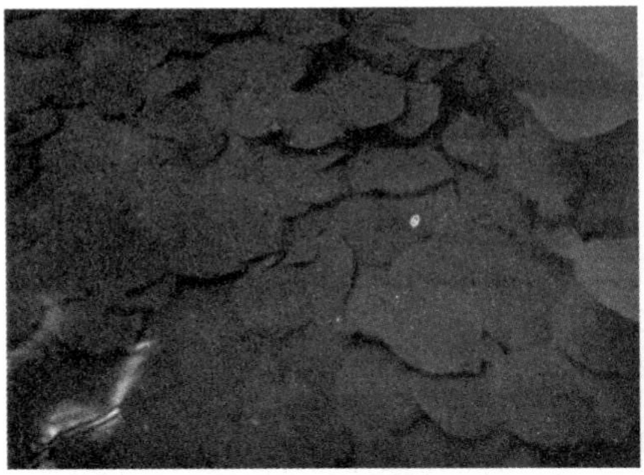

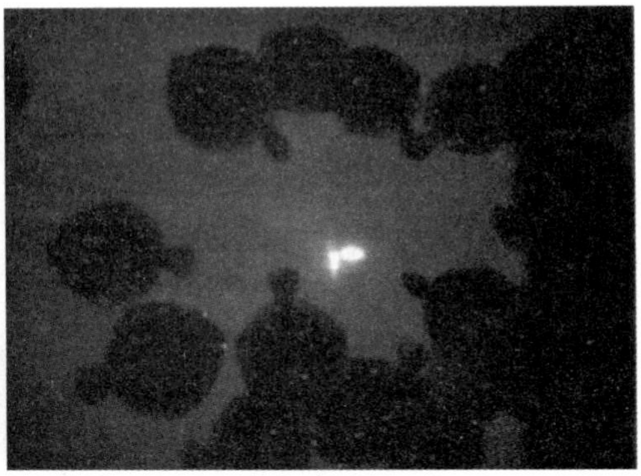

图 10　雌核发育大菱鲆耐高温家系/群体筛选

2.7　牙鲆雌核发育胚胎转录组分析及牙鲆、大菱鲆多倍体研究进展

牙鲆雌核发育胚胎转录组分析:利用转录组分析了牙鲆雌核发育单倍体,同质/异质雌核发育胚胎,结果显示 notch 和 Wnt 信号通路的下调是导致单倍体综合征的重要原因,而染色体的加倍导致 complement C3, formin-2 等基因的上调表达,促进其孵化和成活,但一些免疫和能量代谢相关基因下调导致雌核发育胚胎发育的延迟。

牙鲆与大菱鲆多倍体进展:构建和鉴定了牙鲆二倍体与三倍体肌卫星细胞系,均已经达到 60 余代并显示其都具有增殖与分化的能力,可以为鱼类三倍体生长差异提供研究材料和依据。

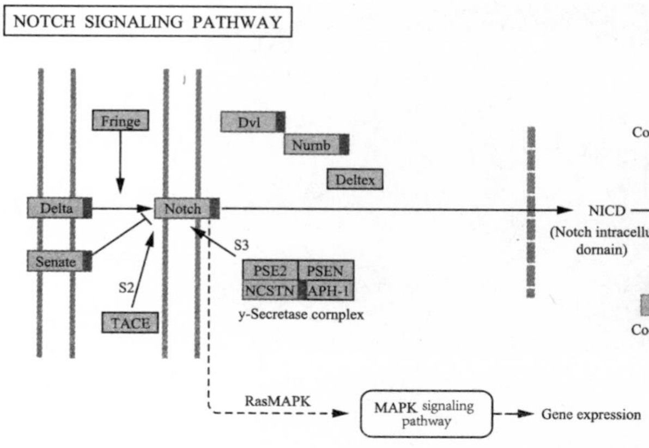

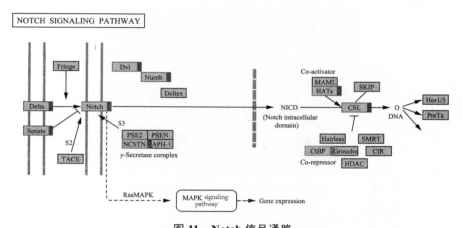

图 11 Notch 信号通路

绿色代表在雌核发育单倍体中下调表达，红色代表上调表达

2.8 牙鲆性别表型形成相关进展

Wnt4 蛋白是哺乳动物性腺分化的关键调控因子，我们对 *wnt4* 在牙鲆性腺分化期、性腺发育期中的差异表达进行了分析，结果表明 *wnt4* 是性别相关基因，在牙鲆不同发育时期起着不同的作用。

2.9 牙鲆生殖细胞胚后迁移及精巢分化发育的组织学观察

通过组织学观察系统研究了牙鲆生殖细胞胚后迁移及精巢分化发育的详细过程。结果发现，孵化后 2 d，20 个左右原始生殖细胞出现在中肾管下方和体腔靠下部位。随后单个生殖细胞向体腔上后方迁移，于 15 d，在肾脏下方和体腔上方，几个生殖细胞排列一起，一侧被体细胞包绕，形成原始生殖脊。最终于 22 d，生殖细胞两侧完全被体细胞包绕，形成细条形原始性腺。

22 d～75 d，生殖细胞缓慢增殖并分散排列于性腺中，形成少量 2 细胞的生殖囊。80 d 后，在性腺的背侧，体细胞之间形成许多缝隙状输精管原基，同时生殖细胞开始加速增殖，于性腺腹侧形成多细胞的腺泡状生殖细胞团。120 d 时，性腺体积变大，加长，呈三角状，腺泡状生殖细胞团增多，密集排布。位于腺泡状生殖细胞团中央的体细胞相互远离，形成中间空隙。8 个月后，腺泡状生殖细胞团逐渐伸长与背部输精管相互汇通，形成发育完整的精巢结构。成熟的精巢中，同步发育的生殖细胞以精小囊分布于精小叶的周围，生殖细胞经过分裂，发育形成成熟的精子，精小囊破裂，精子释放到精小叶腔并汇入生殖管中，等待排出体外。

2.10 大菱鲆减数分裂标记基因 *sycp3* / *Amh* 的表达规律研究

大菱鲆在发育成熟过程中存在一段体重显著增长期，对二龄以下大菱鲆进行取样，检测两个减数分裂标记基因（*sycp3* / *Amh*）在精巢中的相对表达量。结果表明，*sycp3* 的表达量在大菱鲆体重 232.2 g～416.7 g 期间急剧升高，证明在这个阶段大菱鲆精巢进入减数分裂阶

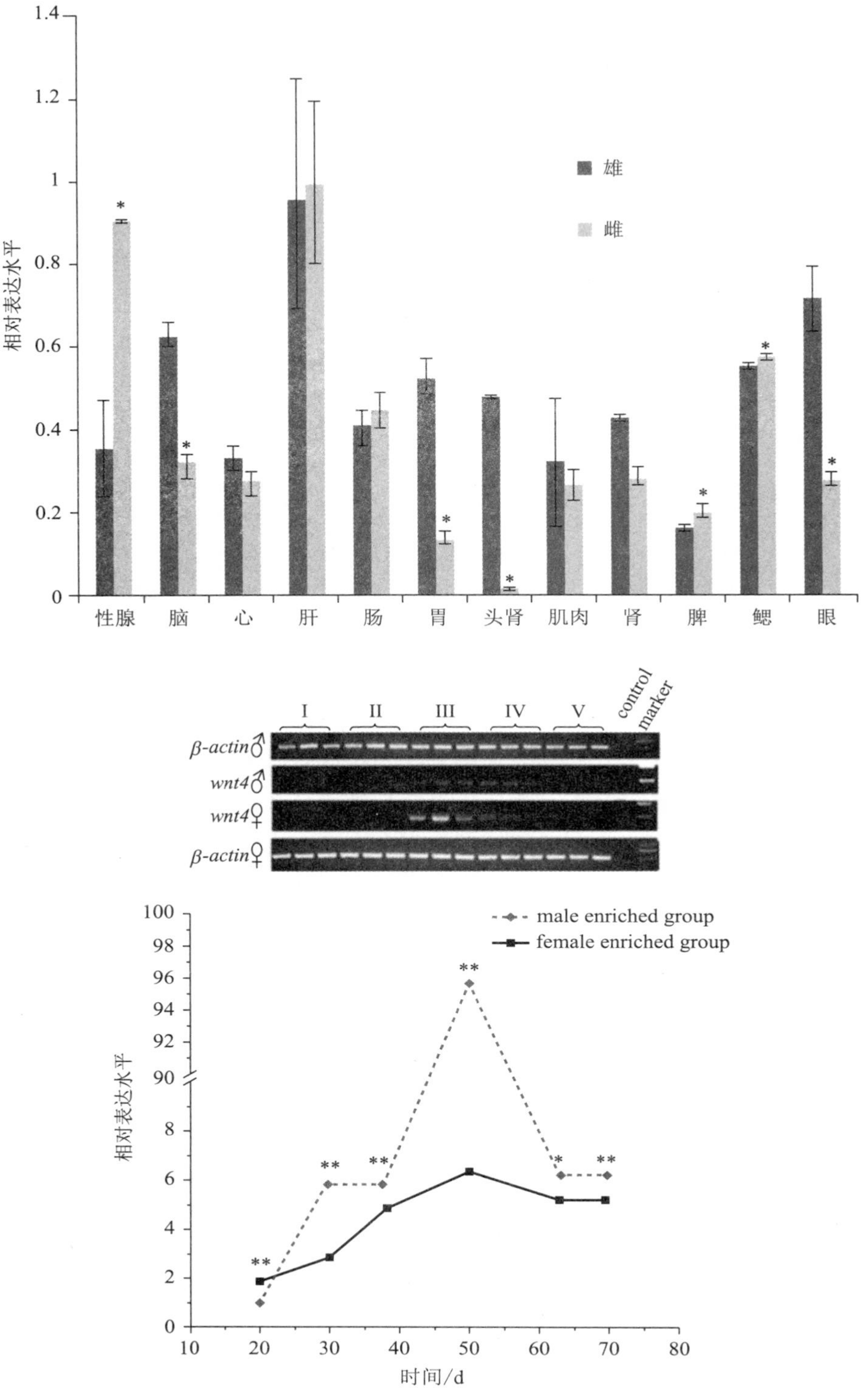

图 12　牙鲆 *wnt4* 基因的差异表达图示

上：成鱼不同组织；中：性腺不同分化期；下：性腺不同发育时期

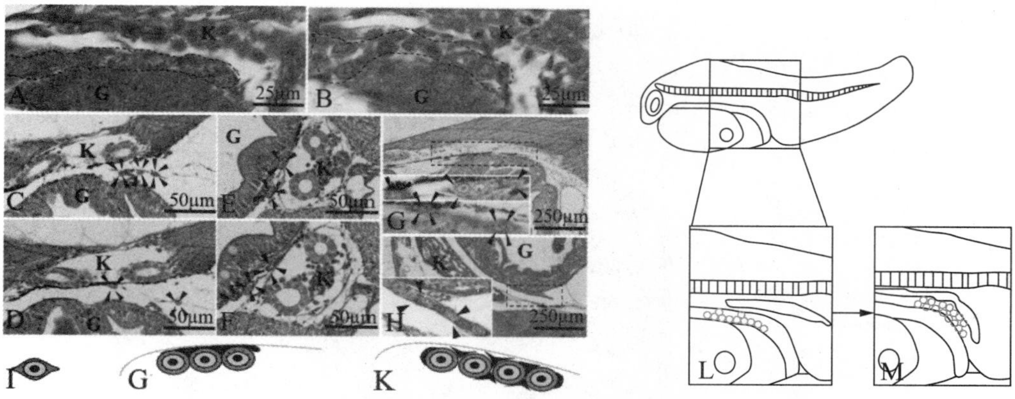

图 13　牙鲆生殖细胞胚后迁移组织学观察

G: 肠道; K: 肾脏; 箭头: 生殖细胞

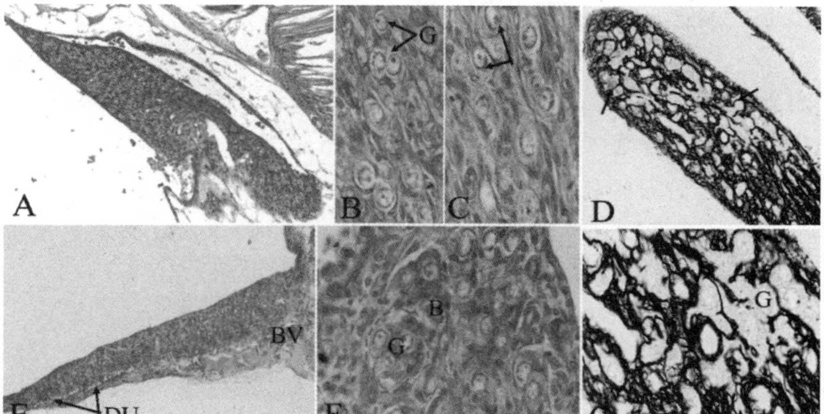

图 14　70～80 天牙鲆精巢组织学结构

G: 生殖细胞; DU: 输精管; B: 血细胞; BV: 血管

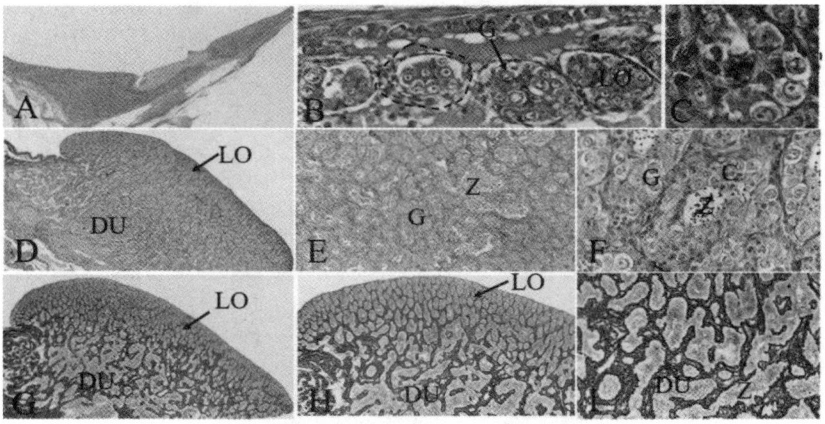

图 15　120 天～8 个月牙鲆精巢组织学结构

G: 生殖细胞; LO: 精小叶; DU: 输精管; Z: 精子细胞; C: 精母细胞

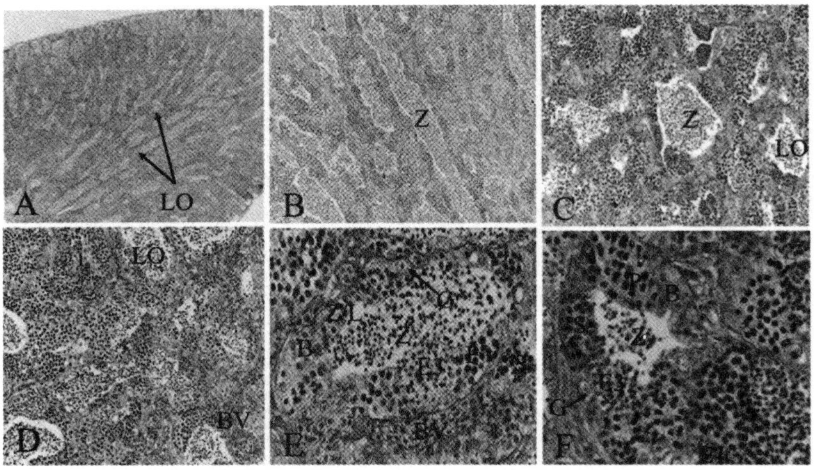

图 16　成熟牙鲆精巢组织学结构

ZL: 偶线期 / 细线期精母细胞; E3 晚期精细胞; P: 粗线期精母细胞; S: 次级精母细胞;
LO: 精小叶; Z: 精子; BV: 血管; G: 精原细胞; B: B 型精原细胞

段。同时发现 *Amh* 的表达量也在大菱鲆体重为 232. 2 g 这一阶段明显升高。

比较研究这一时期孕酮受体 *pgr* 及 *mpRα* 的相对表达量发现: *pgr* 在大菱鲆体重为 17. 2 g ~ 119. 5 g 这个阶段的表达量持续较低,在 119. 5 g ~ 232. 2 g 这一阶段明显升高,之后持续高表达,在体重为 972. 5 g 这个阶段表达量急剧下降。*mpRα* 的表达量和体重或是精巢进入减数分裂期没有明显的相关性。

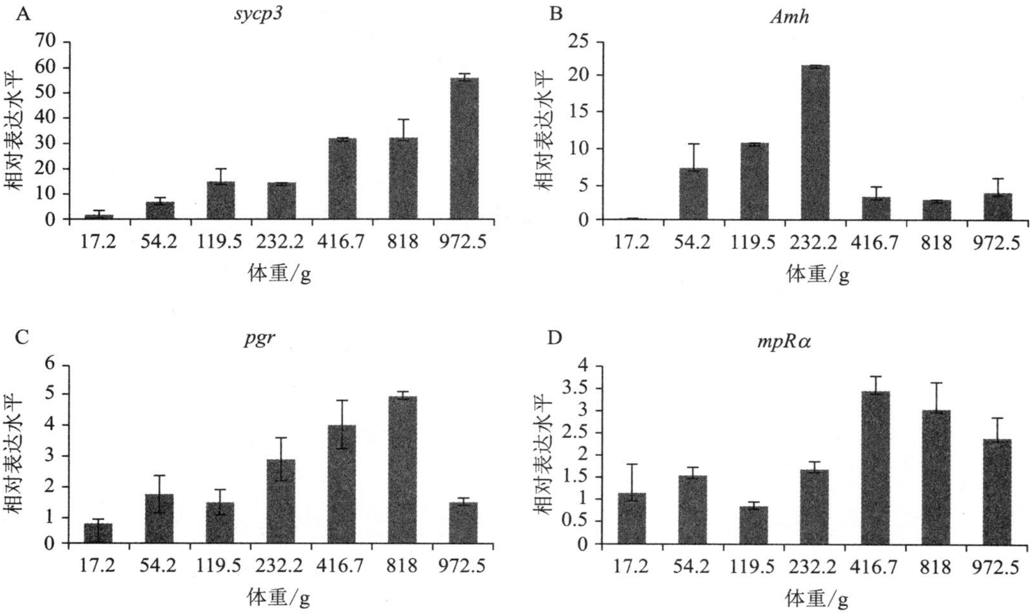

图 17　大菱鲆减数分裂标记基因 *sycp3* / *Amh*、孕酮受体 *pgr* 及 *mpRα* 的相对表达量

2.11　大菱鲆生殖轴上游基因 *GnRH-II* 的克隆及其在繁殖期的表达特征分析

采用 RACE 技术，克隆了大菱鲆下丘脑-垂体-性腺轴的上游调控基因 *GnRH-II* cDNA 全长序列（全长 623 bp），运用 RT-PCR 和荧光定量 PCR 方法，以 *β-actin* 和 *UBQ* 基因为内参基因，检测了大菱鲆 *GnRH-II* 基因在 1 龄幼鱼各组织的表达情况。RT-PCR 结果显示，大菱鲆 *GnRH-II* 基因在雌、雄脑组织呈现高表达的模式，肾脏有极其微弱的表达。荧光定量结果显示，*GnRH-II* 基因在眼、肾脏、肌肉、鳍条、皮肤的表达量较少，与脑组织的表达量呈现显著性差异，而在其他组织几乎检测不到表达。

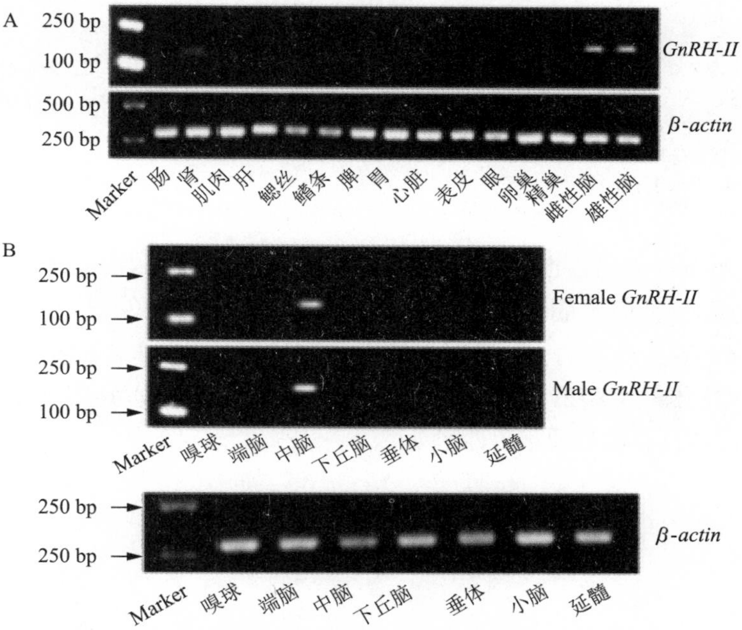

图 18　RT-PCR 半定量分析大菱鲆 *GnRH-II* 组织表达（A）和在脑组织的表达（B）

以 *β-actin* 为内参基因，RT-PCR 检测 1 龄大菱鲆 *GnRH-II* 在雌、雄脑组织的表达量，结果显示，*GnRH-II* 只在中脑有表达，其余各部分脑组织均未检测到表达。据已有的报道哺乳动物和硬骨鱼类，如臭鼩（*Soricidae*）、鲇鱼（*Clarias gariepinus*）、条斑星鲽和金鱼（*Carassius auratus*），*GnRH-II* 细胞体位于中脑，而其神经纤维辐射到除垂体之外其他脑区，*GnRH-II* 作为神经传递物质或者神经传导物质存在于脑。本研究表明，大菱鲆 *GnRH-II* 的分布区域同上述物种一致，暗示了其发挥的作用相似，可能都具有神经传递物质或者神经传导物质的作用。对牙鲆的研究显示 *GnRH-II* mRNA 只在垂体有表达，对于鲤鱼（*Cyprinus carpio*）的研究显示，*GnRH-II* mRNA 不仅在中脑、垂体有高表达，而且在下丘脑、端脑等脑区也有表达，暗示 *GnRH-II* 有刺激其他激素分泌的作用。本研究中，*GnRH-II* 在垂体表达的缺失暗示其间接参与大菱鲆的繁殖。对 *GnRH-II* mRNA 的组织特异性表达分析结果显示，*GnRH-II* mRNA 在雌、雄大菱鲆的脑组织大量表达，其次在肾脏、皮肤、鳍条也有微量表达。

在大菱鲆繁殖季节,随着性腺的发育,*GnRH-II* mRNA 的表达量在雌、雄大菱鲆脑组织中呈现相同的波动趋势。起始精巢、卵巢发育的 II,*GnRH-II* mRNA 的表达量均处于较高水平。随着精巢、卵巢的发育,至性腺发育到 III 期时,*GnRH-II* mRNA 的表达量有显著性的降低,而在随后性腺发育至 IV 期(产卵、产精前)时,表达量又有显著性升高。产卵产精的 V 期时,表达量又有显著性回落。产卵产精结束,性腺退化,*GnRH-II* mRNA 的表达量升高。

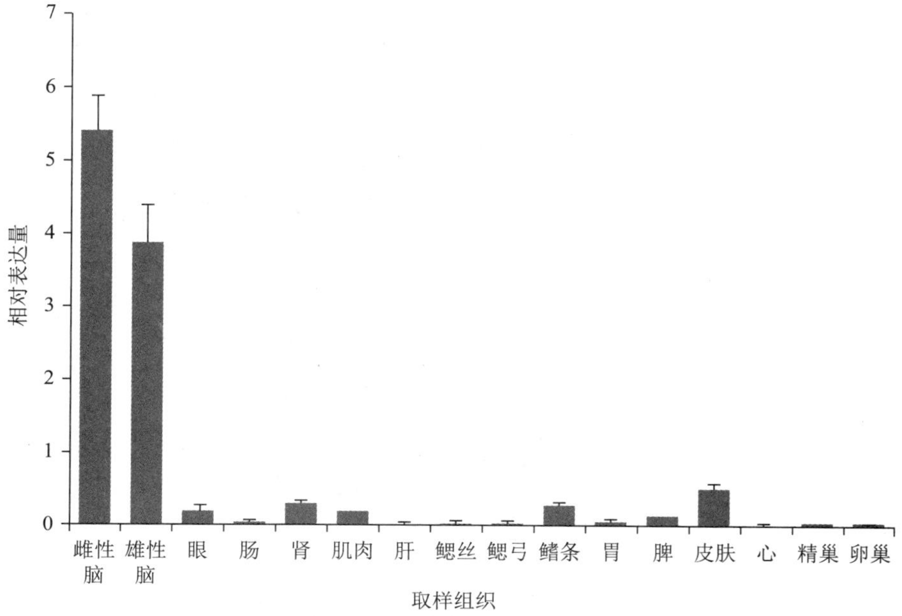

图 19　荧光定量分析大菱鲆 *GnRH-II* 组织表达

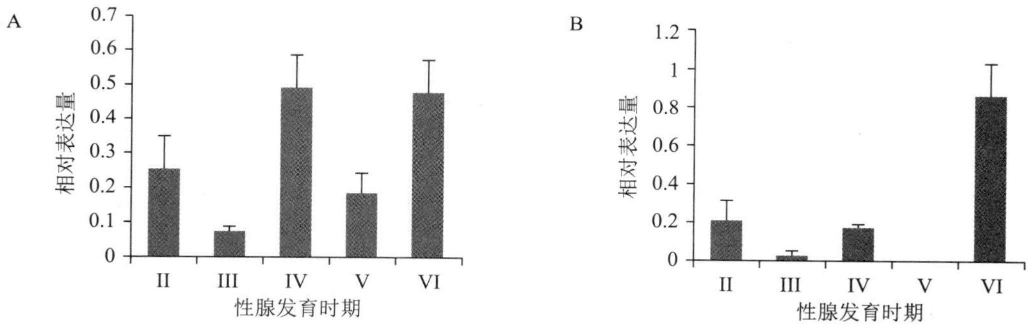

图 20　大菱鲆繁殖期季节 *GnRH-II* 在雌(A)、雄(B)性腺的表达量检测分析

大量的研究表明,*GnRH-II* 具有刺激性腺的成熟及排卵的作用。对革胡子鲇(*Cyprinus carpio*)用 200 μg/kg 剂量的 *GnRH-II* 处理时,对于诱发排卵有明显的促进作用并且实现革胡子鲇 100%排卵。大菱鲆繁殖期 *GnRH-II* 的表达量检测发现,繁殖期雌性大菱鲆 *GnRH-II* 在性腺发育的 V 期表达量相对较低,而此时期正是排卵的关键时期,*GnRH-II* 表达量的降低可能是养殖大菱鲆不能自然排卵的一个重要原因。与此同时，雄性大菱鲆在性腺发育的 V 期表达量也相对降低，这可能与大菱鲆精液量少、精子质量低有一定的关系。

2.12　大菱鲆繁殖期血清类固醇激素的周期变化规律

为了阐明大菱鲆精子发育成熟过程中生理水平激素的变化规律，对 II 期初到 VI 期末大菱鲆血清中激素水平进行初步测定。

进一步分析几种主要类固醇激素在四龄大菱鲆繁殖期（III～VI 期）的变化过程发现：睾酮（T）在 IV 期显著升高，V 期达到最大值，之后 VI 期急剧降低；11-酮基睾酮（11-KT）在 III～V 期表达量没有显著变化，VI 期反而升高；孕酮（P）在 III 期表达量较低，之后持续显著升高，V 期达到最大值，VI 期回降；17α- 双羟孕酮（DHP）从 IV 期开始显著升高，高浓度持续维持至 VI 期。

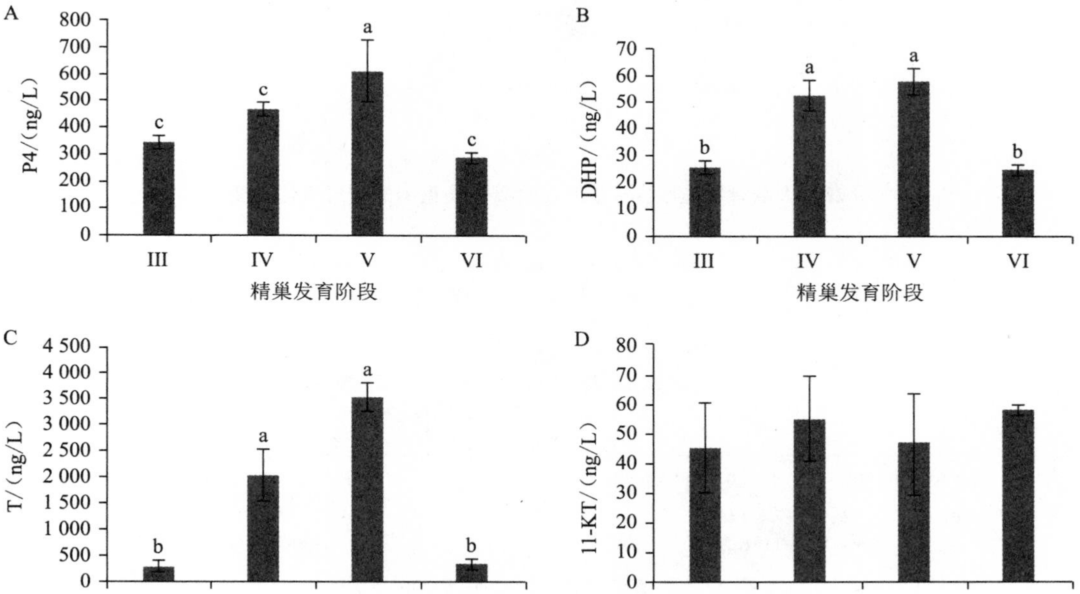

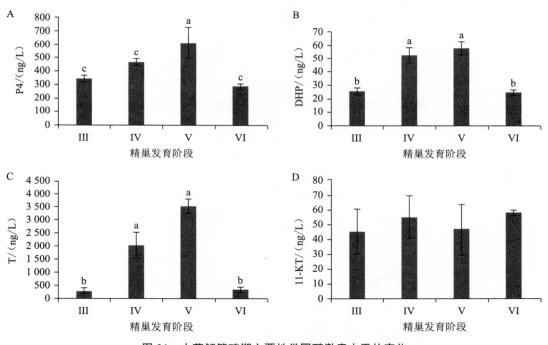

图 21　大菱鲆繁殖期主要性类固醇激素水平的变化

同时发现繁殖期大菱鲆精巢中孕酮核受体（*pgr*）在 III 期表达量较低，IV 期显著升高，V 期达到最大值，VI 期显著降低。

为了进一步分析类固醇激素对精子活力的影响，随机选取精子活力不同的 20 条 V 期大菱鲆为样品，分别检测血清中四种类固醇激素（T、11-KT、P、DHP）的含量。在精子活力不同的大菱鲆血清中，T 的表达量变化没有明显规律。在精子活力低的个体中，11-KT 的表达量高于精子活力高的个体。相反，P 和 DHP 都是在精子活力高的个体上表达量高。

2.13 大菱鲆繁殖期生殖系体细胞标记基因 *Amh*/*Sox9* 及精子变态成熟相关基因的研究

对雄性大菱鲆繁殖期生殖系体细胞的变化规律进行研究。通过同源克隆获得 Sertoli 细胞特异性分子标记 *Sox9* cDNA 全长 3 160 bp，包括 5′UTR 234 bp，3′UTR 1 486 bp，开放阅读框 1 440 bp，编码 479 个氨基酸；获得 *Cx43* cDNA 全长 1 758 bp，包括 5′UTR 190 bp，3′UTR 401 bp，开放阅读框 1 767 bp，编码 388 个氨基酸。

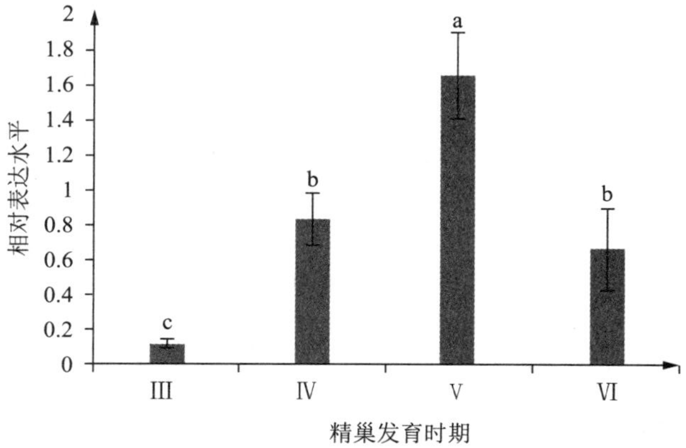

图 22 大菱鲆繁殖期 *pgr* 基因在大菱鲆精巢中的相对表达量变化

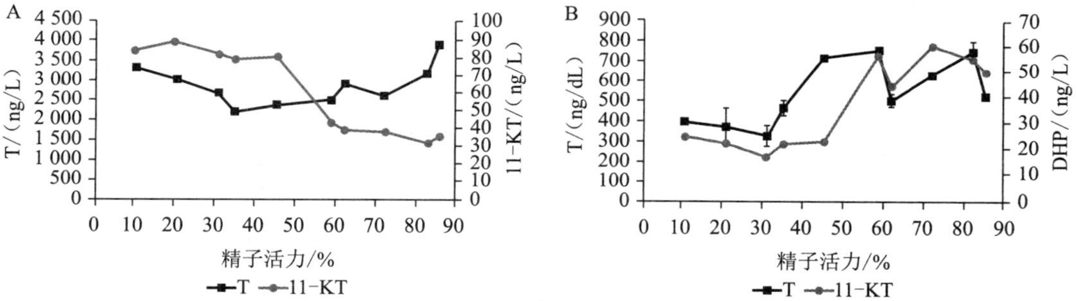

图 23 性类固醇激素和精子活力的相关性

比较了 *Amh*/*Sox9* 在不同组织间的表达差异，发现 *Amh* 在精巢表达较高，在雌、雄性的脑组织中也有部分表达。*Sox9* 在鳃、眼、雌、雄垂体和精巢中都有较高的表达，尤其在软骨组织丰富的鳃中表达量最高。*Amh*/*Sox9* 在雌、雄性腺中的表达存在显著差异，是较好的 Sertoli 细胞标记基因。

分析比较了不同发育时期，雄性大菱鲆精巢中 *Amh*/*Sox9* 基因 mRNA 的表达差异。*Amh*/*Sox9* 从 II 期初到 VI 期末经历了先升高、后下降、再升高的变化。在 II 期(对应 30 g 体重)精巢中的表达量最高，与这一时期快速增殖的 Sertoli 细胞相对应。随着精巢的发育成熟，其内部生殖细胞迅速增多，对应的 Sertoli 细胞比例相对下降，在 IV 期最低。随着 V 期精子的排出和精巢的衰退，生殖细胞数量锐减，Sertoli 细胞所占比例逐步升高。

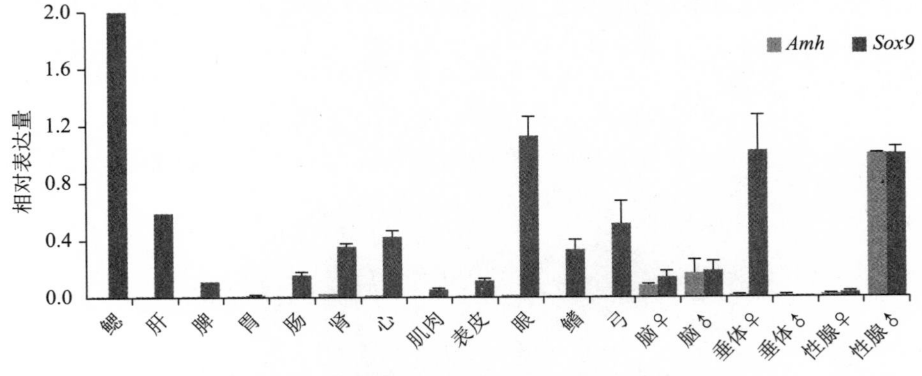

图 24　大菱鲆不同组织间 *Amh* / *Sox9* 的表达

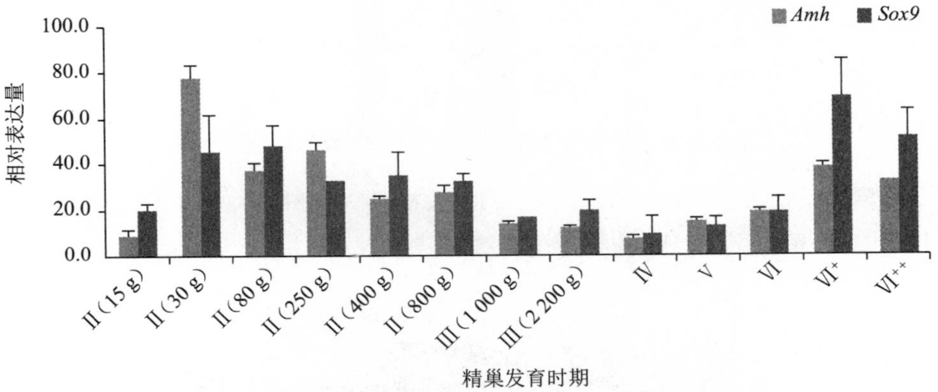

图 25　大菱鲆精巢不同发育时期 *Amh* / *Sox9* 的表达

分析比较了不同发育时期，雄性大菱鲆精巢中 *AR* / *Cx43* 基因 mRNA 的表达差异。*AR* 在 II 期有较高的表达，与该时期精原细胞增殖直接相关，在 III 期表达量较低。随后 IV～V 期逐步升高，与其他文献中报道的雄激素影响精子细胞成熟相关，直到精子排出后（VI 期）又有所回落。*Cx43* 的最高表达出现在 V 期，暗示了其在精子最终成熟阶段可能发挥重要作用。

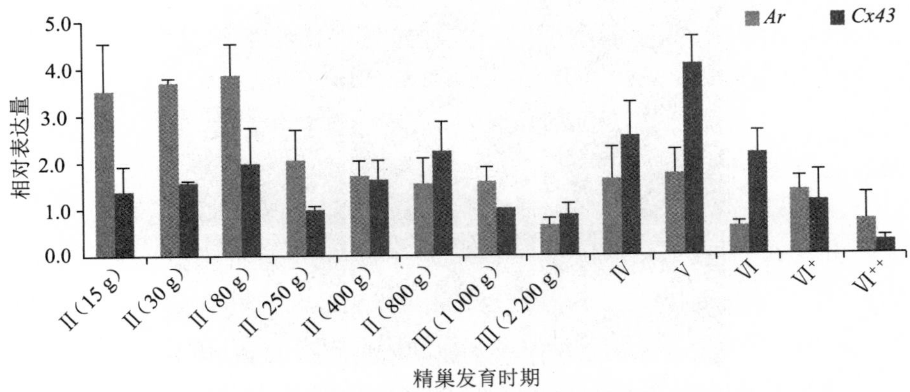

图 26　大菱鲆精巢不同发育时期 *AR* / *Cx43* 的表达

分析比较了不同发育时期，雄性大菱鲆精巢中 *LC3/BECN1* 基因 mRNA 的表达差异。*LC3/BECN1* 从 II 期初到 VI 期末表达量变化趋势一致，在 IV 期达到高峰，与该时期活跃的精子细胞变态过程紧密相关。

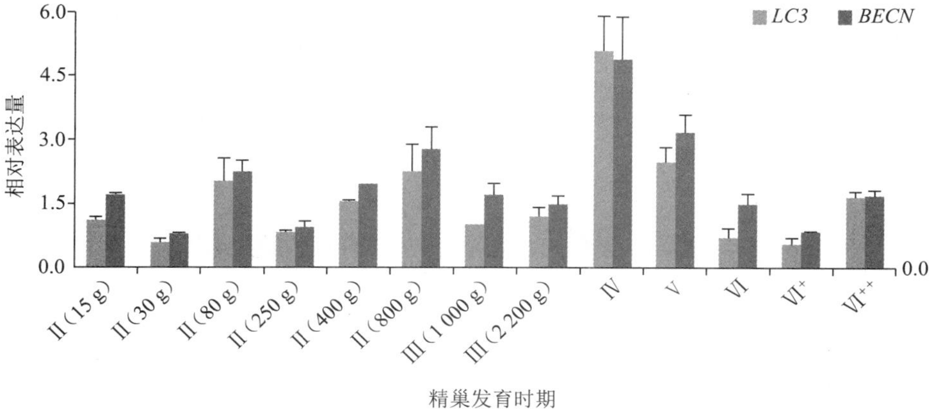

图 27 大菱鲆精巢不同发育时期 *LC3/BECN1* 的表达

通过原位杂交研究了 *Cx43/Sox9* 在不同发育时期大菱鲆精巢中的定位规律，发现二者主要在 II 期的 Sertoli 细胞及 VI 期精子细胞时期共表达。

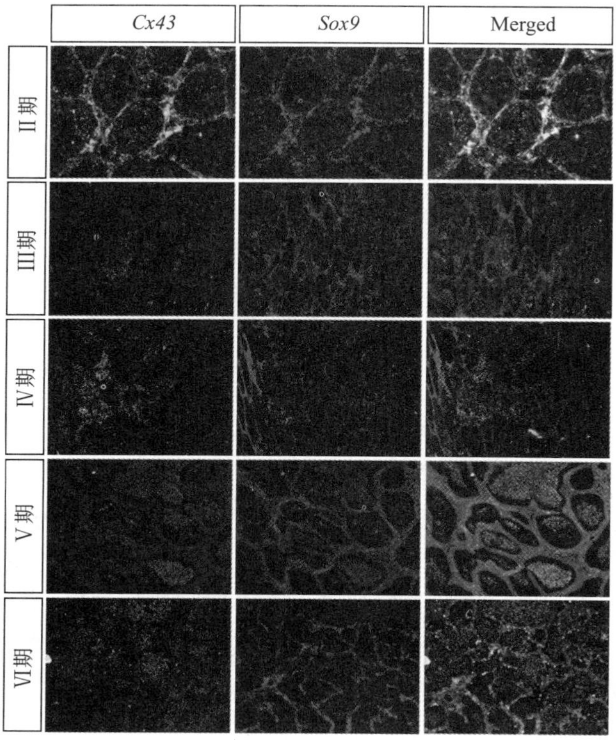

图 28 大菱鲆精巢不同发育时期 *Cx43/Sox9* 基因的定位

2.14　大菱鲆性腺发育成熟过程中 *aqp1a* 的作用研究

通过 RACE 克隆技术获得大菱鲆 *aqp1a* 基因全长 1 280 bp，通过 *aqp1* 和 Na^+-K^+-ATP 酶（Na^+-K^+ 泵）对大菱鲆精巢和卵巢发育过程进行双色原位检测，结果表明 *aqp1a* 在卵巢中主要在卵原细胞和初级卵母细胞中表达，随着雌性生殖细胞的发育，信号逐渐减弱，说明 *aqp1a* 可能参与了卵子的早期发生过程。在精巢中主要在雄性生殖细胞中表达。而 Na^+-K^+-ATP 酶呈泛表达模式，在生殖细胞和周边的体细胞（支持细胞和间质细胞）中均有表达，说明 *aqp1a* 和 Na^+-K^+-ATP 酶协同调控了精子成熟过程。

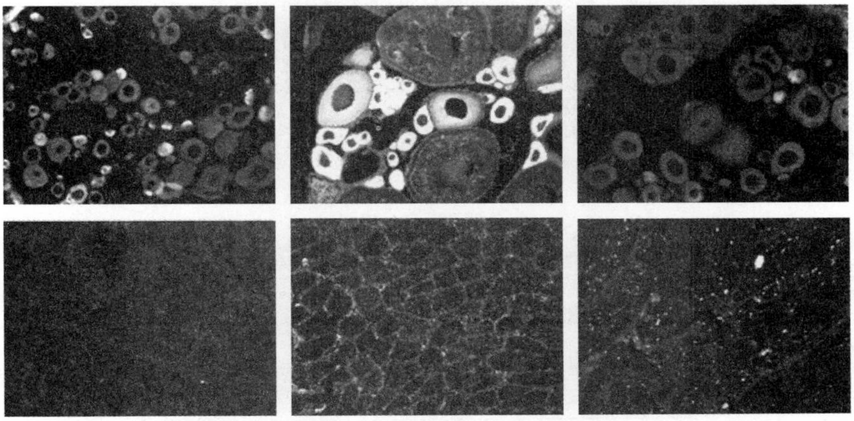

图 29　性腺发育过程中 *aqp1* 和 Na^+-K^+-ATP 酶的双色原位杂交

对大菱鲆的精巢和卵巢进行双色原位实验检测 *Aqp1* 和 *cftr*（一种跨膜蛋白质，又称囊性纤维化跨膜传导调节因子，是一种重要的 HCO_3^-/Cl^- 离子通道），结果表明，*cftr* 主要在早期卵原细胞和初级卵母细胞中定位，说明其可能参与了早期卵原细胞的增殖过程。

通过 AQP1 抗体对大菱鲆成熟期Ⅴ期精巢进行 WB 验证及组化定位，结果表明 AQP1 蛋白的相对分子质量约 28×10^3，主要定位在精子头部上，可能参与了精子激活过程中渗透压的调节。

通过 *aqp1*、Na^+-K^+-ATP 酶及 *cftr* 对大菱鲆精巢发育过程进行荧光定量研究，结果表明 *aqp1* 和 Na^+-K^+-ATP 酶的表达趋势呈正相关，在精巢从 II～IV 期发育过程中都呈现上升的趋势，在 IV 期达到峰值，在 V～VI 期开始下降，说明 *aqp1* 和 Na^+-K^+-ATP 酶在雄性生殖细胞从增殖分化到成熟过程中具有协同效应，主要在成熟过程中起作用。而 *cftr* 基因主要在 II 和 VI 期表达量最高，说明其可能在精原细胞的增殖和分化过程中起作用。

综上，在精巢发育过程中，*aqp1* 和 Na^+-K^+-ATP 酶在精细胞成熟过程中协同起作用，而且 AQP1 蛋白定位在精子头部，可能参与了精子激活过程中渗透压的调节，而 *cftr* 主要在精巢启动发育期和退化期起作用，可能参与了雄性生殖细胞的增殖过程。

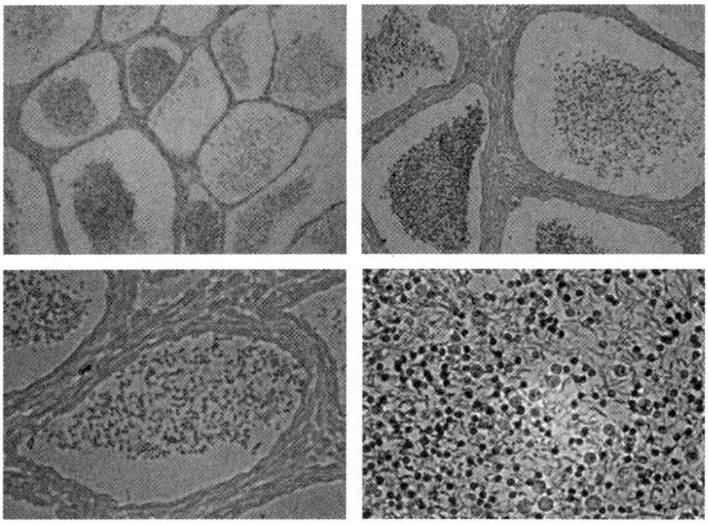

图 30 AQP1 蛋白 WB 验证及在精巢成熟期的组化定位

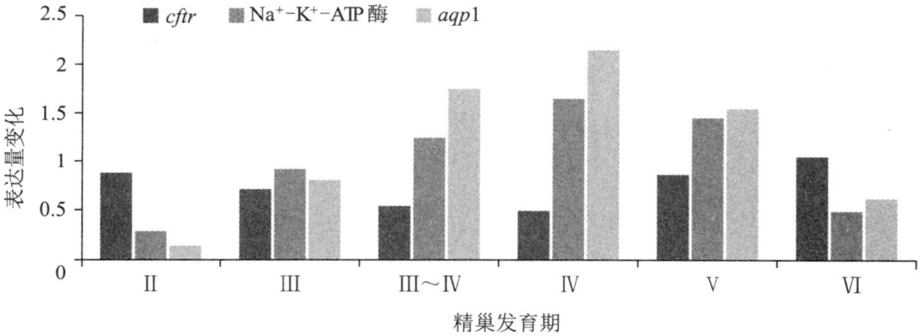

图 31 大菱鲆精巢发育过程中 *aqp1*、Na^{+}-K^{+}-ATP 酶及 *cftr* mRNA 的表达量变化

2.15 雌性大菱鲆繁殖期卵黄蛋白原的生成及其组织特异性分布研究

从大菱鲆肝脏中克隆得到 3 种卵黄蛋白原基因，如图 33 所示，相对荧光定量结果显示，3 个卵黄蛋白原基因在雌性大菱鲆繁殖周期中都呈先上升后下降的趋势表达，但具体的表达量差异显著。*vtgAb* 在 II 期表达量很低，进入 III 期后表达量开始上升，之后迅速攀升，在 V 期达到整个繁殖周期中的表达峰值，最后在 VI 期迅速回落至 II 期水平。*vtgAa* 和 *vtgC* 的表达趋势与 *vtgAb* 基本一致，但明显的区别是这 2 个基因的表达量峰值出现在Ⅳ期。在这 3 个卵黄蛋白原基因中，表达量最为丰富的是 *vtgAb*，而表达量最低的是 *vtgC*。在同一个发育期之内，*vtgAb* 的表达量最高分别达到 *vtgAa* 和 *vtgC* 的 16 倍（II 期）和 50 倍（VI 期），而在Ⅳ期时 *vtgAa* 与 *vtgAb* 差距缩小到最小的 2 倍。

对雌性大菱鲆繁殖周期的肝体系数和性体系数变化分析发现，肝体系数折线存在 2 个峰值，分别出现在 III 期和 V 期，总体看来肝体系数的变化范围不大。性体系数在 II 到 IV 期缓步提升，之后在 V 期迅速攀升达到峰值，最后进入 VI 期后急剧下降至原始水平，这与肝脏中卵黄蛋白原基因的表达规律基本一致，呈先上升后下降的趋势。因此，性体系数和肝脏中卵黄蛋白原基因的表达趋势可以正确反映雌性大菱鲆卵巢卵黄积累的水平。

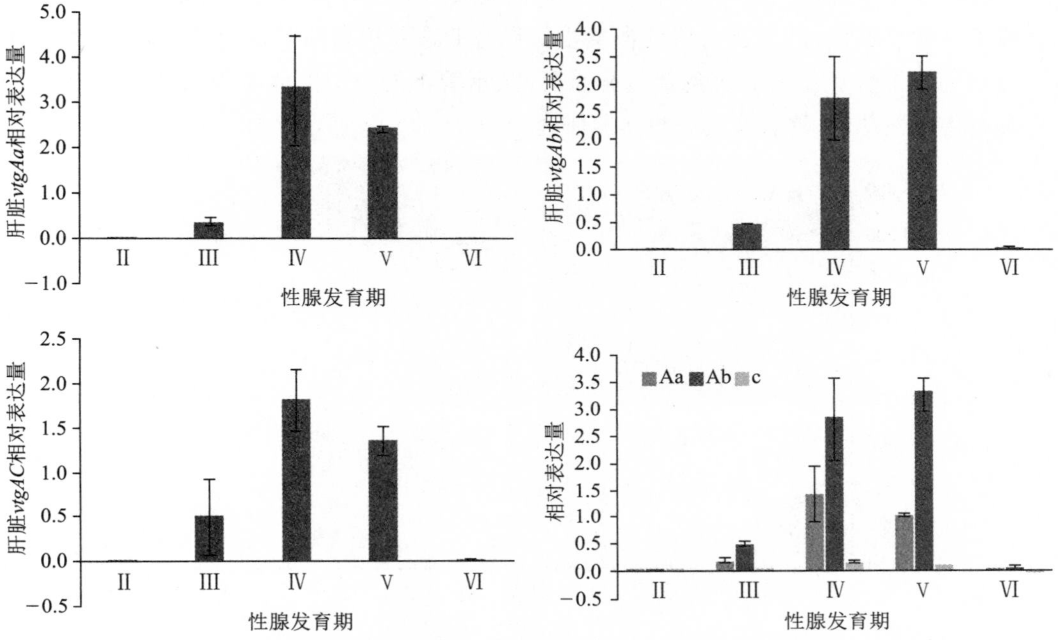

图 32　雌性大菱鲆生殖周期中 3 个 *vtg* 基因在肝脏中的表达变化

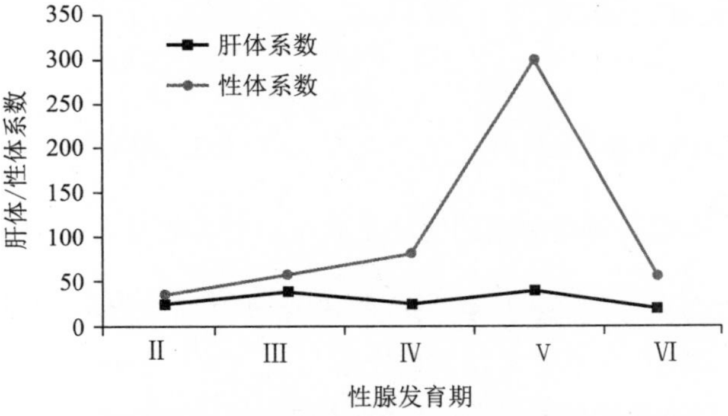

图 33　雌性大菱鲆生殖周期的肝体系数和性体系数变化

相对荧光定量结果显示，3个*vtg*基因在雌性大菱鲆肝脏中的表达量远远高于其他组织。在非肝脏组织中，皮肤、性腺（即卵巢）、脾脏、鳃、肾脏、脑、垂体以及眼睛等组织中均有卵黄蛋白原基因的表达，但其各自表达基因种类及其表达量有所差异。性腺中 3 个基因均有表达，*vtgAb* 的表达量最高，但也仅仅是肝脏相应基因表达量的 1 / 3 2000 00，而 *vtgAa* 和 *vtgC* 的表达量相对较低。皮肤中表达量最高的也是 *vtgAb*，其表达量稍低于性腺，但皮肤中 *vtgAa* 的表达高于除肝脏外的其他组织，是肝脏表达量的 1 / 350 000。在非肝脏组织中，*vtgC* 表达量最高的组织是垂体，是肝脏中表达量的 1 / 22 500。对于鳃和肾脏这两个组织，二者均有一

定量的 *vtgC* 表达,但其他 2 个卵黄蛋白原基因则表达量极低。在脑组织中,3 个基因均有表达,*vtgAb* 和 *vtgC* 的表达量是 *vtgAa* 的两倍。在眼睛中,*vtgAa* 和 *vtgAb* 的表达量明显高于 *vtgC*。由此可知,雌性大菱鲆卵黄积累很有可能主要来源于外源合成途径的肝脏合成,而内源合成相对微乎其微,或卵巢合成的卵黄蛋白原有非营养积累的其他功能。同时,3 个卵黄蛋白原基因在非肝脏组织中表达量差异显著,预示着不同类型的卵黄蛋白原在不同组织中可能具有特异性功能,这有待进一步实验研究。

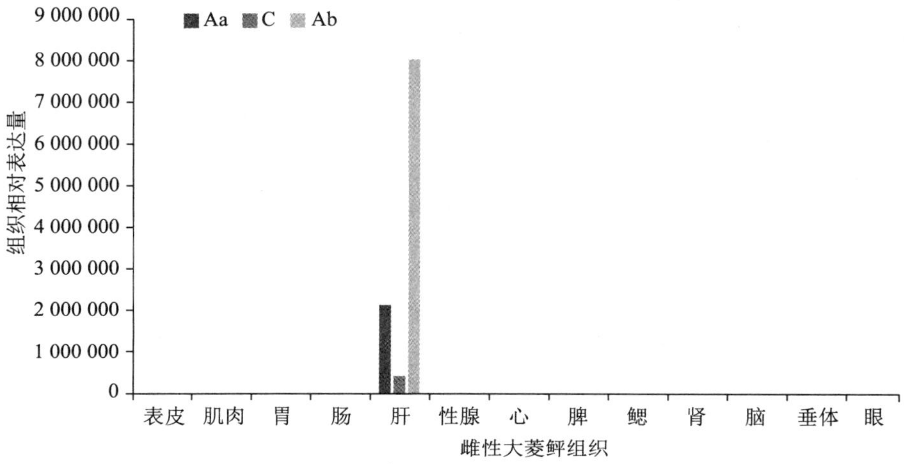

图 34　3 个 *vtg* 基因在包括肝脏在内的 13 个组织中的分布情况

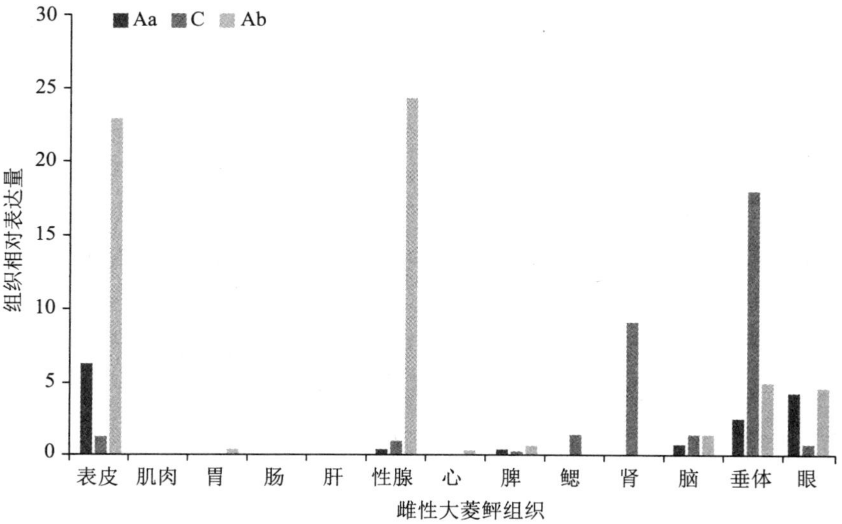

图 35　3 个 *vtg* 基因在非肝脏的 12 个组织中的分布情况

(岗位科学家　李　军)

鲆鲽类循环水养殖系统与关键装备研发进展

工厂化养殖设施设备岗位

1　鲆鲽类封闭循环水养殖综合标准化研究

我国鲆鲽类养殖业是在海水养殖的第四次浪潮中脱颖而出迅速发展起来的。从横向产业结构角度看，该产业是我国海水鱼类养殖的重要品种，是推动我国海水养殖第四次浪潮的主力军之一。根据国家农业部的数据，到2015年时其产量已经达到13.18万吨，居海水鱼类养殖的第二位，而且仅比居第一位的大黄鱼低0.8%。纵向发展趋势上看，首先，其发展速度相对较快。在2005年时，该产业产量仅为鲈鱼的72.5%，而2011年已经上升到了后者的98.5%，两者的距离已经非常小，完全有可能成为我国海水鱼类养殖业之首位。但是，在快速发展的同时各种影响产业健康发展的问题也浮现出来。2006年“多宝鱼”事件一直到目前仍在持续发酵，鱼价一蹶不振，渔民苦不堪言。山东、辽宁等地区由于用水过度，导致地下水枯竭甚至地面沉降。滥用药物使得鱼类种质退化，导致抗病能力下降，等等。因此，建立标准体系表，明确体系框架结构和制定方向，对于促进和推动鲆鲽类产业健康、有序发展具有积极、深远的意义。

1.1　鲆鲽类封闭循环水养殖标准体系表编制原则

鲆鲽类封闭循环水养殖标准体系的建立以市场需求为动力，以技术和技术发展为纲，以具有自主知识产权的核心技术为中心，以传统标准化为基础，站在一定的高度上看待农业生产技术创新，提高工业化农业技术创新的起点、水平和效率，提高设施生产技术、产品和相关农产品的国际竞争力。标准体系既要符合WTO/TBT规则，又要按照《标准体系表编制原则和要求》（GB/T 13016—91）进行框架的构建，具备提纲挈领、效益显著、范围明确、完整可分、集合多能、协调相关、动态可扩等能力，先进、简单、实用，能带动水产养殖生产力进步，促进相关产业升级、减少浪费、维护生态平衡。

（1）“目标明确”原则。总体目标是构建一个以促进鲆鲽类封闭循环水养殖规范化发展，推动鲆鲽类养殖生产方式转变，实现鲆鲽类健康养殖为主要目标的鲆鲽类封闭循环水养殖标准体系,同时要与国际接轨、并适合我国水产业的发展。实现国家标准、行业标准与地方标准间的相互协调,以有效保证鲆鲽类养殖及其产品质量安全,保障消费者健康和权益,推动产业技术进步,提升全国鲆鲽类产业国内外市场竞争力。

(2)"全面成套"原则。全面成套体现在两个方面。一是在采纳标准的级别上全面成套,包含了国家标准、行业标准和地方标准;二是在采纳标准的类型上全面成套,包含了基础标准、设计标准、设备标准、方法标准、安全标准等不同类型的标准。

(3)"层次适当"原则。根据标准的适用范围,使共性标准和个性标准分别在不同层次上形成有机整体。

(4)"划分清楚"原则。按照鲆鲽类封闭循环水养殖所涉及的系统、设备、投入品、生产管理等多个方面进行类别划分,保证了分类清晰不产生交叉。

1.2 鲆鲽类封闭循环水养殖标准体系表

按照《标准体系表编制原则和要求》,鲆鲽类封闭循环水养殖标准体系表对标准化对象、功能、专业、行业门类和层次进行了归类和梳理。如图1所示,标准体系表下列4个子体系,分别为基础、装备与工程、生产管理和产品。

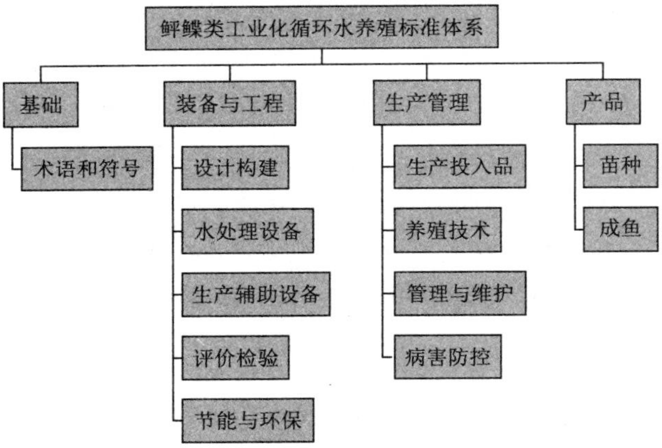

图1 鲆鲽类封闭循环水养殖标准体系表

1.3 标准体系信息统计

整理鲆鲽类封闭循环水养殖相关技术标准120项,基础子体系标准5项,装备与工程子体系标准55项,生产管理子体系标准12项,产品子体系标准1项,建议制订、修订标准43项。

2 臭氧混合添加技术研究与装备研发

针对国内目前养殖企业在臭氧添加和管理方面缺少专用装备技术的问题,开展了样机研发和性能参数优化试验,并在天津天世农养殖有限公司进行了初步试运行。

2.1　臭氧混合装置研发

装置研发从节能性、稳定性和安全性三个方面进行了重点考量。从节能性角度考虑，为了避免不必要的能耗增加，反应装置设计工作在常压状态下，通过孔板滴淋的方式增加气液接触面积，提高气液混合效率。从稳定性角度考虑，装置设计采用双层孔板布水，第一层孔板网孔较粗，主要用以缓冲水流，第二层孔板网孔较细，用以实现均匀布水，提高气液混合效率。由于臭氧混合装置主要设计用于集成了养殖系统水处理支路上，一般采用水泵提水进入反应器，过高的流速会导致布水层流态紊乱，导致臭氧气体外泄逃逸，降低混合效率。双层孔板布水能够很好的解决该问题。从安全性角度考虑，臭氧气体具有非常强烈的氧化性，长时间暴露在高浓度臭氧条件下对人体、鱼体都会产生一定危害。有鉴于此，课题组专门设计研制了残余气体回收装置，利用文丘里效应在反应器出水管路上形成负压，将残余臭氧吸入出水管路进行二次混合，避免臭氧气体的外泄。

2.2　臭氧混合添加技术研究

试验在渔业机械仪器研究所水净化技术实验室内进行。使用一个直径 2 m 的圆形鱼池作为蓄水池，将实际运行中的养殖系统水体抽入池内备用，水深 50 cm。通过水泵将蓄水池内水体泵入臭氧混合器，混合器出水回入蓄水池。通过监测混合器出水水质指标，反应装置处理性能。实验常规参数设置为：水温 13 ℃～15 ℃，pH 为 7.0～7.7，臭氧混合器进水流量 20 m^3/h，臭氧进气量 3 L/min。实验对比了 50%、75%、100% 3 个不同臭氧发生功率条件下池内水质变化情况。

结果如图 2 所示，50% 功率条件下，在实验前 30 分钟时间，池内亚硝酸盐浓度从 0.1 mg/L 降低至 0.001 mg/L，而后无限接近于 0。ORP 在前 30 分钟变化缓慢，而后从 285.6 快速上升至 578.4。水色基本上从 70 线性下降至 41。溶解氧浓度变化符合双模理论，但是由于时间过短，没有显示出最终饱和浓度，从趋势上看基本上在 22 mg/L～23 mg/L 之间。

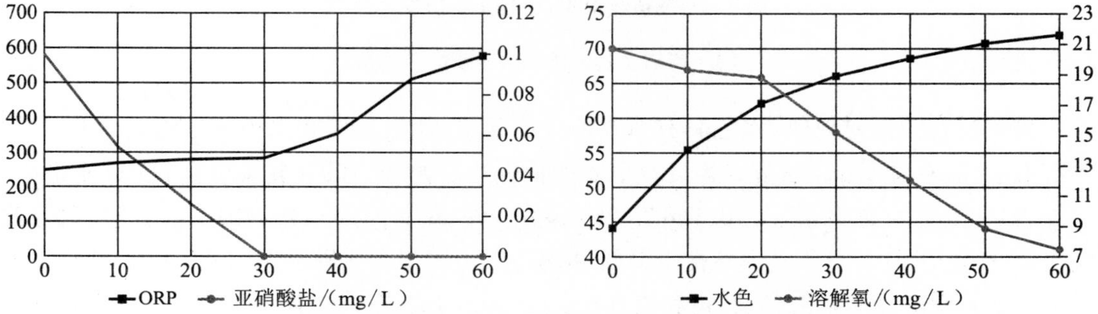

图 2　臭氧功率 50%条件下水质变化情况

如图 3 所示，75% 功率条件下，在实验第 10 min 到 70 min 时间段，亚硝酸盐浓度从 0.563 mg/L 线性下降至 0.027 mg/L，而后无限趋近于 0。ORP 在前 80 min 变化不显著，而后从 302. 7 快速上升至 591. 8。水色指标基本呈线性，从 32 降低至 7。溶解氧浓度变化符合双膜理论，饱和浓度在 22. 8 mg/L 左右。

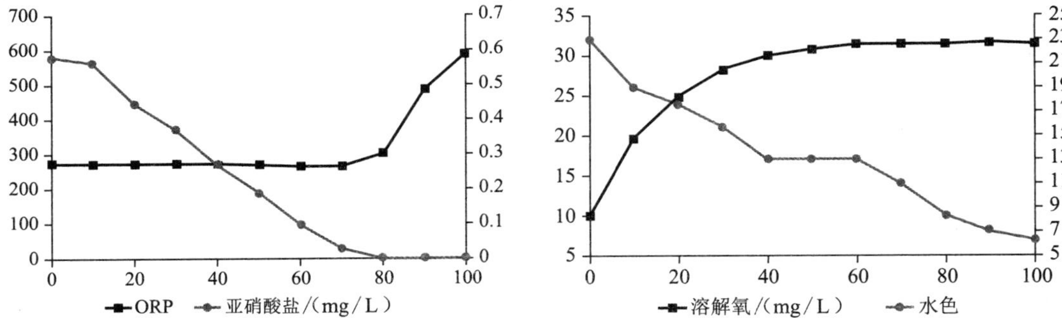

图 3　臭氧功率 75%条件下水质变化情况

如图 4 所示，100%功率条件下，在实验前 30 min 亚硝酸盐浓度无明显变化，第 30 min 到 60 min 时间段，亚硝酸盐浓度从 0. 501 mg/L 线性下降至 0. 013 mg/L，而后无限趋近于 0。ORP 在前 60 min 变化不显著，而后从 260. 2 快速上升至 503。水色指标基本呈线性，从 42 降低至 13。溶解氧浓度变化符合双膜理论，饱和浓度在 22. 8 mg/L 左右。

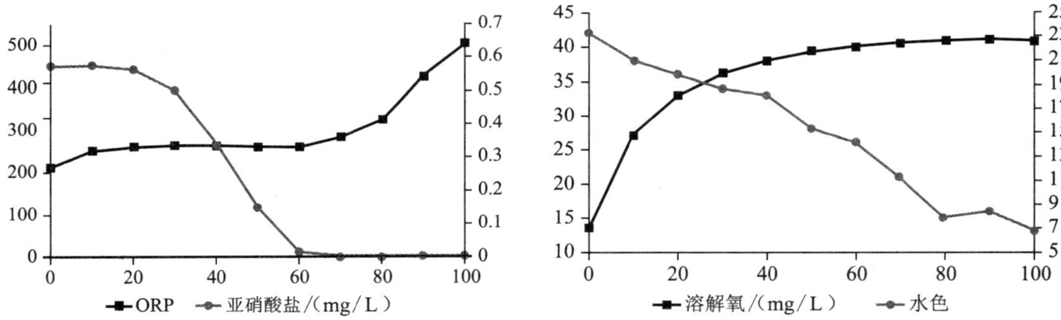

图 4　臭氧功率 100%条件下水质变化情况

2. 3　结论

臭氧混合装置对亚氮、水色具有明显的去除效果，能明显提高水中的溶解氧和 ORP 等水质指标。装置对亚氮的去除效率受臭氧添加量影响显著。在 100%臭氧功率下去除率最高，每添加 1 g 臭氧可去除亚氮约 0. 1 g；在 75%臭氧功率条件下较低，每添加 1 g 臭氧可去除亚氮约 0. 07 g；在 50%臭氧功率条件下最低，每添加 1 g 臭氧可去除亚氮约 0. 04 g。装置增氧效果几乎不受臭氧功率的影响，饱和浓度在 22. 8 mg/L 左右。装置对水色指标的去除基本呈线性。水体 ORP 指标的变化受亚硝酸盐浓度影响较大，当亚硝酸盐浓度接近于 0 时才会开始出现显著上升趋势。

3　生物流化床技术熟化研究

为进一步优化完善生物流化床应用技术，开展了滤器结构优化设计，并通过中试实验深入研究了滤器挂膜规律、水处理性能以及滤料表面的微生物群落结构。

3.1　挂膜规律及水处理性能研究

对传统的流化床生物滤器进行优化，研发出一种可清洗的流化床生物滤器，其装置结构图如图5所示，总高1.6 m，该装置包括床层膨胀区和悬浮物沉淀区两部分。其中，床层膨胀区直径0.5 m，高1.2 m，内装有0.7 m高的滤料，本区域设有自清洗组件，包括一组“米”字型吸料管及一台增压泵，用于对上层滤料进行清洗，经清洗后的滤料通过底部切向进水管进入环形布水腔，进而重新进入床层膨胀区。悬浮物沉淀区位于床层膨胀区上方，直径0.8 m，高0.4 m，主要作用于沉淀自清洗组件分离出的悬浮物或滤料，而后通过排污口将其排出本装置，经沉淀排出的滤料可再次利用。装置底部设有两个成180°的切向进水管，该方式能促进滤器原有的旋转水流，避免了单向进水出现死角区域等问题，进而优化了滤器内部的流态。

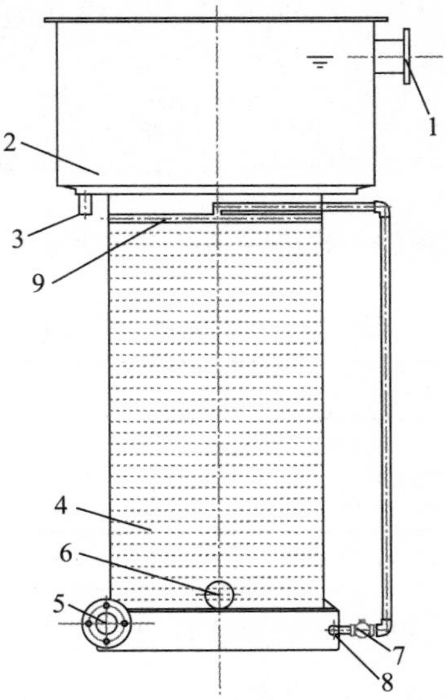

图5　流化床生物滤器结构图

1. 排水口；2. 沉淀区；3. 排污口；4. 膨胀区；5. 进水口；6. 排砂口
7. 增压泵；8. 自清洗装置进水口；9. “米”字形吸料管

对该生物滤器采用自然流水法进行挂膜，经过1个月后，滤料的颜色自下而上变成深灰色，此时经该生物滤器处理后的TAN浓度低于0.26 mg/L，NO_2^--N浓度低于0.03 mg/L，说明生物膜已开始成熟。对滤器中填料进行电镜扫描观察，挂膜前后填料表面发生了一定的变化。挂膜前载体表面无附着物，相对比较光滑。生物膜成熟后，载体表面被微小的悬浮物所包裹，因载体的比表面积巨大，故载体表面的微生物种类非常丰富，加大了固液之间的传质。

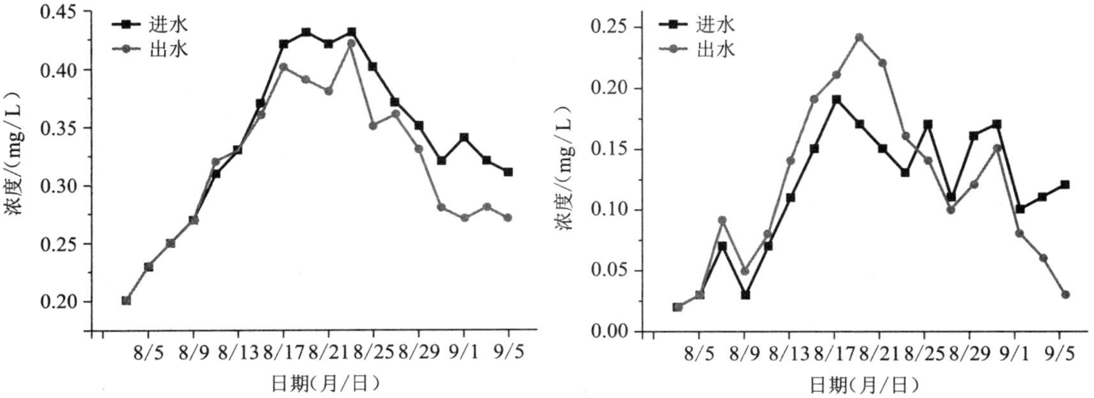

图 6　新型流化床生物滤器挂膜期间 TAN 和 NO_2^--N 的变化情况

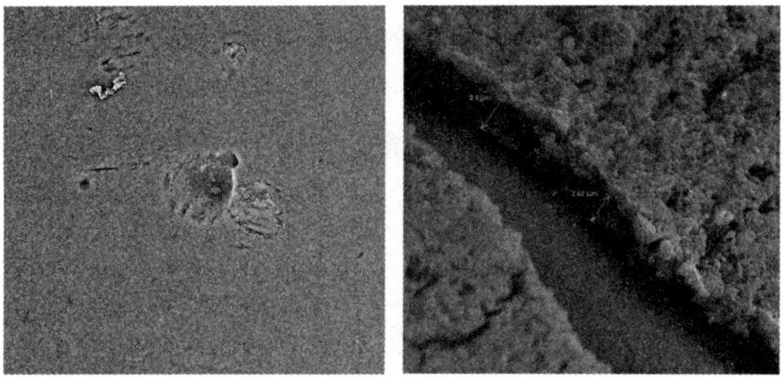

A　　　　　　　　B

图 7　挂膜前（A）、挂膜后（B）载体表面电镜扫描照片

表 1　流化床生物滤器的水处理性能

水质指标	TAN	BOD5	SS
进水浓度/(mg/L)	0.28 ± 0.05	5.22 ± 1.11	25.21 ± 2.37
出水浓度/(mg/L)	0.08 ± 0.04	2.13 ± 0.23	15.11 ± 1.87
去除率/%	61.9 ± 5.23	59.19 ± 4.98	40.06 ± 10.21

流化床生物滤器挂膜成功后，开启自清洗装置，控制床层膨胀率为150%，该生物滤器对不同营养盐的去除情况如表 1 所示。由该表可知，滤器对 TAN 保持了较高的去除率，其平均去除率达到了(61.9 ± 5.23)%，在该工况下，滤器对 TAN 的去除负荷达到了(245.2 ± 50.5) g/(m^3·d)。对 BOD5 的平均去除率达到了(59.19 ± 4.98)%，出水 BOD5 的浓度保持在(2.13 ± 0.23) mg/L，说明此时异养细菌和自养细菌的数量已共存于滤料的表面，其数量也已达到稳定平衡。由于滤器上部悬浮物沉淀区的存在，本装置对 SS 去除率较高，达到了(40.06 ± 10.21)%，有效减轻了系统对悬浮物的去除压力。有学者研究表明，高有机物负荷不利于硝化反应的进行，而本装置拥有较为完善的物理过滤结构，降低了养殖水体中的有机负荷，有助于滤器的硝化作用。

3.2　流化床生物滤器自清洗技术研究

流化床生物滤器在运行过程中，滤器中的滤料因养殖水体中悬浮物的附着逐渐变轻，特别是表层滤料，故滤器在运行过程中床层会逐渐升高。图 8 显示了当自清洗装置启、停时流化床床层的增高情况。由该图可知，当关闭自清洗装置时，随着滤器膨胀率的增加，床层逐渐升高，当膨胀率为 180% 时，滤器的床层周增高达到了(2.5 ± 0.5) cm。而当开启自清洗装置时，随着膨胀率的增加，滤器的床层基本不升高，维持在一定的水平，其增高数值显著低于关闭自清洗装置工况。当开启自清洗装置时，表层滤料表面的附着物被去除，改变了滤料的重量，故当进水量一定时，滤器的床层高度能基本稳定。

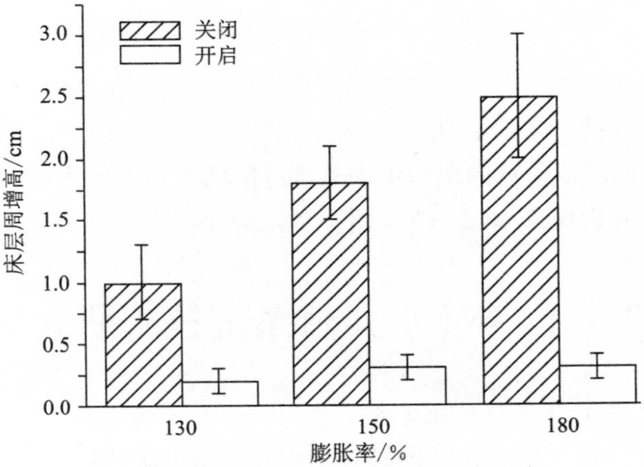

图 8　不同工况下流化床床层增高情况

图 9 显示了不同工况下滤器中滤料的流失情况，该部分滤料主要流失至床层膨胀区，经清洗后可回收使用。由该图可知，随着床层膨胀率的增加，滤料的流失量逐渐增加。在膨胀率为 150% 工况下，当关闭自清洗装置时，滤料的周流失量为(192 ± 12) g，约占滤料总体积的 0.1%，当开启自清洗装置时，滤料每周沉淀至床层膨胀区的重量为(38 ± 5) g，约占滤料总体积的 0.02%，而滤料从滤器中每周流失的重量仅为(8 ± 2) g，集成自清洗装置及颗粒物收集技术可将滤料的流失量降低 95.8%，显著提高了装置的使用效率。

3.3　流化床生物滤器微生物群落结构分析研究

在流化床生物滤器挂膜期，表层区域的优势细菌为绿弯菌门(29.31%)和酸杆菌门(29.65%)，而底层区域的优势细菌为拟杆菌门(54.03%)。生物膜成熟后滤器稳定运行工况下，载体表面门水平的微生物种类分布更加丰富与均衡，共删选出 31 个门，包括变形菌门(29.2%～38.05%)，拟杆菌门(16.81%～33.76%)，绿弯菌门(6.2%～24.18%)，等。通过与 silva 数据库进行比对，共鉴定出 490 个微生物属。其中表层区域的优势细菌

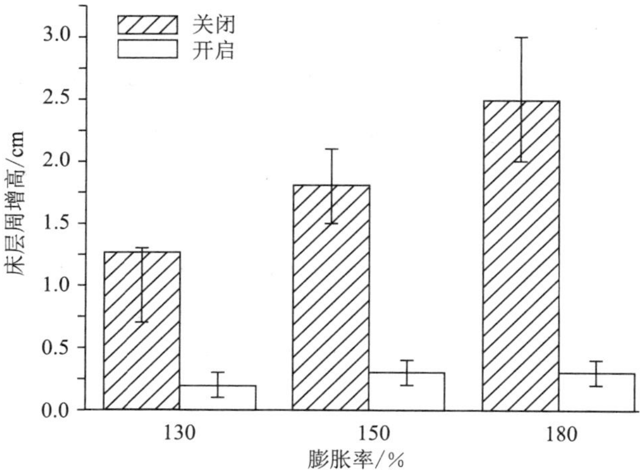

图 9　不同工况下流化床滤料流失情况

基本维持不变,主要包括厌氧绳菌科(8.4%～28%)、黄杆菌科(1.1%～32%)、红杆菌科(2.3%～17%)等。而底层区域的优势细菌随着时间的推移有所变化,主要包括硝化螺菌属(12.45%～17.06%)、微丝菌属(2.6%～8.8%)、*Muricauda*(4.8%～6.3%)等。

4　珍珠龙胆工厂化循环水养殖系统技术研究

珍珠龙胆的人工繁育在国内外已取得突破,而海南、广东、福建沿海地区近年相继开展了网箱和池塘养殖,辽、冀、津、鲁、江、浙等省市也开始引进和建设工厂化循环水试养。为尽快实现集约化生产,在山东莱州明波水产有限公司开展了珍珠龙胆工厂化循环水人工养殖试验研究。

4.1　珍珠龙胆工厂化循环水养殖系统构建

养殖车间为单元化标准设计,每个单元2排,各16个养鱼池,配备一套200 m^3循环水水处理系统。养鱼池为方形抹角水泥池,面积40 m^2,平均水深0.95 m,池底向中心倾斜5%,排水口设置于底部中央,周边相向对流进水并通入液态纯氧,排水口末端设阀门及活动摇臂器控制水位。32个养鱼池有效养殖水面1 280 m^2,养殖水体1 216 m^3(图10)。

工厂化循环水养殖系统的生物承载量(纯氧)≥ 60 kg/m^3。水质指标:水温25 ℃～28 ℃,盐度29～32,DO(养殖池出水)≥ 10 mg/L,pH 7.8～8.5,NH_4^+-N ≤ 0.5 mg/L,NO_2^--N ≤ 0.02 mg/L,COD ≤ 3 mg/L,SS ≤ 10 mg/L,光照度 ≤ 5 000 lx,光照节律与自然光相同。24 h水循环8～24次(可调),系统自维护回流水量15%～30%,24 h新水补充添加量 ≤ 系统水量的5%。

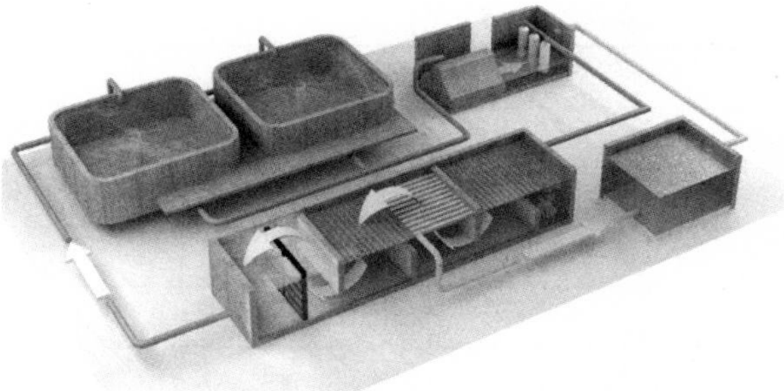

图 10　工厂化循环水养殖示意图

4.2　珍珠龙胆工厂化循环水养殖系统运行效果研究

表 2 数据表明，养殖用水经循环系统处理后，各项水质指标均在设计指标范围之内，且能满足珍珠龙胆正常生长需求。

表 2　养殖车间水质检测指标

	COD /(mg/L)	SS /(mg/L)	NO_2^--N /(mg/L)	NH_4^+-N /(mg/L)	pH	DO /(mg/L)
设计指标	≤3.0	≤10	≤0.02	≤0.5	7.8～8.5	≥10
1	1.9	1.6	0.012	0.15	8.2	10.60
2	2.5	1.5	0.012	0.30	8.1	11.00
3	2.1	1.2	0.014	0.22	8.9	12.20
4	2.1	1.4	0.010	0.20	7.8	11.50
平均值	2.15	1.4	0.012	0.22	8.0	11.32

循环水系统的进水与出水的水质指标相差很小，基本上在 10% 以内

表 3 为 2015 年 3 月至 2016 年 1 月珍珠龙胆养殖生长情况。从中可知采用工厂化循环水养殖珍珠龙胆，10 个月后平均体重为 650 g，达到商品鱼规格。经统计，1 216 m^3 养殖水体共放养 50 g 规格苗种 10 万尾，在水温 25 ℃～28 ℃、盐度 29～32、pH 7.8～8.5、DO（养殖池出水）≥ 10 mg/L 的条件下，经 10 个月培育，平均体重 0.65 kg，共收获成品珍珠龙胆成鱼 9.65 万尾，养殖单产 51.6 kg/m^3，总成活率为 96.5%。

表 3　珍珠龙胆养殖数据统计表

项目	2015 年	2016 年
养殖池鱼数量/尾	100 000	96 500
养殖鱼平均体重/g	50	650
养殖密度/(kg/m^3)	4.1	51.6
水温/℃	27.5	25.5
日投喂次数/次	2～3	1～2

（续表）

项目	2015 年	2016 年
日投饵量 / %	5	3
成活率 / %	100	96. 5
总产量 / kg	5 000	62 725
平均月增重 /（克 / 尾）	50. 0	60. 0

5　船载舱养系统设计研究

根据深蓝渔业发展战略，以 8 万吨级散货船为基础船型，利用深远海优越的气候、资源和水质条件，开展经济性海水鱼类船载舱养系统设计研究。

5. 1　舱型设计与布局

设计利用船体纵、横舱壁，将内部空间划分为 12 个养鱼水舱，用以进行成鱼养殖。由于平台线型关系，共设计 3 种舱型。其中，标准舱 8 个，35. 55 m × 20. 2 m × 13. 7 m，长宽比 1 : 2；尾舱 2 个，10. 48 m × 19. 76 m × 13. 7 m，长宽比 1 : 1。首舱 2 个，为异型舱。

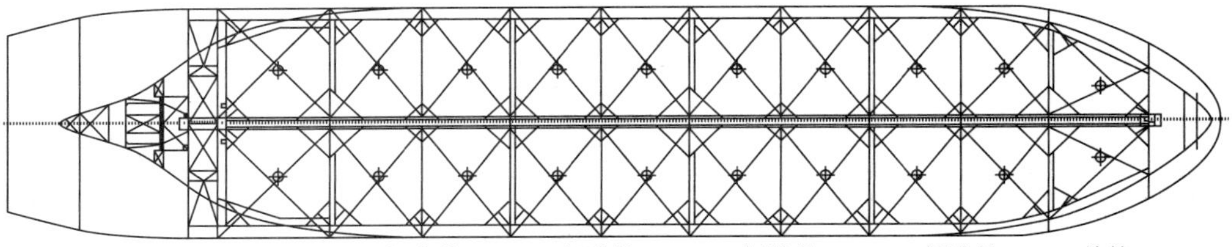

图 11　舱型布置平面图

5. 2　标准舱流态设计与模拟研究

标准舱占所有养殖舱的 75%，是模式设计的主要研究方向。研究采用计算机 CFD 模拟技术对比了不同进出水条件下，标准舱内的水体流态情况。设定换水率 4 次 / 天，进水流速 1. 33 m / s。结果显示：无论是采用“切向进 + 底出”方式（图 12）、“侧上进 + 侧下出”方式（图 13）还是“侧上进 + 右侧出”方式（图 14），舱内水体基本上无法形成一个相对稳定、均匀的流态，进而对养殖舱的集排污效果产生不利的影响。

针对这一问题，课题组参考借鉴多单元跑道式养殖系统（MCR）对标准舱结构和进排水方式进行了针对性设计，以优化舱内的水体流态。具体方式为：在舱体长度方向中间部位设置导流板，形成 2 个舱养单元，通过水流使养殖舱内形成 2 个相互独立的漩涡状水体流态。图 15 和图 16 显示了采用 2 种导流板设置条件下，“切向进 + 切向出”方式形成的水体流态。

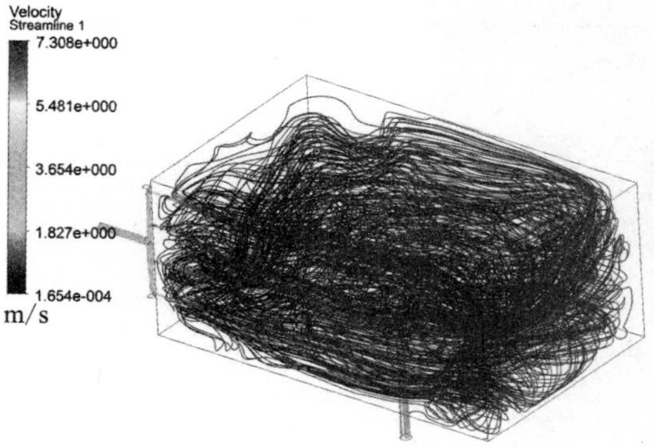

图 12　切向进底出

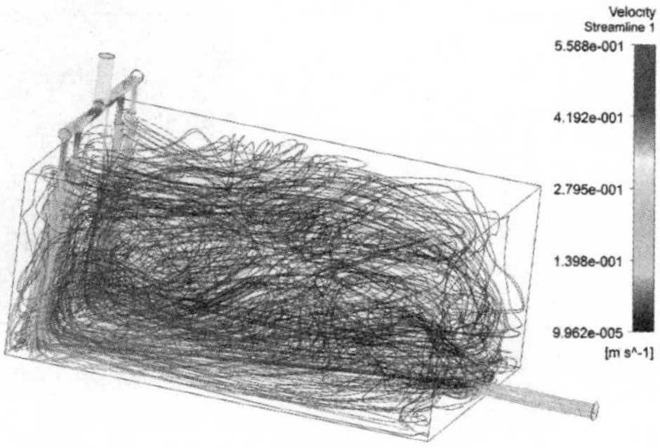

图 13　侧上进侧下出

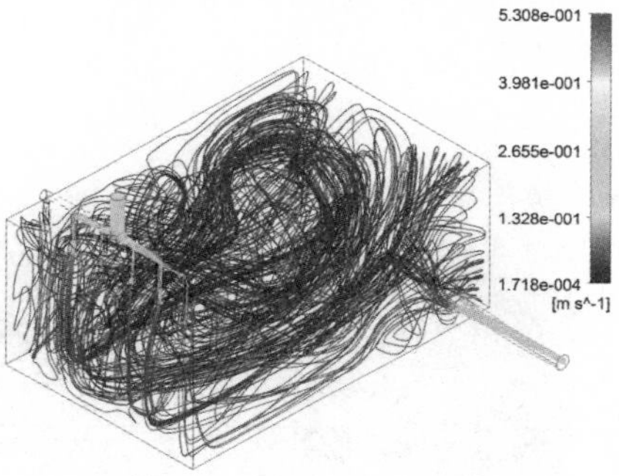

图 14　侧上进右侧出

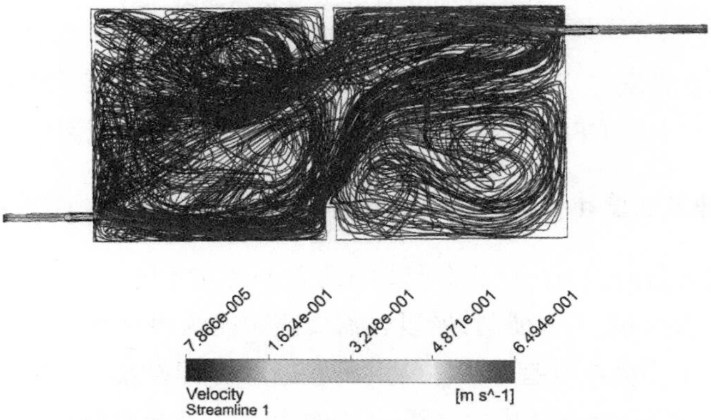

图 15　导流板较短条件下切向进出水方式流态模拟

从中可以看出，导流板长度对流态的形成有较大的影响，当导流板达到一定长度后就有可能形成一个比较均匀的流态。

在经过多次深入研究和优化后，设计方案调整为在舱体长度方向中间部位设置导流板，形成 2 个舱养单元，每个舱养单元底部设计为锥形，同时在锥底设 1 个排水口，通过管路向

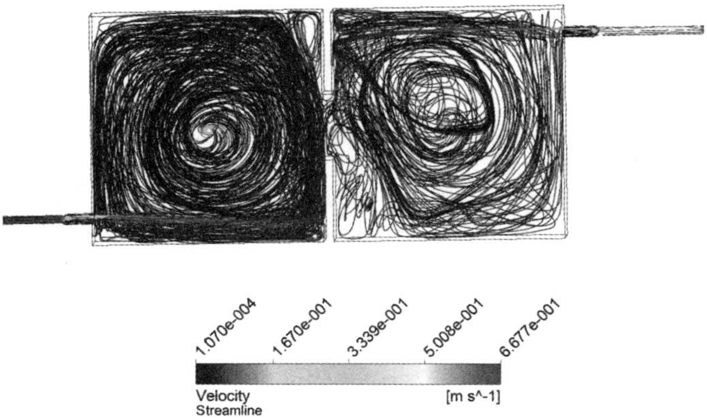

图 16　导流板较长条件下切向进出水方式流态模拟

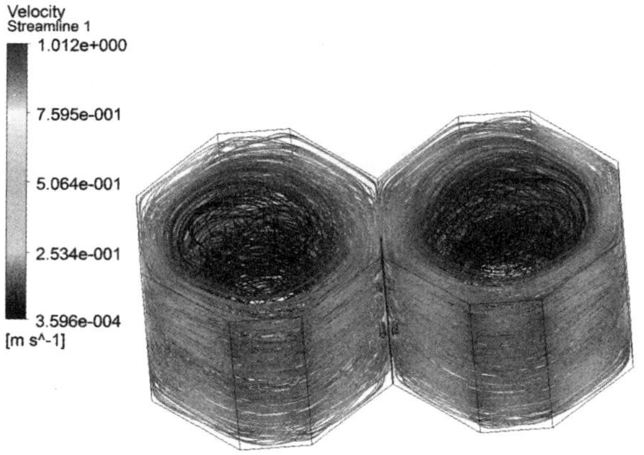

图 17　标准舱 CFD 流态模拟图

外排水。鱼舱进水点设置在 2 个舱养单元中部，同时增设 4 台螺旋桨推流装置(或潜水射流泵)。图 17 所示为标准舱 CFD 流态模拟图，从中可以看出，整体流态非常稳定。

5.3　标准舱水处理工艺设计

养殖舱水处理工艺一方面可以充分借助深远海优越的水质条件，另一方面，由于停泊地点远离大陆，补给运输不是非常便利，所以也需要兼顾能源的充分利用。项目设计采用“常流水 + 内循环”方式实现养殖舱内整体的水质调控，如图 18 所示。其中，“常流水”主要作用是在“二次流”作用下，将水体中的残饵、粪便等固形物排出养殖舱，同时通过外源海水的添加，不断稀释水体中溶解性营养盐和有机物的浓度。“内循环”即采用循环水方式对舱内水体进行处理，主要作用为分离去除水体中残余的悬浮颗粒物，同时，利用生物过滤将氨氮分解为硝氮。

根据养殖舱水处理工艺，结合船体形状和结构，将标准养殖舱设计为图 19 所示。养殖舱上下两侧舱壁处安装气提腔，布置气提管，用以实现内循环过程中的提水，气提区外侧安

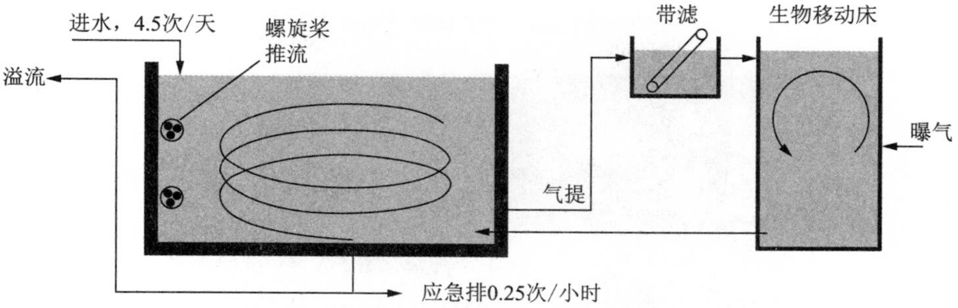

图 18　养殖舱水处理工艺流程图

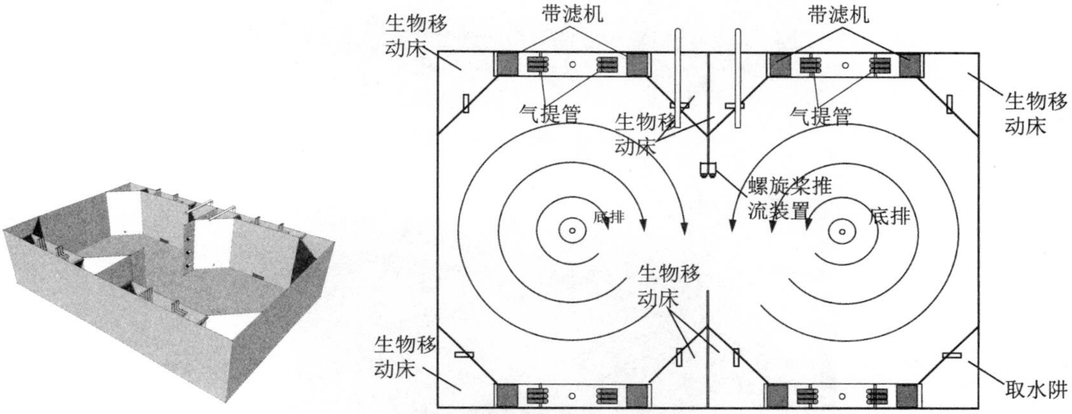

图 19　标准舱结构设计图

装带式微滤机对悬浮颗粒物进行筛滤处理。每个舱养单元四个角落加装 1 块斜板，形成 7 个生物过滤腔，在腔体内部装填浮性滤料进行生物过滤去除养殖过程产生的氨氮。

5.4　养殖负荷

通过整体水处理工艺设计和物质平衡计算，本设计方案海水鱼类养殖密度可达到 22.8 kg/m^3，换水率 4.5 次/天。

6　全自动渔用双罐真空吸鱼泵技术研究

为推进生产管理的自动化水平，针对鱼类养殖过程的自动分级难题，鲆鲽类产业技术体系工厂化养殖设施设备岗位团队研发了全自动鱼苗双罐真空吸鱼泵，采用真空负压传输、高精度四通密封转阀，以及 PLC 集成控制等技术，实现活鱼高效、无损、自动吸捕与转运。罐体采用透明压克力材质，便于观察吸捕的鱼群在罐内的状态。全自动渔用双罐真空吸鱼泵，主机功率：5.5 kW；罐体容积：0.4 m^3 × 2；进口管直径：100 mm；吸捕量：81.7 t/h（鱼水混合，鱼水比 1∶1）。2016 年 12 月 23 日，进行了吸捕活鱼试验。测试鱼品种：鲫鱼；规格：背高 60 mm;数量:100 尾。在吸捕过程中,吸捕鱼群表面未发现有物理损伤痕迹。

图 20　双罐吸鱼泵

图 21　吸捕鱼群试验

(岗位科学家　倪　琦)

鲆鲽类网箱设施与养殖技术研发进展

专用养殖网箱岗位

2016年，鲆鲽类专用养殖网箱岗位围绕离岸网箱养殖及配套设施与技术，开展了福建省近海传统网箱升级改造模式研究、升降式网箱养殖牙鲆试验、离岸大型浮绳式围网的研制、升降式网箱水动力特性试验以及鲆鲽鱼类耐流性试验研究，主要进展如下。

1　福建省近海传统网箱升级改造模式研究

针对福建省近海内湾型传统渔排网箱养殖升级改造产业需求，本年度，专用养殖网箱岗位团队在深入调研与研讨的基础上，确定了福建省近海传统网箱升级改造模式研究方案，并按照年度任务规划实施，主要开展了以下工作：

（1）选取宁德三都澳近海网箱养殖区和升级改造示范区各1个，完成2个区域的环境调查取样（图1、图2），为构建近海传统网箱升级改造模式提供基础数据支撑。近海传统网箱养殖区设置站位3个，升级改造示范区设置站位2个（其中一个为远离养殖区的对照站位）。调查的主要环境因子包括：表层和底层的流速、温度、盐度、溶解氧、pH、浊度、营养盐（包括总氮、总磷、硝酸氮、氨氮、亚硝酸氮、活性磷酸盐、硅酸盐）、悬浮颗粒物浓度、COD、总有机碳、叶绿素a、石油类及沉积物（沉积厚度、硫化物、总有机碳、总氮、总磷、细菌生物量）等。

（2）优化HDPE塑胶网箱结构，研制传统木质养殖网箱替代品。优化确定了HDPE网箱的框架结构、网箱尺度、渔排布局方式和安装工艺，网箱试制工作正在进行中。

图1　环境调查取样

图2　海区流速测定

2 升降式网箱养殖牙鲆试验

设计了一种新型升降式网箱,并进行海上牙鲆养殖试验,以验证升降式网箱养殖鲆鲽鱼类的安全性、可行性、养殖操作技术以及牙鲆养殖的生长情况,为牙鲆等鲆鲽鱼类的升降式网箱养殖提供参考。

2.1 养殖示范基地及海区条件

网箱养殖的试验地点为青岛市黄岛区薛家岛海域,海区水深 20 m 左右,试验期间海水温度 17 ℃～25 ℃。试验网箱为新型 HDPE 升降式网箱,网箱周长 40 m,直径 12.7 m。网箱顶部设有网盖,底部为 2 mm 的无结网,减少网衣和鱼体摩擦。网箱浮架由直径 250 mm 的内外管组成,采用双管自动排气－进水和充气－排水的方式控制网箱的沉降(图 3、图 4),网箱可沉降至海底。

图 3 网箱沉降过程

图 4 网箱完全沉没

2.2 养殖试验主要结果

本次试验时间为 2016 年 5 月 26 日～2016 年 11 月 17 日,历时 175 d,初期投放 700 尾规格为 140 g 左右的牙鲆(图 5),分别于 5 月 26 日、9 月 9 日和 11 月 17 日进行抽样,每次随机抽取 35 尾进行体长体重数据采集。试验结果如下:5 月 26 日～9 月 9 日,经过 107 d 养殖,试养牙鲆的平均体重由 145.7 g 增长到 223.7 g,平均体重增加 53.5%,平均日增加体重 0.73 g,平均体长则由 25 cm 增长到 27 cm,平均体长生长率为 8%,平均日增长 0.02 cm;9 月 9 日～11 月 17 日,经过 68 d 养殖,平均体重由 223.7 g 增长到 268.3 g,平均体重增加 19.9%,平均日增加体重 0.66 g,平均体长由 27 cm 增长到 30.1 cm,平均体长生长率为 11.5%,平均日增长 0.045 cm;5 月 26 日～11 月 17 日,经过 175 d 养殖,平均体重从 145.7 g 增长到 268.3 g,平均体重增加 84.6%,平均日增加体重 0.7 g,平均体长从 25 cm 增长到 30.1 cm,平均体长生长率为 20.4%,平均日增长 0.03 cm。养殖成活率为 98.9%。从养殖结果来看,在前半段,海域水温逐渐增加至牙鲆适宜生长水温,生长速度快,而随着水温的降低，生长速度变缓，所以前半段（5 月 26 日至 9 月 9 日）牙鲆体重增加速度稍快于后半段（9 月 9 日至 11 月 17 日），但是前半段体长增加慢于后半段（图 6）。

图 5　牙鲆陆海接力养殖

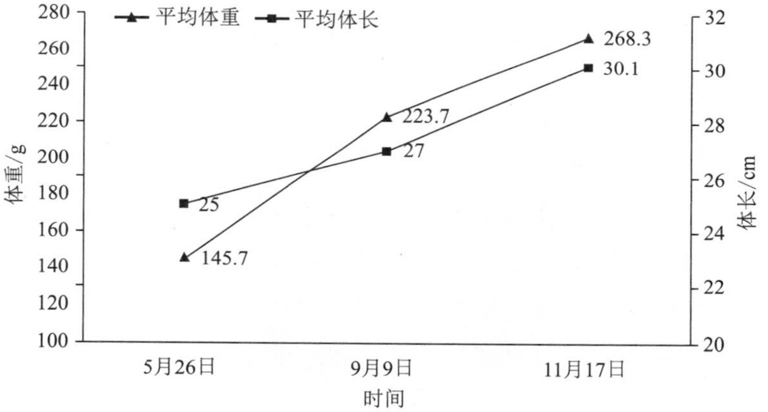

图 6　牙鲆陆海接力养殖生长情况

3　离岸大型浮绳式围网的研制

海水养殖由近岸向离岸深远海区域拓展成为“十三五”渔业发展的重要方向。离岸大型围网作为一种新型养殖模式，可实现大水面规模化健康养殖，提高鱼类品质，通过配套自动投饲、机械化采收、实时监控、在线监测、活鱼运输等装备，可实现精准化生产。为此，鲆鲽类体系专用养殖网箱岗位、高效养殖模式岗位和莱州综合试验站联合，开展了离岸大型浮绳式围网的研制工作。经过多次研讨和设计方案的修改，重点解决了大型围网网底固定、网衣随波浪和潮汐起伏、锚泊敷设、网形固定、精准打桩等技术与施工难题，首次在莱州湾开阔海域建成主尺度为 100 m × 50 m × 12 m，单个围网养殖水体达 60 000 m^3 的大型浮绳式围网（图 7）。该阶段性研究成果于 2016 年 11 月 18 日通过了专家现场验收。

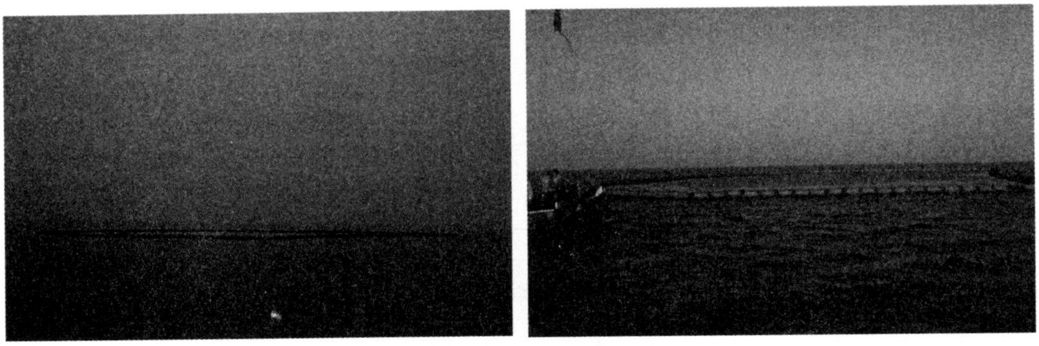

图 7　离岸大型浮绳式围网养殖

4　升降式网箱水动力特性试验

研究不同升降状态下升降式网箱的水动力特性，尤其是网底和网箱整体的稳定性、锚绳受力特性、网衣变形特性，探究下潜深度对网箱受力、变形的影响等，为升降式网箱的设计与选型提供理论依据。

4.1　试验条件及模型布设

试验在大连理工大学海岸和近海国家重点实验室进行。实验波流水槽长 60.0 m，宽 4.0 m，深 2.5 m，工作水深 0.2 m～2.0 m，波浪周期范围 0.5 s～5.0 s，配备液压伺服规则波、不规则波造波系统，微机控制及数据采集系统，2 台 1.0 m^3/s 轴流泵的双向流场模拟系统。采用 1∶20 的模型比尺，对不同升降状态下的网箱水动力特性进行了系统研究，模型网箱布设如图 8 所示。试验设计了纯流、波浪以及波流组合三种作用荷载，主要对漂浮以及不同下潜深度工况下网箱整体的稳定性、网箱的运动特性、锚绳的受力特性以及浮架倾角等进行了研究。

4.2　主要试验结果及分析

升降式网箱模型的波流水槽试验中，网箱处于漂浮状态下的浮架运动量、不同下潜水深时的浮架运动量、下潜状态下的各锚绳受力及漂浮和下潜状态下的锚绳受力对比等结果如图 9 所示。

5　鲆鲽鱼类耐流性试验研究

本试验使用垂直循环造流水槽（图 10）测定了大菱鲆、牙鲆和半滑舌鳎在不同流速下的续航游泳时间、血清生化指标、最大顶流速度和游泳行为观察，测定了沉性饲料在不同流速下的沉降偏移距离。

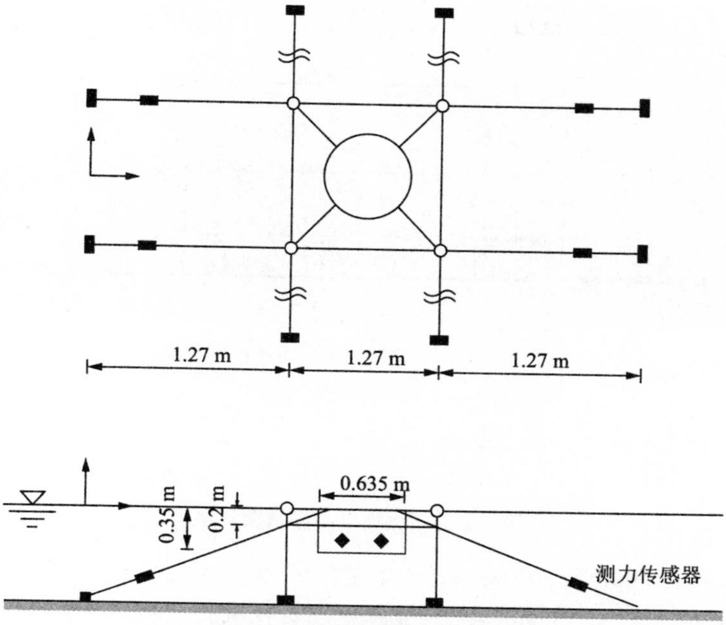

图 8　实验模型布置示意图

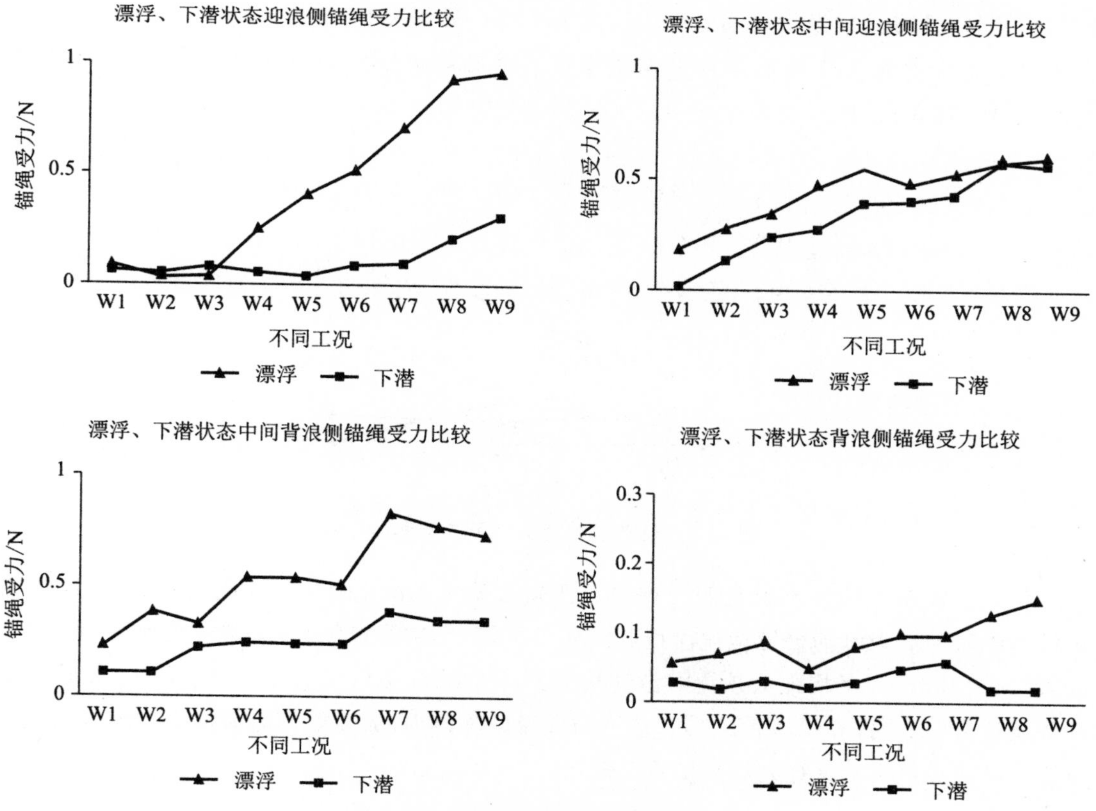

图 9　漂浮、下潜状态锚绳力比较

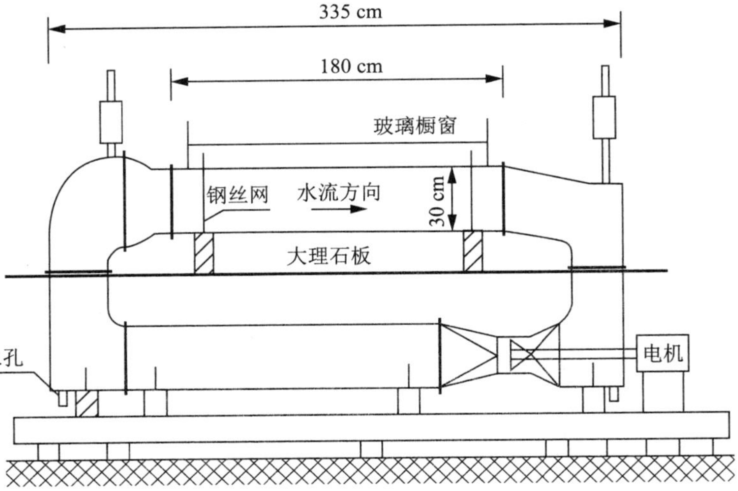

图 10　鱼类耐流性试验垂直循环水槽

试验选取体重为 500 g ± 100 g 的健康大菱鲆、牙鲆和半滑舌鳎，于垂直循环造流水槽中设定了 15 cm/s、25 cm/s、35 cm/s 三种流速工况进行试验，观察并记录三种鱼类的持续游泳时间。由图 11 可见，随着流速的增大大菱鲆续航游泳时间先增大后减小，在流速 35 cm/s 时伏底不动；牙鲆续航游泳时间随着流速增大先减小后增加；半滑舌鳎续航游泳时间随着流速的增大呈幂函数性的增大，说明半滑舌鳎对流速刺激较为敏感，牙鲆和大菱鲆对流速刺激相对半滑舌鳎影响较小。

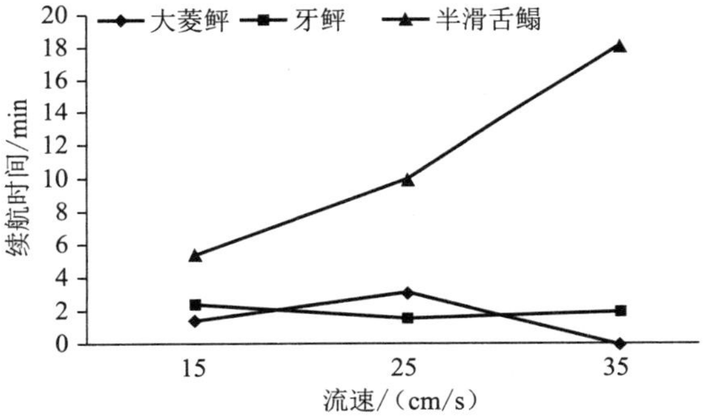

图 11　三种鱼类的续航时间与流速的关系

最大顶流游泳速度是设定的三种鱼类在垂直循环造流水槽中正常游动时不能前进，也不被水流带动后退时的稳定造流流速。由表 1 可知，三种鲆鲽鱼类，牙鲆最大顶流游泳速度最大，大菱鲆最小，平均最大顶流游泳速度牙鲆和半滑舌鳎相差不大。由此可见，试验的三种鲆鲽鱼类中，牙鲆和半滑舌鳎具有较强的顶流游泳能力，大菱鲆顶流游泳能力较弱。流速胁迫下鲆鲽类鱼类游泳行为观察如表 2 描述。

表 1　三种鲆鲽鱼类最大顶流游泳速度

种类	体重 / kg	最大顶流游泳速度 / (cm / s)	平均最大顶流游泳速度 / (cm / s)
大菱鲆	0.531	29.90	29.78
	0.605	31.10	
	0.597	28.34	
牙鲆	0.571	38.00	40.30
	0.605	42.50	
	0.575	40.40	
半滑舌鳎	0.584	39.92	37.91

表 2　10 min 内受流速胁迫下鲆鲽类鱼类游泳行为观察

种类 流速 / (cm/s)	大菱鲆	牙鲆	半滑舌鳎
10	游泳姿态悠闲，不受刺激伏底不动	游泳姿态悠闲，不受刺激伏底不动	游泳姿态悠闲，不受刺激伏底不动
15	顺流时，尾巴不翘起，游泳轻松，伏底	顺流时，尾巴不翘起，游泳轻松，伏底	顺流时，尾部翘起，游泳轻松，伏底
25	游泳吃力，伏底时身体不会上下起伏	游泳吃力，伏底时身体会轻微上下起伏后鳍翘起，顺流尾部翘起	游泳吃力，伏底时间比游泳时间长
35	正常游动会被水流带至后挡网，不能顶流直线前进	游动会被水流带至后挡网，猛力游动才能前进	频繁游动，游动可以前进
45	游动时身体侧翻，伏底被水流带到最后方，靠身体尾部支撑挡网	游动时鱼身倾斜，伏底被水流带到最后方，靠身体尾部支撑挡网	频繁游动，伏底时会被水流带到最后方，靠身体尾部支撑挡网
55	不游动，靠尾部身体一侧支撑挡网，一游动身体侧翻紧贴挡网，无法离开	受流速刺激剧烈游动，身体侧翻，被水流带到后挡网，身体紧贴挡网，无法离开	猛力游动才可前进，前进时侧翻紧贴后方挡网，翻转，出现擦伤

6　年度进展小结

（1）在福建宁德三都澳海域，选取近海网箱养殖区和升级改造示范区各 1 个，完成了 2 个区域的环境调查取样。

（2）在青岛薛家岛离岸海域采用升降式网箱进行了牙鲆养殖试验。

（3）自主设计并制作离岸大型浮绳式围网 1 套，圈养水体达 60 000 m^3，阶段性成果通过专家现场验收。

（4）对升降式网箱漂浮以及不同下潜深度工况下网箱整体的稳定性、网箱的运动特性、锚绳的受力特性以及浮架倾角等进行了系统的物模试验和数值模拟研究。

(5) 开展了主养鲆鲽鱼类耐流性试验,测定了大菱鲆、牙鲆和半滑舌鳎在不同流速下的续航游泳时间、血清生化指标、最大顶流速度,并进行了游泳行为观察,测定了沉性饲料在不同流速下的沉降偏移距离。

(岗位科学家　关长涛)

鲆鲽类工程化池塘养殖技术研发进展

池塘养殖工程岗位

2016年，池塘养殖工程岗位开展了牙鲆岩礁池塘工程化高效养殖示范、工程化池塘循环水养殖系统优化与推广、陆海接力高效养殖技术开发、鲆鲽类生殖调控机制、养殖牙鲆肠道生理健康评价等系列研究，取得的研究进展总结如下。

1　岩礁池塘工程化高效养殖模式优化与养殖示范

2016年，池塘养殖工程岗位从两方面对岩礁池塘进行了工程化设计与优化：① 在岩礁池塘底部设置了集污减排控制系统（在池塘底部中央设置集污沟，与池塘进水方向一致），内铺设塑料篷布，用于收集沉淀，包括残饵、粪便在内的大颗粒废弃物。集污沟内安置可移动式吸污泵，定期将集污沟内的残饵和粪便吸出，减少养殖废水对外海直接排放。② 在岩礁池塘内部设置了增氧环流系统，在养殖池塘两条对角线的四角各安装水轮式增氧机 1 台，除对养殖水体进行增氧外，增氧机叶轮转动还可对养殖水体的流动形成推动作用，可大大提高养殖效率。

在青岛贝宝海洋科技有限公司利用一口进行过工程化改造的面积为10亩的岩礁池塘开展了牙鲆高效养殖示范。2016年5月，放养全长12 cm～14 cm的牙鲆苗种60 000尾，放养水温15 ℃以上。周年养殖条件：水温3 ℃～31 ℃，盐度28～32，日换水率为80%。日投喂饵料2次，按照体重2%～3%投喂。养殖过程中，按照本岗位制定的鲆鲽类工程化池塘养殖技术规范进行生产。本年度养殖过程中，遭遇历史罕见高温，水温28 ℃～31 ℃持续期长达30多天，对养殖鱼的存活和生长造成了较大影响。至2016年11月，养成鱼平均体重达到426.4 g，养殖单产为2 057 千克/亩，养殖期成活率为80.4%。目前，养殖牙鲆已进入越冬期，将采用增氧和零投喂的方法安全越冬，保证越冬成活率。

2　工程化池塘循环水养殖模式优化与推广

2016年，本岗位在辽宁庄河基地推广构建了一套面积为17亩的工程化池塘循环水养殖系统一套，用于河鲀池塘养殖试验与示范。按照本岗位研发的工程化池塘循环水养殖系统设计工艺，庄河基地的池塘循环水养殖系统主要包括回水处理池1个和养殖池塘4个，单个养殖池塘面积为3亩(水深2 m),采用串联方式组合排列,并设置了合理的地势差,从而养殖用水在养殖池塘间形成自流。同时,设置了独立的进排水系统,安装了轴流泵、增氧造流装置机、水质检测系统等设施设备,形成了完善的工程化池塘养殖系统,进行了试运转。

设置的池塘专用增氧造流装置由底座、潜水泵、支架、管道系统、远程控制开关等组成。装置运转时,潜水泵可将底层水抽取到中上水层,实现池塘内上下养殖水层的交换,有利于保持池塘内水环境的均匀并可有效改善养殖池塘底部水质条件。同时,空气通过管道系统被潜水泵负压吸入,射流分散成微小气泡并高速进入养殖水体中,实现了对池塘内中下部养殖水体的高效增氧,还达到了节能的目的。养殖过程中,在养殖池塘底部四角各安装1台内循环增氧装置,将水体通过潜水泵的作用形成高速的流动,推动池塘内养殖水体形成有效的内循环流动,有利于将残饵、粪便等代谢废弃物在池塘底部集中起来清除,可有效改善底质环境,防止烂塘和水质恶化的情况出现,大大提升了池塘养殖生产的效率。

3 陆海接力养殖模式构建与技术开发

3.1 黄条鰤海陆接力养殖技术

2016年度,按照岗位任务要求,建立了大洋性经济鱼类新资源——黄条鰤的海陆接力养殖技术,在庄河基地共养成3龄以上黄条鰤亲鱼3 000余尾,2龄后备亲鱼10 000尾,1龄鱼10 000余尾。海陆接力养殖模式包括深水抗风浪网箱养殖与陆基工厂化养殖两个阶段:

(1)深水抗风浪网箱养殖:网箱为10 m × 10 m × 6 m的方形金属抗风浪网箱,放养密度随鱼的年龄和生长逐渐由1龄以内时的3 000尾/箱~5 000尾/箱降低至2龄及2龄以上时的1 000尾/箱。养殖使用饵料为闲杂鱼与配合饲料,投喂率3%~5%。黄条鰤生长速度快,当年苗种(全长5 cm~8 cm),经3个月的养殖,体重即可生长至300 g。

(2)海陆转运:当秋末自然海水温度下降至18 ℃时,将养殖鱼由海上抗风浪网箱向陆基工厂化养殖车间转移越冬。采用活鱼船与汽车结合的方式搬入陆基工厂化养殖车间,帆布桶充氧运输,运输成活率100%。

(3)工厂化车间越冬养殖:使用面积为50 m^2的水泥池,越冬期间养殖密度为20尾/平方米,养殖水温12 ℃~18 ℃。越冬期间饵料为玉筋鱼或鲐鱼,日投喂1次,投喂量为体重的1%~2%。日换水1次,清底保持养殖池洁净。越冬期间因水温低和饵料投喂减少,苗种生长慢(图1)。翌年5月,海水温度上升至15 ℃以上时,再转移到网箱基地养殖。通过海陆接力养殖方式,黄条鰤苗种经2年养殖(每年适宜生长时期6月~10月),即可生长至体重4 kg~5 kg,表现出快速的生长特性。

3.2 黄条鰤形态度量与内部结构特征

利用传统测量方式、框架测度法、几何形态测量法和解剖学方法,观察和测度了黄条鰤

海上网箱养殖

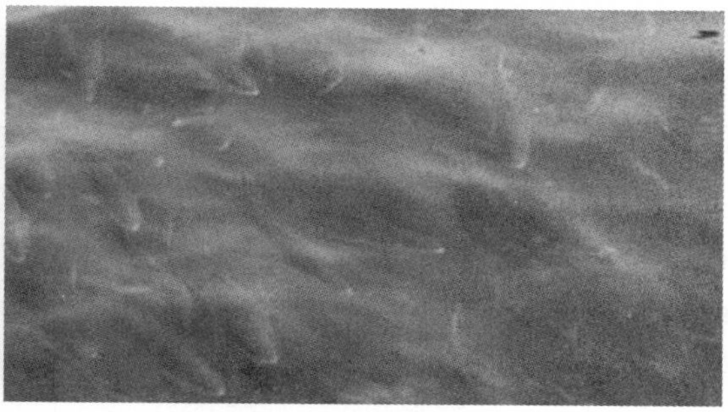

养殖黄条鰤

海陆接力运输船舶

网箱养殖鱼入养殖车间

图 1　黄条鰤海陆接力养殖模式

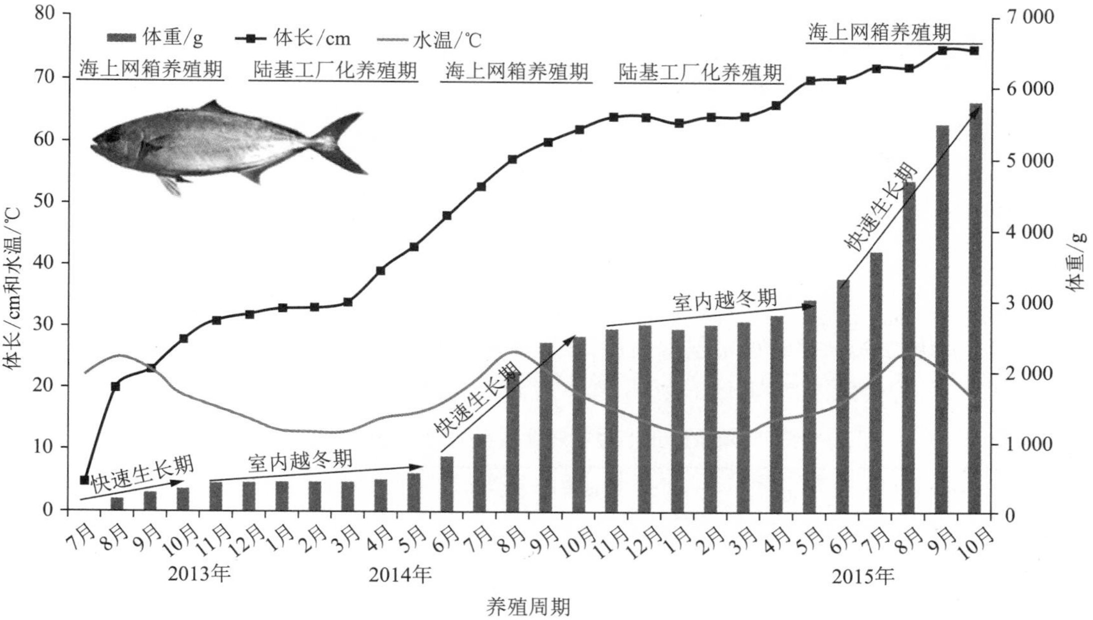

图 2　黄条鰤海陆接力养殖生长

可量与可数性状及内部结构特征,并模拟构建了其形态性状度量框架图。黄条鰤全长与体长比值、下颌长与上颌长比值、尾柄长与尾柄高比值变异较小,表明这些性状关联密切。建立了黄条鰤全长(TL)与体重(BW)间的关系模型 $BW = 2.1652TL^2 - 140.35TL + 2479.9$ ($R^2 = 0.9812$),体高(BH)与体重间的关系模型 $BW = 0.7575BH^{3.0059}$ ($R^2 = 0.9816$)。在12个可量形态性状中,除眼径外的其他11个形态性状间均存在显著相关关系。通径分析揭示体高和体长性状是影响体重的两个关键因素,其对体重的决定程度分别达41.34%和13.11%。黄条鰤比肠长为0.62~0.69,脊椎骨数量为23~25,总出肉率可达75%。这些结果为黄条鰤种质判别、系统分类及人工繁育与养殖技术开发提供形态认知依据。

表1 黄条鰤可数性状

可数性状	指标	可数性状	指标
背鳍数	Ⅰ,Ⅵ,Ⅰ-31~35	侧线鳞数	175~210
臀鳍数	Ⅱ,Ⅰ-19~22	侧线上鳞数	26~31
胸鳍数	20~22	侧线下鳞数	49~55
腹鳍数	Ⅰ-5	鳃弓数	8
尾鳍数	17	鳃耙数	6~9+14~22

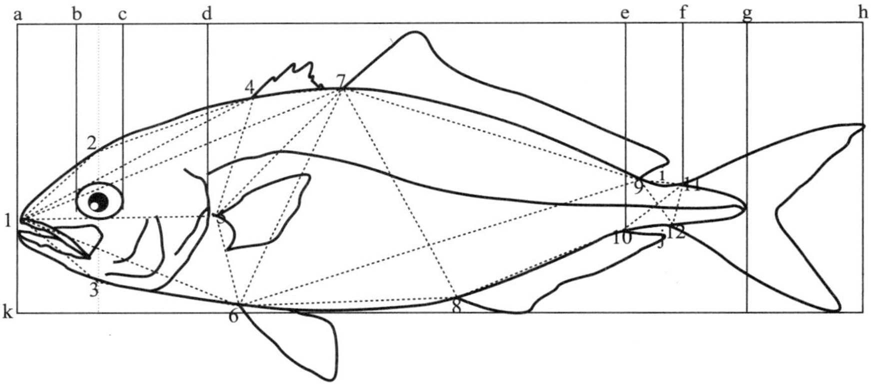

图3 黄条鰤表观形态度量特征

ah:全长;ag:体长;ak:体高;ad:头长;cd:眼后头长;bc:眼径;ef:尾柄长;ij:尾柄高

1:吻端起点;2:过眼睛中点垂直线与背部边缘交叉点;3:过眼睛中点垂直线与腹部边缘交叉点;4:第1背鳍起点;5:胸鳍起点;6:腹鳍起点;7:第2背鳍起点;8:臀鳍起点;9:第2背鳍终点;10:臀鳍终点;11:尾鳍背部起点;12:尾鳍腹部起点

表2 黄条鰤形态性状对体重影响的通径分析结果

性状	相关系数(r)	直接作用(Pi)	间接作用			共线性诊断		决定系数	
			Σ	BH	SL	Tol	VIF	BH	SL
BH	0.976**	0.643**	0.592	—	0.592	0.264	6.557	0.4134	0.4288
SL	0.953**	0.362**	0.333	0.333	—	0.153	5.874	—	0.1311

** 表示差异极显著($P < 0.01$)

3.3　黄条鰤种质特性分析

探明了黄海种群黄条鰤的染色体核型，结果表明：黄条鰤有48条染色体，核型公式为$2^n = 48 = 6sm + 4st + 38t$（图4），其染色体臂的数量为54，单倍体染色体总长度大约为51.97 μm。黄条鰤染色体核型比较独特，具有3对亚中部染色体和2对亚端部染色体，第一对染色体有次缢痕和随体，不同于以往报道的其他鰤属鱼类的染色体核型特征，结果可为黄条鰤遗传鉴定及地理种群区划提供技术依据。

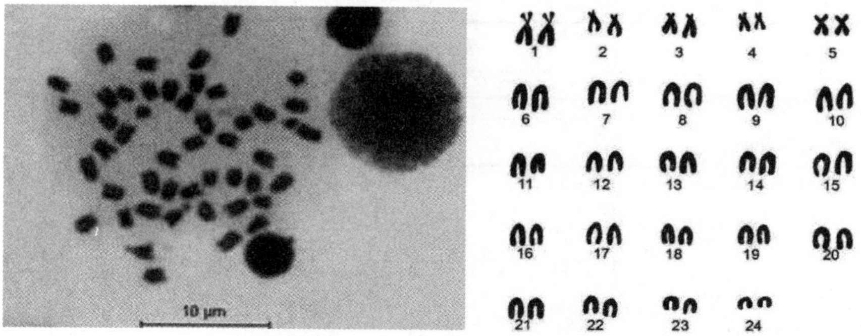

图4　黄条鰤中期分裂相形态及核型图谱

3.4　黄条鰤肌肉营养特性评价

评价了黄条鰤的营养与可食用价值，测定了其肌肉水分、蛋白质、粗脂肪、灰分、氨基酸、脂肪酸、矿物质等营养成分，并对养殖及野生个体的肌肉营养成分组成进行了比较分析（表3～表7）。肌肉中蛋白质含量较高，必需氨基酸和鲜味氨基酸含量丰富，完全符合FAO/WHO推荐的理想蛋白质标准，是一种人类优质蛋白供给源。根据AAS和CS分值，黄条鰤肌肉的第一限制氨基酸为蛋氨酸，第二限制氨基酸为缬氨酸。黄条鰤肌肉的脂肪含量高于三文鱼、金枪鱼、石斑鱼等，且肌肉中含有丰富的不饱和脂肪酸，不饱和脂肪酸与饱和脂肪酸比值较高，特别是EPA和DHA含量高，因此具有较优的口感鲜度和较高的营养价值。另外，黄条鰤肌肉中含有多种机体新陈代谢所需的矿物质。比较发现，养殖黄条鰤较野生黄条鰤具有相对较高的脂肪含量和较低的水分含量，比能值（EP）分别为13.44 kJ/g和8.68 kJ/g，其他营养成分无显著差异，表明养殖与野生黄条鰤肌肉营养价值相似。黄条鰤口感鲜美，营养成分丰富，肌肉蛋白和脂肪质量较高，肌肉蛋白符合理想的优质蛋白标准。

表3　养殖和野生黄条鰤肌肉基本营养组成成分（%，湿重）

营养成分	养殖黄条鰤	野生黄条鰤
水分	56.10	68.10
粗蛋白	20.30	21.90
粗脂肪	21.40	8.80

(续表)

营养成分	养殖黄条鰤	野生黄条鰤
灰分	1.10	1.10
总糖	1.10	0.10
能值(kJ/g)	13.44	8.68
EP	66.21	39.64

表4 养殖和野生黄条鰤肌肉氨基酸组成及含量比较(%,湿重)

氨基酸	养殖黄条鰤	野生黄条鰤
* 苏氨酸 Thr	0.78	0.87
* 缬氨酸 Val	0.96	1.07
* 蛋氨酸 Met	0.60	0.66
* 异亮氨酸 Ile	0.90	0.98
* 亮氨酸 Leu	1.65	1.82
* 苯丙氨酸 Phe	0.84	0.93
* 赖氨酸 Lys	1.79	1.96
# 天门冬氨酸 Asp	1.65	1.82
# 谷氨酸 Glu	1.33	1.47
# 甘氨酸 Gly	1.09	1.23
# 丙氨酸 Ala	1.68	1.89
酪氨酸 Tyr	0.62	0.70
脯氨酸 Pro	0.51	0.55
丝氨酸 Ser	1.07	1.18
& 组氨酸 His	1.18	1.28
& 精氨酸 Arg	1.18	1.30
必需氨基酸(EAA)	7.52	8.29
非必需氨基酸(NEAA)	7.95	8.84
半必需氨基酸(SEAA)	2.36	2.58
鲜味氨基酸(DAA)	5.75	6.41
总氨基酸(TAA)	17.83	19.71
W_{EAA}/W_{TAA}(%)	42.20	42.10
W_{EAA}/W_{NEAA}(%)	94.60	93.80
W_{DAA}/W_{TAA}(%)	32.20	32.50

* 为必需氨基酸;& 为半必需氨基酸;# 为鲜味氨基酸

表 5　黄条𫚕肌肉中必须氨基酸含量与评价

必需氨基酸	黄条𫚕		FAO/WHO标准	鸡蛋蛋白标准	AAS		CS	
	养殖	野生			养殖	野生	养殖	野生
苏氨酸 Thr	243	248	250	292	0. 97	0. 99	0. 83	0. 85
缬氨酸 Val	300	305	310	441	0. 96	0. 99	0. 68	0. 69
蛋氨酸 Met	187	188	220	386	0. 85	0. 86	0. 49	0. 49
异亮氨 Ile	281	280	250	331	1. 12	1. 12	0. 85	0. 84
亮氨酸 Leu	515	519	440	534	1. 17	1. 18	0. 96	0. 97
苯丙氨酸+酪氨酸（Phe+Tyr）	456	465	380	565	1. 19	1. 22	0. 81	0. 82
赖氨酸 Lys	559	559	340	441	1. 64	1. 65	1. 27	1. 27
合计 Total	2 540	2 566	2 250	3 089				

表 6　养殖和野生黄条𫚕肌肉脂肪酸组成比较（%，湿重）

脂肪酸	养殖黄条𫚕	野生黄条𫚕
*C14:0	2. 88	2. 51
*C15:0	0. 42	0. 5
*C16:0	22. 25	21. 23
*C17:0	0. 43	0. 84
*C18:0	4. 82	7. 69
&C16:1	6. 58	5. 52
&C17:1	0. 34	0. 53
&C18:1	22. 15	27. 82
&C22:1n9	3. 64	0. 46
#C18:2n6	1. 61	0. 84
#C18:3n3	1. 00	0. 48
#C18:4n3	2. 21	0. 54
#C20:4n6	—	1. 32
#C22:5n3	1. 09	2. 10
#EPA	6. 60	4. 10
#DHA	13. 63	16. 67
饱和脂肪酸（SFA）	30. 8	32. 77
单不饱和脂肪酸（MUFA）	32. 71	34. 33

(续表)

脂肪酸	养殖黄条鰤	野生黄条鰤
多不饱和脂肪酸(PUFA)	26.14	26.05
UFA/SFA	1.91	1.84

* 为饱和脂肪酸;& 为单不饱和脂肪酸;# 为多不饱和脂肪酸;—表示低于检出限

表7 养殖和野生黄条鰤肌肉矿物质含量的比较(mg/kg,湿重)

矿物元素	养殖黄条鰤	野生黄条鰤
钾 K	1 810.00	1 950.00
钠 Na	241.00	534.00
钙 Ca	96.00	125.00
镁 Mg	268.00	307.00
磷 P	918.00	1 010.00
铜 Cu	< 0.50	< 0.50
锌 Zn	< 0.50	< 0.50
铁 Fe	5.19	4.50
锰 Mn	< 0.10	< 0.10
铬 Cd	< 0.50	< 0.50

4 鲆鲽类生殖调控机制研究

4.1 半滑舌鳎促性腺激素抑制激素(GnIH)克隆和表达特性研究

新发现了半滑舌鳎负向生殖调控因子(GnIH),获得了全长cDNA序列(GenBank登录号:KU612223)为918 bp,其中5′UTR为24 bp,3′UTR为309 bp,ORF为585 bp,编码194个氨基酸的前体多肽,相对分子质量为21.73×10^3,等电点为6.52。半滑舌鳎GnIH前体多肽的信号肽为前24个氨基酸,经过酶切加工可产生2个GnIH成熟肽(图5)。

鲀形目鱼类、欧洲海鲈及半滑舌鳎GnIH前体只编码2种成熟肽,然而其他鱼类编码3种成熟肽。与鲀形目鱼类、欧洲海鲈与斜带石斑鱼类似,半滑舌鳎GnIH-1和GnIH-2多肽的C末端分别为-MPMRF或者-MPQRF基序。然而,鲤形目鱼类GnIH-1和GnIH-2多肽的C末端分别为-LPLRF或者-LPQRF基序(图6)。

同源性分析表明,半滑舌鳎GnIH前体多肽与鲈形目鱼类和青鳉相似度较高(53.48%~63.21%),其次为鲀形目鱼类(46.11%~50%),与鲤形目鱼类相似度较低(25.71%~27.84%)。

进化分析表明,半滑舌鳎GnIH与其他鱼类聚为一支,不同于四足类(图7)。组织分布

```
-24                                                    atggggtccaccaagacggcccc
1     ATGTTGAGCACCGTGTTCCTGTCAGTGCTCCTGATGCTGGGAGGTCCTGGAGGAGCGGCG
1      M  L  S  T  V  F  L  S  V  L  L  M  L  G  G  P  G  G  A  A
61    GCAGCGGCTGACTTCCAGGTCTATGGGAAATCAGCCTACAGTGACAAAAGTTTGCCGAGC
21     A  A  A  D  F  Q  V  Y  G  K  S  A  Y  S  D  K  S  L  P  S
121   AGCGAGGAGGGGAGACACACGGTGAGGAGGCAGCCGCTCCAGCAGGCTAAAGCTGTAACC
41     S  E  K  G  R  H  T  V  R  R  Q  P  L  Q  Q  A  K  A  V  T
181   CGCAGGAGTTTGGACTTGGAGAGGCTCAACATGCGTGTGACCCCGACCGCCAGCAAAAGC
61     R  R  S  L  D  L  E  R  L  N  M  R  V  T  P  T  A  S  K  S
241   AGCCTTCCCACCATCATCAAGCTGTACCCTCCTACAGTCAACCCCCACATACACGCCAAC
81     S  L  P  T  I  I  K  L  Y  P  P  T  V  N  P  H  I  H  A  N
301   ATGCCCATGCGGTTTGGAAGAGAGGTCGAGCCGGAAGACGACCAGTCTCACAACACTCCC
101    M  P  M  R  F  G  R  E  V  E  P  E  D  D  Q  S  H  N  T  P
361   AACATGCCCCAGAGGTTTGGACGGGCCTGGACATTCAACCGAGTGTGTGTGAAGTGTCGA
121    N  M  P  Q  R  F  G  R  A  W  T  F  N  R  V  C  V  K  C  R
421   GGAGACGCCGACCAAGTTCTGCCTGGAACCAGTCTGTACTGGAGTCTCATCAACTCTCTG
141    G  D  A  D  Q  V  L  P  G  T  S  L  Y  W  S  L  I  N  S  L
481   GCTGTTGAACAGTTCTTCAACACTGACTTGCACTGGGCTGAAGATTTTGACTTGACGGCC
161    A  V  E  Q  F  F  N  T  D  L  H  W  A  E  D  F  D  L  T  A
541   AATGCAGAAGAGTTGGAAATGGAGCGAACGTCCTTTAAGGGATAAagacacagtgtctgc
181    N  A  E  E  L  E  M  E  R  T  S  F  K  G  *
601   aatagcctataaatataacatcttttagcccttcaaaacacatttgatgtcaaatacaat
661   aaacatagataacttacagtatatgttattattattaaggcatgggaaataataaaagtt
721   tcaaacaagttgcatttcgctctatctatctatctatctatctatctatctatctatcta
781   tctatctatctatctatctatctatctatctatctatctatctatctatctatctatcta
841   tctataaaaattaagaaaataaaaattgcacaaaaaaaaaaaaaaaaaaaaaaa
```

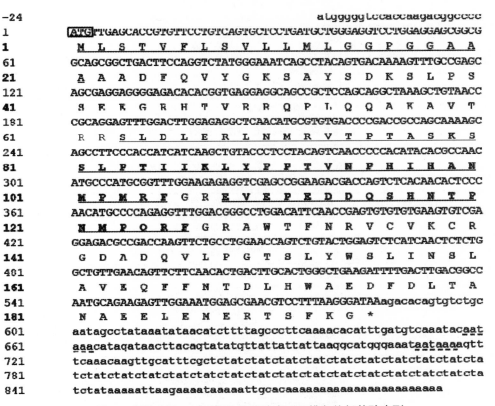

图 5　半滑舌鳎 GnIH 前体核苷酸序列及推断的氨基酸序列

结果显示半滑舌鳎 *gnih* mRNA 主要在脑中表达，其次为卵巢、心脏、胃，在垂体及其他外周组织中表达量较低。半滑舌鳎卵巢发育共分为 5 个时期：卵黄生成前期（II 期）、卵黄生成期（III 期）、卵黄生成后期（IV 期）、成熟期（V 期）以及排卵后期（VI 期）。相对于 II 期，脑中 *gnih* 表达水平在 III、IV 期显著增加，在 V 期表达量降低，然而在 VI 期达到峰值。垂体 *gnih* 表达水平除了在 V 期显著升高外，在卵巢发育过程基本保持稳定。相反，卵巢 *gnih* mRNA 在 III 期显著降低，并且一直持续到 VI 期（图 8）。

4.2　半滑舌鳎 GnIH-R 分子克隆、组织分布及系统进化

半滑舌鳎 GnIH-R 全长 cDNA 序列（GenBank 登录号：KX839491）为 2 201 bp，其中 5′UTR 为 60 bp，3′UTR 为 776 bp，ORF 为 1 365 bp，编码 454 个氨基酸的蛋白。半滑舌鳎 GnIH-R 属于 G 蛋白偶联受体，有 7 个跨膜区、5 个糖基化位点和 7 个磷酸化位点（图 9）。

半滑舌鳎 GnIH-R 氨基酸序列与其他鱼类比对结果见图 10。半滑舌鳎 GnIH-R 与鲈形目鱼类相似度较高，其中与斜带石斑鱼和尼罗罗非鱼相似度分别为 72.41% 和 71.90%，其次为鲀形目鱼类（65.32%～66.22%）。然而，与鲤形目鱼类（金鱼和斑马鱼）3 种 GnIH-R 相似度相对较低（47.33%～57.81%）。

进化分析表明半滑舌鳎 GnIH-R 与其他鱼类聚为一支，不同于四足类（图 11）。半滑

```
goldfish       --MSYFTLVFLALGTLSSFM-LREV---------TALRWPLPDDSDPDRFTWGQFLENAQE
zebrafish      --MSYFALLSLALGILSSFM-LSEV---------TALRLPLSGERDLNGFTWGQFSENAQE
green puffer   MLVTLFLAMLLMIAGLGKAA-VSDLQVTGKVN---DRTLGSREGR-HNMRKELHHQVKNN
tiger puffer   MLVTAFLAMLLMIAGIGGAA-ETDLQVNGKLN---DRTLSSREGR-HNVRKQLRHQIKSN
grass puffer   MLVTAFLAMLLMIAGIGGAA-ETDLQVNGKLN---DRTLSSREGR-HNVRKQLRHQIKSN
tongue sole    MLSTVFLSVLLMLGGPGGAAAAADFQVYGKSAYSDKSLPSSEEGR-HTVRRQPLQQAKAV
sea bass       MLTTVFLSTLLMLGGLGGAA-VSDLQVYGKSIHSDKTLLSSDDGR-HTVRKQPHQQAKGE
grouper        MLTTVFLSALLMLGGLGGAA-AFNRQVHGKSIHSDKTLPGSNDET-HTVRTQPHQQTKNE
Nile tilapia   MLVTMILSALLMLRGLGGS----DVHVFGKSVRSGKTLLSSNDGT-YSVRKQPHQETKNE
medaka         MLTMMMLSVLLVLGGLGGAA-ASDLHVFGKSFHGDDPLESSHDSQLNMLRKQLHQQTKRG
                   :    * :    .       :                    :     .      :

                                                                   LPXRFa-1
goldfish       IPRSLEIEDFTLNVAPTSGRVSSPTILRLHPKITKPTHLHANLPLRFGRDTENTPRERA-
zebrafish      IPRSLEIQDFTLNVAPTSGGASSPTILRLHPIIPKPAHLHANLPLRFGRDAQPGTGDRAP
green puffer   ILRSLDLESLNIHVSPTS-KISFPTIIRLYPPTPQPPLRHINMPMRFGRNSFHGD-DHNP
tiger puffer   ILRSLDMERINIQVSPTSGKVSLPTIVRLYPPTLQPRHQHVNMPMRFGRDGVQGG-DHVP
grass puffer   ILRSLDMERINIQVSPTSGKVSLPTIVRLYPPTLQPHHQHVNMPMRFGRDGVQGG-DHVP
tongue sole    TRRSLDLERLNMRVTPTASKSSLPTIIKLYPPTVNP-HIHANMPMRFGREVEPED-DQS-
sea bass       IRRSLDLESFNIHVSPTTSKISLPTIIRLYPPTAKPLHLHANMPMRFGRDSNPGD-DRSP
grouper        ISRSLDLESFSVHVTPTTSKISLPSIIRLFPPTAKPFQLHANMPMRFGRESVPGD-DSAP
Nile tilapia   IHRSLDLESFNIRVAPTTSKFSLPTIIRFYPPTVKPLHLHANMPLRFGRQS---D-ERTP
medaka         IRRSLDLESFNIRVTPTSSKLNLPTIIKLYPPTAKPLHMHANMPLRFGRESSASD-DRVS
                ***::: :.:.*:**:  . *:*:::.*  :*  * *:*:****:        :

                LPXRFa-2                     LPXRFa-3
goldfish       KSNINLPQRFGRSCT------MCARSGTGLSATLPQRFGRRNIFPLDPFRALTLYKRTPE
zebrafish      KSTINLPQRFGRSCT------MCARSGTGPSATLPQRFGRRNIFALDPLRALALYTRTPE
green puffer   NSAPNMPQRFGRSWKKIQLCEGCYEAHR----ILKHRVKHARNGQR-FISTLLNA-----
tiger puffer   NLNPKMPQRFGRSWKVIRLCEDCSKVQG----VLKHQVRYGRNGQS-LIRTLVNA-----
grass puffer   NLNPNMPQRFGRSWKVIRLCEDCSKVQG----VLKHQVRYGRNGQS-LIRTLVNA-----
tongue sole    HNTPNMPQRFGRAWTFNRVCVKCRGDAD-------QVLPGTSLYWS-LINSLAVE-----
sea bass       NSTPNMPQRFGRSWEVFQMCAECPGVQEPP--KSPQGLRRTSLYWS-LLRTLANA-----
grouper        NSTPNMPQRFGRSWEMCAKCLNVREAQ-NP--ILPQRLGRRSPYWS-LFRTLDNT-----
Nile tilapia   NSSPNLPQRFGRSWEAIRVCAECPSVRRAPNQLLSQRFERNSPYWK-LLRTVASE-----
medaka         NSSPNMPQRFGRAWEVLRMCGGCRSVREAPSPVLPQRFGRNTPHWG-FLNTLANE-----
               .   ::******:                     : .           :  ::

goldfish       SP-FPKERTQVHDYMLETVEDSVEETVKNKDYTVLD
zebrafish      SPSFPKERTQVHDYMFETVEDS-EETVKNTDYTALD
green puffer   -QLLKSALHW--------------------------
tiger puffer   -QLLKTGLHW--------------------------
grass puffer   -QLLKTGLHW--------------------------
tongue sole    -QFFNTDLHWAEDFDLTANAEEL-E-MERTSFKG--
sea bass       -ELLNTGLHWAEDIDFTTSSEEM-Q-MQEKTFNG--
grouper        -WLLNTGLHWAEDFDFTTSSEEV-E-MQEKSFKG--
Nile tilapia   -QLLNTGLHWAENFDSTTTSDEETE-LEEKKSKE--
medaka         -QLLNPELRW--------------------------
                 :
```

图 6　不同鱼类 GnIH 前体氨基酸序列比对

舌鳎 GnIH-R mRNA 主要在脑中表达,其次为垂体,在卵巢等其他外周组织中表达量较低(图 12)。

4.3　半滑舌鳎 GnIH 对下丘脑生殖相关基因表达的影响

利用离体孵育方法研究了 GnIH 对半滑舌鳎下丘脑生殖相关基因表达的影响(图 13):GnIH-1 显著促进了 *gnrh2* mRNA 的表达水平,GnIH-1 处理组 *gnrh2* mRNA 的表达量是对照组的 1.85 倍;而 GnIH-2 使 *gnrh2* mRNA 表达量有所降低,但与对照组相比无显著差异。GnIH-2 显著抑制了 *gnrh3* mRNA 的表达水平,GnIH-2 处理组 *gnrh3* mRNA 表达量为

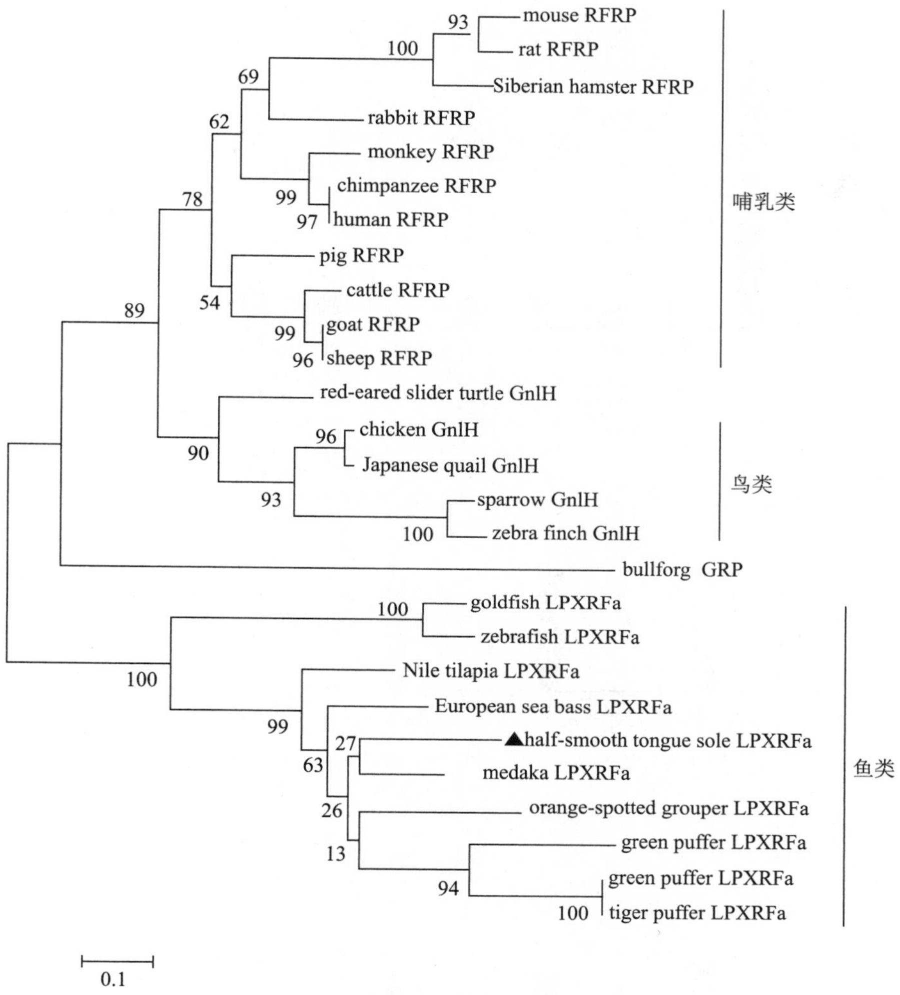

图 7　脊椎动物 GnIH 系统进化分析

对照组的 39%；然而，GnIH-1 不影响 *gnrh3* mRNA 的表达水平。GnIH-1 与 GnIH-2 均不影响 *kiss2* mRNA 的表达水平。GnIH-1 显著促进了 *gnih* mRNA 的表达水平，GnIH-1 处理组 *gnih* mRNA 表达量为对照组的 1.61 倍；然而，GnIH-2 不影响 *gnih* mRNA 的表达水平，GnIH-2 处理组 *gnih* mRNA 表达量为对照组的 83.53%。

4.4　半滑舌鳎 GnIH 对垂体激素合成与分泌的影响

利用原代垂体培养的方法，研究了 GnIH 对半滑舌鳎垂体激素合成与分泌的影响。GnIH-1 显著性地促进了 *gh* mRNA 水平，然而对 *gthα*、*lhβ* 及 *fshβ* 表达量无影响；然而，GnIH-2 均不影响 *gh*、*gthα*、*lhβ* 及 *fshβ* 的表达水平（图 14）。此外，GnIH-1 和 GnIH-2 均显著性地降低了垂体 LH 分泌，然而二者均不影响 GH 和 FSH 分泌（图 15）。

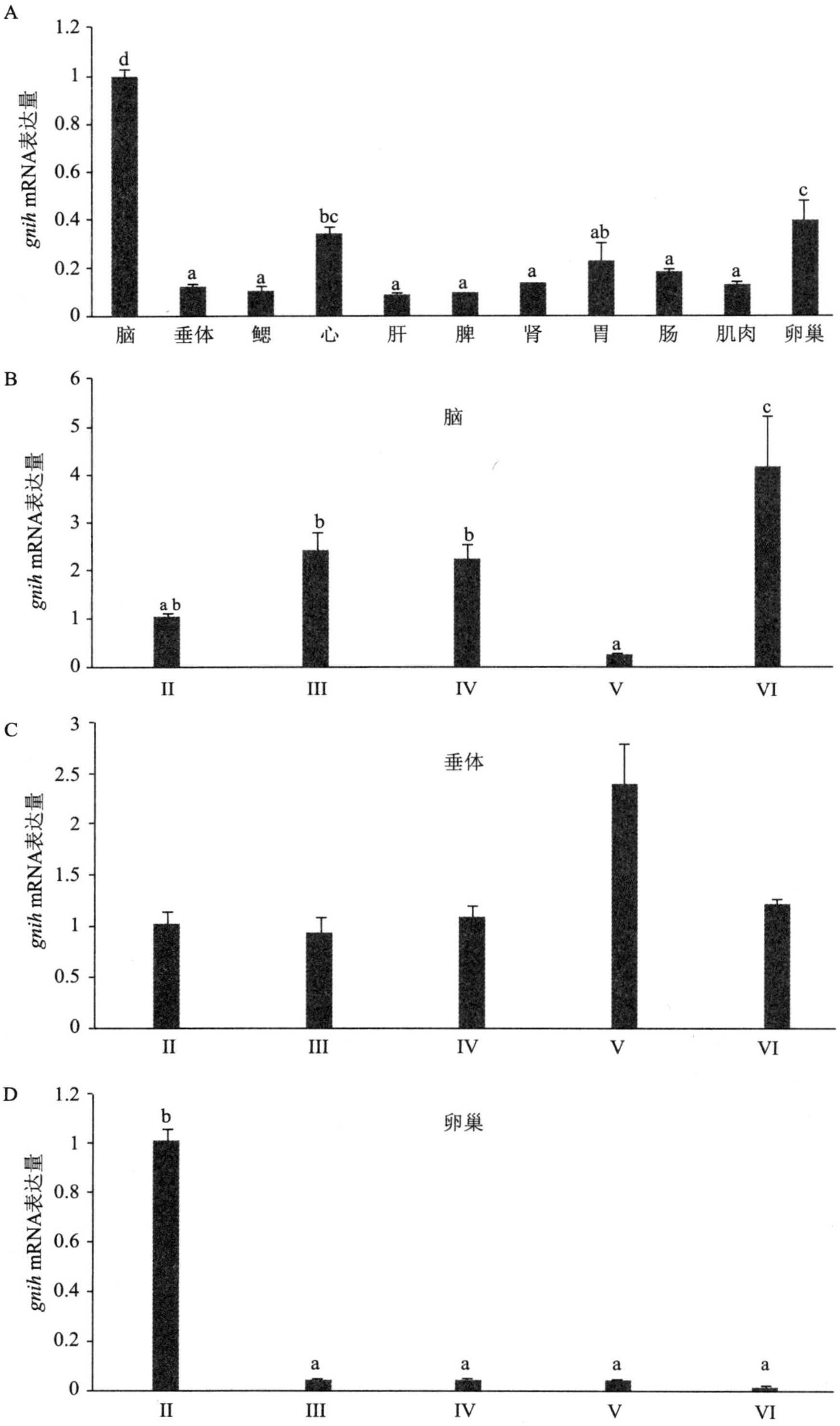

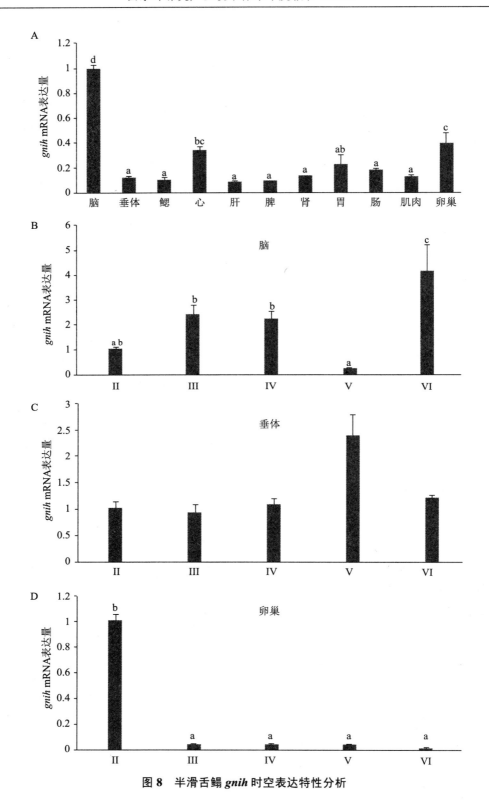

图 8　半滑舌鳎 *gnih* 时空表达特性分析

```
-60    acatggggacaacggcaggtaaatcatttcacaacaattgagctctgcttcagatcaaca
1      ATGGAGTTACTGGATGGAATGATGGGGCAGGAAATGGAGTTTGAGGGCTCTGCCACAGCT
1       M  E  L  L  D  G  M  M  G  Q  E  M  E  F  E  G  S  A  T  A

61     GCAGCACATAAAAAAAACATCAGCACTGACATTGTCACCAACGTCACCAACGTTACATTT
21      A  A  H  K  K  N  I  S  T  D  I  V  T  N  V  T  N  V  T  F

121    TCTCCCTACTATCAGCACTCTCTGTACGTGGCCGCAAGTTACATCTTGGCTTATTTCTTC
41      S  P  Y  Y  Q  H  S  L  Y  V  A  A  S  Y  I  L  A  Y  F  F

181    ATCTTCCTGCTGTGTATGGTTGGGAACATCCTGGTTTGTTTGATTGTACTGGAAAACCAT
61      I  F  L  L  C  M  V  G  N  I  L  V  C  L  I  V  L  E  N  H
                      TM 1
241    CGTATGCGCACAGTCACAAACCTCTTCATCCTAAACCTGGCCATTAGTGACCTCCTGGTG
81      R  M  R  T  V  T  N  L  F  I  L  N  L  A  I  S  D  L  L  V
                                                      TM 2
301    GGCATCTTCTGCATCCCTACCACACTGGTGGACAACCTCATCACAGGCTGGTCCTTTTCT
101     G  I  F  C  I  P  T  T  L  V  D  N  L  I  T  G  W  S  F  S

361    AACACTGTCTGCAAGATGAGTGGATTTGTTCAAGGCGTGTCTGTGTCTGCCTCGGTCTTC
121     N  T  V  C  K  M  S  G  F  V  Q  G  V  S  V  S  A  S  V  F
                                                      TM 3
421    ACCTTGGTGGCCATTGCTGTGGAACGATTTCGCAGCATCATGTATCCTCTGCAACCCAAA
141     T  L  V  A  I  A  V  E  R  F  R  S  I  M  Y  P  L  Q  P  K

481    GCAACTGTACTCGTGGCCAAACTAGCCATCTTCTTTATCTGGGTGTTAGCAGTAGCGATG
161     A  T  V  L  V  A  K  L  A  I  F  F  I  W  V  L  A  V  A  M
                                                      TM 4
541    ATGTGTCCTGCTGTTATGGCTCTCACTGTGAAAAAATTTCCTTCCCATTACATTGTCTAC
181     M  C  P  A  V  M  A  L  T  V  K  K  F  P  S  H  Y  I  V  Y

601    GATGATGACTTCAACCACACGCTCCCTGTTTACATGTGCTACGAGAACTTTGACAAGCCC
201     D  D  D  F  N  H  T  L  P  V  Y  M  C  Y  E  N  F  D  K  P

661    CAGATGAGAAAGATCTACACTGCAGTTCTGTTCGCTCACATCTACCTGGTGCCACTGACC
221     Q  M  R  K  I  Y  T  A  V  L  F  A  H  I  Y  L  V  P  L  T
                                                      TM 5
721    CTCATCACTGTCATGTACGGATGCATTGGCATCAAACTGTGTACTTCTGTGGTGCCAAAC
241     L  I  T  V  M  Y  G  C  I  G  I  K  L  C  T  S  V  V  P  N

781    AGAGAGCCGCAGTTGGATAACGCCACTGTGGCGGTGGGTAGAAGAGCCCCTCAGCCAATG
261     R  E  P  Q  L  D  N  A  T  V  A  V  G  R  R  A  P  Q  P  M

841    ATATCGCAGAAGAAGATCAAGGCCATAAAAATGCTCACTCTGGTGGCTCTGCTTTTCATG
281     I  S  Q  K  K  I  K  A  I  K  M  L  T  L  V  A  L  L  F  M
                                                      TM 6
901    TTGTCCTGGTTGCCCCTCTGGACATTAATGATGATAACTGACTACGCAGGCTGGGACATG
301     L  S  W  L  P  L  W  T  L  M  M  I  T  D  Y  A  G  W  D  M

961    GACCAGCTTGATCTTCTGACCAGTTACATGTTTCCCTTTGCTCACTGGCTGGCTTTCTCC
321     D  Q  L  D  L  L  T  S  Y  M  F  P  F  A  H  W  L  A  F  S
                                                      TM 7
1021   AACTCAAGCATCAACCCCATTATTTATGGCTACTACAATGAAAACTTCAGGCGAGGCTTC
341     N  S  S  I  N  P  I  I  Y  G  Y  Y  N  E  N  F  R  R  G  F

1081   AAGGCAGTGTGCAAGTCCCAGCAATTGTGTTGCAGCTTAGGAGAGACAACTGTGCATCAC
361     K  A  V  C  K  S  Q  Q  L  C  C  S  L  G  E  T  T  V  H  H

1141   CACCCTGTTGTTGTAAGGCTGACCAATCGTGTTCATAATGACAACGAGCTCAAAGACACG
381     H  P  V  V  V  R  L  T  N  R  V  H  N  D  N  E  L  K  D  T

1201   GCTAATAACAACAGAAGTGCACCAGTGGGCAACCCCATGGTCCTCACACAGAGATGTCCT
401     A  N  N  N  R  S  A  P  V  G  N  P  M  V  L  T  Q  R  C  P

1261   TCAAACCAAGTCATGGAAATGGCATCATTGCCCAAGAATAGCTCTGCAGAATCAGAGAGG
421     S  N  Q  V  M  E  M  A  S  L  P  K  N  S  S  A  E  S  E  R

1321   GTCGGCTCACTGGCTGTGTCGGTGCATCAGGCGTGGGAAAACTGAaactccacttctagt
441     V  G  S  L  A  V  S  V  H  Q  A  W  E  N  *
1381   tcaagagaatttggtagaaccatagtgttgtactttataccaggtaatctatagtgcact
1441   taaatgagatgtttttgcatagaccattttcagaacaaaaagccacatgcaacaatttga
1501   cactgtttcacagggtggggttgacatttagttcgtactaaaagacaacaagagttattt
1561   tagacagaatcaaatgtaataaatttgatgggcatacggtgaaagagagctaacagaaat
1621   gctaatttatttttgttgttaaatattcaaataaagaacaacaacattctactggttga
1681   tctcctacctaacccaacccacttgttcaggtttactctattgaaagtctataagaagta
1741   atgcacacccaaatgtaaacatgttgtatttcaagcaaaagaacagtgtttacagtggca
1801   gagtgattagttgttgtttttttttcatttgttgtttctttattgtttaatgtgattct
1861   caacgtgatcattcaacacctgtgtttgagaaatggcgattaatgtccgtgtggatttga
1921   ttcccacaggaaccacatgcttgattgttaatcctcctttgtaagtcgcttcgaataaaa
1981   gcgtctgctaaatgcatacatgtaaataaaacggtggcatcgaaacacgagtattgtgta
2041   attggaacattccttttaagtattttgttaacagtcttgtgtaatcagtgtgttgatgt
2101   taaacattgattttctaaaccgaaaaaaaaaaaaaaaaaa
```

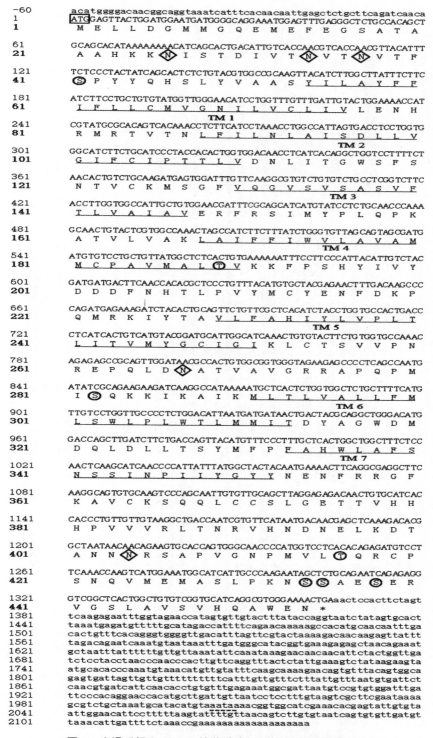

图 9 半滑舌鳎 GnIH-R 核苷酸序列及推断的氨基酸序列

TM:跨膜区;菱形标示糖基化位点;圆圈标示磷酸化位点

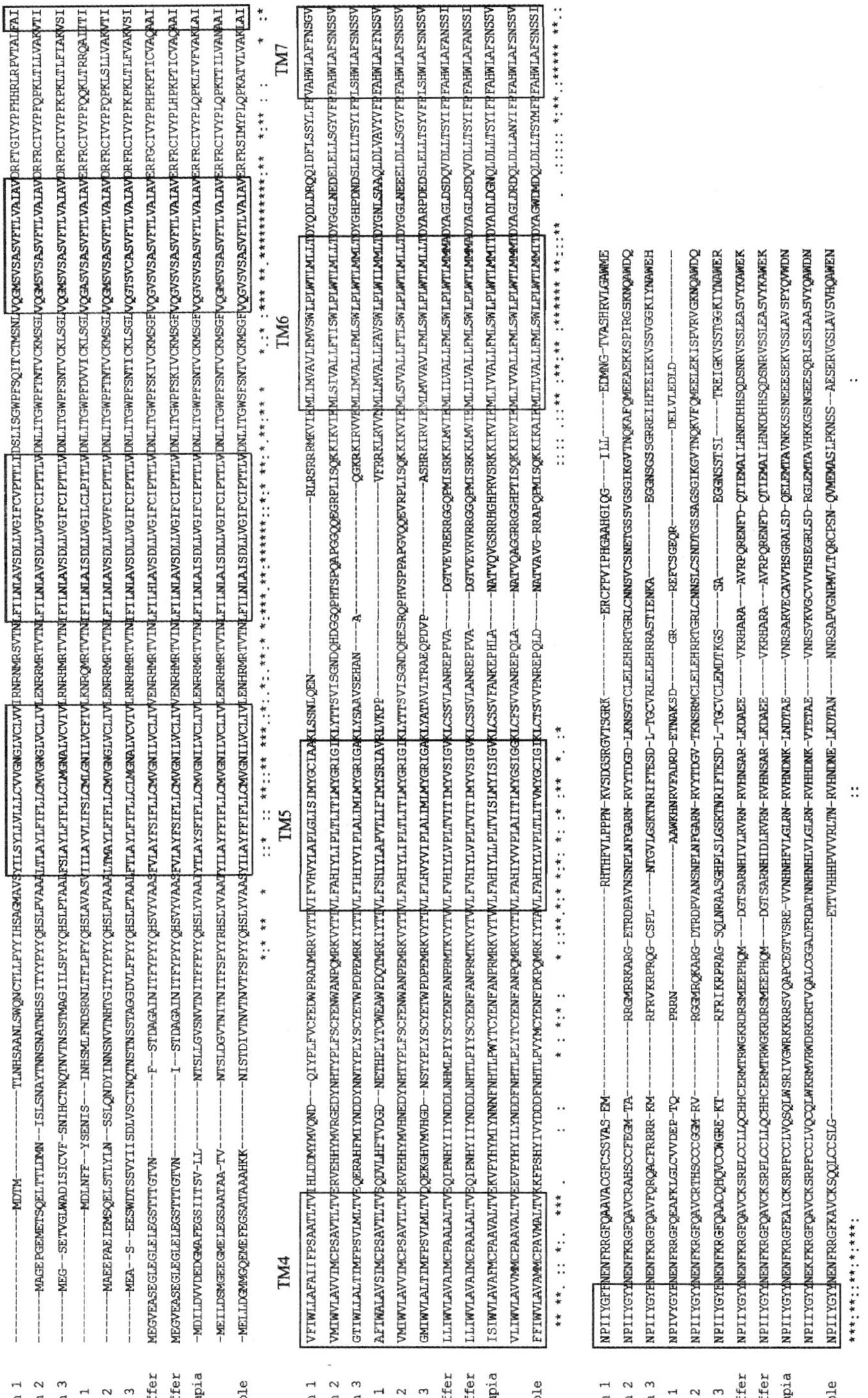

图 10　不同鱼类 GnIH-R 氨基酸序列比对

TM: 跨膜区

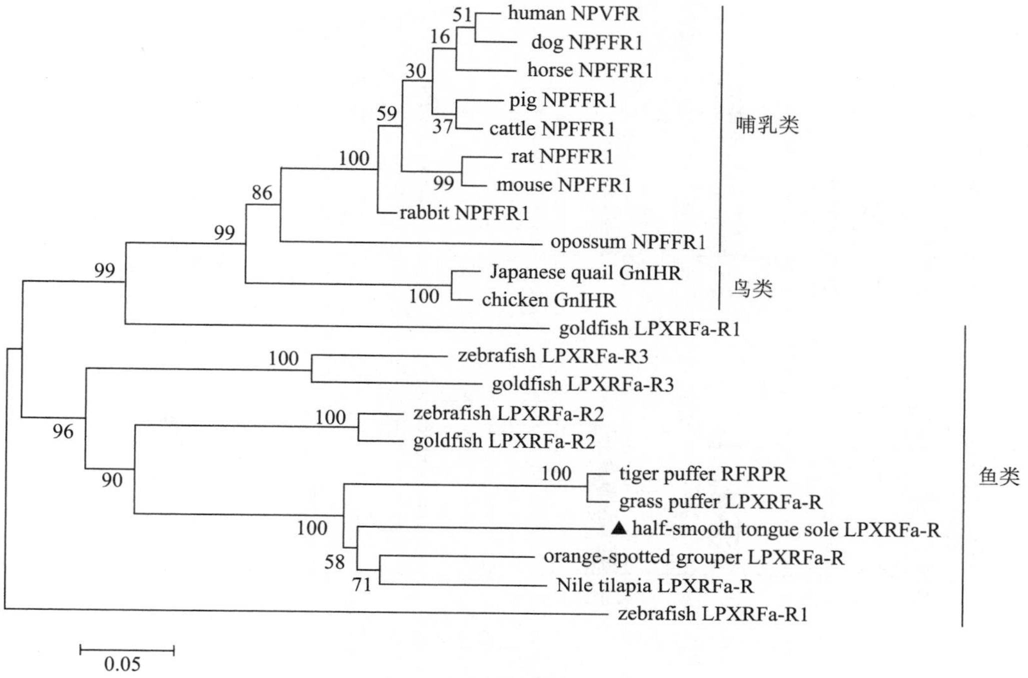

图 11　脊椎动物 GnIH-R 系统进化分析

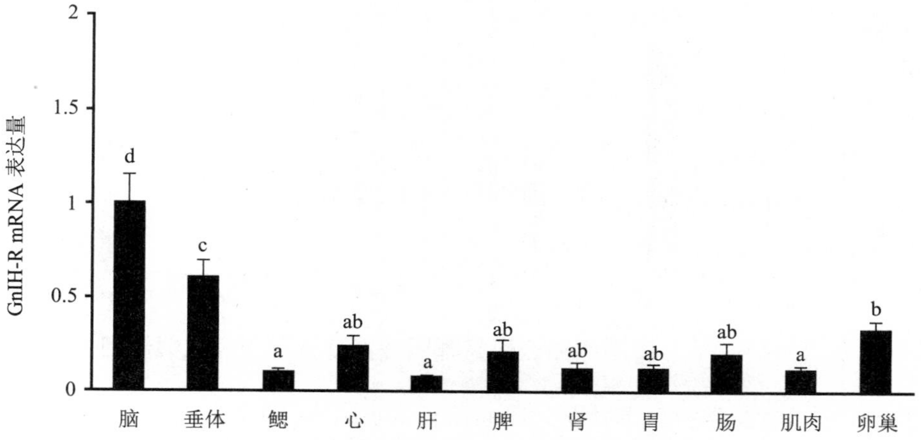

图 12　半滑舌鳎 GnIH-R 组织分布表达特性分析

5　养殖牙鲆肠道菌群发育与结构演替分析

构建了工厂化条件下牙鲆仔、稚、幼鱼等 6 个不同发育时期样品的 16*S* rRNA 基因测序文库，共获得 7 462 个 OTU，分类为 42 个菌门 972 个菌属。随着仔、稚、幼鱼日龄的增加，肠道菌群物种多样性逐渐增加(图 16)。

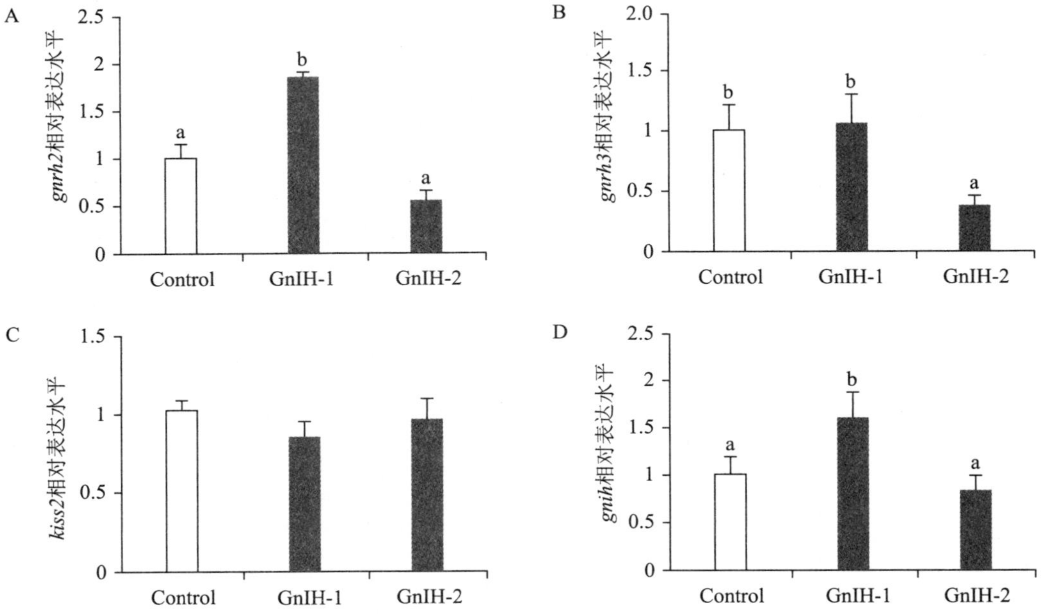

图 13　GnIH 对半滑舌鳎下丘脑生殖相关基因表达的影响

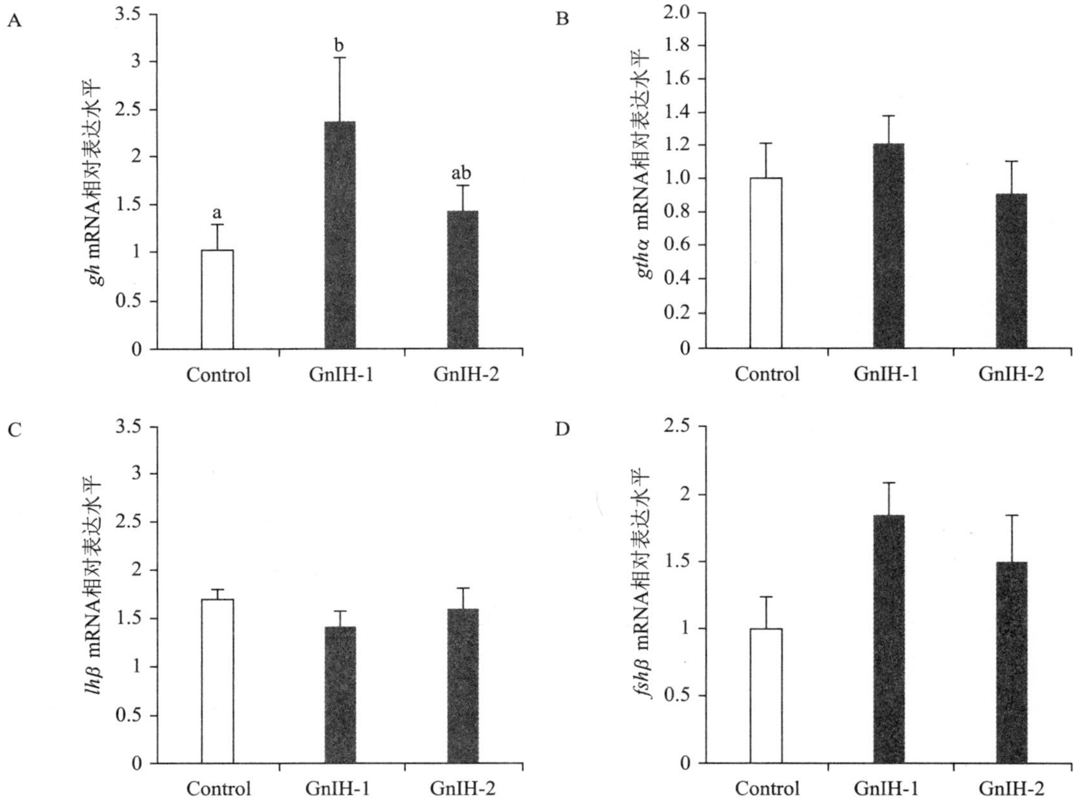

图 14　GnIH 对半滑舌鳎垂体 GH 和 GTH 表达的影响

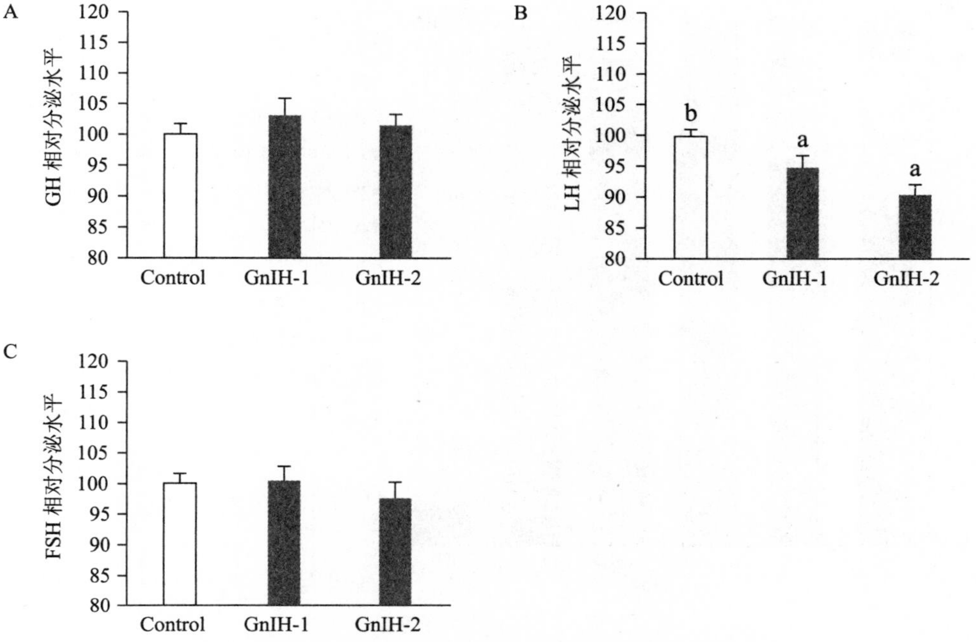

图 15　GnIH 对半滑舌鳎垂体 GH、LH 和 FSH 分泌的影响

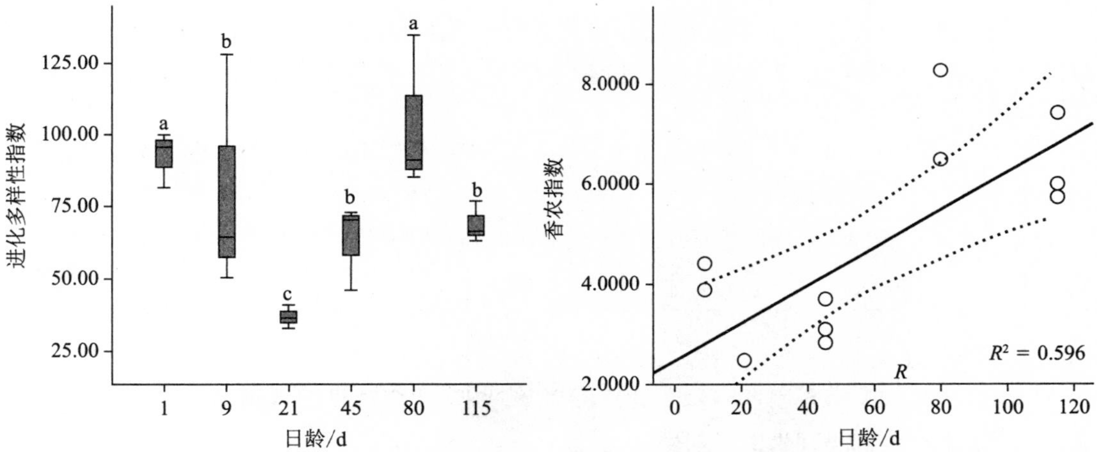

图 16　牙鲆肠道菌群 Alpha 多样性指数分析

初孵仔鱼菌群多样性丰富，优势菌为变形菌门、厚壁菌门和拟杆菌门。在 9 日龄和 21 日龄摄食轮虫和卤虫幼体样品中，肠道的优势菌群结构单一，优势菌群为变形菌门。45 日龄摄食配合饲料后，变形菌门相对丰度显著降低，厚壁菌门和拟杆菌门的相对丰度明显增大，成为肠道菌群的优势菌群（图 17）。

属水平上，仔、稚、幼鱼肠道优势菌群的种类和数量都发生了较大变化。在 9 日龄和 21 日龄时期，弧菌属相对丰度最高，到 45 日龄后丰度锐减。拟杆菌属和普氏菌属在 80 日龄后

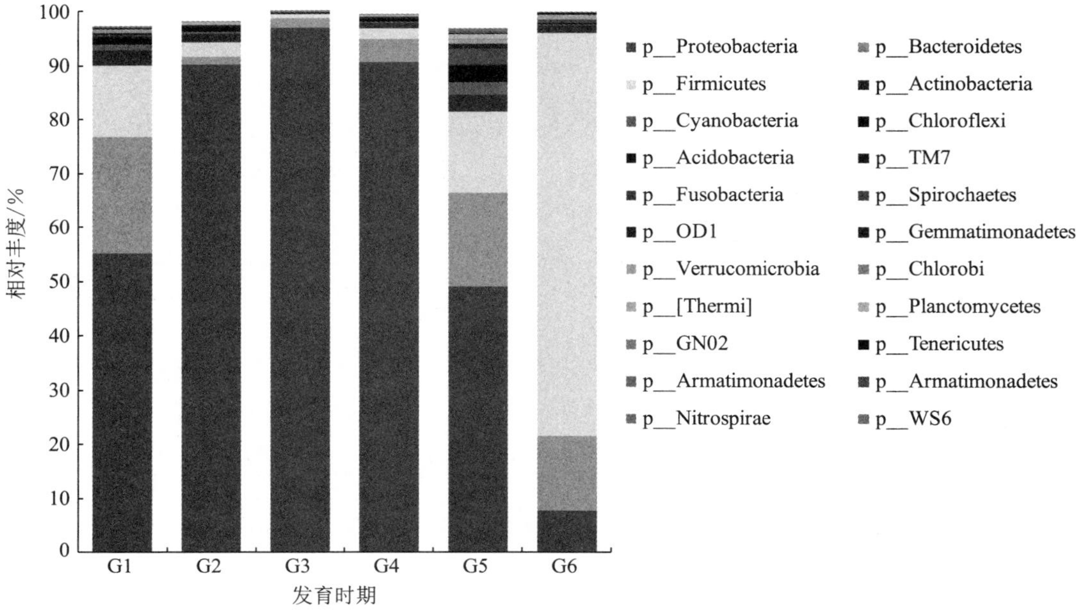

图 17　门水平分类肠道菌群组成

达到较高水平,成为肠道优势菌属。厚壁菌门的 8 个菌属在 80～115 日龄时期均发展成为优势菌属,定植于肠道(图 18)。

摸清了 120 日龄幼鱼肠道主要优势菌属丰度的发育变化,肠道优势菌属主要来自厚壁菌门(8 个属)、拟杆菌门(2 个属)和变形菌门(2 个属),与成鱼基本相似。各优势菌属随着

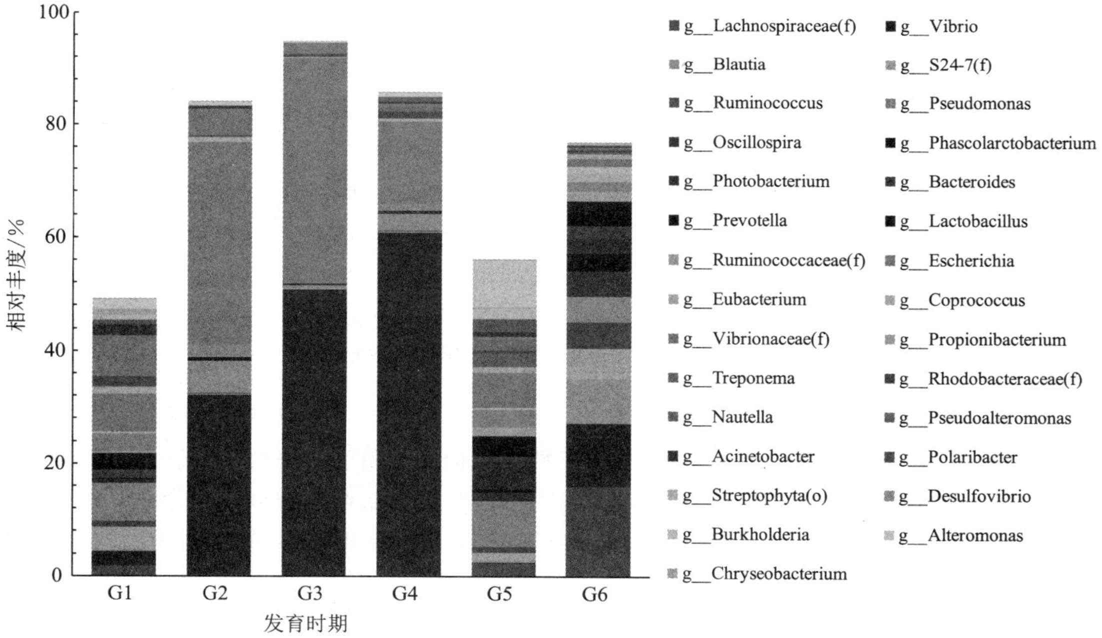

图 18　属水平分类肠道菌群组成

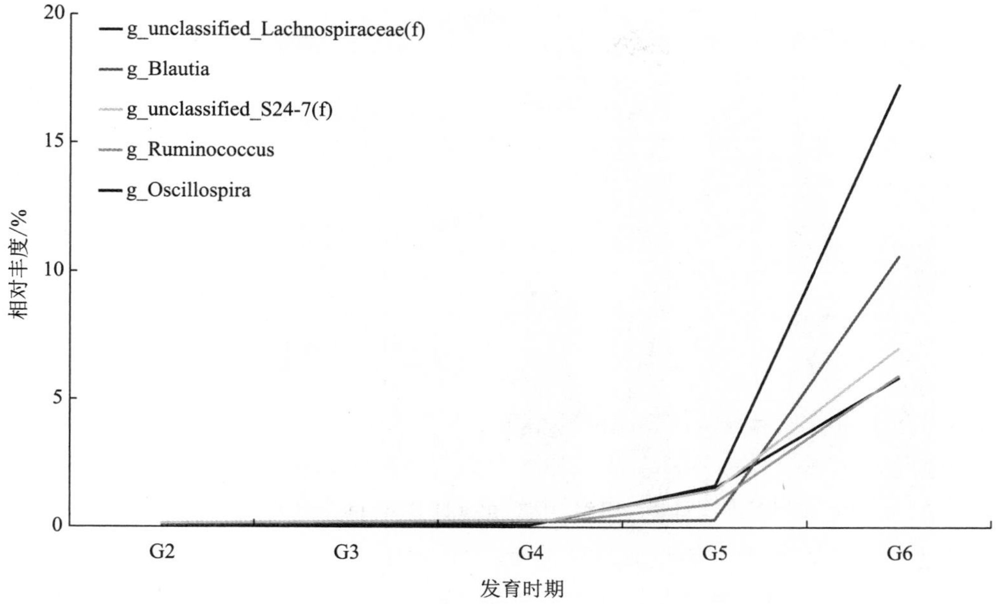

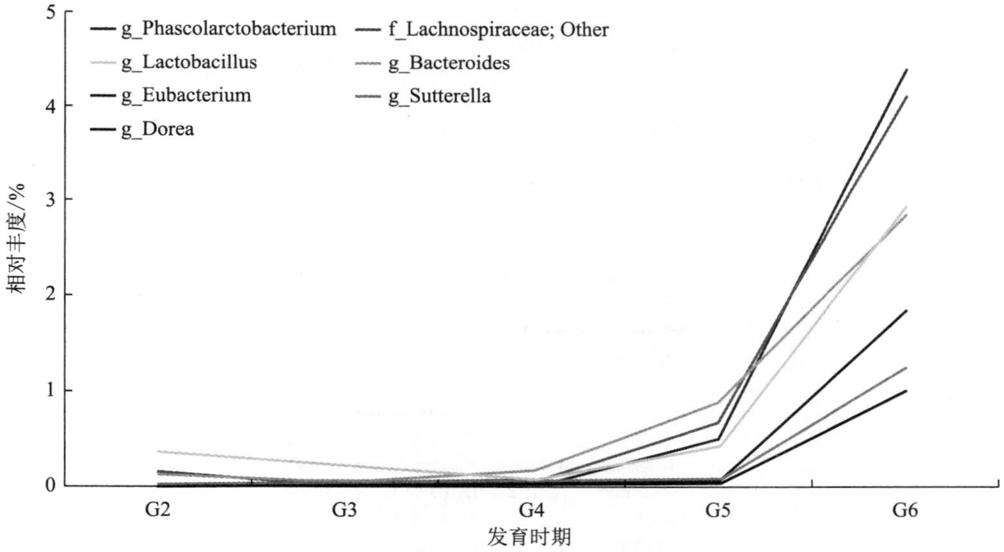

图 19　G6 时期优势菌属丰度的发育变化

摄食与生长丰度逐渐升高，并在 G6 期达峰值（图 19）。

牙鲆仔鱼开始摄食后，受饵料的菌群影响密切。就弧菌丰度看，卤虫菌群中弧菌丰度高，12～40 日龄是投喂卤虫的阶段，此阶段肠道菌群中弧菌丰度最高。同时，水中弧菌丰度也高。认为 45 日龄前肠道中弧菌丰度受饵料和水环境影响大。转为配饵后，饵料和水中弧菌丰度小，其肠道菌群中弧菌丰度迅速下降（图 20）。

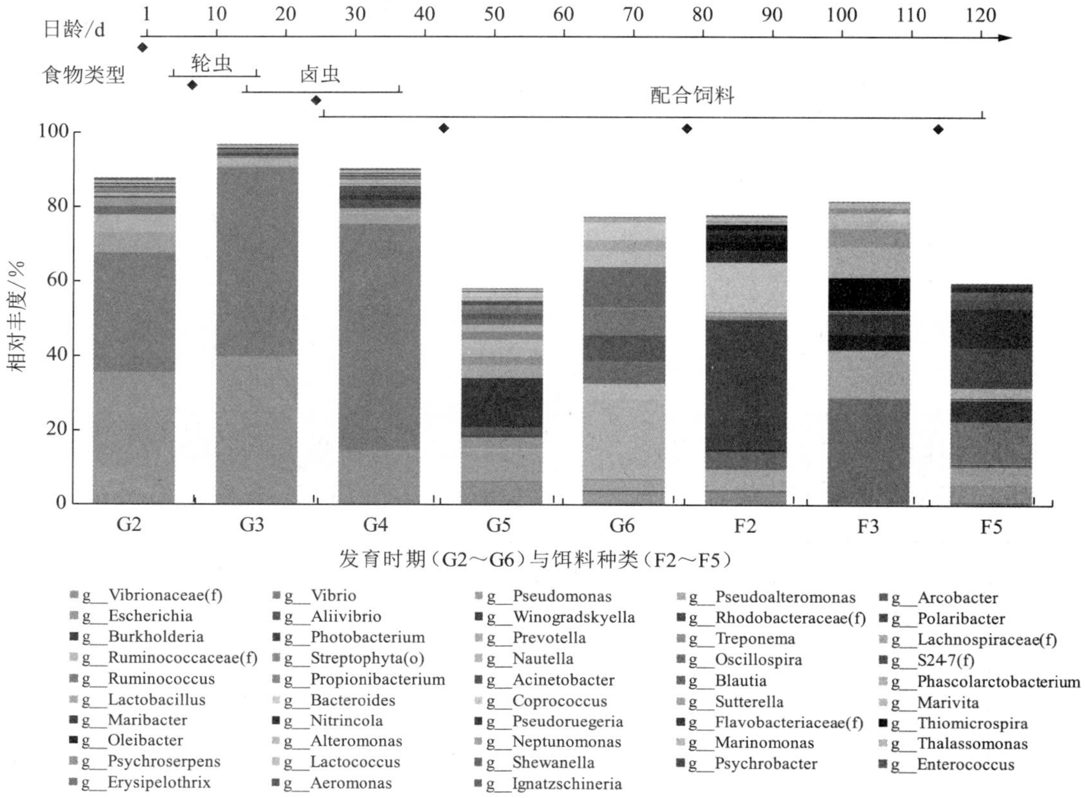

图 20　饵料菌群对不同发育时期肠道优势菌属丰度的发育变化

（岗位科学家　柳学周）

鲆鲽类疾病防控疫苗技术研发进展

疾病防控岗位

针对鲆鲽类两种重要细菌性疾病弧菌病(鳗弧菌)和腹水病(迟钝爱德华氏菌),2016年度疾控岗位的技术研发任务重点围绕细菌性疾病疫苗如下任务目标组织实施相关研发工作:① 实施提交大菱鲆弧菌病鳗弧菌基因工程活疫苗产品注册证申报材料,并为接受中国兽医药品监察所的复核检查做好各项准备工作;② 向农业部提交申请大菱鲆腹水病迟钝爱德华氏菌弱毒活疫苗生产批文;③ 向国家农业部申请鲆鱼弧菌病鳗弧菌灭活疫苗临床试验批件,并适时开展相关试验;④ 制订建立大菱鲆疫苗的生产性免疫接种操作规程,并进行应用示范推广。根据上述技术研发工作目标,积极推进单个疫苗和联合疫苗的临床试验、新兽药注册、生产文号申报,推进技术标准和产业化应用规程制订,同时进行相关产业化示范,加快推进疫苗的产业化进程,各项工作取得了如下进展。

1　鲆鲽类弧菌病疫苗研制

1.1　大菱鲆弧菌病鳗弧菌基因工程活疫苗产品注册申报

在2015年11月25日顺利通过农业部兽药评审中心初审审查的基础上,根据评审中心审查意见和新版兽药典(2015年版)新规定,多次修订并完善了《大菱鲆鳗弧菌基因工程活疫苗(MVAV6203株)制造及检验规程(试行)》《大菱鲆鳗弧菌基因工程活疫苗(MVAV6203株)质量标准(试行)》和使用说明等3项技术标准与规程,完善并向农业部兽药评审中心提交了《大菱鲆弧菌病鳗弧菌基因工程减毒活疫苗新兽药注册申报》修订材料,并按照制造规程与质量标准完成了待检产品制备和自检等各项准备工作,迎接中监所的复核检验,为推进注册证早日获批奠定坚实基础。

1.2　鲆鱼弧菌病灭活疫苗(EIBVA1株)临床试验申报

为今后开发多联疫苗以应对水产养殖生产中面临的多病原挑战,本年度在前期临床前研究及免疫佐剂筛选优化的基础上,在兽用生物制品GMP生产资质企业灭活疫苗生产线完成了5批次临床用GMP产品中间试制,制订了《鲆鱼弧菌病灭活疫苗制造及检验规程(草案)》与《鲆鱼弧菌病灭活疫苗质量标准(草案)》各1项,并在烟台综合试验站、葫芦岛综合试验站、天津综合试验站协助下与3家鲆鲽养殖企业签订了临床协议,形成临床试验申报材

图 1　疫苗效力与安全性自检

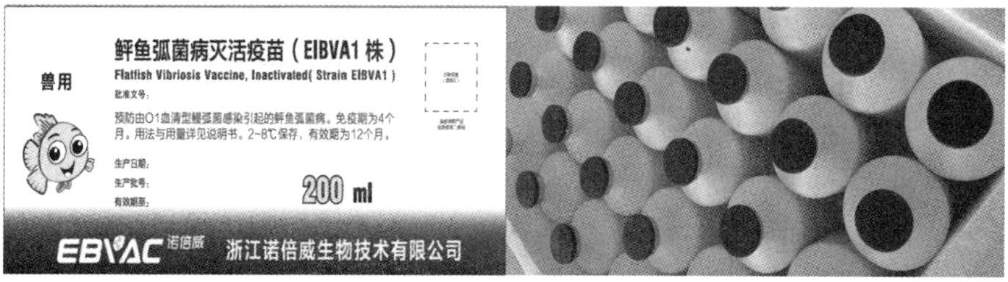

图 2　鲆鱼弧菌病灭活疫苗临床试制产品

料提交农业部审核，临床批件即将获批。

2　鲆鲽类爱德华氏菌病疫苗研制

2.1　迟钝爱德华氏菌基因工程活疫苗(WED 株)临床前研究

2014 年 4 月迟钝爱德华氏菌基因工程活疫苗(WED 株)获得农业部转基因生物安全证书(农基安证字[2013]第 267 号，生产应用)，为以此疫苗株为基础进行临床申报或构建多价载体疫苗奠定了可靠的国家行政许可基础。

本年度在前期工作基础上完成了疫苗临床试验申报所需的 10 批次 GMP 中间试制工作

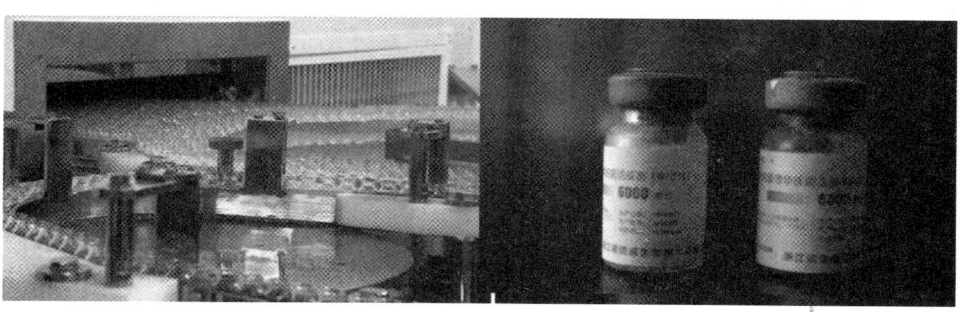

图 3　迟钝爱德华氏菌减毒活疫苗(WED 株) GMP 中间试制

和各项临床前试验，初步制定了《大菱鲆腹水病迟钝爱德华氏菌基因工程活疫苗（WED株）制造及检验规程（草案）》和《大菱鲆腹水病迟钝爱德华氏菌基因工程活疫苗（WED株）质量标准（草案）》。

2.2　迟钝爱德华氏菌弱毒活疫苗（EIBAV1株）生产许可申报

针对大菱鲆迟钝爱德华氏菌弱毒活疫苗（EIBAV1株），在2015年获得了农业部国家一类新兽药注册证后，配合两家兽用疫苗生产企业分别于2016年4月和6月申报获得水产用细菌活疫苗生产线兽药GMP认证，并向农业部提交了生产文号申报书，于2016年12月14日获得我国首例海水鱼活疫苗生产文号（兽药生字110576037），为推进该产品的商用化提供了可靠的GMP生产制造基础和行政许可支持。

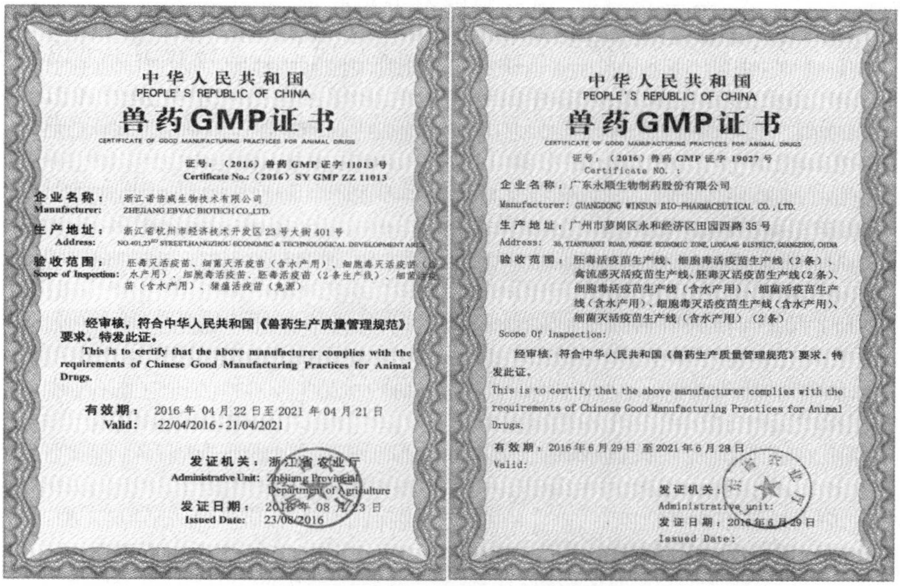
中华人民共和国
PEOPLE'S REPUBLIC OF CHINA
兽药GMP证书
CERTIFICATE OF GOOD MANUFACTURING PRACTICES FOR ANIMAL DRUGS
证号：（2016）兽药GMP证字11013号
Certificate No.:（2016）SY GMP ZZ 11013
企业名称：浙江诺倍威生物技术有限公司
Manufacturer: ZHEJIANG EBVAC BIOTECH CO.,LTD.
生产地址：浙江省杭州市经济技术开发区23号大街401号
经审核，符合中华人民共和国《兽药生产质量管理规范》要求。特发此证。
This is to certify that the above manufacturer complies with the requirements of Chinese Good Manufacturing Practices for Animal Drugs.
有效期：2016年04月22日至2021年04月21日
Valid: 22/04/2016 - 21/04/2021
发证机关：浙江省农业厅
Administrative Unit: Zhejiang Provincial Department of Agriculture
发证日期：2016年08月23日
Issued Date: 23/08/2016

中华人民共和国
PEOPLE'S REPUBLIC OF CHINA
兽药GMP证书
CERTIFICATE OF GOOD MANUFACTURING PRACTICES FOR ANIMAL DRUGS
证号：（2016）兽药GMP证字19027号
Certificate NO.:
企业名称：广东永顺生物制药股份有限公司
Manufacturer: GUANGDONG WINSUN BIO-PHARMACEUTICAL CO., LTD.
生产地址：广州市萝岗区永和经济区田园西路35号
经审核，符合中华人民共和国《兽药生产质量管理规范》要求。特发此证。
This is to certify that the above manufacturer complies with the requirements of Chinese Good Manufacturing Practices for Animal Drugs.
有效期：2016年6月29日至2021年6月28日
Valid:
发证机关：
Administrative unit:
发证日期：2016年6月29日
Issued Date:

图4　水产疫苗生产线GMP证书

3　鲆鲽类疫苗生产性应用示范与产业化推广

为推进大菱鲆疫苗的产业化进程，全面系统进行了《大菱鲆疫苗生产免疫接种操作规程（试行）》制订和完善，进行了针对这些疫苗的工厂化养殖规模化应用示范、推广和疫苗接种规程技术培训，制定了鲆鲽类"计划免疫"的标准接种规程，2016年度疾控岗位联合产业体系烟台综合试验站、山东综合试验站、辽宁综合试验站、天津综合试验站等，以爱德华氏菌弱毒疫苗和鳗弧菌基因工程活疫苗为先导产品进行了产业化应用示范和推广，覆盖我国鲆鲽主产区辽宁、山东、天津等省份的多家重点鲆鲽养殖企业，总计推广示范规模累积达到注射接种幼鱼（体长10 cm以上）约20万尾，浸泡接种商品苗（稚鱼，体长3 cm～5 cm）20万尾。

中华人民共和国农业部
兽药产品批准文号批件

批件号：07040020161021-123

企业名称：浙江诺倍威生物技术有限公司

根据《兽药管理条例》、《兽药产品批准文号管理办法》等有关规定，该产品经审查符合要求，准予生产。

生产地址：浙江省杭州市经济技术开发区23号大街401号

通用名称：大菱鲆迟钝爱德华氏菌活疫苗（EIBAV1株）

商品名称：渔康倍

兽药产品批准文号：兽药生字110576037

含量规格：2000尾份/瓶；3000尾份/瓶；5000尾份/瓶；6000尾份/瓶；10000尾份/瓶

有效期：2016-12-14至2021-12-13

发证日期：2016-12-14

图 5　大菱鲆迟钝爱德华氏菌活疫苗(EIBAV1 株)生产批文

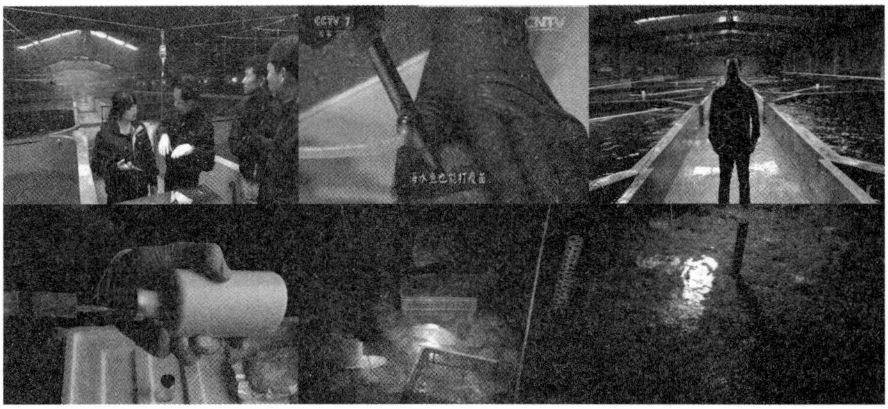

图 6　大菱鲆疫苗生产性应用示范与产业化推广

主要示范企业有山东东方海洋科技股份有限公司、天津奕鸣泉水产养殖有限公司、大连鹤圣丰水产有限公司、大连颢霖水产有限公司、大连天正实业产有限公司等。接种规程的易用性和企业生产养殖规程操作的紧密契合度受到了养殖企业的充分肯定和欢迎，培训和示范活动受到了当地广大鲆鲽养殖企业的热烈响应和欢迎，也得到了当地渔业主管部门的大力支持和积极肯定，为进一步在我国鲆鲽主产区普及养殖病害免疫防控理念及有序推广疫苗接种生产应用起到了积极示范推动作用，也进一步推动了岗位疫苗成果的商用化进程。

4　流行病学调查与新型候选疫苗研发

4.1　鲆鲽养殖主产区流行病学调研

为了全面系统摸清我国鲆鲽类主产区主要病害的流行病学特征与蔓延分布情况，并就养殖企业和用户疫苗应用意愿进行统计调查，在烟台综合试验站、辽宁综合试验站、河北综合试验站、葫芦岛综合试验站、天津综合试验站、山东综合试验站和北戴河综合试验站协助下，本年度5～11月间在鲆鲽主产区域内进行了较为系统的养殖病害情况调研及病原采集工作。总计采集样品150余份，鉴定病原63株，主要病原种类为细菌性疾病，其中绝大多数为弧菌(39株，61.9%)，其次为迟钝爱德华氏菌(9株，14.3%)、各种单胞菌(12株，19.0%)和其他病原菌(3株，4.8%)等。这些工作为将来疫苗的研究提供了对照病原，积累了我国鲆鲽类主产区病害季节性流行资料，为疫苗的生产应用提供区域流行病学参考。

图7　鲆鲽养殖主产区流行学调研

同时，为第一时间更好地了解和监控我国鲆鲽主产区疫情发生情况，与病害高发区相关企业和养殖场签订了病害监控及病原菌采集与临床诊断技术服务合作协议，提供了疫苗及免疫接种示范推广等技术服务等。

4.2　新型候选疫苗产品研发储备

本年度针对临床分离病原如哈维氏弧菌、杀鲑气单胞菌、溶藻弧菌的疫苗开发工作也已取得一定进展。首先建立了攻毒感染模型，根据鱼种、病原种类确定待研疫苗种类，并基于疫苗种类确定研发思路，采用多种策略例如配合佐剂、免疫增强剂等开发疫苗，考察所研疫苗对实验动物及靶动物的安全及效力，并进行一系列免疫效果评价分析。通过以上实验，筛选获得有效疫苗候选株进行临床试验。目前，已建立杀鲑气单胞菌和哈维氏弧菌的感染模型，其灭活疫苗配合商业佐剂后免疫大菱鲆的免疫保护效果均达到50%以上，后续筛选优化工作正在顺利开展。

研发了相关细菌载体疫苗、亚单位疫苗和DNA疫苗的候选株。将费氏弧菌群体感应基因与多种体内诱导调控元件相整合，分别构建了受“铁-细菌密度”或“阿拉伯糖-细菌密度”双重信号调控的体内诱导表达系统(图8)，将编码嗜水气单胞菌的GAPDH编码基因分别引入两个表达系统，转入迟钝爱德华氏菌，构建多价细菌载体疫苗。结果显示，接种两种载体疫苗的大菱鲆针对迟钝爱德华氏菌分别获得57%和65%的免疫保护力，说明基于体内诱导表达系统研发的迟钝爱德华氏菌细菌载体疫苗对预防由该菌引起的病害具有良好的免疫保护作用。

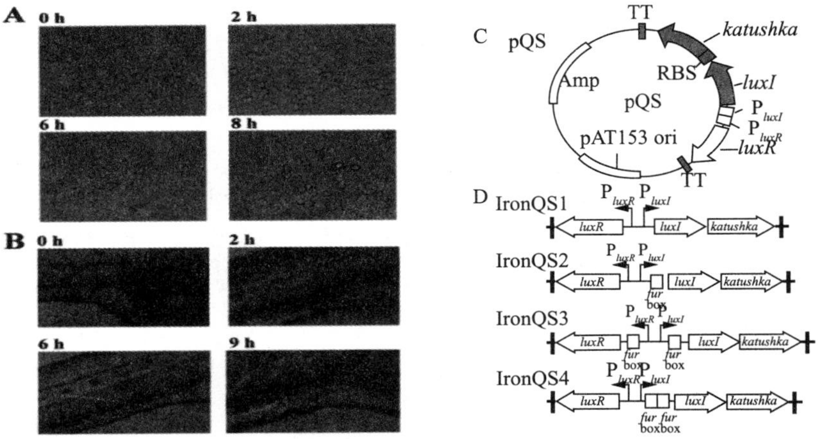

图 8 受双重信号调控的体内诱导的抗原表达系统

利用生物信息学方法预测并获得迟钝爱德华氏菌抗原蛋白 FBA,该蛋白与鳗弧菌、溶藻弧菌、嗜水气单胞菌、哈维氏弧菌的 FBA 高度同源。将 FBA 作为广谱型保护性抗原用于迟钝爱德华氏菌多价疫苗研发。结果显示,接种迟钝爱德华氏菌抗原蛋白 FBA 可在一定程度上激发大菱鲆血清产生针对迟钝爱德华氏菌、鳗弧菌、溶藻弧菌、嗜水气单胞菌、哈维氏弧菌的抗体。接种以上 5 种病原菌抗原蛋白 FBA 的斑马鱼针对以上病原可获得良好免疫保护效果,并产生一定的交叉免疫保护作用,说明基于抗原蛋白 FBA 研发的迟钝爱德华氏菌亚单位疫苗对由该菌引起的病害具有良好的免疫保护作用。

综上所述,2016 年度中由疾控岗位自主研发的大菱鲆迟钝爱德华氏菌活疫苗(EIBAV1 株)已获得国家新兽药注册一类证书并获得生产文号,大菱鲆鳗弧菌基因工程活疫苗(MVAV6203 株)已获得转基因生物安全证书、临床试验批件,正在申报新兽药注册证书,鲆鱼鳗弧菌灭活疫苗(EIBVA1 株)正在申报临床试验批件,迟钝爱德华氏菌减毒活疫苗(WED 株)获得农业转基因生物安全证书,针对杀鲑气单胞菌、哈维氏弧菌、溶藻弧菌等的一系列疫苗产品的研发工作也在有序开展。

为了加快我国海水养殖疫苗的产业化进程,使更多疫苗进入产业化推进阶段,本岗位系统继续采集国际上美国、日本、欧洲等水产疫苗产业发达国家企业和政府管理部门的水产疫苗的生产规程、质量标准、临床管理法规和管理规程,并且同国内各级政府主管部门、学术专家和相关疫苗企业进行了广泛交流与咨询,继续在水产疫苗研发、临床实验、药证申报、产业化示范到成果孵化等各环节完善总结可操作的标准、规范和规程,为该标准化方案进入政府征询意见、纳入行业标准以及国家标准奠定坚实基础。

(岗位科学家 王启要)

鲆鲽类营养与饲料技术开发研发进展

营养与饲料岗位

在体系资助下本年度共发表（含接收）论文8篇，其中SCI论文4篇，核心期刊论文4篇。新申请国家发明专利1项。培养博士后1人，博士研究生2人，硕士研究生5人。组织或参加相关会议、培训等100余场，提交工作日志151篇。另外，分别收集10位鲆鲽类从业人员的详细个人信息、38种仪器设备数据以及29家水产营养与饲料国外相关研究单位概况信息。总体上，完成了任务书规定的年度工作任务。

首先，针对任务书提出的对鲆鲽类生理代谢研究的要求，本年度共开展了涉及大菱鲆幼鱼营养研究（包括鱼类脂肪代谢调控、脂肪酸和添加剂的研究）、仔稚鱼营养生理与微颗粒饲料开发以及亲鱼营养研究及专用饲料开发等方面的7个相关实验。

其次，通过结合大菱鲆替代蛋白源、脂肪代谢调控和饲料添加剂等的相关研究，优化了大菱鲆和半滑舌鳎配合饲料配方。与饲料企业积极合作，又开发出鲆鲽类仔稚鱼和成鱼微颗粒饲料2种，饲料效率160%，生产的鲆鲽类仔稚鱼饲料在价格、稳定性等方面较进口饲料有明显优势。生产推广专用系列鲆鲽类人工微颗粒配合饲料350 t，产值达900余万元。

此外，国家鲆鲽类产业技术体系健康养殖与综合研究室举办了“鲆鲽类健康养殖科学讲堂”，完成各类技术推介、宣传和培训100余次，培训技术推广人员、相关从业人员和科技示范户3 500余人次，编写并发放实用技术宣传手册／资料15 000份。培训养殖户采用经济科学的、可持续发展的方式提高鲆鲽类养殖经济效益。

表1　各项任务及考核指标完成情况对照表

项目	计划任务与研发内容	考核指标	执行情况	完成情况
重点任务	优质高效微颗粒饲料研发	开发出1种优质高效鲆鲽类仔稚鱼（半滑舌鳎）微颗粒饲料，其饲料效率达120%～180%	开发了鲆鲽类仔稚鱼微颗粒饲料1种，饲料效率160%，生产的鲆鲽类仔稚鱼饲料在价格、稳定性等方面较进口饲料有明显优势。生产推广专用系列鲆鲽类人工微颗粒配合饲料350 t，产值达900余万元	超额完成
基础性工作	国家鲆鲽类产业技术研发中心基础数据库	提供营养与饲料团队的论文、著作、专利、成果奖励、新品种、标准规范信息	发表18篇科技论文，并收集了鲆鲽类从业人员10人的详细个人信息。另外，收集了38种仪器设备数据和29个水产营养与饲料国外相关研究单位概况	完成

（续表）

项目	计划任务与研发内容	考核指标	执行情况	完成情况
基础性工作	鲆鲽类营养需要参数及饲料原料生物料利用率基础数据库	蛋白质等营养素营养需要参数建立，完成3种鱼类3种以上营养需求参数数据收集	完成了大菱鲆5种营养物质需求、漠斑牙鲆1种营养物质需求、半滑舌鳎4种营养物质需求、美洲拟鲽1种营养物质需求、塞内加尔鳎2种营养物质需求、星斑川鲽3种营养物质需求数据收集工作	超额完成
前瞻性研究	鲆鲽类系列饲料加工工艺优化	大菱鲆、半滑舌鳎优质高效环保型配合饲料的普及率达到20%以上	0 g～50 g大菱鲆、半滑舌鳎专用配合饲料的普及率达100%，50 g～150 g普及率70%，150 g以上普及率25%～30%。进行了一系列大菱鲆颗粒饲料的相关实验，开发鲆鲽类成鱼饲料1种	完成
	研究影响亲鱼繁殖性能的主要营养因素	年度未具体要求	开展了2个实验，已经完成	超额完成
应急性任务		年度未要求	根据体系的统一布置和要求开展了工作	完成
跨体系任务		年度未要求	根据跨体系的统一布置和要求拟定了跨体系核心技术与实施内容	完成
专利论著标准人才	专利	申请专利1项	申请国家发明专利1项	完成
	专著/论文	完成学术论文3篇	发表（含接收）论文8篇，其中SCI论文4篇，核心期刊论文4篇	超额完成
	标准与规范	年度未要求	无	
	人才培养	培养硕士生2名，博士生2名	培养博士后1名，博士研究生2名，硕士研究生5名	超额完成
	宣传培训	技术推介、宣传和培训2次，累计培训技术推广人员、相关从业人员和科技示范户600人次	完成各类技术推介、宣传和培训100余次，培训技术推广人员、相关从业人员和科技示范户3 500余人次	超额完成
	宣传册/资料	编写并发放实用技术宣传手册/资料150份	编写2册，发放15 000份	超额完成

2016年根据项目要求，共设计了7个实验，具体内容如下。

1 裂壶藻粉在大菱鲆仔稚鱼微颗粒饲料中的应用研究

1.1 裂壶藻粉对大菱鲆仔稚鱼成活率、生长、脂肪酸组成、肠道发育和脂肪代谢相关基因表达量的影响

本实验旨在研究微颗粒饲料中添加裂壶藻粉对大菱鲆仔稚鱼成活率、生长、消化酶活

性、脂肪酸组成和肠道发育的影响。在基础饲料中分别添加0%、5%、10%和15%的裂壶藻粉配制4种等氮等能的试验饲料。另外，选择一种商业料作为对照组。结果表明，饲料中添加裂壶藻粉对大菱鲆仔稚鱼的成活率影响不显著，5%组仔稚鱼终末体重显著高于15%组和对照组，而与0%和10%组间差异不显著。饲料中添加裂壶藻粉对仔稚鱼的特定生长率无显著性影响，对照组显著低于其他组，5%组仔稚鱼终末体长显著高于0%、15%组和对照组，而与10%组差异不显著。添加15%裂壶藻粉组肠道及肠道刷状缘LAP活力显著低于10%组，而与其他组之间差异不显著。15%组肠道及肠道刷状缘AKP活力显著低于5%组，而与0%和10%组间差异不显著。5%和10%饲料组肌肉DHA、*n*-3HUFA含量显著高于0%和15%组，随着添加比例由0%增加到15%，肌肉中EPA、*n*-6HUFA含量逐渐降低，其中，5%、10%和15%组显著低于0%组，5%、10%和15%组间差异不显著。裂壶藻粉对肠道褶皱高度、肠上皮细胞高度和微绒毛高度无显著影响。大菱鲆肝脏PPARα和CPT1的表达量随着裂壶藻粉添加量的增加而显著升高，而FAS的表达量显著降低，且PPARα和CPT1在15%添加组显著高于对照组，FAS在15%添加组显著低于对照组。裂壶藻粉对肝脏LPL和SREBP1的表达量没有显著影响。总之，饲料中添加10%裂壶藻粉能显著提高大菱鲆仔稚鱼的生长，添加15%裂壶藻粉对大菱鲆仔稚鱼生长和消化酶活力不利。

1.2　裂壶藻粉替代鱼油在大菱鲆仔稚鱼微颗粒饲料中的应用

本实验旨在研究微颗粒饲料中裂壶藻粉替代鱼油对大菱鲆仔稚鱼成活率、生长、消化酶活性、脂肪酸组成、肠道发育和脂肪代谢相关基因表达量的影响。在基础饲料中，用裂壶藻粉分别替代0%、25%、50%、75%、100%和150%的鱼油DHA，配制6种实验饲料。另外，选择一种商业料作为对照组。结果显示，饲料中裂壶藻粉替代鱼油对大菱鲆仔稚鱼成活率、终末体重、特定生长率（SGR）和终末体长无显著影响。对照组仔稚鱼SGR显著低于0%组，而与其他组之间差异不显著；对照组仔稚鱼终末体长显著低于0%、50%和150%组，而与其他组之间差异不显著；0%组仔稚鱼体蛋白含量显著高于25%和150%替代组。裂壶藻粉替代鱼油对肠道、肠道刷状缘的AKP和LAP活性无显著影响，对照组的肠道及肠道刷状缘消化酶活力显著低于替代组。裂壶藻粉替代鱼油对肌肉*n*-3HUFA含量无显著影响。随着替代比例由0%增加到150%，肌肉中EPA、C20:4*n*-6含量显著降低，而DHA/EPA比率显著升高。替代比例为0%～150%时，对DHA含量无显著影响，但150%替代组肌肉DHA含量有升高的趋势。随着替代比例由0%增加到100%，肌肉中C18:3*n*-3、C18:2*n*-6和*n*-6HUFA含量显著增加，而150%组显著低于100%组。对照组肌肉*n*-6HUFA与0%组差异不显著，但显著低于其他组，而EPA、DHA、*n*-3HUFA含量显著高于其他组。裂壶藻粉替代鱼油对肠道褶皱高度、肠上皮细胞和微绒毛高度无显著影响。大菱鲆肝脏CPT1、LPL和FAS表达量在0%～100%替代组之间差异不显著，LPL和FAS表达量在150%替代组显著低于其他组，CPT1表达量在150%替代组显著高于75%组。裂壶藻粉替代鱼油对大菱鲆肝脏PPARα和SREBP1的表达量没有显著影响。总之，在本实验条件下，饲料中裂壶藻粉替代150%鱼油DHA对大菱鲆仔稚鱼的成活率、生长、体成分和消化酶以及肠道组织无显著性影响。综上所述,裂壶藻可以替代大菱鲆仔稚鱼微颗粒饲料中150%鱼油DHA。

2 大菱鲆幼鱼饲料添加剂研究

2.1 茶多酚对大菱鲆幼鱼降脂作用的研究

高脂饲料以及豆油替代鱼油会引起大菱鲆鱼体脂肪异常沉积。本实验旨在探究上述饲料中添加茶多酚对缓解大菱鲆幼鱼脂肪异常沉积的影响及作用方式,寻找一种高效安全的降脂添加剂。在基础高脂饲料中添加0%、0.05%、0.2%、0.5%的茶多酚,配制出4种等氮等能(粗蛋白50%,脂肪18%)的实验饲料。每组饲料随机投喂3组实验鱼,饱食投喂,养殖周期10周。

目前,养殖实验已经结束,相关分析工作正在进行中。

2.2 饲料中添加奇力素和赛乐硒对大菱鲆幼鱼生长、体组成、免疫功能、抗氧化酶活力、肠道结构、抵御嗜水气单胞菌攻毒以及免疫相关基因表达的影响

本实验旨在探究奇力素和赛乐硒对大菱鲆存活、生长、体组成、消化酶活性、肠道结构、免疫反应及免疫相关基因的表达水平上的影响,寻找安全、无毒、无副作用的抗生素替代品。在基础饲料中添加0%、0.4%、0.8%、1.6%和3.2%的Actigen和0 mg/kg、0.15 mg/kg、0.30 mg/kg的Sel-Plex(美国奥特奇公司),共配制出10种等氮等能(粗蛋白50 %,脂肪12%)的实验饲料。每组饲料随机投喂3组实验鱼,饱食投喂,养殖周期10周。养殖实验结束后,进行嗜水气单胞菌的攻毒实验。

目前,养殖实验已经结束,相关分析工作正在进行中。

2.3 饲料中有机微量元素替代无机微量元素对大菱鲆幼鱼生长性能、组织残留、粪便元素排泄量、免疫功能、抗氧化酶活力的影响

本实验旨在探究有机微量元素替代无机微量元素对大菱鲆存活、生长、粪便排泄、体组织矿物残留等方面的影响,探究更高效更环保的矿物元素利用。有机微量元素(OTM)(铁、铜、锰、锌和硒为百乐和赛乐系列)替代基础饲料中25%、50%、75%和100%的无机微量元素。配制出6种等氮等能(粗蛋白50%,脂肪12%)的实验饲料。每组饲料随机投喂3组实验鱼,饱食投喂,养殖周期10周。

目前,养殖实验已经结束,相关分析工作正在进行中。

3 亲鱼营养研究及专用饲料开发

3.1 饲料中不同水平的花生四烯酸(ARA)对不同发育时期的半滑舌鳎亲鱼性类固醇激素合成量及合成过程的影响

本实验旨在研究饲料中不同水平的花生四烯酸(ARA)对不同发育时期的半滑舌鳎亲鱼性类固醇激素合成量及合成过程的影响。配制了3种等氮等脂(脂肪含量13%)的实验饲料,分别含有不同梯度水平的ARA:0.72%(不额外添加ARA精制油的对照组,C),5.14%(添加低水平ARA的处理组,ARA-L)及15.44%(添加高水平ARA的处理组,ARA-H)(各数值均为占总脂肪酸的比例)。每组饲料投喂3个实验桶,每桶放25尾3龄大菱鲆亲鱼(雌雄比例约为2:3)。养殖实验在室内流水系统内进行,每天饱食投喂2次,养殖周期为3个月。养殖结束后,分别取发育前期的雌鱼、发育后期的雌鱼、发育后期的雄鱼测血清雌二醇和睾酮的含量,并检测性腺中性类固醇激素合成相关蛋白的基因表达量以及脂肪酸组成。结果表明:ARA的添加降低了雌鱼性腺中雌二醇的含量,但是升高了雄鱼性腺中睾酮的含量。ARA的添加显著降低了卵巢中芳香化酶的基因表达量,但是显著提高了精巢中3β-羟化类固醇脱氢酶的基因表达量。在成熟的卵巢中,ARA-L显著降低了促卵泡激素受体、3β-羟化类固醇脱氢酶及17β-羟化类固醇脱氢酶基因表达量,但是在发育前期的卵巢中,ARA-L显著提高了促卵泡激素受体、固醇合成急性调节蛋白、胆固醇侧链裂解酶、3β-羟化类固醇脱氢酶及17β-羟化类固醇脱氢酶基因表达量。在所有类型的性腺中,17α-羟化酶受饲料ARA影响的方式跟其他性类固醇激素合成相关蛋白都不同。脂肪酸组成分析显示,ARA优先沉积在性腺中,其相对含量在成熟的精巢中最高,在未成熟的卵巢中次之,在成熟的卵巢中最低。与雌鱼相比,雄鱼性腺中含有更高的DHA含量,而肝脏中含有更低的DHA含量。总上,ARA对性类固醇激素的合成以及ARA在性腺中的沉积与半滑舌鳎亲鱼的性别以及发育阶段都有关系,饲料ARA对雄性亲鱼中的重要性高于对雌性亲鱼,在发育前期的雌性亲鱼中的重要性要高于在发育成熟的雌性亲鱼中。

3.2 饲料中花生四烯酸对大菱鲆亲鱼性类固醇激素合成的影响

为了研究饲料中不同水平的ARA对发育前期的大菱鲆亲鱼性类固醇激素合成量及合成过程的影响,配制了3种等氮等脂(脂肪含量13%)的实验饲料,分别含有不同梯度水平的ARA:0.72%(不额外添加ARA精制油的对照组),5.63%(添加低水平ARA的处理组,ARA-L)及15.03%(添加高水平ARA的处理组,ARA-H)(各数值均为占总脂肪酸的比例)。每组饲料投喂3个实验桶,每桶放25尾3龄大菱鲆亲鱼(雌雄比例约为1:1)。养殖实验在室内流水系统内进行,每天饱食投喂2次,养殖周期为5个月。养殖结束后,分别取未发育雌雄鱼测血清雌二醇和睾酮的含量并检测性腺中性类固醇激素合成相关蛋白的基因表达量。结果表明:与对照组相比,饲料中高水平的ARA显著降低了雌鱼血清中雌二醇的含量,而雄鱼血清中睾酮的含量在ARA-L组显著降低。在卵巢中,饲料中高水平的ARA显著降低了促卵泡激素受体的mRNA表达量,但是饲料中低水平的ARA显著升高了17α-羟化酶的mRNA表达量。在精巢中,饲料中低水平的ARA显著降低了固醇合成急性调节蛋白以及17α-羟化酶的mRNA表达量,但显著升高了芳香化酶的mRNA表达量。饲料中ARA提高了组织中ARA的含量,但降低了二十碳五烯酸(EPA)的含量,卵巢中的ARA累积量高于精巢。综上所述,饲料中一定量ARA的添加抑制了发育前期大菱鲆亲鱼雌二醇和睾酮的合成,在卵巢中,这种抑制作用可能是通过对促卵泡激素受体表达的抑制来实现,而在精巢中,可能是通过对固醇合成急性调节蛋白以及17α-羟化酶的抑制来实现。

(岗位科学家　麦康森)

鲆鲽类产品质量安全与加工技术研发进展

加工与质量控制岗位

1　水产品可追溯体系研究进展

1.1　编制鲆鲽类产品物理溯源技术规范 1 项

在既有工作基础上，整理鲆鲽类产品产地溯源各阶段工作模块，结合条码选择打印、二维码编制生成、鱼体挂标、信息录入以及条码扫描查询等内容，初步建立溯源技术体系的框架，编制鲆鲽类产地溯源操作规范。

本操作规范是通过标识技术将二维码标签标记在鱼体上，通过养殖、销售等环节的信息实时录入，终端二维码扫描查询，建立鲆鲽类的产地溯源技术体系，实现养殖鲆鲽类的全程可追溯。设计的二维码数字码段为符合国际 EAN/UCC 编码体系规定的 128 码，由企业代码（7～9 位）、商品代码（4 位）和产品批次代码（4～8 位）组成，根据二维码尺寸及预添加企业名称等信息，确定标签的规格为 5 cm × 1 cm，选择韧性较好的聚对苯二甲酸乙二醇酯（PET）材料和树脂碳带作为标签材料。

采用钉射嵌挂法进行加标：将胶针和钉射枪针头用 75% 酒精浸泡消毒后，使用 12 mm 胶针将标签钉射在鱼体的背部肌肉处，同时为促进伤口愈合，避免运输过程中的感染损伤，产品需养殖 1 个月后方可上市。标记过程中一次性挂标率可达 95% 以上，挂标 1 个月内脱标率低于 2%，鱼体死亡率低于 3%。标识用钉射枪、胶针、标识过程和标识后的鱼体实例如图 1 所示。

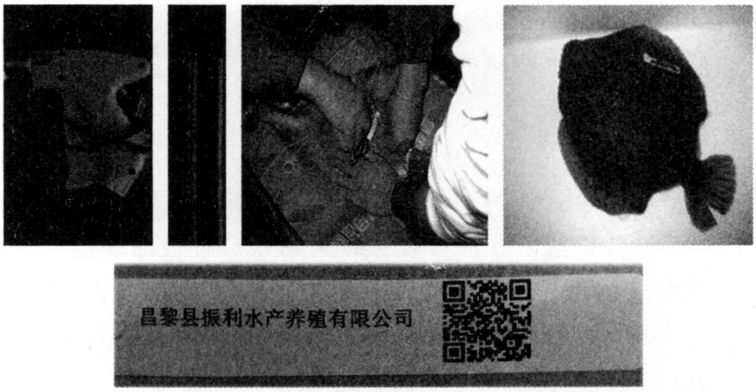

图 1　标识用钉射枪、胶针、标识过程和标识后的鱼体实例以及二维码追溯标签

追溯信息的采集包括固定信息和实时信息的采集，包括养殖企业在鲆鲽鱼类质量安全监管与追溯平台(http://www.safefood.gov.cn/quyu/hhs/)注册用户、录入企业基本信息、产品信息、苗种信息、饵料信息、病害及用药情况、养殖过程、销售过程，通过智能手机扫描二维码可实现鲆鲽鱼类的全程可追溯。

1.2 完善二维码追溯标签内容、优化追溯展示界面

二维码标签规格为 5 cm × 1 cm，在标签上增加生产企业名称，以提供直观印象，结合后期二维码扫描追溯，保证产品信息的完整查询。已在秦皇岛市昌黎县振利水产养殖有限公司对 1 000 尾大菱鲆进行二维码追溯标签的挂标试点，并且，挂标产品已在秦皇岛市场上进行销售。通过与山东省标准化研究院合作，对二维码扫描展示信息界面进行了优化，从原有的网页式转变为现有的图片式，提供更友善的展示界面，使追溯信息更加明确易于读取。

2 水产品质量安全管理与控制技术的研究

2.1 鲆鲽类嘌呤化合物的安全性评价

确立了反相高效液相色谱测定鲆鲽类中腺嘌呤、鸟嘌呤、次黄嘌呤和黄嘌呤含量的方法，对大菱鲆不同组织的嘌呤含量进行了测定，并将此方法推广到其他水产品中进行了测定。同时，探究了嘌呤化合物在水产品贮藏和加工过程中的变化规律，以期为人们的健康饮食提供科学的指导基础。研究发现：水煮、超声和微波等加工方式可以有效辅助水产品中嘌呤溶出进而消减水产品中嘌呤含量，微波脱除率 > 水煮脱除率 > 超声脱除率，采用上述技术对水产品中的嘌呤进行消减，10 min 时嘌呤脱除率分别为 80%、60%、25%。

通过建立持续、稳定的高尿酸血症鹌鹑动物模型，长期摄入腺嘌呤、鸟嘌呤、次黄嘌呤和黄嘌呤，发现高尿酸血症和正常鹌鹑的血尿酸水平均会下降。摄入的嘌呤使个体腺苷脱氨酶酶活力降低，导致外源性嘌呤生成的次黄嘌呤含量下降，同时，生物体内存在大量内源性的腺嘌呤及其衍生物，肝脏腺苷脱氨酶酶活力及 mRNA 表达水平的降低，也将显著影响内源性腺嘌呤的转化。因此，尿酸前体物质嘌呤的长期摄入反而使个体血尿酸水平下降，明确了鲆鲽类嘌呤化合物并不是造成痛风的主要原因。

2.2 X 射线检测海水鱼片中鱼刺

通过 X 光机设备参数优化、鱼刺分布规律总结及冷冻结合 X 光机技术 3 个方面对 X 射线检测鱼刺效果进行研究，确定了 5 种鱼对应的 X 射线最优检测参数(表 1)。发现鱼片在冷冻状态下鱼刺的检出率均高于解冻状态下鱼刺检出率，且冷冻状态下人工读 X 射线图判断鱼刺更接近于实际情况。进一步提高了残留鱼刺检出率，一定程度上解决了海水冷冻鱼片产品鱼刺残留问题。

结果表明大部分海产鱼片产品的最优检测参数为：电压范围 30 kV ～ 45 kV、电流范围 3 000 μA ～ 6 000 μA、亮度范围 4 ～ 6（感官指数 0 ～ 10）。

表 1　不同鱼种对应的 X 射线最优检测参数

参数	竹荚鱼	狭鳕	鲐鱼	好吉鱼	红鱼	秋鲑
电压 / kV	45	30 ～ 40	30 ～ 35	30 ～ 40	35 ～ 40	30 ～ 40
电流 / μA	5 500 ～ 6 000	4 200 ～ 5 200	5 000 ～ 6 000	4 500 ～ 5 000	3 200 ～ 4 700	4 000 ～ 5 500
亮度	4 ～ 5	4	4 ～ 5	4	4 ～ 6	4

2.3　氧化石墨烯在水中对氟喹诺酮类药物的吸附和解析特性的研究

探究影响氧化石墨烯（GO）在水中对氟喹诺酮类药物的吸附和解析特性的影响因素，并用高效液相色谱进行药物浓度的检测研究。结果表明，GO 在 pH 3 ～ 11 范围内都具有较好的吸附特性，且 0.01 mol/L 的 NaOH 具有较好的洗脱效果，洗脱率达到 90.5% ～ 108.5%。大菱鲆鱼蛋白对 GO 吸附药物的能力影响较大，但是可以通过增大体系中 GO 的含量来吸附药物。接下来将进一步探究 GO 应用于水产品中实际样品药物残留检测中前处理回收率的检测。

2.4　组织完成企业内部禁用药物自查示范

协助日照国际海洋城淘宝渔业专业合作社建立企业内部的“养殖大菱鲆中禁用药物残留日常检测制度”。培训快速检测技术工作人员 1 名，完成 4 次企业内部自检，均未检出硝基呋喃类代谢物。

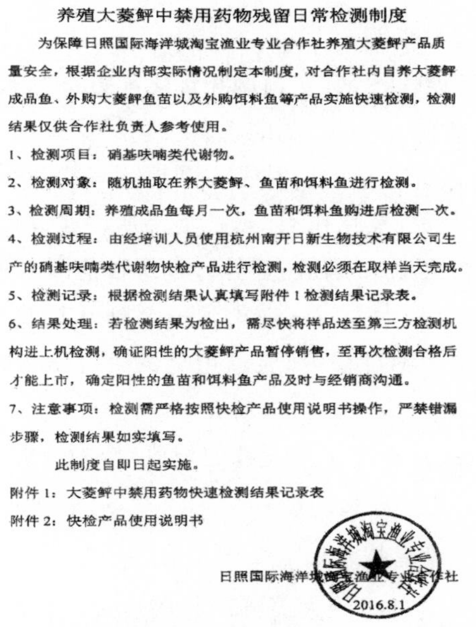
养殖大菱鲆中禁用药物残留日常检测制度

为保障日照国际海洋城淘宝渔业专业合作社养殖大菱鲆产品质量安全，根据企业内部实际情况制定本制度，对合作社内自养大菱鲆成品鱼、外购大菱鲆鱼苗以及外购饵料鱼等产品实施快速检测，检测结果仅供合作社负责人参考使用。

1、检测项目：硝基呋喃类代谢物。

2、检测对象：随机抽取在养大菱鲆、鱼苗和饵料鱼进行检测。

3、检测周期：养殖成品鱼每月一次，鱼苗和饵料鱼购进后检测一次。

4、检测过程：由经培训人员使用杭州南开日新生物技术有限公司生产的硝基呋喃类代谢物快检产品进行检测，检测必须在取样当天完成。

5、检测记录：根据检测结果认真填写附件 1 检测结果记录表。

6、结果处理：若检测结果为检出，需尽快将样品送至第三方检测机构进上机检测，确证阳性的大菱鲆产品暂停销售，至再次检测合格后才能上市，确定阳性的鱼苗和饵料鱼产品及时与经销商沟通。

7、注意事项：检测需严格按照快检产品使用说明书操作，严禁错漏步骤，检测结果如实填写。

此制度自即日起实施。

附件 1：大菱鲆中禁用药物快速检测结果记录表

附件 2：快检产品使用说明书

日照国际海洋城淘宝渔业专业合作社

2016.8.1

图 2　养殖大菱鲆中禁用药物残留日常检测制度

3 鲆鲽鱼类深加工及高值化综合利用技术研究

3.1 开发大菱鲆“一鱼多吃”产品

将大菱鲆做成鱼头、鱼块、鱼骨、鱼皮、鱼柳、鱼片等适于电子商务的“一鱼多吃”冰鲜产品(图3),使每条鱼的价值大大增加。“一鱼多吃”产品不仅可以改善目前大菱鲆价格持续低迷的现状,也为鲆鲽类深加工产品的进一步发展提供了新的思路。2016年已在两家酒店完成了“一鱼多吃”产品的食用推广,得到酒店方面和消费者的一致好评,今后会进一步加强此类产品的研发工作。

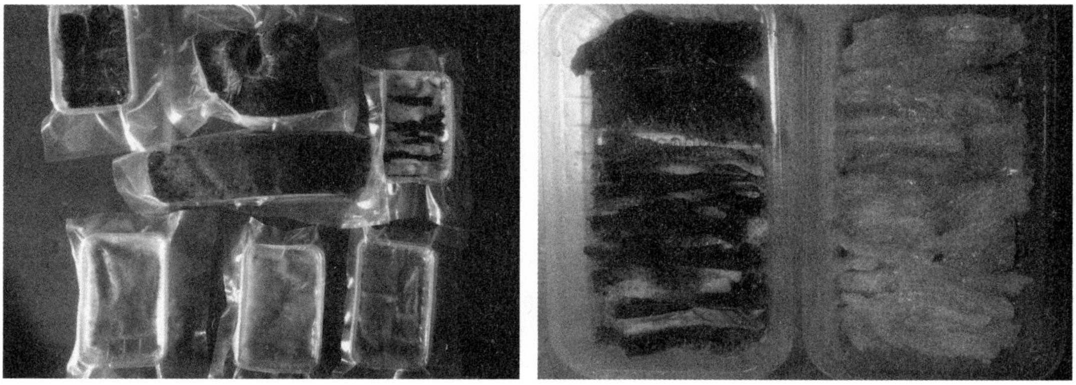

图3 “一鱼多吃”产品图

3.2 鲆鲽类加工产品的品质保持、风味调制及包装流通技术研究

3.2.1 大菱鲆熏制预调味技术研究

研究了大菱鲆的熏制预调味技术,在预调味优化配方下经熏制后鱼肉品质完整结实、色泽金黄明显、具有熏制特有香味,有着广阔的消费市场。烟熏是一种特殊的传统食品加工技术,在食品干燥而减少其中水分含量的同时吸收烟熏特有的芳香气味和风味烟气,赋予制品特殊的熏香味,并有防腐杀菌、抗氧化作用,延长货架期。通过探讨大菱鲆的熏制技术,可以扩大海水鱼的加工利用渠道,提高鱼肉加工利润,延缓目前海水鱼价格的进一步下滑。

该项技术的具体工艺流程:大菱鲆→前处理→剖割→盐渍→脱盐→调味→干燥→熏制→冷却→保存。在调味优化关键过程中,通过正交试验对调味影响最大的食盐、白糖、料酒和酱油比例进行优化。

通过对大菱鲆的特色熏制,提高了大菱鲆的价值,作为一种休闲即食鱼片,营养丰富、风味独特、食用方便,值得推广和产业化生产。

3.2.2 研究大菱鲆产品包装流通技术

与山东潍坊龙威实业有限公司就大菱鲆产品的包装制作和包装方法进行了交流,探讨了礼品盒装大菱鲆分割产品的加工方法和销售前景，先后考察了莱西信智、光生源等几家包装材料企业，并对进口牛卡纸箱强度等进行了实地测试，后期针对大菱鲆加工包装材料进行设计和交流，经过筛选和调研，初步确定了以进口牛卡纸箱为主体的外包装（图 4）。

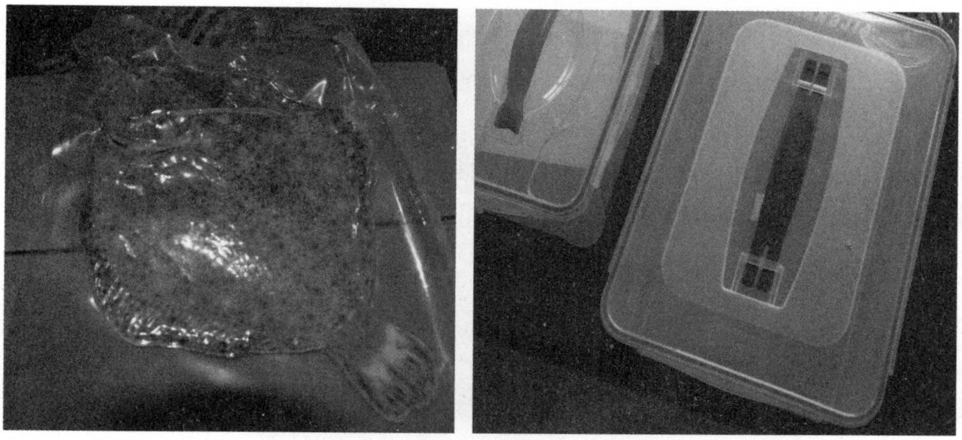

图 4　大菱鲆产品包装及流通材料

3.3　指导建设以鲆鲽类为主的海水鱼加工生产线

指导山东潍坊龙威公司进行海水鱼加工生产线的建设（图 5），进行了鱼糜生产线的试制和鱼丸、鱼糕、鱼肠等鱼糜系列产品的生产，制定了鱼糜及制品生产工艺流程和操作要点。

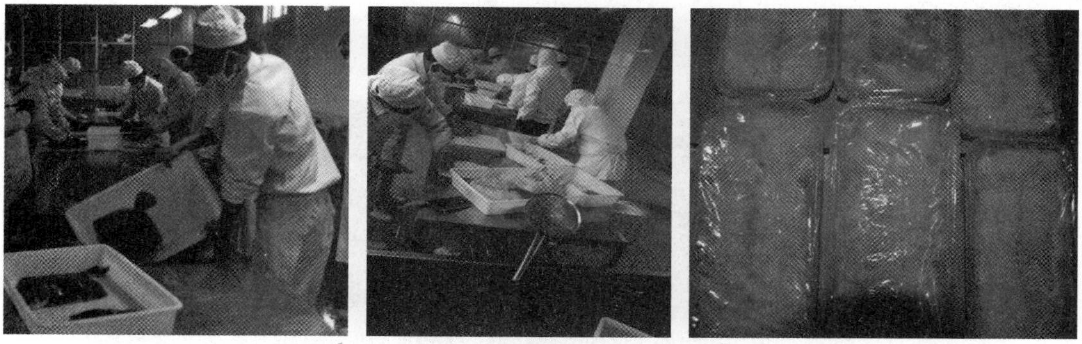

图 5　大菱鲆加工生产线

3.4　研发鲆鲽类成鱼加工产品工艺流程

以大菱鲆为主要原料，通过预处理、漂洗、去腥、软化、烘干、油炸、调味等工艺制成鱼排，研究了食盐、白糖和味精等主要影响因素的添加量对鱼排品质的影响，结果表明：使用 4% 味精、9% 食盐和 4% 白糖作为主要调味料调制的鱼排品质最佳，油炸鱼排在 110 ℃～150 ℃时颜色、风味和组织等较理想，质量较为稳定。通过科学利用鱼类加工下脚料，达到了提高附加值、降低主导产品成本的目的。加工流程：原料→处理→漂洗→鱼排→去腥→软化→烘干→油炸→调味→包装→成品。通过改变工艺参数、调整配方等措施生产调味鱼排产品,具有一定的实用价值。

3.5 多宝鱼系列产品的初步探索

3.5.1 腊制多宝鱼

腌制腊鱼是我国传统水产加工食品,其风味独特,耐贮藏,在我国南方的饮食文化中有着重要地位。在腌腊鱼的加工过程中,由于微生物和原料鱼中酶类的作用,发生硝酸盐的还原、蛋白质的水解、氨基酸的脱羧和脱氨、脂质的水解与氧化等生化变化,使鱼体具有独特的风味和营养。多宝鱼营养价值极高,经过腌制晒干后的多宝鱼制品,储藏期将大大增加,本身独特的风味加上腌制风干后的特殊香味成分,将成为一种新的半成品加工方式。

3.5.2 半干多宝鱼

从传统意义上来说,半干的腌鱼也属于干货制品,主要是因为腌制的时间较短,水分并没有被完全的晾干,或者是为了保持一定的柔软度,而致使腌鱼处于半干的状态。半干的鱼与完全风干的鱼一样,都会由于酶类等物质的作用,产生不同于普通做法的特殊风味,半干制品更加方便烹饪加工,对于快节奏生活方式的人群来说具有一定的市场价值。

3.6 鲆鲽类模拟空运无水保活技术研究

针对鲆鲽类活鱼运输的现状,研究了一种无水保活运输技术,采用模拟空运状态对大菱鲆、牙鲆的运输情况进行研究分析,并改进无水保活工艺(图6),为今后无水保活运输技术提供了相应的技术支持。

图6 模拟空运无水保活技术

采用无水保活工艺:鲆鲽类→挑选→停食→暂养→降温→起鱼→装箱→无水运输→补冰→实验→验活。

选用购自烟台开发区天源水产有限公司的大菱鲆和牙鲆,选择有活力、无损伤、体重为1.5 kg～2.3 kg的大菱鲆、牙鲆,先行停食1 d,活体运输前夜置于养殖池中进行梯度降温,降温速度 1 ℃/h～2 ℃/h。第二天清晨起鱼加冰装箱，每种鱼 20 箱，湿物覆盖鱼体，模拟空运条件进行，运输过程 4 ℃～8 ℃。2 h 后到达山东潍坊龙威实业有限公司加工车间加冰，以保持温度 4 ℃以下，平稳放置，隔 2 h 观察鱼体状态，记录死亡时间和数量。将大菱鲆、牙鲆放置于海水中，通过观察经多次刺激后鳃盖和鱼体的反应来确定是否是活体，当死亡率达 50%时结束试验。结果显示，大菱鲆存活达到 1.5 d，牙鲆当天就全部死亡。

4　鲆鲽类化学危害物快速检测技术研究进展

针对水产品中恩诺沙星的残留，建立了一种水产品中恩诺沙星免疫层析毛细管检测技术（CICA），并对其检测前处理过程进行初步探究，以适应快速现场检测的要求。

以直接竞争的实验原理，通过优化检测区抗原、质控区二抗、金标一抗检测浓度初步建立免疫层析毛细管检测方法如图 7 所示，并进一步确定其检测灵敏度、特异性、重复性等及前处理的优化操作。该方法检测恩诺沙星的视觉检测限为 5 ng/mL，半定量检测限为 1.29 ng/mL。在大菱鲆、鲅鱼、舌鳎、南美白对虾阴性样品中的添加回收率在 75%～130%。除对环丙沙星外，对其他相似物的特异性较好。环丙沙星的视觉检测限为 10 ng/mL，半定量检测限为 4.37 ng/mL，满足国家标准检测二者低于 100 ng/mL 的限量要求。同时，针对 CICA 的前处理方法进行初步探究，最终确定采用常温条件下磺基水杨酸的方法，检测用时不超过 25 min。恩诺沙星残留免疫层析毛细管检测技术具有简单、快速、重复性好的优势，为食品药物残留检测领域提供了检测思路新方法。

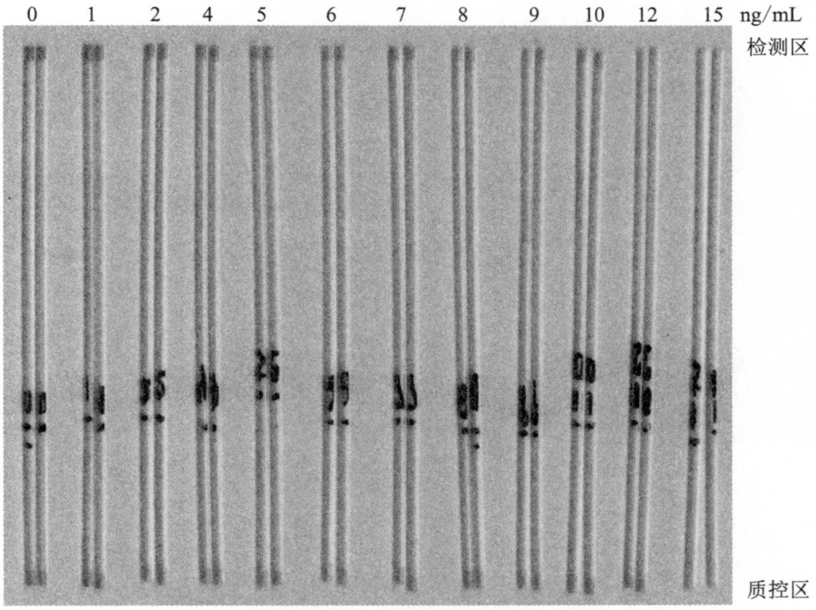

图 7　免疫层析毛细管检测方法展示图

5 基础性科技资料的整理与分析

2016年度采集鲆鲽类加工行业信息数据、风险因子信息数据统计均达到50条以上。加工信息数据包括加工厂占地面积、冷冻及恒温库容量、年度加工产量、出口创汇、利润、成本、产值、不同鱼类的加工量、鲜鱼及加工制品销售市场占比、加工产品种类等。风险因子信息数据包括企业目前厂区建设布局情况,加工设备和技术的更新情况,管理方法存在的相关问题,技术人员的培训,加工、贮藏和运输中微生物控制,流通和销售渠道等。

(岗位科学家　林　洪)

鲆鲽类高效养殖模式技术研发进展

高效养殖模式岗位

从全球水产养殖发展现状来看，尽管中国的渔业发展速度较快，但从技术、经济上，从产业化和工业水平上，等等，与发达国家相比，仍存在较大的差距。如何利用经济规律，尊重自然发展来建立完整科学的养殖模式，将水产养殖推向可持续性发展方向，合理利用养殖水域和自然资源，保护养殖水域生态平衡，成为水产养殖业迫切需要解决的问题。为此，2016 年，鲆鲽类高效养殖模式岗位围绕工厂化健康养殖技术、健康养殖技术规范、鲆鲽类健康指标体系构建等方面开展了系统研究和养殖示范，并取得了一定的研究进展。

1　鲆鲽类工厂化健康养殖技术研究

在国家供给侧结构调整和社会消费需求变化背景下，针对鲆鲽类养殖产业供给增长过快和大菱鲆突发事件冲击等因素的影响导致的鲆鲽类销售价格持续走低，大批养殖企业和养殖户亏损的局面，高效养殖模式岗位根据体系鲆鲽类养殖提质增效关键技术研发与集成示范任务，以规范养殖工艺、保障产品优质安全为重点，开展了大菱鲆工厂化健康养殖技术方面的研究，分别在工厂化养殖条件下养殖密度对大菱鲆生长、固有免疫与氧化应激水平的影响研究和大菱鲆应对关键水质因子胁迫的生理相应研究等关键技术环节进行了一系列较深入的研究。

1.1　工厂化循环水养殖条件下，养殖密度对大菱鲆生长、固有免疫与氧化应激水平的影响研究

通过研究养殖密度对不同大小规格的大菱鲆影响，从生长和福利水平评估大菱鲆适宜的养殖密度，为大菱鲆工厂化循环水养殖提供参考依据。

研究结果显示，不同养殖密度对总氨氮（TAN）和化学耗氧量（COD）的影响较大，对 NO_2^--N 和 PO_4^{3-}-P 以及细菌总数没有影响。在三个养殖阶段后期，高密度组 TAN 和 COD 含量显著高于低密度组。高密度养殖条件下，大菱鲆生长受到抑制，并产生应激反应，伴随着皮质醇、葡萄糖和乳酸等生理学参数的升高，显著不利于大菱鲆鱼类福利状况。研究还表明不同大小规格的大菱鲆对密度的适应不同。在本研究条件下，70 g 左右的大菱鲆密度不可超过 14.3 kg/m^2，180 g 左右的大菱鲆密度不可超过 25.7 kg/m^2，510 g 左右的大菱鲆密

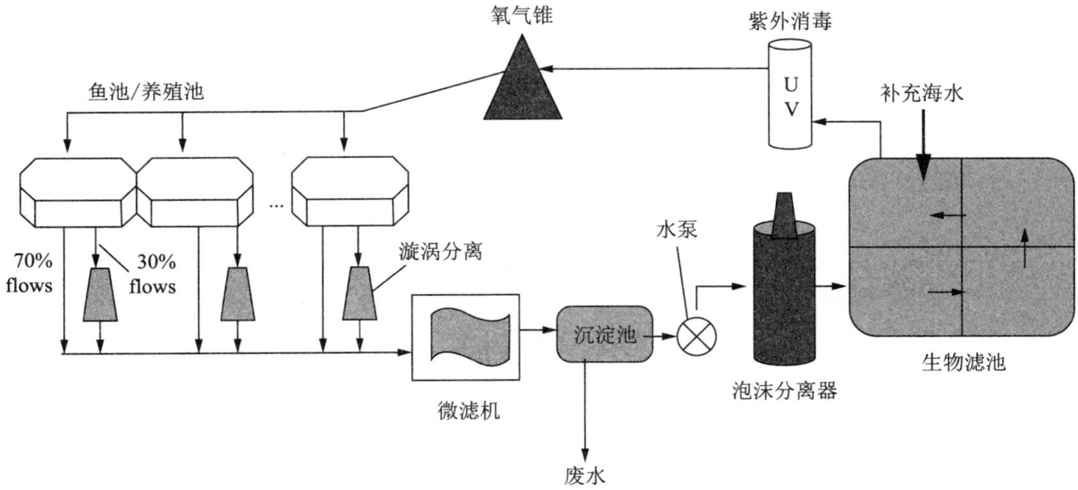

图 1 循环水系统示意图

水流方向为:养殖池—旋涡分离器—微滤机—沉淀池
—水泵—泡沫分离器—生物滤池—紫外杀菌设备—纯氧添加装置—养殖池

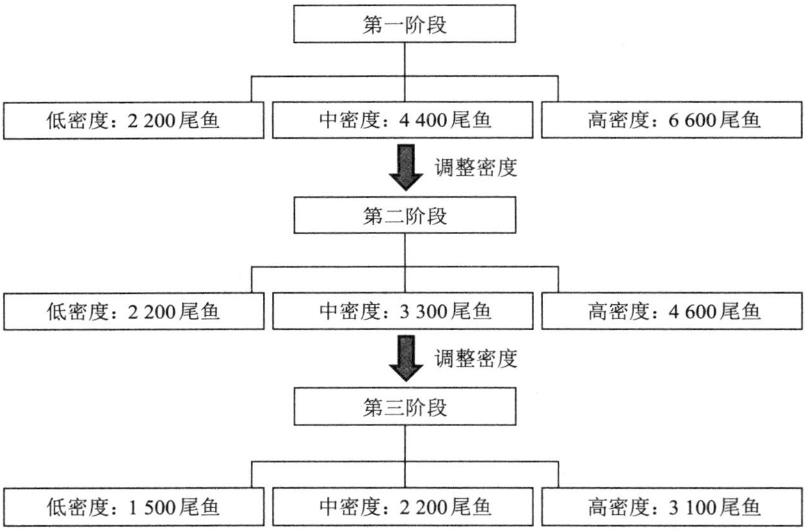

图 2 实验养殖密度设置

度不可超过 52. 3 kg/m^2。

1. 2 大菱鲆应对关键水质因子胁迫的生理相应研究

1.2.1 亚硝酸盐对大菱鲆氧化应激、凋亡和免疫能力的影响研究

在循环水养殖系统中,由于过高的养殖密度、过量的投喂、消化细菌和反硝化细菌的失衡等因素导致水体中亚硝酸盐含量的升高。长期暴露于高浓度的亚硝酸盐会导致大菱鲆生长减慢甚至死亡。因此本研究从生理应激、抗氧化状态、细胞凋亡以及免疫反应等方面评估

亚硝酸盐胁迫对大菱鲆的影响。

该研究结果显示，血液中亚硝酸盐胁迫改变离子平衡，促使高铁血红蛋白（MetHb）产生，导致缺氧应激反应，伴随着皮质醇、葡萄糖（Glu）、热休克蛋白（HSP）、谷草转氨酶（GOT）、谷丙转氨酶（GPT）、补体C3、补体C4和碱性磷酸酶（ALP）水平的升高，免疫参数溶菌酶（LZM）和免疫球蛋白（IgM）的下降。在鳃组织中，亚硝酸盐引起氧化应激，导致抗氧化物质（SOD、CAT、GPx、GSH等）含量下降和丙二醛（MDA）含量升高，诱导凋亡发生，使凋亡相关基因（JUK1、P53、caspase-3、caspase-7和caspase-9）表达上调，引起细胞内应激响应、炎症反应和免疫毒性，并导致HSP70、HSP90、MT、TNF-1α、TLR-3和IL-1β基因表达上调，LZM和IGF-1基因表达下调。

1.2.2 氨氮对大菱鲆应激反应和免疫能力的影响

研究评估了氨氮暴露对大菱鲆HPI轴、GH/IGF-1轴、抗氧化应激和免疫功能的影响，旨在为大菱鲆及其他海水鱼养殖中氨氮胁迫的防控提供科学依据。结果显示，高浓度的TAN引起大菱鲆激素水平的改变，激活HPI轴，抑制GH/IGF-1轴。随着暴露时间和TAN浓度的增加，大菱鲆肝组织中发生氧化应激反应，引起SOD和过氧化氢酶（CAT）活力的升高、GSH含量的消耗。

2 鲆鲽类健康养殖技术规范研究

为了完成体系鲆鲽类养殖提质增效关键技术研发与集成示范任务，规范养殖工艺、保障产品优质安全，高效养殖模式岗位以鱼类健康养殖理念为指导，根据我国鲆鲽类养殖的技术需求和鲆鲽类主产区气候、环境、养殖模式等差异化因素，按照辽宁、天津-河北、山东-江苏三大区域板块，根据工厂化流水养殖模式与区域特点，结合鲆鲽类体系“十二五”的研究基础，初步制定了“大菱鲆（区域性）工厂化流水健康养殖技术标准”（初稿）3项。

3 鲆鲽类健康指标体系构建

开展了鲆鲽鱼类不同生理状态同表观性状、血液理化因子、内分泌调控和功能基因等相关指标的筛选，初步构建了工厂化养殖模式下大菱鲆生理生化与免疫指标体系。下一步准备通过长期实地跟踪监测大菱鲆养殖实际状况，根据养殖鱼类生长发育状态，结合临床症状等方面研究，利用经典生理生态学方法和现代生物多组学分析方法定期监测其应激、免疫、代谢等生理指标，通过大数据分析和试验验证，对可准确反映大菱鲆健康状况的生理生态学指标因子进行筛选评定。建立基于实际养殖生产条件下鲆鲽鱼类“体检机制”和快速评价方法。

表 1　鱼类健康指标体系结构

序号	健康指标分类	健康指标范围	敏感影响因子
一、养殖鱼类表观指标			
1	生长速率		养殖密度、水质环境等
2	代谢速率		养殖密度、水质环境等
3	鱼类行为		DO、养殖密度、pH、氨氮浓度等
4	体表黏液		各种胁迫应急
二、养殖鱼类血液指标			
1	红细胞数量		DO
2	白细胞数量		水温、氨氮浓度、pH、养殖密度等
3	血红蛋白含量		
4	血浆中葡萄糖和乳酸水平		
5	血清中的超氧化物歧化酶(SOD)		
6	碱性磷酸酶		
7	溶菌酶		
8	转氨酶		
9	细胞脆性		
10	细胞直径		
三、养殖鱼类神经中枢激素			
1	血浆中皮质醇		养殖密度等各种胁迫应急
2	甲状腺激素		
3	生长激素		
4	性腺激素等和大脑中 5-羟色胺		
5	5-羟吲哚乙酸		
四、生物标志物(基因)			
1	HSP 家族		
2	AMPK 基因		

4　鲆鲽工厂化健康养殖示范点建设

分别在山东文登海合水产育苗有限公司和江苏连云港际洲水产有限公司建立鲆鲽类工厂化健康养殖示范点 2 个，其中文登海合示范面积 1.5 万平方米，养殖大菱鲆 60 万尾，连云港际洲示范面积 0.45 万平方米，养殖大菱鲆 15 万尾。推广与实施工厂化健康养殖技术标准，用以精准指导和规范鲆鲽类各主产区的养殖生产，构建企业标准化养殖生产技术体系，以期减少病害发生，杜绝违禁药物滥用，确保养殖产品品质与安全，实现提质增效。

5　年度进展小结

（1）通过研究养殖密度对不同大小规格的大菱鲆影响，从生长和福利水平评估大菱鲆适宜的养殖密度，70 g 左右的大菱鲆密度不可超过 14.3 kg/m^2，180 g 左右的大菱鲆密度不可超过 25.7 kg/m^2，510 g 左右的大菱鲆密度不可超过 52.3 kg/m^2，为大菱鲆工厂化循环水养殖提供参考依据。

（2）从生理应激、抗氧化状态、细胞凋亡以及免疫反应等方面评估亚硝酸盐胁迫对大菱鲆的影响。结果显示：血液中亚硝酸盐胁迫改变离子平衡，促使高铁血红蛋白产生，导致缺氧应激反应；免疫参数溶菌酶和免疫球蛋白下降；在鳃组织中抗氧化物质含量下降和丙二醛含量升高；诱导凋亡发生，使凋亡相关基因表达上调；引起细胞内应激响应、炎症反应和免疫毒性，并导致相关基因表达发生变化。

（3）评估了氨氮暴露对大菱鲆 HPI 轴、GH/IGF-1 轴、抗氧化应激和免疫功能的影响，结果显示，高浓度的氨氮引起大菱鲆激素水平的改变，激活 HPI 轴，抑制 GH/IGF-1 轴。随着暴露时间和氨氮浓度的增加，大菱鲆肝组织中发生氧化应激反应，引起超氧化物歧化酶和过氧化氢酶活力的升高、GSH 含量的消耗。研究为大菱鲆及其他海水鱼养殖中氨氮胁迫的防控提供科学依据。

（4）根据我国鲆鲽类养殖的技术需求和鲆鲽类主产区气候、环境、养殖模式等差异化因素，按照辽宁、天津－河北、山东－江苏三大区域板块，根据工厂化流水养殖模式与区域特点，结合鲆鲽类体系“十二五”的研究基础，初步制定了“大菱鲆（区域性）工厂化流水健康养殖技术标准”（初稿）3 项。

（5）开展了鲆鲽鱼类不同生理状态同表观性状、血液理化因子、内分泌调控和功能基因等相关指标的筛选，初步构建了工厂化养殖模式下大菱鲆生理生化与免疫指标体系。

（6）在山东文登和江苏连云港共建立鲆鲽类工厂化健康养殖示范点 2 个，示范面积 1.95 万平方米，养殖大菱鲆 75 万尾，推广与实施工厂化健康养殖技术标准，用以精准指导和规范鲆鲽类各主产区的养殖生产，构建企业标准化养殖生产技术体系。

（岗位科学家　黄　滨）

第二篇

鲆鲽类主产区调研报告

天津综合试验站产区调研报告

1　示范县(市、区)鲆鲽类养殖现状

本综合试验站下设五个示范县(市、区),分别为:天津市塘沽区、天津市大港区、天津市汉沽区、浙江省舟山市普陀区、温州市苍南县。其育苗、养殖品种、产量及规模见附表1

1.1　育苗面积及苗种产量

1.1.1　育苗面积

五个示范县育苗总面积为28 500 m^2,其中大港区1 200 m^2、汉沽区27 300 m^2。按品种分:大菱鲆育苗面积11 700 m^2、半滑舌鳎10 800 m^2、牙鲆6 000 m^2。

1.1.2　苗种年产量

五个示范县共计21户育苗厂家,总计育苗2 830万尾,其中:大菱鲆1 250万尾、半滑舌鳎1 180万尾、牙鲆400万尾。各县育苗情况如下:

大港区:2户育苗企业,天津海升水产养殖有限公司、天津立达海水资源开发有限公司,生产半滑舌鳎苗种180万尾,用于本区养殖。

汉沽区:19户育苗厂家,生产牙鲆苗种约400万尾,生产大菱鲆苗种1 250万尾,生产半滑舌鳎苗种1 000万尾,用于天津地区养殖及供应山东、河北、辽宁。

1.2　养殖面积及年产量、销售量、年末库存量

1.2.1　工厂化养殖

养殖方式有工厂化循环水养殖、工厂化非循环水养殖,养殖企业共有18家,工厂化养殖面积110 080 m^2,年总生产量为1 364. 68 t,销售量为744. 06 t,年末库存量为620. 02 t。其中:

塘沽区:1户,养殖面积35 000 m^2,养殖半滑舌鳎35 000 m^2,年产量353. 78 t,销售151. 49 t,年末库存202. 29 t。

大港区:3户,养殖面积9 880 m^2,半滑舌鳎9 880 m^2,年产量124. 96 t,销售84. 96 t,年末库存40 t。

汉沽区:13户,养殖面积60 200 m^2大菱鲆23 700 m^2,年产量389. 78 t,销售32. 03 t,年末库存357. 75 t;牙鲆9 100 m^2,年产量127. 48 t,销售127. 48 t,年末库存0 t;半滑舌鳎

27 400 m^2,年产量 248.68 t,销售 228.1 t,年末库存 20.58 t。

苍南县:仅 1 户,半滑舌鳎 5 000 m^2,年产量 120 t,销售 120 t,年末库存 0 t。

1.2.2　池塘养殖

只有浙江省舟山市普陀区采用池塘养殖的方式,种类为牙鲆,采用池塘单养池塘或者牙鲆与三疣梭子蟹池塘混养,养殖户 1 户,养殖面积 20 亩,年总生产量为 5 t,销售量为 5 t,年末库存量为 0 t。

1.3　品种构成

品种养殖面积及产量占示范县养殖总面积和总产量的比例见附表 2。

统计五个示范县鲆鲽类养殖面积调查结果,各品种构成如下:

工厂化育苗总面积为 28 500 m^2。其中牙鲆为 6 000 m^2,占总育苗面积的 21.05%;半滑舌鳎为 10 800 m^2,占总面积的 37.90%;大菱鲆为 11 700 m^2,占总面积的 41.05%。

工厂化育苗总出苗量为 2 830 万尾。其中牙鲆为 400 万尾,占总出苗量的 14.13%;半滑舌鳎为 1 180 万尾,占总出苗量的 41.70%;大菱鲆为 1 250 万尾,占总出苗量的 44.17%。

工厂化养殖总面积为 110 080 m^2。其中牙鲆为 9 100 m^2,占总养殖面积的 8.27%;半滑舌鳎为 77 280 m^2,占总养殖面积的 70.20%;大菱鲆为 23 700 m^2,占总养殖面积的 21.53%。

工厂化养殖总产量为 1 310.94 t。其中牙鲆为 185.19 t,占总养殖产量的 14.13%;半滑舌鳎为 738.32 t,占总养殖产量的 56.32%;大菱鲆为 387.43 t,占总养殖产量的 29.55%。

池塘养殖总面积为 20 亩,全部为浙江舟山本地牙鲆。

池塘养殖总产量为 5 t,全部为浙江舟山本地牙鲆。

从以上统计可以看出,在五个示范县内,半滑舌鳎、牙鲆、大菱鲆三个品种养殖面积和产量都占绝对优势。

2　示范县(市、区)科研开展情况

2.1　科研课题情况

试验站依托单位天津渤海水产研究所积极申请鲆鲽类产业相关项目,做好产业技术集成与示范,通过地方体系与国家体系对接,更好地完成产业体系的示范工作。以天津市鲆鲽类产业技术体系技术集成与示范为科技成果的“鲆鲽类高效生态养殖技术的集成与示范项目”,来源为国家农业科经成果转化资金,项目在 2016 年通过了验收,在天津市滨海新区鲆鲽类养殖示范基地进行转化及产业化推进,通过研究、集成、组装、配套等技术措施,重点完成海水工厂化养殖优质高产、模式升级、设施优化、产品质量安全保障等技术体系建设任务。通过循环水养殖系统与装备的研究与示范,优化集成循环水处理系统自动运行、故障报警、数据远程传输和水质在线监控技术。

2.2　发表专利和论文情况

1. 海水工厂化养殖水质在线监测预警系统

软件著作权，2016SR131956；天津渤海水产研究所。

2. 一种用于鲆鱼和舌鳎类亲鱼精卵采集的辅助装置

实用新型，ZL2015210707004；张博、刘克奉、贾磊、郑德斌、肖广侠。

3. 一种鲆鲽类亲鱼多层保活运输专用袋

实用新型，ZL2015210703978；张博、宋文平、刘克奉、贾磊、郑德斌、肖广侠。

4. 一种舌鳎类成鱼高效保活无水运输专用袋

实用新型，ZL201521063475.1；张博、宋文平、刘克奉、贾磊、郑德斌、肖广侠。

5. 一种用于鲆鲽活鱼血液采集的辅助装置

实用新型，ZL201521082046.9；张博、贾磊、刘克奉、郑德斌、马超、肖广侠。

6. 一种鲆鲽鱼类装袋辅助装置

实用新型，ZL201521117634.1；张博、贾磊、郑德斌、刘克奉、马超、肖广侠。

7. 悬浮物对褐牙鲆肌肉抗氧化酶活性及相关基因表达的影响

海洋科学进展，2016年34卷4期，542-552；肖广侠、徐文远、贾磊、郑德斌、张博、马超。

8. 汉沽养殖区海水重金属含量分布特征及潜在生态危险评价

安徽农业科学，2016年44卷20期，66-68；韩现芹、陈春秀、贾磊。

3　鲆鲽类产业发展中存在的问题

3.1　养殖品种与模式

海水工厂化养殖水平参差不齐，多数养殖场选择的品种（大菱鲆、牙鲆等）"易养难卖"价格低，养殖效益低，甚至亏损，当前唯一具有较高效益的工厂化养殖品种半滑舌鳎因发病率达到50%以上，部分企业无法继续养殖。导致多数海水工厂化养殖场车间闲置，海水养殖总体技术水平急需提升。

3.2　产业结构尚需优化，水产产业经济研究几乎空白

目前存在水产品加工滞后，产业链条短，产品附加值低；渔业一、二、三产业比例不合理，融合发展程度低；品种少，更新换代慢，品种结构、养殖模式单一等问题。渔业的产业结构调整、发展动态、国内市场变化、趋势预测、宏观调控决策缺乏系统研究。

3.3　养殖设施与装备技术

封闭式循环水养殖设施的资金投入问题。封闭式循环水养殖可以节水，节省能源，并提高养殖产量，可是，前期需要投入大量资金用于完善设施设备，并且对企业员工的知识

水平和操作能力也要求很高,完全由企业出资进行设备设施改造有一定困难,需要政府出台支持政策和配套融资措施。

3.4 鲆鲽类病害防控技术

各种养殖品种病害依然是影响养殖成活率的主要因素,目前大菱鲆、牙鲆的病害还没彻底解决,半滑舌鳎新的病害又不断出现,严重影响养殖企业的养殖热情。

3.5 鲆鲽类营养与饲料

养殖企业普遍认为进口饲料比国产饲料营养价值高,鲆鲽鱼类生长良好,并且出口药检没问题,但是进口饲料价格昂贵,成本难以承受。国产人工配合饲料各厂家质量差异大,同一厂家质量也不稳定,投喂后生长效果不好,担心饲料或饲料原料被污染导致药检出问题。另外很多企业投喂冰鲜小杂鱼,虽然养殖鱼生长效果表现良好,但是饲料系数高,饲料成本随着有限的渔业资源的过度开发越来越高,而且对养殖水质的污染严重。

4 当地政府对产业发展的扶持政策

本试验站与示范区多家海水工厂养殖企业签订科技合作协议,为养殖企业提供科技服务,协助养殖企业完成鲆鲽类市级、国家级良种场建设工作,协助养殖企业完成天津海水工厂化循环水养殖车间建设项目建设工作。积极帮助鲆鲽类养殖企业多渠道争取资金支持,发展鲆鲽类产业。同时通过合作关系,能够更好地把体系成果应用到本区域示范企业中去。

5 产业发展建议

5.1 推进养殖技术与模式升级

通过引进适合我市海水工厂化养殖的高档海水养殖新品种,优质、高效、环保、低成本饲料配制,免疫增强剂研制开发,病害综合防治技术研发,苗种培育技术示范,循环水处理工艺的提升等措施解决工厂化养殖车间闲置和养殖效益低等问题。

5.2 增加投入,加快疫病防控体系建设

国家应设置专项资金用于水产品质量监管体系的建设、检测仪器设备的购置和配备、水产品质量检测和抽查、执法队伍建设,组织科研部门研究健康养殖相关技术。制定优惠政策,全力支持研制开发高效、速效、长效水产疫苗,以此保障产业素质和发展质量。加快完善水产养殖动物防疫体系,开发一批防治药物、免疫制剂以及快速检测技术与产品。重点建设一批国家级、省部级和县级水产养殖动物防疫基础设施及疫病参考实验室。

5.3 健全追溯机制，推行产业链监管，建立协调有效的认证认可体系

从养殖者、养殖环境、养殖品种、流通过程等方面入手，实现“从产地到餐桌”的全程质量监控，开展对无公害水产品的认证管理，确保水产品质量安全。实现水产品“生产有记录、流向可追踪、质量可追溯、责任可界定、违者可追究”。一旦出现问题，可以准确无误地追究相关企业责任，也为有效实施不安全水产品召回制度打下良好基础。

附表 1　2016 年度天津综合试验站示范县鲆鲽类育苗及成鱼养殖情况统计

		塘沽区	大港区		汉沽区			普陀区	苍南县
		半滑舌鳎	大菱鲆	半滑舌鳎	大菱鲆	牙鲆	半滑舌鳎	牙鲆	半滑舌鳎
育苗	面积 / m^2	—	—	1 200	11 700	6 000	9 600	—	—
	产量 / 万尾	—	—	180	1 250	400	1 000	—	—
工厂养殖	面积 / m^2	35 000	—	9 880	23 700	9 100	27 400	—	5 000
	年产量 / t	353.78	—	124.96	389.78	127.48	248.68	—	120
	年销售量 / t	151.79	—	84.96	32.03	127.48	228.1	—	120
	年末库存量 / t	202.29	—	40	357.75	0	20.58	—	0
池塘养殖	面积 / 亩	—	—	—	—	—	—	20	—
	年产量 / t	—	—	—	—	—	—	5	—
	年销售量 / t	—	—	—	—	—	—	5	—
	年末库存量 / t	—	—	—	—	—	—	0	—
户数	育苗户数	—	—	2	8	4	7	—	—
	养殖户数	1	—	3	5	2	6	1	1

附表 2　天津综合试验站 5 个示范县养殖面积、养殖产量及品种构成

项目 \ 品种	年产总量	大菱鲆	牙鲆	半滑舌鳎
工厂化育苗面积 / m^2	28 500	11 700	6 000	10 800
工厂化出苗量 / 万尾	2 830	1 250	400	1 180
工厂化养殖面积 / m^2	110 080	23 700	9 100	77 280
工厂化养殖产量 / t	1 364.68	389.78	127.48	847.42
池塘养殖面积 / 亩	20	—	20	—
池塘年总产量 / t	5	—	5	—
网箱养殖面积 / m^2	—	—	—	—
网箱年总产量 / t	—	—	—	—
各品种工厂化育苗面积占总面积的比例 / %	100	41.05	21.05	37.90

(续表)

品种 项目	年产总量	大菱鲆	牙鲆	半滑舌鳎
各品种工厂化出苗量占总出苗量的比例/%	100	44.17	14.13	41.70
各品种工厂化养殖面积占总面积的比例/%	100	21.53	8.27	70.20
各品种工厂化养殖产量占总产量的比例/%	100	28.56	9.34	62.10
各品种池塘养殖面积占总面积的比例/%	100	—	100.00	—
各品种池塘养殖产量占总产量的比例/%	100	—	100.00	—

(天津综合试验站站长　贾　磊)

河北综合试验站产区调研报告

1　示范县(市、区)鲆鲽类养殖现状

河北综合试验站下设5个示范县(市、区),分别为:昌黎县、乐亭县、滦南县、丰南区、黄骅市。2016年育苗、养殖品种、产量及规模见附表1。

1.1　育苗面积及苗种产量

1.1.1　育苗面积

5个示范县育苗总面积为12 000 m^2,其中昌黎县7 000 m^2、黄骅市5 000 m^2。按品种分:半滑舌鳎育苗面积5 000 m^2、牙鲆7 000 m^2。

1.1.2　苗种年产量

5个示范县共计5户育苗厂家,年育苗量400万尾,其中:半滑舌鳎100万尾、牙鲆300万尾。各示范县育苗情况如下:

昌黎县:共有牙鲆育苗厂家2个,半滑舌鳎育苗厂家2个,育苗水体共计7 000 m^2其中牙鲆育苗水体2 000 m^2,年生产牙鲆苗种200万尾;半滑舌鳎育苗水体5 000 m^2,年生产半滑舌鳎苗种100万尾。

黄骅市:共有牙鲆育苗厂家1个,育苗水体5 000 m^2,年生产牙鲆苗种100万尾。

丰南区、滦南县和乐亭县:没有鲆鲽类育苗厂家。

1.2　养殖面积及年产量、销售量、年末库存量

1.2.1　工厂化养殖

5个示范县共有工厂化养殖户96家,养殖面积808 000 m^2,年总生产量2 454. 87 t,年销售量3 386. 7 t,年末库存量6 986. 78 t。其中:

昌黎县:65户,养殖面积60. 0万平方米。大菱鲆养殖45户,养殖面积39. 5万平方米,产量1 122. 15 t,销售1 234. 05 t,年末库存3 342. 5 t;牙鲆养殖16户,养殖面积13. 0万平方米,年产量372. 15 t,销售728. 41 t,年末库存2 385. 7 t;半滑舌鳎养殖4户,养殖面积7. 5万平方米,年产量171. 49 t,年销售344. 94 t,年末库存825. 8 t。

丰南区:1户,养殖面积5 000 m^2。牙鲆养殖面积1 000 m^2,年产量16. 86 t,年销售12. 9 t,

年末库存 8.96 t;大菱鲆养殖面积 4 000 m^2,年产量 10.11 t,年销售 18.31 t,年末库存 11.0 t。

滦南县:1 户,牙鲆养殖面积 3 000 m^2,年产量 76.95 t,年销售 54.59 t,年末库存 22.36 t。

乐亭县:29 户,养殖面积 20 万平方米。大菱鲆养殖 27 户,养殖面积 19.8 万平方米,年产量 670.11 t,年销售 937.2 t,年末库存 382.81 t;半滑舌鳎养殖 2 户,养殖面积 2 000 m^2,年产量 15.05 t,年销售 56.3 t,年末库存 7.65 t。

黄骅市:本年度未进行工厂化成鱼养殖。

1.2.2 池塘养殖

本站示范区内 2016 年无池塘养殖。

1.2.3 网箱养殖

本站示范区内 2016 年未进行网箱养殖。

1.3 品种构成

每个品种养殖面积及产量占示范县养殖总面积和总产量的比例见附表 2。

统计 5 个示范县鲆鲽类育苗、养殖情况,各品种构成如下:

工厂化育苗总面积为 12 000 m^2其中牙鲆 7 000 m^2,占育苗总面积的 58.33%;半滑舌鳎 5 000 m^2,占总面积的 41.67%。

年总出苗量为 400 万尾。其中牙鲆为 300 万尾,占总出苗量的 75%;半滑舌鳎为 100 万尾,占总出苗量的 25%。

工厂化养殖总面积为 808 000 m^2 其中大菱鲆为 597 000 m^2,占总养殖面积的 73.89%;牙鲆为 134 000 m^2,占总养殖面积的 16.58%;半滑舌鳎为 77 000 m^2,占总养殖面积的 9.53%。

工厂化养殖总产量为 2 454.87 t。其中大菱鲆为 1 802.37 t,占总量的 73.42%;牙鲆 465.96 t,占总量的 18.98%,半滑舌鳎为 186.54 t,占总量的 7.6%。

从以上统计数据可以看出,5 个示范县内,大菱鲆工厂化养殖面积和产量都占绝对优势,其次是牙鲆,半滑舌鳎养殖较少。

2 示范县(市、区)科研开展情况

2.1 科研开展情况

2016 年河北省鲆鲽类项目 5 个:

“全封闭循环海水工厂化养殖技术示范与推广”,省科技厅项目,承担单位河北省海洋与水产科学研究院,项目资金 30 万元。采用机械过滤、生物过滤器、紫外消毒机、增氧机等手段,使养殖废水循环利用,单产达到 20 kg/m^2,当年推广面积 3 万平方米。

“河北省现代农业产业体系海产品创新团队海水鱼健康增养殖及配套技术研究岗位”,

依托单位河北省海洋与水产科学研究院，2016年度资金30万元。开展全雌牙鲆“北鲆1号”和野生牙鲆大规模苗种培育对比实验，建立全雌牙鲆养殖示范区1个，繁育全雌牙鲆苗种30万尾，养殖单产20 kg/m^2，比普通牙鲆平均体重增长16.5%以上。

“牙鲆良种亲本引进更新”，承担单位黄骅市海富源水产养殖有限公司，资金20万元。

“野生半滑舌鳎亲鱼人工驯养培育与苗种繁育”，承担单位黄骅市水产技术推广站，资金50万元，主要进行野生半滑舌鳎采集、驯化及设备添加。

“半滑舌鳎全封闭式循环水养殖技术研究与集成示范”，承担单位秦皇岛鼎盛海洋生态科技开发有限公司，资金100万元。主要进行了循环水设施、设备更新改造。

2016年河北综合试验站通过与基层农技推广体系对接，在昌黎示范县进行全封闭循环海水工厂化养殖模式示范与推广，目前已有7个养殖大户采用了全封闭循环海水工厂化养殖模式，水体达到11万平方米。其中秦皇岛江鹏水产科技开发公司4.5万平方米、秦皇岛鼎盛海洋生态水产养殖有限公司2万平方米、秦皇岛粮丰海洋生态科技开发股份有限公司4.5万平方米。2016年，我站与河北省现代农业产业技术体系特色海产品创新团队海水鱼健康增养殖及配套技术研究岗位密切合作，通过实验研究和工厂化循环海水养殖技术推广，提高了鲆鲽类单位面积产量，减少了养殖废水排放，产生了较好的示范效果。

2.2　发表论文情况

2016年5个示范县发表鲆鲽类研究论文共计4篇，分别是：

（1）大菱鲆人工育苗关键技术要点

孙玉华、丁军；沧州市水产技术推广站；河北渔业，2016年9期。

（2）黄粉虫粉替代鱼粉对大菱鲆幼鱼生长性能的影响

李垚垚、郭冉；河北农大海洋学院；河北渔业，2016年5期。

（3）水中Mn（Ⅱ）对大菱鲆幼鱼生长及碱性磷酸酶和超氧化物歧化酶活性的影响

何忠伟、宫春光、殷蕊、孙桂清、于骞、符冬林；河北农业大学海洋学院、河北省海洋与水产科学研究院；水产科学，2016年7期。

（4）河北沿海焦氏舌鳎资源现状分析

胡保存、齐遵利、高文斌；河北农业大学海洋学院、河北省海洋与水产科学研究院；河北渔业，2016年12期。

3　鲆鲽类产业发展趋势及存在问题

3.1　产业发展趋势

3.1.1　苗种生产情况

2016年苗种生产厂5个，育苗水面12 000 m^2，比2015年减少了14.29%；育苗量400

万尾,比 2015 年减少了 55.06%,见表 1。分析育苗厂数量减少及苗种生产量降低的主要原因是:① 大菱鲆成鱼价格下降,使育苗者对苗种生产获利期望值下降;② 饲料价格略有增加,育苗盈利空间减少。

表 1　育苗规模对比表

年份	育苗厂家	育苗水面 / m^2	育苗量 / 万尾
2016 年	5	12 000	400
2015 年	9	14 000	890

3.1.2　鲆鲽类养殖情况

2016 年鲆鲽类养殖厂家 96 家,较 2015 年减少了 3.03%,养殖面积和产量分别为 808 000 m^2 和 2 454.87 t,分别比上年减少 1.25% 和 48.72%,商品鱼年末存量 6 986.78 t,比上年减少 11.70%,说明 2016 年鲆鲽类养殖面积和养殖产量均呈下降趋势。养殖面积、产量及产品鱼库存下降的原因主要是:① 2016 年上半年大菱鲆销售价格低迷,1 龄鱼养殖数量减少;② 10 月以后,大菱鲆销售价格回升,成品鱼销售量增加,造成产品库存减少(表 2)。

表 2　养殖规模对比表

年份	养殖厂数量	养殖面积 / m^2	年产量 / t	存量 / t
2016	96	808 000	2 454.87	6 986.78
2015	99	818 200	4 787.25	7 912.21

3.2　产业存在问题

河北省的鲆鲽类养殖,多为大排大放的初级工厂化养殖模式,一方面,对外界水环境依赖性大,另一方面,养殖废水排放也加剧了外界环境的恶化,同时导致水资源和能源大量消耗。近年来港口建设、石油开发、工业排污、生活污水排放等给河北沿海水域造成严重污染,致使水产养殖水质恶化,病害频发,在这种背景下,开展封闭式循环水养殖,不仅可以节约水之源,提高产量,还可以减少养殖业本身对海洋的污染。群众有开展封闭式循环海水养殖的需求,但由于前期投入太高,完全由企业出资进行设备设施改造困难很大,需要政府出台政策支持和配套融资措施。

河北省的鲆鲽类养殖业所需苗种,主要来源于野生种鱼的人工繁育苗种,经过累代繁殖,种质退化严重,表现为生长速度慢、发病率高、抗逆性差,导致养殖周期加长,养殖成本提高,养殖效益下降。群众迫切需要生长快、高抗病的优质鲆鲽类苗种。

各示范县鲆鲽类养殖存在的具体问题如下:

(1) 半滑舌鳎成品鱼价格相对较高,保持在 60～110 元 / 斤,渔民养殖的积极性较高,但因半滑舌鳎雄性苗种率较高,急需雌性比率较高的优质苗种及半滑舌鳎苗种培育技术。

(2) 因自然海水水质较差且夏季温度偏高,多采用地下卤水养殖,地下卤水铁、锰含量较高，养殖业者多用沉淀、砂滤方式去除铁、锰，既耗时，去除效果也不理想；在去除铁、锰的同时，地下水温度升高，利用价值降低。

（3）河北省即将对全省地下水超采进行综合治理，到2017年年底，压采地下水42亿立方米。地下水开采也将逐步实行收费制度，这将增加鲆鲽鱼类养殖成本，对鲆鲽类养殖将会造成不良影响。

4　当地政府对产业发展的扶持政策

河北沿海鲆鲽类养殖产业发展较快，成为渔民致富的重要途径，是沿海各县的支柱产业之一，特别是昌黎县、乐亭县尤为突出。2016年河北省响应国家产业结构调整及绿色发展要求，用原来燃油补贴经费的一部分，进行秦唐沧沿海地区工厂化养殖水处理设施改造、循环水养殖设备购置，使产区鲆鲽类养殖系统得到初步更新。

5　鲆鲽类产业技术需求

5.1　全封闭循环海水养殖水质净化技术

造价低廉的海水处理设备、全封闭循环水系统维护技术。

5.2　病害预防与控制技术

预防鲆鲽类疾病的安全、有效药物，急需预防鲆鲽类疾病发生的疫苗。

5.3　优质鲆鲽类全价饲料

国内各饲料厂家生产的人工配合饲料质量差异较大，同一种饲料的质量也不是很稳定，投喂后生长效果不好。有些企业还存在投喂冰鲜杂鱼现象，造成水质污染和对渔业资源的浪费，急需优质鲆鲽类全价饲料，代替目前使用的冰鲜杂鱼，节约资源，降低污染，提高产量及产品品质。

5.4　优质高抗鲆鲽类新品种

5个示范县对生长快、抗病力强的鲆鲽类新品种需求迫切。

附表 1　2016 年河北综合试验站示范县鲆鲽类育苗及成鱼养殖情况统计表

品种		昌黎县			丰南区			滦南县		乐亭县			黄骅市		
		大菱鲆	牙鲆	半滑舌鳎	大菱鲆	牙鲆	半滑舌鳎	牙鲆	半滑舌鳎	大菱鲆	牙鲆	半滑舌鳎	大菱鲆	牙鲆	半滑舌鳎
育苗	面积 / m^2	—	2 000	5 000	—	—	—	—	—	—	—	—	—	5 000	—
	产量 / 万尾	—	200	100	—	—	—	—	—	—	—	—	—	100	—
工厂化养殖	面积 / m^2	395 000	130 000	75 000	4 000	1 000	—	3 000	—	198 000	—	2 000	—	—	—
	年产量 / t	1 122. 15	372. 15	171. 49	10. 11	16. 86	—	76. 95	—	670. 11	—	15. 05	—	—	—
	年销售量 / t	1 234. 05	728. 41	344. 94	18. 31	12. 90	—	54. 59	—	937. 2	—	56. 3	—	—	—
	年末库存量 / t	3 342. 5	2 385. 7	825. 8	11. 00	8. 96	—	22. 36	—	382. 81	—	7. 65	—	—	—
池塘养殖	面积 / 亩	—	—	—	—	—	—	—	—	—	—	—	—	—	—
	年产量 / t	—	—	—	—	—	—	—	—	—	—	—	—	—	—
	年销售量 / t	—	—	—	—	—	—	—	—	—	—	—	—	—	—
	年末库存量 / t	—	—	—	—	—	—	—	—	—	—	—	—	—	—
户数	育苗户数	—	2	2	—	—	—	—	—	—	—	—	—	1	—
	养殖户数	45	16	4	1	—	—	1	—	27	—	2	—	—	—

附表 2　河北综合试验站五个示范县养殖面积、养殖产量及品种构成

项目＼品种	年产总量	大菱鲆	牙鲆	半滑舌鳎
工厂化育苗面积 / m^2	12 000	—	7 000	5 000
工厂化出苗量 / 万尾	400	—	300	100
工厂化养殖面积 / m^2	808 000	597 000	134 000	77 000
工厂化养殖产量 / t	2 454. 87	1 802. 37	465. 96	186. 54
池塘养殖面积 / 亩	—	—	—	—
池塘年总产量 / t	—	—	—	—
各品种工厂化育苗面积占总面积的比例 / %	100	—	58. 33	41. 67
各品种工厂化出苗量占总出苗量的比例 / %	100	—	75	25
各品种工厂化养殖面积占总面积的比例 / %	100	73. 89	16. 58	9. 53
各品种工厂化养殖产量占总产量的比例 / %	100	73. 42	18. 98	7. 60

（河北综合试验站站长　赵海涛）

北戴河综合试验站产区调研报告

1　示范县(市、区)鲆鲽类养殖现状

北戴河综合试验站下设五个示范县(市、区),分别为:河北省唐山市曹妃甸区、秦皇岛市山海关区,辽宁省丹东东港市,江苏省南通市如东县,福建省漳州市东山县。五个示范县鲆鲽类养殖模式、品种等各有不同,如曹妃甸示范县以工厂化养殖大菱鲆、牙鲆和半滑舌鳎为主,山海关示范县以工厂化养殖大菱鲆为主,东港示范县以池塘养殖牙鲆为主,如东示范县以工厂化养殖半滑舌鳎、漠斑牙鲆,池塘养殖漠斑牙鲆为主,东山示范县以工厂化和池塘网箱养殖牙鲆、大菱鲆为主。

1.1　育苗面积及苗种产量

1.1.1　育苗面积

五个示范县育苗总面积为108 999 m^2,其中曹妃甸区89 000 m^2、山海关区2 000 m^2、东港市6 000 m^2、东山县11 999 m^2,山海关区、东港市和东山县育苗品种为牙鲆,曹妃甸区育苗品种包括牙鲆、大菱鲆和半滑舌鳎。

1.1.2　苗种年产量

五个示范县共计23户育苗厂家,总计培育苗种5 460万尾。各县育苗情况如下:

曹妃甸区:11户育苗厂家。其中牙鲆3户,生产苗种1 120万尾;半滑舌鳎4户,生产苗种1 490万尾;大菱鲆4户,生产苗种800万尾。苗种主要用于工厂化养殖。

山海关区:仅秦皇岛海鑫水产养殖科技开发有限公司1家,生产牙鲆苗种50万尾(5 cm～8 cm),全部用于增殖放流和海洋牧场建设。

东港市:6户育苗厂家,生产牙鲆苗种1 500万尾(3 cm左右),用于本市池塘养殖及增殖放流。

如东县:无育苗厂家。

东山县:5户育苗厂家,生产牙鲆苗种500万尾(5 cm左右),全部用于本县工厂和网箱养殖。

1.2　养殖面积及年产量、销售量、年末库存量

五个示范县成鱼养殖厂家共220家,包括工厂化养殖和池塘养殖。河北省的曹妃甸区

和山海关区只有工厂化养殖，没有池塘养殖。

1.2.1　工厂化养殖

五个示范县均为开放式养殖，养殖面积 330 500 m^2，年总生产量为 3 029. 3 t，销售量为 1 208 t，年末库存量为 1 952. 7 t。其中：

曹妃甸区：牙鲆养殖户 5 家、半滑舌鳎养殖户 4 家、大菱鲆养殖户 3 家，养殖面积 277 000 m^2，全年生产量 2 834. 3 t，全年销售量 1 041. 7 t，年末库存 1 792. 6 t。其中牙鲆养殖面积 38 000 m^2，全年生产量 367. 05 t，全年销售量 101 t，年末库存 266. 05 t；半滑舌鳎养殖面积 200 000 m^2，全年生产量 2 264. 75 t，全年销售量 787. 75 t，年末库存 1 477 t；大菱鲆养殖面积 39 000 m^2，全年生产量 202. 5 t，全年销售量 152. 95 t，年末库存 49. 55 t。

山海关区：养殖厂家 4 家，其中企业 2 家，个体 2 家。养殖面积 32 000 m^2，均养殖大菱鲆，全年生产量 130 t，全年销售量 110 t，年末库存 57. 1 t。

东港市：养殖面积约 16 000 m^2，仅作为待售牙鲆暂养、当年苗种越冬或苗种培育之用，年末库存 75 t。

如东县：成鱼养殖厂共 4 家，养殖面积 2 500 m^2，全年生产量 16 t，全年销售量 22 t，年末库存 13. 3 t。其中漠斑牙鲆养殖面积 1 000 m^2，全年生产量 12 t，年销售量 16 t，年末库存 13. 3 t；半滑舌鳎养殖面积 1 500 m^2，全年生产量 4 t，年销售量 6 t，没有年末库存。

东山县：工厂化养殖品种只有牙鲆一种，养殖面积 3 000 m^2，全年生产量 49 t，全年销售量 34. 3 t，年末库存 14. 7 t。

1.2.2　池塘养殖

五个示范县池塘养殖面积为 2 100 亩和 53 728 m^2 的网箱（网箱放池塘中），年产量 1 132. 3 t，年销售量 808. 9 t，年末库存 324 t。其中大菱鲆池塘养殖面积为 45 728 m^2 网箱，全年生产量 780 t，年销售量 546 t，年末库存 234 t；牙鲆池塘养殖面积 2000 亩和 8 000 m^2 的网箱，全年生产量 350 t，年销售量 260 t，年末库存 90 t；漠斑牙鲆池塘养殖面积 100 亩，全年生产量 2. 3 t，年销售量 2. 9 t。

东港市：养殖户 150 家，养殖面积 2 000 亩，全部养殖牙鲆，全年生产量 200 t，全年销售量 200 t，没有年末库存。

如东县：池塘养殖面积 100 亩，养殖品种为漠斑牙鲆，养殖年产量 2. 3 t，全年销售量 2. 9 t，没有年末库存。

东山县：使用在池塘中加网箱的养殖方式，网箱规格为 3 m × 4 m × 2. 5 m。养殖大菱鲆和牙鲆，养殖厂家共 50 户。养殖网箱面积 53 728 m^2，养殖年产量 930 t，年销售量 606 t，年末库存量 324 t。其中大菱鲆养殖网箱面积 45 728 m^2，全年生产量 780 t，年销售量 546 t，年末库存 234 t。牙鲆养殖网箱面积 8 000 m^2，全年生产量 150 t，年销售量 60 t，年末库存 90 t。

1.3 品种构成

每品种养殖面积及产量占示范县养殖总面积和总产量的比例见附表 2。

统计 5 个示范县鲆鲽类养殖面积调查结果,各品种构成如下:

工厂化育苗总面积为 108 900 m^2其中大菱鲆为 15 000 m^2,占总养殖面积的 13.78%;牙鲆为 43 900 m^2,占总养殖面积的 40.31%;半滑舌鳎为 50 000 m^2,占总养殖面积的 45.91%。

工厂化育苗总出苗量为 5 460 万尾。其中大菱鲆为 800 万尾,占总出苗量的 14.65%;牙鲆为 3 170 万尾,占总出苗量的 58.06%;半滑舌鳎为 1 490 万尾,占总出苗量的 27.29%。

工厂化养殖总面积为 330 500 m^2其中大菱鲆为 71 000 m^2,占总养殖面积的 21.48%;牙鲆为 57 000 m^2,占总养殖面积的 17.25%;半滑舌鳎为 201 500 m^2,占总养殖面积的 60.97%;漠斑牙鲆为 1 000 m^2,占总养殖面积的 0.3% 。

工厂化养殖总产量为 3 029.3 t。其中大菱鲆为 332.5 t,占总量的 10.98%;牙鲆 416.05 t,占总量的 13.73%;半滑舌鳎为 2 268.75 t,占总量的 74.89%;漠斑牙鲆为 12 t,占总量的 0.4%。

池塘养殖总面积为 2 100 亩。其中牙鲆为 2 000 亩,占总养殖面积的 95.24%;漠斑牙鲆为 100 亩,占总养殖面积的 4.76% 。

池塘养殖总产量为 202.3 t。其中牙鲆 200 t,占总量的 98.86%;漠斑牙鲆为 2.3 t,占总量的 1.14%。

网箱养殖总面积为 53 728 m^2其中大菱鲆为 45 728 m^2,占总养殖面积的 85.11%;牙鲆为 8 000 m^2,占总养殖面积的 14.89%。

网箱养殖总产量为 930 t。其中大菱鲆为 780 t,占总量的 83.87%;牙鲆 150 t,占总量的 16.13%。

从以上统计数据可以看出,五个示范县内,牙鲆的育苗面积和出苗量占比最高,分别为 40.31%和 58.06%。工厂化养殖面积和产量半滑舌鳎的占比最高,分别为 60.97%和 74.89%。池塘养殖面积和产量牙鲆的占比最高,分别为 95.24%和 98.86%。网箱养殖面积和产量大菱鲆的占比最高,分别为 85.11%和 83.87%。

从成品鱼价格来看,半滑舌鳎最高,在 70～100 元/斤,大菱鲆价格在 11～22 元/斤,牙鲆价格在 20～25 元/斤。规格不同价格差别较大,如第四季度山海关示范县 1～1.5 斤/尾的大菱鲆,单价 20 元/斤,1.5～2.0 斤/尾的单价 19 元/斤,2.0 斤/尾以上的单价 13 元/斤。又如曹妃甸示范县 1.5～2.0 斤/尾的半滑舌鳎单价 80 元/斤,2.0 斤/尾以上的单价 100 元/斤。随季节和地区不同价格变化也较大,福建东山示范县第二、四季度牙鲆价格最低,为 15 元/斤,河北曹妃甸第三、四季度牙鲆价格最高,为 25 元/斤。大菱鲆价格整体比较低,山海关第二季度大菱鲆价格低至 11 元/斤,第三、四季度有所回升。如东和曹妃甸半滑舌鳎的价格差别比较大,如东半滑舌鳎平均每斤低 20 元。综上所述,2016 年大菱鲆价格整体较低，且规格越大，价格越低。半滑舌鳎规格越大价格越高。五个示范县曹妃甸牙鲆价格最高，福建东山牙鲆价格最低。

2　示范县(市、区)科研开展情况

五个示范县今年只有福建东山县实施了一个省级科研项目：牙鲆鱼南方工厂化育苗与养成。项目金额15万元，参与研发人员5人，获省科技进步二等奖。

3　鲆鲽类产业发展中存在的问题

3.1　市场需求和价格变动幅度大

今年牙鲆商品鱼价格在19～24元/斤浮动，养殖成本一般不会高于15元/斤，半滑舌鳎价格一般在75元/斤以上，养殖成本在30～50元/斤不等，这两种鱼都在盈利状态，但养殖户也认为鱼价不可预测，浮动太大，今年盈利，明年就可能赔钱。而大菱鲆情况很糟，今年10月前价格在9～12元/斤之间，养殖成本却在17～22元/斤，养殖户能挺住的都在亏本经营，租厂房和资金有限的大都撤摊了。造成大菱鲆价格低迷的原因，一是高档酒店需求的急剧减少，二是舆论对大菱鲆“药残”事件的过度渲染，甚至在微信朋友圈广泛流传着一篇题为《这是一条致癌鱼》的文章。

3.2　加工与销售

鲆鲽类一直被认为是高档消费鱼类，销售渠道主要为鲜活市场和饭店。目前除了半滑舌鳎外，牙鲆和大菱鲆的价格已下降到完全可以走上普通老百姓的餐桌，即使半滑舌鳎价钱很高，也可以在节日的饭桌上来一条。造成大菱鲆价格低迷的原因，除“药残”事件外，产能过剩也是原因之一，但许多水产界以外的人根本不认识这些鱼，即使认识也不知怎么吃，因此需要加大宣传力度，拓展销售渠道，开创新的销售模式。

4　当地政府对产业发展的扶持政策

五个示范县基本沿用以前的优惠政策，2016年没有出台新的扶持政策。

5　鲆鲽类产业技术需求

五个示范县鲆鲽类养殖在养殖模式、病害和新品种等核心问题上均有不同程度的需求。

5.1 养殖模式方面

河北省鲆鲽类主要是工厂化养殖，养殖模式是利用地下淡水和卤水井进行勾兑进行养殖。目前由于养殖规模的扩大，地下水资源过度消耗，取水位下降厉害，政府部门已经认识到这不是一种可持续发展模式，都在推荐循环水养殖模式，但循环水存在占地多、成本高、效率低等缺点，虽然政府给予了一些补贴，但杯水车薪，尤其是大菱鲆价格上不去，更不可能使用循环水。

5.2 病害方面

牙鲆疾病：一是纤毛虫病，发病时间在牙鲆体重 200 g 以下；二是淋巴囊肿病，被公认为牙鲆养殖中危害最大的疾病，发生时间一般在牙鲆半成鱼的时期，养殖成本花了不少，但没有一点养殖价值，因此特别需要抗淋巴囊肿新品种，以便从根本上杜绝淋巴囊肿病的发生；三是牙鲆出血病，比较严重，发病时在尾鳍、背鳍、臀鳍等部位有鲜血渗出，没药可治，会引起陆陆续续的死亡；四是腹水症，主要以控制养殖用水和饵料新鲜度来控制。大菱鲆相比牙鲆病害较少，主要是肠炎比较严重，能引起大量死亡。其次是皮疣病，虽然不致死，但影响外观，降低了销售价格。半滑舌鳎疾病：一是烂底、腐皮病，主要症状为底板鳃盖下出现溃疡出血，表皮鱼鳞脱落出血，养殖全期皆有发生；二是断尾症，主要出现在 5 cm～15 cm 的苗种期；三是寄生虫病，主要是车轮虫、鞭毛虫、杯体虫及其他纤毛虫类，养殖期全程都有发生；四是肠炎、腹水病，主要症状为肠道积水，肝脏坏死，养殖全期都有发生。

5.3 新品种方面

半滑舌鳎新品种：由于半滑舌鳎雌大雄小的特性，苗种的进量及优选是养殖成败的基础。通过几年来的摸索，在当前半滑舌鳎育苗的技术手段下，同一批鱼苗中有养殖价值的雌鱼比例仅为30%，实际雌鱼的比例可能稍高一些。在30%优选后的鱼苗中，只有不超过21%的鱼 12 个月可以从 7 cm～8 cm 长到 1 000 g 以上，55%左右的鱼在 500 g 以上，24%的鱼在 350 g～500 g 之间。以上数据加上养殖中损耗，可作为当前养殖过程中苗种数量选购的依据。如果想提高养殖效果，那么优选苗种的比例就要进一步缩小，如此苗种成本则会进一步增加。比如：当前情况下以放养 1 万尾 6 cm～8 cm 舌鳎苗，以 2.5 元/尾计算，万尾购苗成本 2.5 万元。优选后的苗种在 3 000 尾左右，有效养殖的尾苗成本为 8.3 元(不包括苗种损耗所造成的成本上升)。如此苗种成本再加上其他成本，半滑舌鳎养殖存在风险要高于其他品种。因此急需一个能提高雌化率的新品种。

牙鲆新品种：抗淋巴囊肿新品种，提高养殖成活率。

其他养殖品种：在调研过程中，调研对象也谈到市场问题之一——产能过剩。养殖场建了，养多了价格下去了，赔钱；不养，养殖场闲着浪费。他们希望能够有新的品种养殖，降低单一品种的产量，增加养殖品种的数量，提高整个鲆鲽类养殖的产量和效益。

附表 1　2016 年度北戴河综合试验站示范县鲆鲽类育苗及成鱼养殖情况统计表

品种		曹妃甸区			山海关区		东港市	如东县		东山县	
		牙鲆	半滑舌鳎	大菱鲆	大菱鲆	牙鲆	牙鲆	半滑舌鳎	漠斑牙鲆	大菱鲆	牙鲆
育苗	面积/m^2	24 000	50 000	15 000	0	2 000	6 000	0	0	0	11 999
	产量/万尾	1 120	1 490	800	0	50	1 500	0	0	0	500
工厂养殖	面积/m^2	38 000	200 000	39 000	32 000	0	16 000	1 500	1 000	0	3 000
	年产量/t	367. 05	2 264. 75	202. 5	130	0	0	4	12	0	49
	年销售量/t	101	787. 75	152. 95	110	0	0	6	16	0	34. 3
	年末库存量/t	266. 05	1 477	49. 55	57. 1	0	75	0	13. 3	0	14. 7
池塘养殖	面积/亩	0	0	0	0	0	2 000	0	100	0	0
	年产量/t	0	0	0	0	0	200	0	2. 3	0	0
	年销售量/t	0	0	0	0	0	200	0	2. 9	0	0
	年末库存量/t	0	0	0	0	0	0	0	0	0	0
网箱养殖	面积/m^2	0	0	0	0	0	0	0	0	45 728	8 000
	年产量/t	0	0	0	0	0	0	0	0	780	150
	年销售量/t	0	0	0	0	0	0	0	0	546	60
	年末库存量/t	0	0	0	0	0	0	0	0	234	90
户数	育苗户数	3	4	4	0	1	6	0	0	0	5
	养殖户数	5	4	3	4	0	150	3	1	40	10

附表 2　北戴河综合试验站五个示范县养殖面积、养殖产量及品种构成

项目 \ 品种	年产总量	大菱鲆	牙鲆	半滑舌鳎	漠斑牙鲆
工厂化育苗面积 / m^2	108 999	15 000	43 999	50 000	0
工厂化出苗量 / 万尾	5 460	800	3 170	1 490	0
工厂化养殖面积 / m^2	330 500	71 000	57 000	201 500	1 000
工厂化养殖产量 / t	3 029. 3	332. 5	416. 05	2 268. 75	12
池塘养殖面积 / 亩	2 100	0	2 000	0	100
池塘年总产量 / t	202. 3	0	200	0	2. 3
网箱养殖面积 / m^2	53 728	45 728	8 000	0	0
网箱年总产量 / t	930	780	150	0	0
各品种工厂化育苗面积占总面积的比例 / %	—	13. 78	40. 31	45. 91	0
各品种工厂化出苗量占总出苗量的比例 / %	—	14. 65	58. 06	27. 29	0
各品种工厂化养殖面积占总面积的比例 / %	—	21. 48	17. 25	60. 97	0. 30
各品种工厂化养殖产量占总产量的比例 / %	—	10. 98	13. 73	74. 89	0. 40
各品种池塘养殖面积占总面积的比例 / %	—	0	95. 24	0	4. 76
各品种池塘养殖产量占总产量的比例 / %	—	0	98. 86	0	1. 14
各品种网箱养殖占总面积的比例 / %	—	85. 11	14. 89	0	0
各品种网箱养殖产量占总产量的比例 / %	—	83. 87	16. 13	0	0

（北戴河综合试验站站长　于清海）

辽宁综合试验站产区调研报告

1　示范县(市、区)鲆鲽类养殖现状

辽宁综合试验站下设五个示范县(市、区),分别为大连市的旅顺口区、甘井子区、瓦房店市、庄河市和盘锦市的大洼县。示范基地7处,分别为大连天正集实业有限公司、大连鹤圣丰养殖场、大连万洋渔业有限公司、大连德洋水产有限公司、大连庄河富谷水产有限公司、大连瓦房店灏霖水产、盘锦光合蟹业有限公司。在示范县和示范基地主要进行鲆鲽类养殖技术的示范和推广工作,各个示范县区的人工育苗、养殖品种、产量及规模见附表1。

1.1　育苗面积及苗种产量

1.1.1　育苗面积

5个示范县育苗总面积为4 000 m^2,其中旅顺口区2 000 m^2,甘井子区2 000 m^2,牙鲆。

1.1.2　苗种年产量

5个示范县共计2户育苗厂家,总计生产牙鲆苗400万尾,各县育苗情况如下:

旅顺口区:1户育苗厂家,生产牙鲆苗200万尾,全部用于完成放流任务。

甘井子区:1户育苗厂家,生产牙鲆苗200万尾,全部用于完成放流任务。

1.2　养殖面积及年产量、销售量、年末库存量

1.2.1　工厂化养殖

5个示范县共计11家养殖户,工厂化养殖面积58 200 m^2,年总生产量为269. 5 t。今年销售量346 t,年末库存量为100 t。其中:

旅顺口区:养殖单位3户,养殖面积19 200 m^2。其中养殖大菱鲆18 000 m^2,全年生产量82 t,销售41 t,年末库存50 t。养殖条斑星鲽200 m^2,圆斑星鲽500 m^2,石鲽500 m^2,产量较少。

瓦房店市:养殖单位2户,养殖种类为大菱鲆,面积5 000 m^2,年产量37 t。销售38 t,年末库存13. 5 t。

甘井子区:养殖单位4户,养殖面34 000 m^2。其中,大菱鲆养殖面积27 000 m^2,年产量

111. 5 t,年销售量 102 t,年末库存 34. 5 t;牙鲆养殖面积 5 000 m^2,年产量 34 t,年销售量 40 t,年末库存 2 t;半滑舌鳎 2 000 m^2,年产量 5 t,年销售量 25 t,年末库存 0 t。

庄河市:养殖单位 1 户,养殖种类为牙鲆,面积 1 000 m^2,年产量 8 t,移到池塘养殖 8 t,年末库存 0 t。

盘锦大洼:养殖单位 1 户,养殖种类为牙鲆,面积 1 000 m^2,年产量 8 t,移到池塘养殖 8 t,年末库存 0 t。

1.2.2 池塘养殖

本示范区只有大洼县和庄河市进行池塘养殖牙鲆,均采用混养方式,池塘养殖总面积为 2 000 亩,产量 84 t,销售量 100 t,年末库存量为 0 t。

大洼县:1 户养殖企业进行池塘养殖,全部养殖牙鲆。养殖面积 1 000 亩,年产量 42 t,销售量 100 t,年末库存量为 0 t。

庄河市:1 户养殖企业进行池塘养殖,全部养殖牙鲆。养殖面积 1 000 亩,年产量 42 t,销售量 100 t,年末库存量 0 t。

1.2.3 网箱养殖

本示范区只有旅顺口区有 1 户养殖企业进行网箱养殖,大菱鲆养殖网箱 5 m × 5 m 规格 400 个,10 m × 5 m 规格 200 个,养殖面积 20 000 m^2,年产量为 100 t。

1. 3 品种构成

经过对本试验站内五个示范县区的鲆鲽类养殖情况的调查统计,每个品种的养殖面积及产量占示范县养殖总面积和总产量的比例(见附表 2)情况如下:

工厂化育苗总面积为 4 000 m^2,其中牙鲆为 4 000 m^2,占总育苗面积的 100%。

工厂化育苗的总出苗量为 400 万尾,其中牙鲆 400 万尾,占总出苗量的 100%。

工厂化养殖的总面积为 58 200 m^2其中牙鲆为 5 000 m^2,占总养殖面积的 8. 59%;大菱鲆为 50 000 m^2,占总养殖面积的 85. 91%;星鲽为 700 m^2,占总养殖面积的 1. 2%;石鲽为 500 m^2,占总养殖面积的 0. 86%;半滑舌鳎为 2 000 m^2,占总养殖面积的 3. 43%。

工厂化养殖的总产量为 269. 5 t。其中牙鲆 34 t,占总量的 12. 62%;大菱鲆为 230. 5 t,占总量的 85. 53%;半滑舌鳎为 5 t,占总量的 1. 85%;条斑星鲽、圆斑星鲽、石鲽极少,忽略不计。

池塘养殖总面积为 2 000 亩,养殖品种为牙鲆,养殖总产量为 84 t。

示范区内鲆鲽类浅海及池塘网箱养殖面积 20 000 m^2,养殖大菱鲆总产量 100 t。

从以上统计可以看出:在 5 个示范县内,育苗以牙鲆为主;工厂化养殖以大菱鲆和牙鲆为主,少量半滑舌鳎,条斑星鲽、圆斑星鲽、石鲽产量极少;池塘养殖品种为牙鲆;网箱养殖以大菱鲆为主。

2　示范县(市、区)科研开展情况

2.1　科研课题情况

大连庄河示范区、旅顺口区示范区、瓦房店示范区、盘锦大洼县示范区进行科研项目2项,承担单位为辽宁省海洋水产科学研究院,分别为"刺参、大菱鲆工厂化健康养殖模式技术研究与示范"和"辽宁省村级综合服务平台建设"。

大连市甘井子区示范区进行科研项目3项,承担单位为大连市天正实业有限公司,3项项目分别为"大连市标准化示范区-名贵海水鱼渔业"、"农业部渔业节能减排示范项目"、"辽宁省地方标准-星鲽工厂化人工繁育及养殖技术规程"。

2.2　发表论文、标准、专利情况

2016年,辽宁综合试验站内各示范区专利1项、标准2项,具体如下:

鲍相渤、苏浩等;辽宁省海洋水产科学研究院申请鲬和李氏鮨的快速鉴别试剂盒及鉴别方法;申请号CN201610072848.4,公开号CN105525020A。

3　鲆鲽类产业发展中存在的问题

辽宁综合试验站各示范县区主养大菱鲆、牙鲆,少量养殖半滑舌鳎、星鲽、石鲽等鱼类。各示范县区养殖条件与品种不同,养殖存在的问题也不同。

3.1　大菱鲆养殖存在的问题

大菱鲆市场价格的大幅下降,造成养殖效益减少,养殖积极性受挫。因此养殖面积及产量下降。如何提振大菱鲆养殖业成为当务之急。同时,大菱鲆工厂化养殖疫苗防治效果较好,但厂家缺少使用知识,应加快推广普及。

3.2　牙鲆养殖存在的问题

牙鲆池塘养殖受市场价格的大幅下降及其他养殖品种发展的影响,养殖面积及产量大幅下降。发展池塘高效及生态养殖,提高产量、降低成本,成为牙鲆养殖发展的出路。

3.3　半滑舌鳎养殖存在的问题

由于受辽宁地区气温较低影响,半滑舌鳎工厂化养殖成本较高,养殖规模不大,发展较慢。如何减低成本,发展节能减排养殖技术,急待解决。

4 当地政府对产业发展的扶持政策

辽宁省和大连市政府对鲆鲽类产业发展采取一定的扶持政策。支持科研机构与企业联合创新,建立多个鲆鲽类原、良种场,大力发展高效、节能工厂化全封闭循环水养殖示范基地,大力支持无公害水产品养殖,制定养殖地方标准。大力开展辽宁省村级服务平台建设,进行技术培训、科技下乡、专家帮扶等活动。

5 鲆鲽类产业技术需求

5.1 陆海接力及池塘高效生态养殖技术

降低养殖成本,进行陆海接力养殖技术。具有高效、造价低廉的池塘养殖海水处理设备、高效生态养殖技术。

5.2 产品质量综合控制技术

优良苗种采购,饲料质量的保证,预防疾病发生,各种药物合理使用技术。

5.3 鲆鲽工厂化提质增效养殖技术

优良品种,循环水养殖设备,疫苗免疫防病,高效饲料综合应用技术。

附表 1　2016 年度辽宁综合试验站示范县鲆鲽类育苗及成鱼养殖情况统计表

品种		旅顺口区				大洼	甘井子区			瓦房店	庄河市
		大菱鲆	星鲽	石鲽	牙鲆	牙鲆	大菱鲆	舌鳎	牙鲆	大菱鲆	牙鲆
育苗	面积 / m^2	—	—	—	2 000	—	—	—	2 000	—	—
	产量 / 万尾	—	—	—	200	—	—	—	200	—	—
工厂养殖	面积 / m^2	18 000	700	500	—	—	27 000	2 000	5 000	5 000	—
	年产量 / t	82	产量少，忽略不计	产量少，忽略不计	—	—	247. 5	5	50	37	—
	年销售量 / t	41	—	—	—	—	111. 5	25	34	38	—
	年末库存量 / t	50	—	—	—	—	102	0	40	13. 5	—
池塘养殖	面积 / 亩	—	—	—	—	1 000	34. 5	—	2	—	1 000
	年产量 / t	—	—	—	—	42	—	—	—	—	42
	年销售量 / t	—	—	—	—	50	—	—	—	—	50
	年末库存量 / t	—	—	—	—	0	—	—	—	—	0
网箱养殖	面积 / m^2	20 000	—	—	—	—	—	—	—	—	—
	年产量 / t	100	—	—	—	—	—	—	—	—	—
	年销售量 / t	200	—	—	—	—	—	—	—	—	—
	年末库存量 / t	0	—	—	—	—	—	—	—	—	—
户数	育苗户数	—	—	—	1	—	1	—	1	—	—
	养殖户数	3	1	—	1	1	3	1	2	2	1

附表 2　辽宁综合试验站五个示范县养殖面积、养殖产量及主要品种构成

项目＼品种	年产总量	牙鲆	大菱鲆	星鲽	石鲽	半滑舌鳎
工厂化育苗面积/m^2	4 000	4 000	—	—	—	—
工厂化出苗量/万尾	400	400	—	—	—	—
工厂化养殖面积/m^2	58 200	5 000	50 000	700	500	2 000
工厂化养殖产量/t	494	58. 5	425. 5	产量少，忽略不计	产量少，忽略不计	5
池塘养殖面积/亩	2 000	2 000	—	—	—	—
池塘年总产量/t	84	84	—	—	—	—
网箱养殖面积/m^2	20 000	—	20 000	—	—	—
网箱年总产量/t	100	—	100	—	—	—
各品种工厂化育苗面积占总面积的比例/%	100	100	—	—	—	—
各品种工厂化出苗量占总出苗量的比例/%	100	100	—	—	—	—
各品种工厂化养殖面积占总面积的比例/%	100	8. 59	85. 91	1. 2	0. 86	3. 43
各品种工厂化养殖产量占总产量的比例/%	100	12. 62	85. 53	产量少，忽略不计	产量少，忽略不计	1. 85
各品种池塘养殖面积占总面积的比例/%	100	100	—	—	—	—
各品种池塘养殖产量占总产量的比例/%	100	100	—	—	—	—
各品种网箱养殖面积占总面积的比例/%	100	—	100	—	—	—
各品种网箱养殖产量占总产量的比例/%	100	—	100	—	—	—

(辽宁综合试验站站长　赫崇波)

葫芦岛综合试验站产区调研报告

1　示范县(市、区)鲆鲽类养殖现状

本综合试验站下设五个示范县(市、区)，分别为：兴城现代渔业园区、兴城沙后所、绥中县、葫芦岛龙港区、福建省宁德市霞浦县。其育苗、养殖品种、产量及规模见附表 1

1.1　育苗面积及苗种产量

1.1.1　育苗面积

五个示范县大菱鲆育苗总面积为 5 000 m^2，在兴城现代渔业园区。

1.1.2　苗种年产量

五个示范县共计 1 户育苗厂家，总计育苗 300 万尾，全部为大菱鲆苗种，用于本市养殖。

1.2　养殖面积及年产量、销售量、年末库存量

1.2.1　工厂化养殖

五个示范县有四个示范县为大菱鲆工厂化养殖，一个示范县为网箱养殖。养殖户 633 户，面积 254 万平方米，年生产量为 21 317 t，销售量为 25 950 t，年末存池量为 17 247 t。其中：

兴城现代渔业园区：养殖户 450 户，养殖面积 160 万平方米，年产量 16 315 t，销售 20 100 t，年末存池量 12 275 t。

兴城沙后所：养殖户 30 户，养殖面积 40 万平方米，年产量 1 850 t，销售量 2 250 t，年末存池量 2 440 t。

绥中县：养殖户 120 户，养殖面积 42 万平方米，年产量 2 282 t，销售 2 870 t，年末存池量 2 002 t。

葫芦岛龙港区：养殖户 33 户，养殖面积 12 万平方米，年产量 870 t，年销售量 730 t，年末存池量 530 t。

1.2.2　网箱养殖

在福建省宁德市霞浦县进行南北接力式网箱养殖大菱鲆。网箱养殖户 35 户，网箱数量 4 000 个(3. 3 m × 3. 3 m × 2 m)，年养殖产量 290 t，年销售量 290 t，年末存量 0 t。

2 示范县(市、区)科研开展情况

2016年1月19日,葫芦岛市举办以“大美葫芦岛,放心多宝鱼”为主题的首届“互联网+水产品”兴城多宝鱼绿色创新发展原产地高峰论坛。来自国家和省、市的多位渔业专家、养殖企业从政策、品牌、养殖、销售等不同环节,探讨了在“互联网+”时代多宝鱼行业的现状、问题、成果和前景,为兴城多宝鱼的发展献计献力。

本届“互联网+水产品”兴城多宝鱼绿色创新发展原产地高峰论坛,由兴城市政府、葫芦岛和兴城两级海洋与渔业部门共同举办,旨在把关注和支持多宝鱼产业发展的各界朋友请进来,把各方信息和建议及最新理念与技术传进来,推动多宝鱼产业健康、可持续发展,把兴城打造成为全国优质多宝鱼产业带示范基地。论坛上,国家鲆鲽类产业技术体系和中国水产科学研究院黄海水产研究所岗位科学家马爱军、省水产养殖协会秘书长杨珍茂以及来自广东、北京、山东等地的专家,共同分析解读了“互联网+”时代多宝鱼行业如何更好地走绿色创新发展之路,引起了广大养殖企业的共鸣。央视《致富经》栏目组、阿里巴巴集团控股有限公司、新华网、中国网、中国日报网等20多家新闻媒体对论坛进行了报道。

当日,举行了首届兴城多宝鱼“鱼王”争霸赛,葫芦岛市10余家国家级、省级无公害产地认证企业带来了加注二维码溯源标签的多宝鱼参赛,并签定了水产品质量安全承诺书。

2016年1～12月,试验站技术依托单位兴城龙运井盐水产养殖有限责任公司承担由全国水产技术推广总站组织的“2016年农业部动物疫情监测与防治——病原菌耐药性普查”项目。该项目通过对大菱鲆发病所做的药敏试验,大菱鲆养殖主要病原微生物耐药性,逐步摸清大菱鲆发病原因及对症的药物治疗。

2016年12月7日,由中国水产流通与加工协会、中国农产品市场协会和中国农村杂志社主办的2016年中国水产品品牌大会在广州隆重召开,“兴城多宝鱼”以综合评价第十名的成绩荣获“2016最具影响力水产品区域公用品牌”。“兴城多宝鱼”地理标志证明商标于2012年注册成功的,经过几年来的品牌应用与运营取得了丰硕成果,为稳定和保持葫芦岛市乃至全国的大菱鲆产业健康可持续发展发挥了重要作用。当前正处于渔业供给侧结构性改革的关键时刻,葫芦岛市将以“兴城多宝鱼”品牌的成功运营为代表,积极推进渔业养殖业绿色发展、提质增效,助推行业稳步前行。

2016年由葫芦岛市水产技术推广站承担的《中草药在防治大菱鲆病害的应用》项目。中草药可预防大菱鲆多种常见病害,对发生的淋巴囊肿病的大菱鲆在使用中草药后治愈率可达93%以上;在大菱鲆肠炎病防治中,治愈率达91%以上,防治效果优于抗生素。同时,与使用西药相比可节约用药成本52%以上,对环境无污染,生态效益显著,经检测产品质量合格率100%。经大菱鲆养殖户使用,效果明显。

2016年由葫芦岛市水产技术推广站引进圆斑星鲽鱼苗1.95万尾,在兴城现代渔业园区示范养殖,通过一年的饲养,葫芦岛地区水温、盐度等各项水质指标,均符合圆斑星鲽生长条件,积累了一定经验,为葫芦岛地区工厂化鲆鲽鱼养殖新的品种提供了保证。

3　鲆鲽类产业发展中存在的问题

3.1　鲆鲽鱼养殖产业发展现状

葫芦岛地区自2001年开展工厂化鲆鲽鱼——大菱鲆养殖以来，经过十六年的发展，在国家鲆鲽产业体系及相关部门的大力支持下，截至2016年底，葫芦岛市有633户养殖企业（户）开展大菱鲆工厂化养殖，工厂化面积254万平方米，年产大菱鲆2.2万吨。工厂化养殖大菱鲆已成为葫芦岛市大农业中一项举足轻重的产业。

葫芦岛地区鲆鲽鱼发展，期间经历了引种试验、高速增长、提高整合和稳定四个阶段。葫芦岛市按照“规范管理、控制规模、提质增效”的思路，严格执行《葫芦岛市大菱鲆 养殖技术操作规程》，建立健全了《水产品养殖行业质量安全可追溯体系》，严格控制违禁药品及添加剂的使用。

工厂化养殖业的发展，不仅解决农民就业和收入的增加，也带动这一区域二、三产业的发展。目前从苗种繁育到生产、销售，从饲养、养殖技术服务到储藏和运输等都逐渐形成一个较完整的产业链，从业人员近万人，拥有一大批经济人才。

3.2　鲆鲽鱼养殖取得的成绩

葫芦岛工厂化养殖始于2001年，2002年经省政府批准，兴城大菱鲆养殖园区成为辽宁省首批39个现代农业园区中唯一的一个渔业园区。

2012年成功注册“兴城多宝鱼”地理标志商标，葫芦岛地区所有养殖的大菱鲆均为“兴城多宝鱼”，成为名扬全国的知名品牌。“兴城多宝鱼”建立了完善的质量溯源系统，实现了从渔场到餐桌，一扫即现，全程实现“来源可追溯、去向可查证、责任可追究”的放心鱼。2016年通过转口贸易，远销美国、马来西亚、韩国、港澳台等国家和地区。

2008年，“兴城多宝鱼”被列入北京奥运会、残奥会指定水产品；2009年，“兴城多宝鱼”被列入上海世博会指定水产品；2014年，兴城被辽宁省评定为辽宁特产“大菱鲆之乡”。2011～2015年，“兴城多宝鱼”连续五年接受农业部抽检，合格率达100%；2016年“兴城多宝鱼”荣获“最具影响力水产品区域公用品牌”。

3.3　大菱鲆养殖存在问题

大菱鲆是一条“好鱼”，无论在改善和提高人民生活质量，还是带动当地群众致富等方面都发挥了积极的作用。但16年来回头看，在养殖区建设、产业发展、环境与资源影响等方面还存在明显问题。归纳起来，突出问题有3个。

3.3.1　产业生存的压力

一方面，园区有养殖行业协会1家，专业合作社4家。不仅合作组织建设起步晚、发展慢、数量少、不规范，而且协会、合作组织的作用没有得到很好地发挥。大多数养殖业户处于散兵游勇、各自为战状态,没有形成集团化组织和有能力的领军企业,防范和抵御市场风险能力差,也很难形成市场价格的控制力,市场一有风吹草动,整个产业就受到冲击和影响。

另一方面,整个产业还处于低端运行状态,深加工、冷链物流还是短板。2016年年底转口贸易刚刚起步,数量有限,年不足几十吨;电商合作也刚刚起步,年销量预计也只有百吨左右。

3.3.2 水资源的压力

目前,养殖区内均采用开放式流水养殖,这种原始粗放养殖方式用水量大,对地下井盐水的依赖性强。长期超负荷开采地下水资源,势必对地下水资源形成严峻挑战。

3.3.3 生态环境的压力

养殖户分散于沿岸一线,点多线长,养殖废水难以集中处理。养殖废水长期直排,势必对生态环境造成破坏。

因此,从长远发展来看,加强行业自律,提高企业准入门槛,规划和调查产业布局,集中扶持和建立几个或多个企业集团和领军企业,势在必行。

4 当地政府对产业发展的政策

本着产业发展与环境保护相结合的原则,葫芦岛市政府高度重视大菱鲆养殖产业环境,强化养殖区环境卫生监管力度,投入资金,组织专人规范垃圾收集、清运工作,减少生活垃圾对环境、水域的污染。同时本着"谁排污,谁负责"的原则收取垃圾清运费,与大菱鲆养殖户签定卫生保证责任书。强化宣传教育,让养殖业主养成良好的卫生习惯,做到不乱扔垃圾。对拒不服从管理,仍随意乱扔垃圾的依法打击处理。

积极推进循环水养殖技术,加强养殖废水可持续循环利用,确保养殖水域环境及时净化处理。指导养殖企业修建废水沉淀过滤池,对养殖废水储存沉淀后排放,可减少残饵、粪便等固体物排放,同时也有利于养殖废水中消毒剂等药物排放前失效。全面推广健康养殖模式,减少药物使用量,实现科学、生态养殖。

5 鲆鲽类产业技术需求

5.1 发展循环水养殖

目前,大菱鲆养殖区内均采用开放式流水养殖,这种原始粗放养殖方式用水量大,对地下井盐水的依赖性强。长期超负荷开采地下水资源,势必对地下水资源形成严峻挑战,因此进行节能减排,发展循环水养殖。

5.2　加强生态环境保护

养殖户分散于沿岸一线，结合生态生境修复项目，利用生物方法，充分利用碱蓬草将养殖废水进行处理，保护养殖环境。

5.3　加快推进产品追溯体系的建立与应用

建立水产品质量安全可追溯体系，将养殖、流通、消费等环节全部纳入二维码溯源体系，从渔场到餐桌，全程实现“来源可追溯、去向可查证、责任可追究”的管理体系，建立产业组织形式，协调生产数量，采取“公司＋农户”的形式，建立与应用产品追溯体系，进一步推进品牌文化建设。

5.4　加快推进大菱鲆疫苗的应用

对已经研发获批的大菱鲆疫苗在已试点示范的基础上，加快推进大菱鲆疫苗的使用。

5.5　引进适合工厂化养殖模式的名、特、优鲆鲽新苗种

针对大菱鲆几年来市场价格情况，加快名、特、优鲆鲽新品种的引进。

5.6　政府引导、制定标准

通过政府引导，协会加强行业自律，提高企业准入门槛，规划和调查产业布局，扶持和建立几个或多个企业集团和领军企业。

加强产品质量监管和查处力度，改进产地检疫检测管理。出台制定标准：政府制定大菱鲆养殖销售地方标准，养殖行业协会制定大菱鲆养殖销售行业标准，养殖企业制定大菱鲆养殖销售企业标准，规范生产、销售行为。

附表1　2016年度葫芦岛综合试验站示范县鲆鲽类育苗及成鱼养殖情况统计表

项目 \ 品种		兴城现代渔业园区	兴城沙后所	绥中县	龙港区	福建省宁德市霞浦县
		大菱鲆	大菱鲆	大菱鲆	大菱鲆	大菱鲆
育苗	面积/m^2	5 000	—	—	—	—
	产量/万尾	300	—	—	—	—
工厂养殖	面积/万平方米	160	40	42	12	—
	年产量/t	16 315	1 850	2 282	870	—
	年销售量/t	20 100	2 250	2 870	730	—
	年末库存量/t	12 275	2 440	2 002	530	—
池塘养殖	面积/亩	—	—	—	—	—
	年产量/t	—	—	—	—	—
	年销售量/t	—	—	—	—	—

（续表）

项目 \ 品种		兴城现代渔业园区	兴城沙后所	绥中县	龙港区	福建省宁德市霞浦县
		大菱鲆	大菱鲆	大菱鲆	大菱鲆	大菱鲆
池塘养殖	年末库存量 / t	—	—	—	—	—
网箱养殖	面积 / m^2	—	—	—	—	40 000
	年产量 / t	—	—	—	—	290
	年销售量 / t	—	—	—	—	290
	年末库存量 / t	—	—	—	—	0
户数	育苗户数	1	—	—	—	—
	养殖户数	450	30	120	33	35

附表 2　葫芦岛综合试验站五个示范县养殖面积、养殖产量及品种构成

项目 \ 品种	年产总量	牙鲆	漠斑牙鲆	半滑舌鳎	大菱鲆
工厂化育苗面积 / m^2	5 000	—	—	—	5 000
工厂化出苗量 / 万尾	300	—	—	—	300
工厂化养殖面积 / m^2	2 540 000	—	—	—	2 540 000
工厂化养殖产量 / t	21 317	—	—	—	21 317
池塘养殖面积 / 亩	—	—	—	—	—
池塘年总产量 / t	—	—	—	—	—
网箱养殖面积 / m^2	40 000	—	—	—	40 000
网箱年总产量 / t	290	—	—	—	290
各品种工厂化育苗面积占总面积的比例 / %	100	—	—	—	100
各品种工厂化出苗量占总出苗量的比例 / %	100	—	—	—	100
各品种工厂化养殖面积占总面积的比例 / %	100	—	—	—	100
各品种工厂化养殖产量占总产量的比例 / %	100	—	—	—	100
各品种池塘养殖面积占总面积的比例 / %	—	—	—	—	—
各品种池塘养殖产量占总产量的比例 / %	—	—	—	—	—

（葫芦岛综合试验站站长　王　辉）

烟台综合试验站产区调研报告

1　示范县(市、区)鲆鲽类养殖现状

本综合试验站下设五个示范县(市、区),分别为:烟台市福山区、海阳市、蓬莱市、长岛县,福建连江县。福山区、海阳市、蓬莱市主要以工厂化养殖鲆鲽类为主,长岛县及连江县以网箱养殖鲆鲽类为主。

1.1　育苗面积及苗种产量

1.1.1　育苗面积

五个示范县育苗总面积为 53 500 m^2,其中海阳市 25 000 m^2、福山区 14 000 m^2、蓬莱市 14 500 m^2。按品种分:大菱鲆育苗面积 38 000 m^2、牙鲆 9 000 m^2、半滑舌鳎 5 000 m^2、其他 1 500 m^2。

1.1.2　苗种年产量

五个示范县共计 31 户育苗厂家,总计育苗 2 660 万尾。其中:大菱鲆 1 280 万尾(5 cm～6 cm),牙鲆 780 万尾,半滑舌鳎 500 万尾(5 cm～6 cm),其他 100 万尾。长岛县及连江县主要是网箱养殖鲆鲽鱼类,因此无育苗业户,所需苗种均为外地购买。各县育苗情况如下:

海阳市:20 户育苗厂家,较大规模的育苗厂家为海阳黄海水产有限公司。大菱鲆育苗面积 18 000 m^2,生产苗种 560 万尾;牙鲆育苗面积 2 000 m^2,生产苗种 180 万尾;半滑舌鳎育苗面积 5 000 m^2,生产苗种 500 万尾。

福山区:共 5 户育苗厂家,主要育苗企业有烟台开发区天源水产有限公司、山东东方海洋科技股份有限公司、仁和水产有限公司。大菱鲆育苗面积 10 000 m^2,生产苗种 420 万尾;牙鲆育苗面积 4 000 m^2,生产苗种 300 万尾。

蓬莱市:8 户育苗厂家,较大规模的为蓬莱宗哲养殖有限公司、蓬莱安源水产有限公司。大菱鲆育苗面积 10 000 m^2,生产苗种 300 万尾;牙鲆育苗面积 3 000 m^2,生产苗种 300 万尾;黄盖鲽育苗面积 1 500 m^2,生产苗种 100 万尾。

1.2　养殖面积及年产量、销售量、年末库存量

1.2.1　工厂化养殖

五个示范县中,海阳市、福山区、蓬莱市均为工厂化养殖,共计 230 家养殖户,养殖面积

495 000 m²,年总生产量为 3 510 t,销售量为 2 859 t,年末库存量为 651 t。其中:

海阳市:现有 160 家养殖业户,养殖面积 367 000 m²。大菱鲆养殖面积 228 000 m²,年产量 2 100 t,销售 1 800 t,年末库存 300 t;牙鲆 30 000 m²,年产量 58 t,销售 50 t,库存 8 t;半滑舌鳎 108 000 m²,年产量 320 t,销售 270 t,库存 50 t。

福山区:现共有 14 家养殖业户,养殖面积 50 000 m²。养殖大菱鲆 45 000 m²,年产量 380 t,销售 280 t,年末库存 100 t;牙鲆 5 000 m²,年产量 40 t,销售 32 t,年末库存 8 t。

蓬莱市:共有 15 个鲆鲽类养殖业户,养殖面积 76 000 m²。养殖大菱鲆 46 000 m²,年产量 490 t,销售 320 t,年末存量 170 t;牙鲆 10 000 m²,年产量 60 t,销售 60 t;半滑舌鳎 10 000 m²,年产量 40 t,销售 32 t,年末存量 8 t;养殖黄盖鲽 10 000 m²,年产量 20 t,销售 15 t,年末存量 5 t。

1.2.2 网箱养殖

在长岛县以深海网箱和浅海筏式网箱的养殖方式进行鲆鲽类养殖,连江县则是浅海筏式网箱养殖,主要养殖品种为牙鲆。

长岛县:鲆鲽类养殖业户有 20 户,网箱养殖面积 12 000 m²,养殖产量 190 t,销售 190 t。

连江县:鲆鲽类养殖业户 20 户,网箱养殖面积 13 000 m²,养殖产量 200 t,销售 200 t。

1.3 品种构成

统计五个示范县鲆鲽类养殖面积调查结果,各品种构成如下:

工厂化育苗总面积为 53 500 m²其中大菱鲆为 38 000 m²,占育苗总面积的 71.03%;牙鲆为 9 000 m²,占育苗总面积的 16.82%;半滑舌鳎为 5 000 m²,占育苗总面积的 9.34%;其他品种为 1 500 m²,占育苗总面积的 4.25%。

工厂化育苗总产量为 2 660 万尾。其中大菱鲆 1 280 万尾,占总产苗量的 48.12%;牙鲆为 780 万尾,占总产苗量的 29.32%;半滑舌鳎为 500 万尾,占总产苗量的 18.80%;其他品种为 100 万尾,占总产苗的 3.76%。

工厂化养殖总面积为 495 000 m²其中大菱鲆为 321 000 m²,占总养殖面积的 64.85%;牙鲆为 45 000 m²,占总养殖面积的 9.09%;半滑舌鳎为 118 000 m²,占总养殖面积的 23.84%;黄盖鲽养殖面积为 10 000 m²,占总养殖面积的 2.02% 。

工厂化养殖总产量为 3 510 t。其中大菱鲆为 2 970 t,占总量的 84.62%,牙鲆为 158 t,占总量的 4.50%;半滑舌鳎为 360 t,占总量的 10.26%;其他品种为 22 t,占总量的 0.63%。

网箱养殖总面积 25 000 m²,养殖总产量 400 t,养殖品种为牙鲆和耐高温大菱鲆。

从以上统计可以看出,在进行工厂化养殖的三个示范县中,大菱鲆为主要养殖品种,面积和产量都占绝对优势。在进行网箱养殖的两个示范县中,受养殖环境影响,主要养殖品种为牙鲆。

2　示范县(市、区)科研开展情况

2016年,正在实施的跨年度课题项目2项,分别是海阳市黄海水产有限公司承担的山东省泰山学者蓝色产业领军人才团队支撑计划项目“分子育种技术在鲆鲽鱼类育种中的研发与产业化”、山东省良种工程项目“大菱鲆‘丹法鲆’良种保种及新品种选育”,目前进展顺利,均已按计划完成年度规定的各项研究和经济指标。烟台开发区天源水产有限公司承担的山东“海上粮仓”建设发展资金项目——“大菱鲆良种基地设施提升建设”,项目总投资200万元,其中专项资金100万元,现已建设完成。

3　鲆鲽类产业发展中存在的问题及产业技术需求

目前存在问题:一是目前存在过多的无证经营的鲆鲽类育苗企业或个体户,影响了鲆鲽类苗种的质量和信誉。二是受市场消费的较大影响,鲆鲽类苗种及成鱼价格降幅过大,短时期难以提升,影响养殖户收益。

建议:一是相关主管部门出台相关政策保护养殖产业,规范水产行业的经营秩序,使其良性发展;二是规范鲆鲽类市场的价格及销售问题,对市场进行宏观调控;三是统一各地质量追溯平台,鲆鲽类产品统一平台查询生产信息。

4　当地政府对产业发展的扶持政策

烟台市政府对海洋产业经济发展非常重视,制定了一系列的政策进行产业扶持。为海水养殖企业创办了质量追溯平台,举办各种技术培训班,推广健康养殖技术,保证食品安全。

附表 1　2016 年度烟台综合试验站示范县鲆鲽类育苗及成鱼养殖情况统计表

品种		海阳市			福山区		蓬莱市				长岛县	连江县
		大菱鲆	牙鲆	半滑舌鳎	大菱鲆	牙鲆	大菱鲆	牙鲆	半滑舌鳎	黄盖鲽	牙鲆	牙鲆、大菱鲆
育苗	面积 / m^2	18 000	2 000	5 000	10 000	4 000	10 000	3 000	—	1 500	—	—
	产量 / 万尾	560	180	500	420	300	300	300	—	100	—	—
工厂化养殖	面积 / m^2	228 000	30 000	108 000	45 000	5 000	46 000	10 000	10 000	10 000	—	—
	年产量 / t	2 100	58	320	380	40	490	60	40	20	—	—
	年销售量 / t	1 800	50	270	280	32	320	60	32	15	—	—
	年末库存量 / t	300	8	50	100	8	170	0	8	5	—	—
网箱养殖	面积 / m^2	—	—	—	—	—	—	—	—	—	12 000	15 000
	年产量 / t	—	—	—	—	—	—	—	—	—	190	210
	年销售量 / t	—	—	—	—	—	—	—	—	—	190	210
	年末库存量 / t	—	—	—	—	—	—	—	—	—	0	0
户数	育苗户数	10	2	6	3	2	5	2	—	1	—	—
	养殖户数	100	10	50	12	2	12	2	1	1	20	20

附表2　烟台综合试验站五个示范县养殖面积、养殖产量及品种构成

项目＼品种	年产总量	大菱鲆	牙鲆	半滑舌鳎	其他
工厂化育苗面积/m^2	53 500	38 000	9 000	5 000	1 500
工厂化出苗量/万尾	2 660	1 280	780	500	100
工厂化养殖面积/m^2	495 000	321 000	45 000	118 000	10 000
工厂化养殖产量/t	3 510	2 970	158	360	22
网箱养殖面积/m^2	25 000	2 000	23 000	—	—
网箱年总产量/t	400	10	390	—	—
各品种工厂化育苗面积占总面积的比例/%	100	71.03	16.82	9.34	4.25
各品种工厂化出苗量占总出苗量的比例/%	100	48.12	29.32	18.80	3.76
各品种工厂化养殖面积占总面积的比例/%	100	64.85	9.09	23.84	2.02
各品种工厂化养殖产量占总产量的比例/%	100	84.62	4.50	10.26	0.63

（烟台综合试验站站长　杨　志）

青岛综合试验站产区调研报告

1 示范县(市、区)鲆鲽类养殖现状

本综合试验站下设五个示范县(市、区),分别为:青岛市黄岛区、烟台市莱阳市、日照市岚山区、威海市环翠区和江苏省赣榆县。其育苗、养殖品种、产量及规模见附表1

1.1 育苗面积及苗种产量

1.1.1 育苗面积

五个示范县区育苗总面积为44 000 m^2,其中黄岛区3 000 m^2、环翠区41 000 m^2,岚山区、赣榆县和莱阳市没有苗种生产。

1.1.2 苗种年产量

五个示范县区2015年总计育苗为7 240万尾,2016年总计育苗为4 150万尾,同比增加42.7%。苗种产量中,大菱鲆共计4 100万尾,占总产量的98.8%,仍为鲆鲽类中苗种产量最大的品种。

1.2 养殖面积及年产量、销售量、年末库存量

1.2.1 工厂化养殖

青岛市黄岛区大菱鲆工厂化养殖面积129 000 m^2,与2015年略有增加,大菱鲆的养殖产量为762 t,比2015年有所减少。半滑舌鳎产量为333 t,比2015年有所增加。

莱阳市大菱鲆工厂化养殖面积500 000 m^2,与2015年相比大幅增加。该地区2016年工厂化养殖的鲆鲽类品种只有大菱鲆,无其他品种,养殖模式也全部为工厂化养殖,无池塘养殖和网箱养殖。

日照岚山区工厂化养殖面积45 000 m^2,较2015年显著增加,养殖品种仍呈现多样化特点,其中大菱鲆85 t、牙鲆111 t、半滑舌鳎38 t。

威海环翠区大菱鲆工厂化养殖养殖面积65 000 m^2较2015年显著减少,均为大菱鲆养殖,产量1 538 t。

赣榆县鲆鲽类工厂化养殖总面积320 000 m^2,与2015年规模相当,养殖品种为大菱鲆,产量2 670 t。

1.2.2　池塘养殖

本示范区仅青岛市黄岛区有池塘养殖，养殖品种为牙鲆，池塘的型式均为岩礁池。岩礁池养殖牙鲆，是黄岛区鲆鲽类养殖的一个特色养殖模式。2016 年总面积为 280 亩，与 2015 年一致，产量为 1 460 t。

1.2.3　网箱养殖

本示范区 2016 年度无鲆鲽类网箱养殖。

1.3　品种构成

每品种养殖面积及产量占示范县养殖总面积和总产量的比例见附表 2。

统计五个示范县鲆鲽类养殖面积调查结果，各品种构成如下：

工厂化育苗总面积为 44 000 m^2。其中大菱鲆为 43 000 m^2，占总育苗面积的 97. 73%；牙鲆为 1 000 m^2，占总面积的 2. 27%。

工厂化育苗总出苗量为 4 150 万尾。其中大菱鲆 4 100 万尾，占总出苗量的 98. 80%；牙鲆为 50 万尾，占总出苗量的 1. 20%。

工厂化养殖总面积为 1 059 000 m^2。其中大菱鲆为 1 020 000 m^2，占总养殖面积的 96. 32%；牙鲆为 1 500 m^2，占总养殖面积的 1. 42%；半滑舌鳎为 24 000 m^2，占总养殖面积的 2. 26%。

工厂化养殖总产量为 8 239 t。其中大菱鲆 7 755 t，占总量的 94. 13%；牙鲆为 111 t，占总量的 1. 35%；半滑舌鳎为 373 t，占总量的 4. 52%。

池塘养殖总面积为 280 亩，全部为牙鲆，总产量为 1 460 t。

无网箱养殖。

从以上统计可以看出，在五个示范县内，大菱鲆育苗、养殖的产量和面积都是最高的，占绝对优势，且相比 2015 年，大菱鲆的占比又有所提高。而池塘养殖和网箱养殖的规模均很小。说明在五个示范县区内，工厂化养殖大菱鲆是鲆鲽类养殖的主要品种和养殖模式。

2　示范县(市、区)科研开展情况

青岛市黄岛区青岛通用水产养殖有限公司是青岛综合试验站的建设依托单位，在持续与岗位科学家协作进行大菱鲆全雌苗种研究，并进行循环水养殖的试验与示范、大菱鲆疫苗免疫试验与示范等研究工作。

2016 年 10 月 13 日，黄海水产研究所邀请有关专家，在青岛市黄岛区通用水产养殖有限公司，对国家鲆鲽类产业技术体系全雌苗种生产岗位“大菱鲆全雌苗种生产技术研究”任务进行了现场验收。

威海环翠区圣航公司在 2016 年继续与科研院所合作开展了大菱鲆选育等研发工作，该公司是鲆鲽类重要的受精卵供应商，亲鱼储备多，有很好的资源进行技术研发。威海市新增大菱鲆国家级良种场一家，预计将来可能会开展大菱鲆品种选育、育苗相关的研究工作。

3 鲆鲽类产业发展中存在的问题

3.1 产业面临转型发展的压力

自2015年，鲆鲽类的主要品种——大菱鲆市场出现价格下跌、销量下降显著的情况，社会经济与消费特点的变化是主要原因，即高端餐饮消费下降，整体餐饮消费理性，海水鱼产品消费结构发生变化，产业从追求量的增加向追求质量的提高转变。同时，鲆鲽类养殖又处在向工业化升级的关键阶段，而大多数养殖业户仍以经验养殖为主。建议体系更多地进行养殖工艺方面的研究，为养殖业户提供更多直接指导生产的技术，推动养殖的科学化、标准化，从而提高养殖稳定性、质量和效益。

3.2 质量安全问题

近年来，随着相关法规的宣传、技术培训、抽检监管等措施的加强，大菱鲆养殖生产中违禁药物的使用已大幅减少，产品合格率显著提高，2016年度各消费地区流通环节的抽检结果也表明了这个情况。但鲆鲽类的质量安全水平仍缺乏长效的保障机制。

3.3 鲆鲽类病害垂直传染的问题亟待重视，相关技术亟待开发

鲆鲽类产业分工越来越细，现已形成亲鱼培育企业、育苗企业和养殖企业的分工。多数没有培育亲鱼，育苗场从亲鱼培育企业购苗受精卵，但现在生产环节中缺乏受精卵消毒环节，导致垂直传染的病原从亲鱼培育企业直接传递到育苗场，再从育苗场通过鱼苗传递到养殖场，病原的传播畅通无阻，可能是鲆鲽类病害多、发病面广的重要因素。育苗企业也有反映，似乎一些育苗过程中的病害与受精卵的来源有显著的相关性。

很多品种的研究表明，受精卵可能携带多种病原。在西方渔业发达国家，受精卵消毒等预防病原垂直传染的技术已是常规的、例行的繁育技术规范。而这方面在我国很多海水鱼繁育实践中尚很欠缺。

3.4 机械化、自动化设备和技术亟待开发

为提高生产效率，降低劳动强度，实现工业化养殖的目标，鲆鲽类养殖产业需要实现养殖生产方面的机械化与自动化。就养殖生产而言，可考虑以下设备与产品的研发：自动投饵、鱼苗筛分和计数等设备，水质监测远程传输系统、工厂化养殖管理软件系统等。

3.5　鲆鲽类文化宣传欠缺

鲆鲽类养殖以工厂化养殖为主，环境可控，不受环境污染、气候变化的影响。鲆鲽类是营养丰富的健康海水鱼。这些涉及鲆鲽类产品、生产方式的特点缺乏基本的宣传，消费者对于鲆鲽类产品不了解，限制了市场的推广。而且，在出现个别质量安全问题时，容易造成质量安全恐慌。所以我们认为，鲆鲽类的宣传与产业文化建设是当前急需的工作之一。

3.6　鲆鲽类生产单位组织化程度低

鲆鲽类养殖生产者的大多数是小型养殖场，以个体经营为主，缺乏龙头企业，缺乏专业合作社组织（已注册的合作社组织多数无实质运营）。这导致盲目生产、产业无序发展、质量安全监管的困境、价格大起大落等等。所以我们认为，促进鲆鲽类产业组织化程度是当前紧迫的工作之一。

2016年，中国水产流通与加工协会成立，各地也成立了较多合作社、协会，有了提高组织化程度的趋势。

3.7　鲆鲽类产品应开拓新的流通方式

鲆鲽类产品主要以活鲜的形式进入市场，流通环节分别为：养殖业户、鱼贩、海鲜批发市场、餐饮酒店，主要以餐饮酒店或宴会消费为主。

近年来，由于整体消费环境变化、食品安全谣传等因素的影响，大菱鲆消费显著下降，造成养殖产品积压。在这种情况下，应积极开拓产品形式（比如冰鲜、冷冻），新的流通渠道（比如超市、电商等）。

3.8　鲆鲽类养殖需向质量效益型转变

传统的鲆鲽类养殖属于粗放型、资源消耗型养殖，在鲆鲽类养殖的发展阶段，产业主要的关注点在如何扩大产量，但是，当前社会经济的转型已深刻影响到鲆鲽类产业，产业面临从追求产量向质量效益型的转变压力。对大菱鲆养殖而言，传统的“深井海水＋温室大棚”的养殖模式需要消耗大量的地下海水资源，地下海水资源的短缺成为养殖发展的瓶颈，水量不足又可能导致病害发生，养殖效益下降，且容易出现滥用药物的情况。所以，鲆鲽类养殖迫切需要采用更先进的养殖模式，降低对大量深井海水的依赖，进行工业化养殖，降低养殖成本，提高养殖产品质量，从而实现向质量效益型的转变。

3.9　鲆鲽类养殖技术体系仍亟待深入发展和提升

经过多年的发展，鲆鲽类养殖技术体系已得到长足发展，已成为海水鱼工厂化养殖的典范。但是，由于过去养殖业户更主要地追求产量，采取不科学的方法降低成本，对于规范化重视不足，仍存在很多技术需提高或推广。向质量效益型的工业化养殖模式发展需要系统的技术体系支撑，这需要从育种、苗种培育、养殖、加工、饲料等各个方面建立标准和规范。

4 当地政府对产业发展的扶持政策

海水鱼养殖仍是示范县区重要的产业之一，各县区渔业部门对于促进鲆鲽类养殖业发展采取各种扶持政策。比如，莱阳市、青岛市黄岛区通过建设网站帮助鲆鲽类产品销售、鼓励合作社的建立、鼓励水产品品牌建设等措施，推动产业的发展。山东省也推出了可追溯技术与平台，鼓励养殖业户实施。定期举办技术培训也是各示范县区均每年进行的扶持措施。

5 鲆鲽类产业技术需求

根据2016年示范县区调研及我站对示范县区产业状况的分析，总结技术需求如下：

（1）大菱鲆育苗阶段腹水病防治技术；

（2）循环水养殖技术：系统的设计建造与运行管理技术；

（3）半滑舌鳎全雌苗种培育技术；

（4）大菱鲆全雌苗种繁育技术；

（5）机械化、自动化设备和技术；

（6）鲆鲽类病害垂直传染类病害防控技术。

附表 1　2016 年度青岛综合试验站示范县鲆鲽类育苗及成鱼养殖情况统计表

品种		青岛市黄岛区			日照市岚山区			江苏省赣榆县		莱阳市	威海市环翠区	
		大菱鲆	牙鲆	半滑舌鳎	大菱鲆	牙鲆	半滑舌鳎	大菱鲆	半滑舌鳎	大菱鲆	大菱鲆	牙鲆
育苗	面积 / m^2	3 000	—	—	—	—	—	—	—	—	40 000	1 000
	产量 / 万尾	500	—	—	—	—	—	—	—	—	3 600	50
工厂养殖	面积 / m^2	114 000	—	15 000	21 000	15 000	9 000	320 000	—	500 000	65 000	—
	年产量 / t	762	—	333	85	111	38	2 670	2	2 700	1 538	—
	年销售量 / t	792	—	264	132	136	51	2 670	2	2 000	1 600	—
	年末库存量 / t	400	—	189	34	36	30	330	0	1 500	600	—
池塘养殖	面积 / 亩	—	280	—	—	—	—	—	—	—	—	—
	年产量 / t	—	1 460	—	—	—	—	—	—	—	—	—
	年销售量 / t	—	1 400	—	—	—	—	—	—	—	—	—
	年末库存量 / t	—	800	—	—	—	—	—	—	—	—	—
网箱养殖	面积 / m^2	—	—	—	—	—	—	—	—	—	—	—
	年产量 / t	—	—	—	—	—	—	—	—	—	—	—
	年销售量 / t	—	—	—	—	—	—	—	—	—	—	—
	年末库存量 / t	—	—	—	—	—	—	—	—	—	—	—
户数	育苗户数	3	4	—	6	7	6	—	—	—	20	1
	养殖户数	35	26	3	20	18	12	60	1	15	30	—

附表 2　青岛综合试验站五个示范县养殖面积、养殖产量及品种构成

项目 \ 品种	年产总量	大菱鲆	牙鲆	半滑舌鳎
工厂化育苗面积 / m^2	44 000	43 000	1 000	—
工厂化出苗量 / 万尾	4 150	4 100	50	—
工厂化养殖面积 / m^2	1 059 000	1 020 000	15 000	24 000
工厂化养殖产量 / t	8 239	7 755	111	373
池塘养殖面积 / 亩	280	—	280	—
池塘年总产量 / t	1 460	—	1 460	—
网箱养殖面积 / m^2	—	—	—	—
网箱年总产量 / t	—	—	—	—
各品种工厂化育苗面积占总面积的比例 / %	100	97. 73	2. 27	—
各品种工厂化出苗量占总出苗量的比例 / %	100	98. 80	1. 20	—
各品种工厂化养殖面积占总面积的比例 / %	100	96. 32	1. 42	2. 26
各品种工厂化养殖产量占总产量的比例 / %	100	94. 13	1. 35	4. 52
各品种池塘养殖面积占总面积的比例 / %	100	—	100	—
各品种池塘养殖产量占总产量的比例 / %	100	—	100	—

(青岛综合试验站站长　张和森)

莱州综合试验站产区调研报告

1　示范县(市、区)鲆鲽类养殖现状

莱州综合试验站下设莱州市、昌邑市、龙口市、招远市、乳山市五个示范县，并把莱州市、昌邑市作为优良半滑舌鳎苗种主推县市，把莱州市作为大菱鲆优质苗种的生产县市，把乳山市作为牙鲆优质苗种的生产县市。2016年，在体系首席专家、岗位专家、功能室及示范县各技术骨干、养殖示范企业、养殖户的支持协作下，体系工作进展顺利，在鲆鲽类价格下降的不利因素下，示范区养殖面积和产量稳定，产业发展合理，并取得了多项验收技术成果。其育苗、养殖品种、产量及规模见附表1

1.1　育苗面积及苗种产量

1.1.1　育苗面积

五个示范县育苗总面积为103 600 m^2，其中莱州市90 000 m^2、乳山县13 600 m^2。按品种分：大菱鲆育苗面积69 600 m^2、半滑舌鳎30 000 m^2、牙鲆4 000 m^2。

1.1.2　苗种年产量

五个示范县共计72户育苗厂家，总计育苗1 099.5万尾，其中大菱鲆430万尾(5 cm)、半滑舌鳎512万尾(5 cm)、牙鲆157.5万尾。各县育苗情况如下：

莱州市：26家育苗企业，其中大菱鲆育苗企业12家、半滑舌鳎育苗企业14家。生产大菱鲆300万尾、半滑舌鳎512万尾。年末存量大菱鲆120万尾、半滑舌鳎133万尾。苗种除自用外，其余主要销往辽宁、河北、山东、江苏、天津等省市。

乳山市：44家育苗企业，其中大菱鲆育苗企业33家、牙鲆育苗企业11家。生产大菱鲆130万尾、牙鲆苗种157.5万尾。年末存量大菱鲆65万尾、牙鲆3万尾。苗种除本市自用外，其余销往山东沿海县市。

1.2　养殖面积及年产量、销售量、年末库存量

试验站所辖五个示范县养殖模式为工厂化养殖，养殖面积总计为1 747 000 m^2，养殖企业共计1 024家，年产鲆鲽类共计5 335 t，销售量为5 308.9 t，年末库存量为5 651.5 t。

莱州市：共有养殖企业465户，工厂化养殖面积887 000 m^2，养殖大菱鲆和半滑舌鳎。

其中大菱鲆养殖面积 722 000 m²,全年产量 2 411 t,销售量 2 307 t,年末库存量 1 822 t;半滑舌鳎养殖面积 165 000 m²,全年产量 188 t,销售量 220 t,年末库存量 704 t。

龙口市:共有养殖企业 53 户,工厂化养殖面积 88 000 m²,养殖大菱鲆和半滑舌鳎。其中大菱鲆养殖面积 74 000 m²,全年产量 247.6 t,销售量 268.1 t,年末库存量 332 t;半滑舌鳎养殖面积 14 000 m²,全年产量 37.9 t,销售量 46.7 t,年末库存量 48.1 t。

招远市:共有养殖企业 38 户,工厂化养殖面积 80 000 m²,养殖大菱鲆和半滑舌鳎。其中大菱鲆养殖面积 70 000 m²,全年产量 112.7 t,销售量 89.7 t,年末库存量 345 t;半滑舌鳎养殖面积 10 000 m²,全年产量 14 t,销售量 8.7 t,年末库存量 23.5 t。

昌邑市:共有养殖企业 344 户,工厂化养殖面积 610 000 m²,养殖大菱鲆和半滑舌鳎。其中大菱鲆养殖面积 530 000 m²,全年产量 1 288 t,销售量 1 282 t,年末库存量 1 250 t;半滑舌鳎养殖面积 80 000 m²,全年产量 784 t,销售量 781 t,年末库存量 801 t。

乳山市:共有养殖企业 124 户,工厂化养殖面积 82 000 m²,养殖大菱鲆和半滑舌鳎。其中大菱鲆养殖面积 64 000 m²,全年产量 200.7 t,销售量 255.7 t,年末库存量 270 t;牙鲆养殖面积 18 000 m²,全年产量 51.1 t,销售量 50 t,年末库存量 56 t。

1.3 品种构成

每品种养殖面积及产量占示范县养殖总面积和总产量的比例见附表 2。

统计五个示范县鲆鲽类养殖面积调查结果,各品种构成如下:

工厂化育苗总面积为 103 600 m²。其中大菱鲆为 69 600 m²,占总育苗面积的 67.18%;半滑舌鳎为 30 000 m²,占总面积的 28.96%;牙鲆为 4 000 m²,占总面积的 3.86%。

工厂化育苗总出苗量为 1 099.5 万尾。其中大菱鲆 430 万尾,占总出苗量的 39.11%;半滑舌鳎为 512 万尾,占总出苗量的 46.57%;牙鲆为 157.5 万尾,占总出苗量的 14.32%。

工厂化养殖总面积为 1 747 000 m²。其中大菱鲆为 1 460 000 m²,占总养殖面积的 83.57%;半滑舌鳎为 269 000 m²,占总养殖面积的 15.40%;牙鲆为 18 000 m²,占总养殖面积的 1.03%。

工厂化养殖总产量为 5 335 t,其中大菱鲆 4 260 t,占总量的 79.85%,半滑舌鳎为 1 023.9 t,占总量的 19.19%;牙鲆为 51.1 t,占总量的 0.96%。

从以上统计可以看出,在五个示范县内,大菱鲆养殖面积和产量最大,其次为半滑舌鳎,牙鲆养殖面积和产量最小。

2 示范县(市、区)科研开展情况

2.1 技术创新

创新养殖模式－离岸大型浮绳式围网建立:2016 年试验站与体系专用养殖网箱岗、高效养殖模式岗进行技术合作创新，突破围网构建装备与工程技术开发、适养品种选择、立体养殖工艺等关键技术，首次在莱州湾开阔海域建立2个离岸大型浮绳式围网（100 m × 50 m × 12 m），新增深远海养殖面积10 000 m^2、养殖水体120 000 m^3，围网抗风浪、耐流性能良好，并开展底层半滑舌鳎等鲆鲽类、中上层游泳性鱼类的立体生态养殖实验，养殖鱼类生长良好，成果通过专家组现场验收。

2.2 成果鉴定

（1）2016年7月5日，“云龙斑（云纹石斑鱼♀ × 鞍带石斑鱼♂）繁育群体、家系建立及优良苗种培育”通过现场验收；

（2）2016年7月5日，“石斑鱼种质细胞冷冻库建立及应用技术”通过现场验收；

（3）2016年7月5日，“褐石斑鱼（♀）× 鞍带石斑鱼（♂）杂交育苗”通过现场验收；

（4）2016年7月5日，“赤点石斑鱼（♀）× 鞍带石斑鱼（♂）杂交育苗”通过现场验收；

（5）2016年7月5日，“清水石斑鱼（♀）× 棕点石斑鱼（♂）杂交育苗”通过现场验收；

（6）2016年7月5日，“驼背鲈（♀）× 鞍带石斑鱼（♂）杂交育苗”通过现场验收；

（7）2016年7月5日，“驼背鲈规模化人工繁育技术”通过现场验收；

（8）2016年9月14日，“云纹石斑鱼（♀）× 鞍带石斑鱼（♂）杂交苗种‘云龙斑’养殖”通过现场验收；

（9）2016年11月18日，“国家鲆鲽类产业技术体系离岸大型浮绳式围网创新养殖模式”通过现场验收。

2.3 发表论文情况

2016年，莱州示范县参与发表海水鱼繁育相关论文6篇：《红鳍东方鲀（*Takifugu rubripes*）雌、雄个体的形态特征比较》、《棕点石斑鱼（*Epinephelus fuscoguttatus* ♀）× 鞍带石斑鱼（*Epinephelus lanceolatus* ♂）F1染色体制备方法及核型分析》、《养殖2龄赤点石斑鱼（*Epinephelus akaara*）的营养价值与畸形原因分析》、《斑石鲷早期发育的异速生长模式》、*The Genome and Transcriptome of Japanese Flounder Provide Insights into Flatfish Asymmetry*、《盐度对云纹石斑鱼（*Epinehelus moara* ♀）× 鞍带石斑鱼（*Epinehelus lanceolatus* ♂）受精卵孵化的影响及杂交仔稚幼鱼形态发育观察》。循环水相关论文2篇：《基于高通量测序的石斑鱼循环水养殖生物滤池微生物群落分析》、《不同浓度臭氧对循环水养殖系统生物膜活性及其净化效能的影响》。

2.4 申请专利、授权情况

2016年申请专利9项，其中授权专利6项。

表1 申请专利

序号	专利名称	类型	申请时间	申请号
1	池塘循环水养殖水处理系统	发明专利	2016年5月	201610332643.5
2	驼背鲈与鞍带石斑鱼的工厂化杂交育苗方法	发明专利	2016年5月	201610332683.X
3	循环水养殖池塘内置处理池	发明专利	2016年5月	201610332599.8

表2 授权专利

序号	专利名称	类型	授权时间	授权号
1	斑石鲷工厂化循环水养殖系统	实用新型专利	2016年10月	ZL201620460747.X
2	成片养殖池虹吸排污装置真空集中控制系统	实用新型专利	2016年10月	ZL201620460731.9
3	循环水养殖保温棚	实用新型专利	2016年10月	ZL201620460695.6
4	循环水养殖池塘紫外线消毒系统	实用新型专利	2016年10月	ZL201620460675.9
5	循环水养殖池塘自动排污系统	实用新型专利	2016年10月	ZL201620460656.6
6	坐落于生物精华池内的循环水分离井	实用新型专利	2016年10月	ZL201620460591.5

3 鲆鲽类产业发展中存在的问题

随着鲆鲽类产业逐渐稳定发展，苗种繁育和养殖技术的成熟，对养殖技术、设施设备、从业人员素质要求的门槛降低，养殖行业逐渐趋向于平衡，即总产量围绕市场需求量之间波动。这样产品的价格与其本身价值的平衡关系也基本确定，即鲆鲽类养殖利润会逐渐被压缩，而进入微利时代，原有的较大的利润空间被压缩。这样要求养殖从业者需要不断优化养殖工艺、升级养殖设施设备、提高养殖技术，从而达到提高养殖成活率、提高养殖效率、降低养殖成本和增加养殖效益的目的。

其次，养殖门槛的降低也可能导致市场无序化，尤其是当出现价格上涨时，如半滑舌鳎在2015年价格到了260元/千克，2016年春天育苗量大幅增加，养殖从业者迅速增加，从试验站辖区示范县统计数据可知，2016年产量比2015年增加了58%。另外一个明显的例子就是大菱鲆，2015年市场不景气导致2016年辖区五个示范县累计养殖量缩减1 388.5 t，比2015年减少24.58%。但随着大菱鲆价格的反弹，养殖量会逐渐增加，比较明显的表现就是大菱鲆苗种单价到了1.8元/尾。但这种较大的波动不利于产业的发展，属于市场的无序化。

此外，由于鲆鲽类养殖企业比较分散，个体养殖企业较多，难以实现规范化管理，养殖水产品质量管理难以控制，呋喃类“药残”事件仍然存在，这对于产业的健康发展极其不利，这也导致2016年大菱鲆价格的波动幅度较大，直接影响养殖企业的效益。因而，如何规范产业发展，实现质量可控，是我们要面对的问题。

4　当地政府对产业发展的扶持政策

山东省是渔业大省，莱州综合试验站下设5个示范县都是山东省海水养殖重点县市，对海水养殖行业推动作用较大。山东省“海上粮仓”战略实施，莱州明波、招远发海、昌邑海丰、乳山科合等企业都有重点项目，各地渔业主管部门出台相应优惠政策，如用海、用地的手续简化，渔业补贴等，刺激渔业的快速发展。

此外，体系与山东省现代农业产业技术体系鱼类创新团队加强合作，加快推进产业发展。

三是各地对水产技术人才的重视，每年由主管部门、企业举办的技术研讨会、基层渔业技术人才培训等，邀请行业专家到场培训指导，为水产行业基层提供良好的人才支撑和智力保障，提升行业技术水平。

5　鲆鲽类产业技术需求

5.1　半滑舌鳎、圆斑星鲽体色改良技术研究

针对目前半滑舌鳎养殖过程中，底板颜色容易黑化的问题，依托现代育种技术，以及通过营养强化、环境控制等调控，改变养殖商品鱼体色，提高商品鱼的附加值。

5.2　大型钢制管桩围网等新型养殖模式的构建

针对渔业发展向深远海拓展的方向，加快钢制管桩围网、浮绳式围网等新型大水体养殖模式的开发，实现立体健康养殖，提升养殖鱼类品质，降低养殖能耗，推进渔业的转型升级。

5.3　开展鱼藻共生养殖

工厂化循环水养殖是“十三五”渔业发展重要内容，也是工厂化养殖的最先进的模式，但循环水养殖长期运行过程中，铵态氮、硝酸氮富集问题，以及少量外排水硝酸氮的浓度比较大的问题，值得讨论。这也是开展鱼藻共生系统研究的目的，利用大型藻类实现硝酸氮的利用转化，降低水体中的硝酸氮，真正实现养殖的可持续。

附表 1　2016 年度莱州综合试验站示范县鲆鲽类育苗及成鱼养殖情况统计表

品种		莱州市		昌邑市		招远市		龙口市		乳山市	
		大菱鲆	半滑舌鳎	大菱鲆	半滑舌鳎	大菱鲆	半滑舌鳎	大菱鲆	半滑舌鳎	大菱鲆	牙鲆
育苗	面积 / m^2	60 000	30 000	—	—	—	—	—	—	9 600	4 000
	产量 / 万尾	300	512	—	—	—	—	—	—	130	157. 5
工厂养殖	面积 / m^2	722 000	165 000	530 000	80 000	70 000	10 000	74 000	14 000	64 000	18 000
	年产量 / t	2 411	188	1 288	784	112. 7	14	247. 6	37. 9	200. 7	51. 1
	年销售量 / t	2 307	220	1 282	781	89. 7	8. 7	268. 1	46. 7	255. 7	50
	年末库存量 / t	1 822	704	1 250	801	345	23. 5	332	48. 1	270	56
池塘养殖	面积 / 亩	—	—	—	—	—	—	—	—	—	—
	年产量 / t	—	—	—	—	—	—	—	—	—	—
	年销售量 / t	—	—	—	—	—	—	—	—	—	—
	年末库存量 / t	—	—	—	—	—	—	—	—	—	—
网箱养殖	面积 / m^2	—	—	—	—	—	—	—	—	—	—
	年产量 / t	—	—	—	—	—	—	—	—	—	—
	年销售量 / t	—	—	—	—	—	—	—	—	—	—
	年末库存量 / t	—	—	—	—	—	—	—	—	—	—
户数	育苗户数	12	14	—	—	—	—	—	—	33	11
	养殖户数	256	209	208	136	35	3	49	4	98	26

附表 2　莱州综合试验站五个示范县养殖面积、养殖产量及品种构成

项目 \ 品种	年产总量	大菱鲆	半滑舌鳎	牙鲆
工厂化育苗面积 / m^2	103 600	69 600	30 000	4 000
工厂化出苗量 / 万尾	1 099. 5	430	512	157. 5
工厂化养殖面积 / m^2	1 747 000	1 460 000	269 000	18 000
工厂化养殖产量 / t	5 335	4 260	1 023. 9	51. 1
池塘养殖面积 / 亩	—	—	—	—
池塘年总产量 / t	—	—	—	—
网箱养殖面积 / m^2	—	—	—	—
网箱年总产量 / t	—	—	—	—
各品种工厂化育苗面积占总面积的比例 / %	100	67. 18	28. 96	3. 86
各品种工厂化出苗量占总出苗量的比例 / %	100	39. 11	46. 57	14. 32
各品种工厂化养殖面积占总面积的比例 / %	100	83. 57	15. 40	1. 03
各品种工厂化养殖产量占总产量的比例 / %	100	79. 85	19. 19	0. 96
各品种池塘养殖面积占总面积的比例 / %	—	—	—	—
各品种池塘养殖产量占总产量的比例 / %	—	—	—	—

（莱州综合试验站站长　翟介明）

山东综合试验站产区调研报告

1 示范县(市、区)鲆鲽类养殖现状

本综合试验站下设五个示范县(市、区),分别为:日照东港、烟台牟平、威海荣成、威海文登、滨州无棣。其中威海荣成是全国大菱鲆苗种的主要产区。各示范县(市、区)其育苗、养殖品种、产量及规模见附表1

1.1 育苗面积及苗种产量

1.1.1 育苗面积

五个示范县育苗总面积为180 200 m^2,其中日照东港13 000 m^2、威海荣成160 000 m^2、威海文登6 000 m^2、滨州无棣12 000 m^2,烟台牟平无育苗生产。按品种分:大菱鲆育苗面积166 000 m^2、半滑舌鳎7 200 m^2、牙鲆7 000 m^2。

1.1.2 苗种年产量

五个示范县共计31户育苗厂家,总计育苗10 325万尾,其中大菱鲆9 600万尾、半滑舌鳎225万尾、牙鲆500万尾。各县育苗情况如下:

日照东港:半滑舌鳎育苗厂家4家,生产半滑舌鳎苗种185万尾;牙鲆育苗厂家6家,生产牙鲆苗种500万尾。

威海荣成:大菱鲆育苗厂家20家,生产大菱鲆苗种9 000万尾。我国目前大菱鲆生产所需苗种主要来自威海荣成。

威海文登:大菱鲆育苗厂家2家,生产大菱鲆苗种600万尾。

滨州无棣:半滑舌鳎育苗厂家仅海城生态科技集团有限公司一家,生产半滑舌鳎苗种39.6万尾。

烟台牟平:无育苗生产。

1.2 养殖面积及年产量、销售量、年末库存量

1.2.1 工厂化养殖

五个示范县均为开放式养殖,共计546家养殖户,养殖面积524 800 m^2,年总生产量为

4 174 t，销售量为 2 581 t，年末库存量为 1 593 t。其中：

日照东港：416 户，大菱鲆、牙鲆、半滑舌鳎养殖面积分别为 220 000 m^2、6 000 m^2、60 000 m^2。大菱鲆产量 2 127 t，销售 538 t，年末存量 1 589 t；牙鲆产量 124 t，销售 35 t，年末存量 99 t；产量 361 t，销售 122 t，年末存量 122 t。

威海荣成：主要为网箱养殖，工厂化养殖所占比例较小，工厂化养殖面积 3 000 m^2，生产大菱鲆 30 t，销售 20 t，年末存量 10 t。

威海文登：文登养殖大棚目前大部分已被政府收回，养殖户进行了转产，目前成鱼养殖 16 户，养殖面积 200 000 m^2，生产大菱鲆 1 039 t，销售 1 029 t，年末存量 10 t。

滨州无棣：2 户，养殖面积 18 000 m^2，生产半滑舌鳎 310 t，销售 219 t，年末存量 91 t。

烟台牟平：2 户，养殖面积 17 800 m^2，生产大菱鲆 183. 9 t，销售 118. 4 t，年末存量 65. 5 t。

1.2.2　网箱养殖

只有威海荣成存在网箱养殖方式。养殖 110 家，养殖面积 96 000 m^2，养殖产量 1 500 t。

1. 3　品种构成

每品种养殖面积及产量占示范县养殖总面积和总产量的比例见附表 2。

统计五个示范县鲆鲽类养殖面积调查结果，各品种构成如下：

工厂化育苗总面积为 180 200 m^2。其中牙鲆为 7 000 m^2，占总育苗面积的 3. 9%；半滑舌鳎为 7 200 m^2，占总面积的 4. 0%；大菱鲆为 166 000 m^2，占总面积的 92. 1% 。

工厂化育苗总出苗量为 10 325 万尾。其中牙鲆 500 万尾，占总出苗量的 4. 2%；半滑舌鳎为 225 万尾，占总出苗量的 2. 8%；大菱鲆为 9 600 万尾，占总出苗量的 93%。

工厂化养殖总面积为 524 800 m^2。其中牙鲆为 6 000 m^2，占总养殖面积的 1. 1%；半滑舌鳎为 78 000 m^2，占总养殖面积的 14. 9%；大菱鲆为 440 800 m^2，占总养殖面积的 84% 。

工厂化养殖总产量为 4 174. 2 t。其中牙鲆 123. 7 t，占总量的 3%；半滑舌鳎为 671. 1 t，占总量的 14. 9%；大菱鲆为 3 379. 4 t，占总量的 80. 9%。

网箱养殖面积为 96 000 m^2。其中牙鲆 64 000 m^2，占 66. 7 %；牙鲆 32 000 m^2，占 33. 3%。

网箱养殖产量为 1 500 t。其中牙鲆 1 000 t，占总量的 66. 7%；牙鲆为 500 t，占总量的 33. 3%。

从以上统计可以看出，在五个示范县内的工厂化养殖，大菱鲆养殖面积和产量都占 80% ，　绝对优势，网箱养殖主要以牙鲆为主。

2 示范县(市、区)科研开展情况

2.1 科研课题情况

东港示范县进行科研项目3项,分别为“星斑川鲽优质苗种繁育及养殖技术开发”“鲆鲽类现代工业化养殖与加工产业化开发”“漠斑牙鲆遗传育种-全雌种苗培育和养殖示范基地”,其中“星斑川鲽优质苗种繁育及养殖技术开发”项目获山东省科技进步二等奖。

2.2 发表论文情况

2000年到2016年文登示范县共拟写有关鲆鲽类养殖方面的论文11篇,分别为:《鲆、鲽鱼类室内水泥池养殖易发疾病及其防治》《半滑舌鳎与中国对虾无公害池塘混养技术》《利用井海水进行菊黄东方鲀越冬试验》《海水池塘鱼、虾、贝生态综合养殖技术》《利用塑料大棚及地下海水养殖刺参》《海水池塘鱼、虾、贝生态综合养殖技术》《大菱鲆与牙鲆盾纤类纤毛虫病的防治技术》《利用大菱鲆养殖设施多茬养殖刺参试验》《半滑舌鳎与中国对虾无公害池塘混养术》《星斑川鲽工厂化养殖技术》。东港示范县发表论文1篇:《地下水再利用养殖牙鲆试验》。

3 鲆鲽类产业发展中存在的问题

3.1 水产品质量安全生产意识有待于进一步加强

2006年11月、2015年7月“大菱鲆”事件给大菱鲆养殖带来毁灭性打击,导致大菱鲆销售价格低于养殖成本。主要原因是从业人员水产品质量安全生产意识淡薄,养殖过程中滥用药物,导致水产养殖品中药物残留超标。因工厂化养殖是高密度、集约化的生产方式,加剧了养殖产品疾病的发生。养殖单位、养殖户为了减少损失,加大了用药;部分养殖单位、养殖户为了过分追求效果,追求低成本,不惜使用一些低价高残留药物;个别制药厂在产品名称上追求立新,一药多名和一名多药,造成养殖品中重复用药,过量用药;没有科学的指导盲目用药现象也加重了养殖中药物残留问题,导致水产品质量安全事故时有发生。

3.2 地下水资源破坏严重,海水倒灌现象严重

以日照东港为例,全区100余万平方米工厂化养殖面积,基本采用地下海水资源进行养殖,按照70%的利用率,养殖面积达到80万平方米,按照最低3个流程计算,每天消耗地下海水资源达到140万吨以上,对地下海水资源破坏严重,海水倒灌严重,部分养殖区水质盐度下降,水质理化指标不稳定,导致养殖病虫害发生频繁。目前,东港、文登示范县均已受到影响。

3.3　养殖种质退化，病害发生频繁

现有养殖品种种质退化现象十分严重，性状表现为性成熟年龄提前，性成熟个体体重变小，生长速度减慢，使达到要求规格的商品鱼养殖周期延长，增加了养殖周期和养殖成本。同时，抗逆性下降，造成水产养殖过程中突发事件经常发生，出现养殖品种大批死亡现象，使不少养殖单位、养殖户遭受巨大损失。

3.4　从业人员素质不高，技术更新缓慢

海水工厂化从业人员大多只有初中及以下学历，几乎没有高端人才。规模小、分布散的行业现状极大制约了高端人才的引进，使养殖人才缺失，与渔业科研高等院所对接困难，严重滞缓了全区海水工厂化养殖实现跨越式发展的进程。同时许多养殖企业对机遇和信息的敏感性不强，缺乏必要的技术储备，对引进的技术和设施（如工厂化循环水养殖）缺乏创新性的消化和吸收，对新技术的接受和应用反应迟钝，致使产品与市场需求不相适应，导致养殖成本居高不下，养殖效益不显著。

3.5　科技支撑力度不够，新品种、新技术更新缓慢

从事渔业技术研究与推广专业技术的人员少，与新形势下面上渔业技术需求不相适应，对渔业发展的科技支撑力度不够。

3.6　信息不灵，市场行情不详

养殖单位、养殖户在养殖过程中有一定的盲目性，一哄而上、一哄而败的现象仍然普遍存在；对市场信息掌握不够，产品销售主要依靠中间人，不能直接对接市场，压价销售现象普遍，一定程度上影响养殖效益。

3.7　缺少品牌意识，市场竞争力不强

多数养殖单位、养殖户还没有把水产养殖品像其他行业产品一样树立品牌，他们参与竞争的意识不强，单纯依靠产品市场价格上扬来增收创收，而不能从加大养殖对象的科技含量、产品整体包装来获取更高的经济效益和社会效益。

4　当地政府对产业发展的扶持政策

俗话说“靠山吃山，靠水吃水”，水产养殖业是各地沿海支柱产业之一，是渔民收入的重要来源。各地政府对其发展采取不同扶持政策。东港示范县通过标准化建设工程、品牌培育工程等，帮助渔民专业合作社登记注册、规范运作，切实提高渔民的组织化程度，大力支持和帮助渔民专业合作社进行产业化、标准化、规模化基地建设，培育和打造农产品品牌，促进

社员增收。通过水产健康养殖活动,给予参与企业水产品健康养殖补贴资金,不断提高水产健康养殖水平。如东示范县大力开展技术培训、送科技下乡、结对帮扶等活动,把各地养殖品种的产前、产中、产后的各种技术送到广大农民手中,使广大养殖户的科学技术水平得到大幅度提高。

5 鲆鲽类产业技术需求

(1) 控制传统养殖模式规模,推进循环水养殖模式发展。

(2) 引导成立水产养殖行业协会,促进水产品的认证认可工作。

(3) 加大执法监督和水产品质量安全培训力度,确保水产品质量安全。

附表 1　2016 年度山东综合试验站示范县鲆鲽类育苗及成鱼养殖情况统计表

品种		东港			荣成		文登	牟平	无棣
		大菱鲆	牙鲆	半滑舌鳎	大菱鲆	牙鲆	大菱鲆	大菱鲆	半滑舌鳎
育苗	面积 / m^2	—	7 000	6 000	160 000	—	6 000	—	1 200
	产量 / 万尾	—	500	185	9 000	—	600	—	39. 6
工厂养殖	面积 / m^2	220 000	600	6 000	3 000	—	200 000	17 800	18 000
	年产量 / t	2 126. 5	123. 7	361. 1	30	—	1 039	183. 9	310
	年销售量 / t	3 293	58. 2	325. 2	20	—	1 029	118. 4	219
	年末库存量 / t	537. 8	99. 4	239. 3	10	—	10	65. 5	91
池塘养殖	面积 / 亩	—	—	—	—	—	—	—	—
	年产量 / t	—	—	—	—	—	—	—	—
	年销售量 / t	—	—	—	—	—	—	—	—
	年末库存量 / t	—	—	—	—	—	—	—	—
网箱养殖	面积 / m^2	—	—	—	32 000	64 000	—	—	—
	年产量 / t	—	—	—	500	1 000	—	—	—
	年销售量 / t	—	—	—	400	900	—	—	—
	年末库存量 / t	—	—	—	100	100	—	—	—
户数	育苗户数	—	6	4	20	—	2	—	1
	养殖户数	312	26	78	110	—	16	2	2

附表 2　山东综合试验站五个示范县养殖面积、养殖产量及品种构成

项目 \ 品种	年产总量	牙鲆	半滑舌鳎	大菱鲆
工厂化育苗面积 / m^2	180 200	7 000	7 200	166 000
工厂化出苗量 / 万尾	10 325	500	225	9 600
工厂化养殖面积 / m^2	524 800	6 000	78 000	440 800
工厂化养殖产量 / t	4 174. 2	123. 7	671. 1	3 379. 4
池塘养殖面积 / 亩	—	—	—	—
池塘年总产量 / t	—	—	—	—
网箱养殖面积 / m^2	—	96 000	64 000	32 000
网箱年总产量 / t	1 500	1 000	—	500
各品种工厂化育苗面积占总面积的比例 / %	100	3. 9	4. 0	92. 1
各品种工厂化出苗量占总出苗量的比例 / %	100	4. 8	2. 2	93. 0
各品种工厂化养殖面积占总面积的比例 / %	100	1. 1	14. 9	84. 0
各品种工厂化养殖产量占总产量的比例 / %	100	3. 0	16. 1	80. 9
各品种网箱养殖面积占总面积的比例 / %	100	66. 7	—	33. 3
各品种网箱养殖产量占总产量的比例 / %	100	66. 7	—	33. 3

（山东综合试验站站长　姜海滨）

日照综合试验站产区调研报告

1　示范县(市、区)鲆鲽类养殖现状

本综合试验站下设5个示范基地(市、区),分别为山东省日照市开发区、山东省潍坊市滨海开发区、山东省青岛市崂山区、浙江省台州市温岭市、日照市涛雒镇。其育苗、养殖品种、产量及规模见附表1。

1.1　育苗面积及苗种产量

1.1.1　育苗面积

5个示范基地育苗总面积为19 000 m^2,其中日照市开发区2 800 m^2、青岛崂山区2 000 m^2,日照市涛雒镇14 200 m^2。按品种分:大菱鲆育苗面积11 100 m^2,牙鲆育苗面积4 900 m^2,半滑舌鳎育苗面积3 000 m^2。浙江台州市温岭市、潍坊滨海开发区无育苗生产。

1.1.2　苗种年产量

5个示范基地共计67户育苗场家,总计育苗3 530万尾,其中大菱鲆1 760万尾,牙鲆870万尾,半滑舌鳎900万尾。浙江台州市温岭市、潍坊滨海开发区无育苗户。各县育苗情况如下:

日照市开发区:大菱鲆育苗面积2 100 m^2,全年生产360万尾,共有育苗户8家;牙鲆育苗面积700 m^2,全年生产130万尾,共有育苗户8家。

青岛崂山区:大菱鲆育苗面积2 000 m^2,全年生产400万尾,共有育苗户7家。

日照市涛雒镇:大菱鲆育苗面积7 000 m^2,全年生产1 000万尾,共有育苗户16家;牙鲆育苗面积4 200 m^2,全年生产740万尾,共有育苗户12家;半滑舌鳎育苗面积3 000 m^2,全年生产900万尾,共有育苗户16家。

1.2　养殖面积及年产量、销售量、年末库存量

日照综合试验站所辖区域主要是工厂化养殖、深水海水养殖。5个示范基地共计461家养殖户,养殖面积606 000 m^2,年总生产量为5 564 t,销售量为5 087 t。其中:

日照市开发区:115户,养殖面积154 500 m^2养殖大菱鲆80 500 m^2,产量810 t,销售量780 t,年末库存量30 t;牙鲆46 000 m^2,产量260 t,销售量220 t,年末库存量40 t;半滑舌鳎

28 000 m^2,产量138 t,销售量125 t,年末库存量13 t;星突江鲽5 000 m^2,产量10 t,销售量10 t,年末库存量0 t。

日照市涛雒镇:252户,养殖面积279 200 m^2养殖大菱鲆142 700 m^2,产量1 988 t,销售量1750 t,年末库存量238 t;牙鲆84 000 m^2,产量950 t,销售量920 t,年末库存量30 t;半滑舌鳎52 500 m^2,产量368 t,销售量323 t,年末库存量45 t;星突江鲽17 000 m^2,产量50 t,销售量50 t,年末库存量0 t。

青岛市崂山区:32户,大菱鲆养殖面积63 500 m^2,产量340 t,销售量315 t,年末库存量25 t。

潍坊滨海开发区:38户,大菱鲆养殖面积80 000 m^2,产量380 t,销售量350 t,年末库存量30 t。

浙江省台州市温岭市:24户,养殖面积35 600 m^2养殖大菱鲆28 800 m^2,产量230 t,销售量204 t,年末库存量26 t;养殖半滑舌鳎6 800 m^2,产量为40 t,销售量40 t,年末库存量0 t。

1.3 品种构成

每品种养殖面积及产量占示范县养殖总面积和总产量的比例见附表2。

统计5个示范基地鲆鲽类养殖面积调查结果,各品种构成如下:

工厂化育苗总面积为19 000 m^2其中大菱鲆为11 100 m^2,占总育苗面积的58.42%;牙鲆为4 900 m^2,占总育苗面积的25.79%;半滑舌鳎为3 000 m^2,占总育苗面积的15.79%。

工厂化育苗总出苗量为3 530万尾。其中大菱鲆1 760万尾,占总出苗量的49.86%;牙鲆870万尾,占总出苗量的24.64%;半滑舌鳎900万尾,占总出苗量的25.50%。

工厂化养殖总产量为5 564 t。其中大菱鲆3 748 t,占总量的67.36%;牙鲆1 210 t,占总量的21.75%;半滑舌鳎546 t,占总量的9.81%;星突江鲽60 t,占总量的1.08%。

工厂化养殖总面积为628 000 m^2其中大菱鲆395 500 m^2,占总量的62.98%;牙鲆130 000 m^2,占总量的20.70%;半滑舌鳎80 500 m^2,占总量的12.82%;星突江鲽22 000 m^2,占总量的3.50%。

从以上统计可以看出,在5个示范基地内,大菱鲆的养殖和产量占绝对优势,其次是牙鲆,半滑舌鳎和星突江鲽所占的比例很少。

2 示范县(市、区)科研开展情况

2.1 专利申请与获得情况

2016年11月24日,申请一项中国发明专利《一种冷冻分割多宝鱼的加工方法》,专利申请号:201611044895.4。

2.2　产业技术宣传与培训情况

2016年1月14日，日照综合试验站在日照举办“养殖大菱鲆使用药物调研及指导”培训班，培训班培训基础技术人员30人，培养养殖户26人。

2016年4月20日，日照综合试验站以山东美佳集团有限公司为依托举办培训会议，培训内容围绕鲆鲽鱼产品的加工工艺、加工过程质量控制以及对鲆鲽鱼类文化的推广，对示范县管理人员、技术骨干、车间主要技术人员进行培训，共培训技术人员80人。

2016年7月19日，加工与质量控制岗位、日照综合试验站和山东省虾蟹创新团队在日照举办“2016年国家鲆鲽类产业技术体系与省级水产创新团队水产品加工与质量安全岗位建设规划研讨会”。

3　鲆鲽类产业发展中存在的问题

大菱鲆屠宰过程中屠宰放血难度较高，现在的屠宰方式是直接静脉、动脉放血，放血不彻底，造成肉质颜色发红，感官品质不良。直接放血，很多血液黏附在鱼体表面、地面、工器具上，容易造成致病菌微生物的生长，对食品安全控制很不利。

消费者对大菱鲆的抵触心理仍然很严重，消费者对冰鲜大菱鲆和速冻大菱鲆的接受度不高，仍然喜欢活鱼。企业加工大菱鲆，综合考虑成鱼的安全性问题、加工成本，最终的冰鲜、速冻产品价格会远远高于市场活鱼价格，消费者难以接受，销售困难。

如何建立完善的养殖-加工的无缝化质量安全控制，是现在亟待解决的问题。

4　当地政府对产业发展的扶持政策

水产养殖业是沿海地区的支出产业之一，是渔民收入的重要来源，各地各级政府对水产养殖业相当重视，并且采取不用的扶持政策，加强对产业的支持。对近海养殖户做好统计管理，帮助养殖企业工商注册登记，发放养殖场养殖执照，海域使用执照等，建立起完善的管理制度。帮助企业做好标准化、规模化基地建设，重点奖励通过“三品一标”认证的养殖企业，加大对名牌建设的力度，努力培育水产品品牌产品，给养殖企业增产增效增值。

开展技术培训、科技下乡等活动，建立定向帮扶，专人负责专企。加强对养殖用药、养殖饲料的管理和喂养指导，政府投资建立检测点，免费检测水产品的质量安全，帮助养殖企业建立科学的养殖观念，坚决杜绝滥用兽药，保证水产品的安全。

5　鲆鲽类产业技术需求

建议实行体系认证制度，建立完善的生产准入制度，设置养殖生产准入，对于规范养殖的养殖户予以发证,由体系进行鉴别和保护,体系对通过认证的养殖户进行培训、抽检,提高养殖规模化、标准化水平,缩小监管范围。在养殖过程中进行指导和监管,发现问题移交相关部门处理。构建对养殖者、贩卖者、加工企业及整个的生产、流通、加工、销售环节的可追溯机制。

加强鲆鲽类养殖产业发展战略研究规划,采取积极政策大力扶持得到认证的合格的养殖户,加大鲆鲽类正面的宣传,引导大众的消费倾向,大力宣传体系推广的先进的养殖技术、控制技术、可追溯二维码,在体系内及消费者中大力的推广,完善可追溯体制并深入人心,确保产品从养殖到餐桌的安全、安心、放心。

针对养殖、加工、销售的相关环节,研究制定系列化、标准化、适合企业实际实施的操作规范,构建完整的质量安全控制体系和溯源体系。

加强与相关专家的沟通联系,探讨新的加工模式,提高副产物的综合利用率。建议体系探讨鲆鲽类新产品加工及市场推广,建立鲆鲽类加工品出口质量安全体系并探讨实施鲆鲽类产品“互联网 +”的销售模式。

附表1　2016年度日照综合试验站示范县鲆鲽类育苗及成鱼养殖情况统计表

品种		育苗 面积/m^2	育苗 产量/万尾	工厂养殖 面积/m^2	工厂养殖 年产量/t	工厂养殖 年销售量/t	工厂养殖 年末库存量/t	池塘养殖 面积/亩	池塘养殖 年产量/t	池塘养殖 年销售量/t	池塘养殖 年末库存量/t	网箱养殖 面积/m^2	网箱养殖 年产量/t	网箱养殖 年销售量/t	网箱养殖 年末库存量/t	户数 育苗户数	户数 养殖户数
日照市开发区	大菱鲆	2 100	360	80 500	810	780	30	—	—	—	—	—	—	—	—	8	55
	牙鲆	700	130	46 000	260	220	40	—	—	—	—	—	—	—	—	8	34
	半滑舌鳎	—	—	28 000	138	125	13	—	—	—	—	—	—	—	—	—	23
	星突江鲽	—	—	5 000	10	10	0	—	—	—	—	—	—	—	—	—	3
青岛市崂山区	大菱鲆	2 000	400	63 500	340	315	25	—	—	—	—	—	—	—	—	7	32
日照市涛雒镇	大菱鲆	7 000	1 000	142 700	1 988	1 750	238	—	—	—	—	—	—	—	—	16	80
	牙鲆	4 200	740	84 000	950	920	30	—	—	—	—	—	—	—	—	12	98
	半滑舌鳎	3 000	900	52 500	368	323	45	—	—	—	—	—	—	—	—	16	60
	星突江鲽	—	—	17 000	50	50	0	—	—	—	—	—	—	—	—	—	14
浙江省台州市温岭市	大菱鲆	—	—	28 800	230	204	26	—	—	—	—	—	—	—	—	—	18
	半滑舌鳎	—	—	6 800	40	40	0	—	—	—	—	—	—	—	—	—	6
潍坊市滨海开发区	大菱鲆	—	—	80 000	380	350	30	—	—	—	—	—	—	—	—	—	38

附表 2　日照综合试验站五个示范县养殖面积、养殖产量及品种构成

项目＼品种	年产总量	牙鲆	半滑舌鳎	大菱鲆	星突江鲽
工厂化育苗面积 / m^2	19 000	4 900	3 000	11 100	—
工厂化出苗量 / 万尾	3 530	870	900	1 760	—
工厂化养殖面积 / m^2	628 000	130 000	80 500	395 500	22 000
工厂化养殖产量 / t	5 564	1 210	546	3 748	60
池塘养殖面积 / 亩	—	—	—	—	—
池塘年总产量 / t	—	—	—	—	—
网箱养殖面积 / m^2	—	—	—	—	—
网箱年总产量 / t	—	—	—	—	—
各品种工厂化育苗面积占总面积的比例 / %	100. 00	25. 79	15. 79	58. 42	—
各品种工厂化出苗量占总出苗量的比例 / %	100. 00	24. 64	25，50	49. 86	—
各品种工厂化养殖面积占总面积的比例 / %	100. 00	20. 70	12. 82	63. 98	3. 50
各品种工厂化养殖产量占总产量的比例 / %	100. 00	21. 75	9. 81	67. 36	1. 08
各品种池塘养殖面积占总面积的比例 / %	—	—	—	—	—
各品种池塘养殖产量占总产量的比例 / %	—	—	—	—	—

（日照综合试验站站长　郭晓华）

第三篇

2016 年度研究论文选编

Immunological characterization and expression of lily-type lectin in response to environmental stress in turbot (*Scophthalmus maximus*)

Zhihui Huang, Aijun Ma, Dandan Xia, Xinan Wang, Zhibin Sun, Xiaomei Shang, Zhi Yang, Jiangbo Qu

Yellow Sea Fisheries Research Institute, Chinese Academy of Fisheries Sciences
Key Laboratory of Sustainable Development of Marine Fisheries, Ministry of Agriculture
Qingdao Key Laboratory for Marine Fish Breeding and Biotechnology, Qingdao, China

Abstract Lectins are a superfamily of carbohydrate-binding proteins that are widely distributed throughout living organisms. In earlier work, we identified lily-type lectin (*Sm*LTL) in the skin mucus of turbot *Scophthalmus maximus*, and we characterized the protein in the present study. Results from qRT-PCR indicated that *Sm*LTL was expressed highly in skin, intestine and gill tissue. Changes in *Sm*LTL expression occurred in these tissues in response to environmental stressors including ciliate infection, high temperature and salinity. Recombinant *Sm*LTL purified from *Escherichia coli* was able to haemagglutinate mouse erythrocytes in the absence of calcium, and was inhibited by *D*-mannose. In addition, *Sm*LTL displayed selective binding to bacterial species including *Edwardsiella tarda and Vibrio anguillarum*, and exhibited toxicity towards *Philasterides dicentrarchi*, with a mortality of over 60% after 24 h at a concentration of only 100 μg / mL.To investigate this toxicity further, we measured binding of *Sm*LTL after incubating the ciliate in FITC-*Sm*LTL solution. Surface fluorescence decreased substantially in the presence of 400 mmol/L *D*-mannose. Together these results suggest that lily-type lectins serve as the first line of defence against microbial attack and play a pivotal role in the mucosal immune system.

Keywords lily-type lectins; haemagglutinate; environmental stressors; anti-ciliate activity; turbot

1 Introduction

The aquatic environment harbours a wide array of biological, physical and chemical hazards, and the mucosal surfaces of the skin, gills and gut constitute the first line of defence in fish (Rajan et al, 2011). These sensitive tissues are invariably associated with pathogen entry and disease progression. Fish skin mucus comprises a series of immune components including lysozyme, IgM, calmodulin, C-reactive protein, proteolytic enzymes and antimicrobial peptides (Alexander et al, 1992; Magnadóttir 2006). In addition to these molecules, lectins are often present in fish skin secretions. Lectins are sugar-binding proteins that are ubiquitous in bacteria, fungi, plants, invertebrates, and vertebrates, and play a vital role in many cellular processes. These proteins are not enzymes are but able to recognize and bind specifically to soluble carbohydrates or sugar moieties on glycoproteins and glycolipids (Zhang et al, 2010). Skin mucus lectins have been purified from fish such as the windowpane flounder *Lophopsetta maculate* (Kamiya et al, 1980), Japanese eel *Anguilla japonica* (Tasumi et al, 2004), conger eel *Conger myriaster* (Kamiya et al, 1988; Shiomi et al, 1989), pufferfish *Fugu rubripes* (Tsutsui et al, 2003), and striped murrel *Channa striatus* (Arasu et al, 2013). Lectins exhibit marked structural diversity and are classified into five distinct families (Shigeyuki et al, 2011); galectin (Tasumi et al, 2004; Muramoto et al, 1992; Muramoto et al, 1999), C-type (Tasumi et al, 2002; Tsutsui et al, 2007), lily-type (Tsutsui et al, 2003; Karla et al, 2009), rhamnose-binding lectin (RBL) (Okamoto et al, 2005) and pentraxin (Tsutsui et al, 2009). Lily-type lectins were initially discovered in the skin mucus of the pufferfish *T. rubripes*. Pufflectins are characterized by a three-fold internal repeat (β-prism architecture) and the consensus mannose-binding motif QXDXNXVXY (Karla et al, 2009). Although there is no sequence homology with any known animal lectins, pufflectins share relatively high homology (29.4% identity) with mannose-binding lectins from monocotyledonous plants such as snowdrop and garlic (Suzuki et al, 2003). Monocot mannose-binding (MMB) lectins are widely distributed in higher plants and are an important tool in plant protection and plant biotechnology because their genes confer resistance against sucking insects and nematodes (Lin et al, 2003). Additionally, MMB lectins have been shown to bind to carbohydrates on the surfaces of retroviruses such as HIV (Balzarini et al, 1992). The interesting properties of MMB lectins has resulted in increasing attention (Ganapati et al, 2010). However, only lily-type lectins from pufferfish *Fugu rubripes* (Suzuki et al, 2003), striped murrel *Channa striata* (Arasu et al, 2013), and half-smooth tongue sole *Cynoglossus semilaevis* have received significant attention (Sun et al, 2016).

Turbot, *Scophthalmus maximus* (Family Bothidae), is economically important in wild fisheries and has stimulated the development of aquaculture for several other species. Since turbot was introduced to China in 1992, breeding and culture technology has been greatly advanced and commercialized (Huang et al, 2011). However, the flatfish farming industry remains limited to certain areas by environment constraints. Water is a perfect medium for bacteria and parasitic microbes, in intensive aquaculture, fish are constantly exposed not only to pathogen attack but also alterations in the aquatic environment, such as temperature, salinity and photoperiod (Tsutsui et al, 2003; Martins et al, 2011). The mucosal innate immune system of aquatic vertebrates plays an important role in protecting against microbial infections and changes in environmental conditions. We previously identified lily-type lectin *Sm*LTL in the skin mucus of turbot *Scophthalmus maximus* using a comparative proteomics approach (Huang et al, 2011; Ma et al, 2013). We determined the full-length cDNA sequence (GenBank accession no. KU199003) but functional studies have not yet been reported. In this study, we investigated the temporal expression of *Sm*LTL following exposure to environment stress, and probed the immunological characteristics. In order to suggest that lily-type lectins serve as the first line of defence against environmental stressors and microbial infections and play a pivotal role in the innate mucosal immune system.

2 Materials and methods

2.1 Fish

In all test, turbot broodstocks were obtained from the Tianyuan Fisheries Co. Ltd. Yantai, China, and held in 60 L circular seawater tanks with a filter-aeration system. Fish were fed twice a day throughout the experiment. The body weight of fish on the sampling day ranged from 80 g～100 g [average 90 g ± 10.2 g (mean ± SD)]. The experiment fish were acclimatized at a normal factor (14 ℃ ± 0.5 ℃ ; salinity: 30 ± 0.4) for two weeks before setting the experiment. Tissue samples including gills, spleen, liver, head kidney, kidney, heart, intestine, skin and muscle were dissected from the killed fish and submerged in RNAlater (Tiangen, Beijing, China) and stored at −80 ℃ until RNA extraction.

2.2 Expression pattern of *Sm*LTL transcripts in different fish tissues

The relative expression of *Sm*LTL in gills, spleen, liver, head kidney, kidney, heart, intestine, skin and muscle were measured by quantitative real-time polymerase chain reaction (qRT-PCR). Total RNA was isolated with the RNAprep Tissue Kit (Tiangen, Beijing, China). One microgram of total RNA was used for cDNA synthesis with the Superscript II reverse transcriptase (TransGen Biotech, Beijing, China). qRT-PCR was carried out using ABI 7500 Real-time Detection System (Applied Biosystems, Foster City, CA, USA) according to the manufacturer's instructions in the SYBR Premix Ex Taq Kit (Takara, Dalian, China), in 20 reaction volume containing 4 μL of cDNA from each tissue, 10 μL of Fast SYBR® Green Master Mix, 2 μL of each of the forward and reverse primers L-YGDL-S-117 and L-YGDL-AS-117 (1 μmol/L). β-actin was used as a reference gene. Sequences are listed in Table 1. Dissociation curve analysis of amplification products was performed after the qRT-PCR to confirm the specificity of PCR products. All samples were analysed in three duplications and the relative *Sm*LTL expression level as exhibited by $2^{-\Delta\Delta Ct}$ was determined for each group and the values were shown as mean ± SD (n = 3).

Table 1 PCR primers used in this study

Primer name	Sequence (5′-3′)
L-YGDL-S-117	CGAGTTCAAAGCCATCTTCC
L-YGDL-AS-117	TTATCCGGTTGCAGAAGGAC
β-actin-F	GTAGGTGATGAAGCCCAGAGC
β-actin-R	CTGGGTCATCTTCTCCCTGT

2.3 Effects of environment stress on the *Sm*LTL response

2.3.1 qRT-PCR analysis of *Sm*LTL gene expression in response to experimental challenge to *Philasterides dicentrarchi*

Ciliates *Philasterides dicentrarchi* generates a severe systemic infection invading internal organs such as brain, gills, liver and intestine that generally results in the death of the host. The ciliates were isolated and cultured in the laboratory under the conditions described by Zhang (Zhang et al, 2013). After 10～15 days, the ciliates were pelleted by centrifugation (650 g, 5 min), resuspended in sterile physiological saline (SPS) and counted in a haemocytometer. Live fish (average 90 g ± 10.2 g (mean ± SD)) were divided randomly in two groups (40 fish/group). The first group was intraperitoneal inoculation which described by A. Parama′ (Paramá et al, 2003), that was injection into abdominal cavity of 200 μL of SPS containing 10^5 ciliates (the fish weighing 20 g～40 g with a dose of 10^4 or 5 × 10^4 ciliates produced 100% systemic infection); and the second group with SPS which served as control. Each group had five fish in each of three replicates (i.e. five per tank and 15 in total). The fishes were sacrificed at 0.5, 1, 2, 3, 4 and 5 days. Gill, intestine, skin were taken aseptically from five fish and used for total RNA extraction, cDNA synthesis and qRT-PCR were performed as described above. The data were expressed as mean ± SD and subjected to ANOVA (two-way analysis of variance) to determine differences among treatments. The differences were considered as significant at $P < 0.05$. All the Statistical analysis was performed using SPSS 19.0 for windows.

2.3.2 qRT-PCR analysis of *Sm*LTL gene expression in response to temperature change

The fish (average 90 g ± 10.2 g (mean ± SD)) were acclimatised at a normal factor (14 ℃ ± 0.5 ℃ ; salinity: 30 ± 0.4) for two weeks, and then were divided into four high-temperature groups and a control group, similar to the experimental scheme of Diegane Ndong (Diegane et al, 2007).

Each group had five fish in each of three replicates (i.e. five per tank and 15 in total). Temperature was programmed as a 3 ℃ increment every 48 h until the desired experimental temperature was reached (14 ℃ ± 0.5 ℃ , 19 ℃ ± 0.5 ℃ , 22 ℃ ± 0.5 ℃ , 25 ℃ ± 0.5 ℃ and 27 ℃ ± 0.5 ℃). The experiment was conducted for 48 h. Gill, intestine, skin were taken aseptically from five fish and used for total RNA extraction, cDNA synthesis and qRT-PCR were performed as described above.

2.3.3 qRT-PCR analysis of *Sm*LTL gene expression in response to salinity change

Following acclimation, turbot (average 90 g ± 10.2 g (mean ± SD)) were transferred directly to salinities of 10, 20, 30 (control) and 40. The experimental salinities were obtained by diluting filtered seawater at ambient salinity (30) with fresh water or adding with artificial sea salt. Salinities were measured with densimeter (Sanli, China). Each group had five fish in each of three replicates (i.e. five per tank and 15 in total). The experiment was conducted for 48 h. Gill, intestine, skin were taken aseptically from five fish and used for total RNA extraction, cDNA synthesis and qRT-PCR were performed as described above.

2.3.4 Statistical analysis

The data which *Sm*LTL expression level in response to temperature and salinity changes, were analyzed with SPSS 19.0 software package for one way ANOVA test followed Dunnett two-sided test to determine differences among treatments. The differences were considered as significant at $P < 0.05$. All the Statistical analysis was performed using.

2.4 Expression of recombinant proteins *Sm*LTL

2.4.1 Construction of pEt-*Sm*LTL

To construct pEt-*Sm*LTL, which express His-tagged recombinant *Sm*LTLthe coding sequences of the protein were amplified by PCR with primer pairs *Sm*LTL F/*Sm*LTL R (Table 1). *Nde*I and *Hin*dIII sits were introduced in the primers, respectively. The PCR products of *Sm*LTL was digested with *Nde*I and *Hin*dIII, and sub-cloned into the plasmids pET-43.1a(+) digested with the corresponding restriction enzymes, resulting in pEt-*Sm*LTL.

2.4.2 Recombinant expression and purification of *Sm*LTL

The recombinant plasmid (pET-43.1a (+)-*Sm*LTL) was transformed into *Escherichia coli* BL21 (DE3) (TransGen, Beijing, China) and subjected to nucleotide sequencing. The transformants were cultured in 200 mL LB medium (containing 100 mg/mL ampicillin) at 37 ℃ with shaking at 195 r/min. The parent vector without an insert fragment was used as negative control. When the culture medium reached OD600 of 0.6～0.8, the β-*D*-1-thiogalactopyranoside (IPTG) with final concentration of 0.5 mmol/L was added to induce the recombinant proteins expression. After growing at 15 ℃ for overnight, the cells were collected by centrifugation at 8 000 g at 4 ℃ for 15 min, resuspended in 50 mL buffer A (pH 8.0) consisting of 50 mmol/L Tris, 150 mmol/L NaCl, 1 mmol/L DTT, 1% Triton X-100, 1 μg/mL Pepstatin A, 1 μg/mL Leupeptin, 20 mmol/L iminazole, and sonicated on ice. The protein was collected by centrifugation at 13 000 g at 4 ℃ for 30 min. The supernatants were filtered through a 0.4 μm filter membrane, and loaded onto a Ni-IDA resin column (GE Healthcare, Piscataway, USA) as recommended by the manufacturer. The column was successively washed with PBS containing 8 mol/L urea and 20 mmol/L imidazole (pH 7.4). The proteins were concentrated with Amicon Ultra Centrifugal Filter Devices (Millipore, Billerica, USA). All the purified proteins were electrophoresed in 12% SDS-PAGE, following stained with Coomassie brilliant blue R-250 and their concentrations were measured according to Bradford, using bovine serum albumin (BSA) as the standard.

2.4.3 Western blot analysis

Western blot analysis was carried out to test *Sm*LTL expression at the protein level and performed as described by Chen Sun (Sun et al, 2013). The proteins were electrophoresed on a 12% SDS-PAGE gel, and electro-transferred onto a PVDF membrane. After blocking with 4% bovine serum albumin (BSA) in Tris-buffered saline (TBS:10 mmol/L Tris and 150 mmol/L NaCl, pH 8.0) at room temperature for 2 h, the PVDF membrane was incubated with anti-His tag mouse monoclonal antibody (CWBIO) diluted 1:3000 with 4% BSA in TBS (pH 8.0) at 4 ℃ overnight. After washed with TBST (0.1% Tween-20 in TBS) four times (10 min/time), the membrane was incubated in peroxidase-conjugated goat anti-mouse IgG (CWBIO) (1:5 000 with 4% BSA in TBS) for 3 h at room temperature. The binding was visualized through peroxidase catalyzed stain reaction.

2.5 Biochemical characterization of recombinant *Sm*LTL

2.5.1 Characterization of haemagglutination activity

Haemagglutination assay was carried out in round-bottomed microtitre plates using 2% (*V*/*V*) trypsinized mouse erythrocytes, according to the method described by Abirami (Arasu et al, 2013). Erythrocytes were washed three times with TBS-Ca buffer (50 mmol/L Tris-HCl, pH 7.5, 150 mmol/L NaCl, 10 mmol/L $CaCl_2$) and then suspended at 2% (*V*/*V*) in TBS-Ca buffer. A total volume of 50 μL was used in each well: 25 μL aliquot of serial two-fold dilutions of the lectin in TBS-Ca buffer were mixed with 25 μL erythrocyte suspension in the microtitre plate. Hemagglutination values were expressed as hemagglutinating unit and defined as the minimal amount (mg) of protein per milliliter able to induce visible agglutinagtion of cells after different time (1 h, 2 h, 4 h, 6 h, 8 h and 10 h) at room temperature. Erythrocytes mixed with serial dilutions of bovine serum albumin (BSA, Sigma, USA) were processed in parallel with the above-described erythrocytes to serve as controls.

To test whether calcium is required for the haemagglutination activity of *Sm*LTL, 12.5 μL of serial dilutions of EDTA in TBS-Ca buffer was mixed with 12.5 μL of *Sm*LTL, 25 μL 2% (*V*/*V*) trypsinized mouse erythrocytes were added, and the mixture was incubated for 1h at room temperature. The method described by Abirami (Arasu et al, 2013). Assays were performed in triplicate. Cells were observed in light microscopy.

2.5.2 Sugar binding specificity of *Sm*LTL

To determine the sugar-binding specificity of *Sm*LTL, 12.5 mL of *Sm*LTL was mixed with an equal volume of 200 mmol/L *D*-glucose, lactose, sucrose, *D*-galactose, *D*-mannose, maltose, *D*-fructose and xylose (Sangon, Shanghai, China), respectively, and incubated for 30 min at 4 ℃. Then 2% trypsinized mouse erythrocytes were added, and the mixture was incubated for 1 h at room temperature and observed for agglutination as described above. This assay was performed in triplicate.

2.5.3 Effect of temperature and pH

The effect of temperature on the *Sm*LTL hemagglutinating activity was evaluated by heating an aliquot of *Sm*LTL during 4 h at 4 ℃, 10 ℃, 25 ℃, 37 ℃, 45 ℃ and 60 ℃.

To determine the optimal pH, the lectin was mix with PBS buffer of different pH (2.0, 4.0, 7.0, 8.0, 9.0 and 10.0) 37 ℃ for 4 h, and the activites of the solutions were measured.

2.6 Immunological activity

2.6.1 Bacterial agglutination assay

Six strains of fish pathogenic bacteria (Yellow Sea Fisheries Research Institute, Chinese Academy of Fisheries Sciences) were tested for agglutination following the methodology of Yang (Yang et al, 2007). The Gram-negative bacteria, *E. coli*, *A. hydrophila*, *Edwardsiella tarda* and *Vibrio anguillarum*. The Gram-positive *Bacillus mycoides* and *Staphylococcus aureus*. Those bacteria were suspended in TBS buffer (50 mmol/L Tris-HCl, 100 mmol/L NaCl, pH 7.5) to 2×10^9 CFU/mL, and were labeled with 0.2 mg/mL fluorescein isothiocyanate (FITC) (Tiangen, Beijing, China). Ten microliters of bacteria were mixed with 40 μL of recombinant *Sm*LTL in serial dilution (5 μg/mL, 25 μg/mL, 50 μg/mL and 100 μg/mL) by TBS as a control, followed by incubation at 4 ℃ for overnight. Agglutination was observed with a fluorescence microscope (Nikon E800, Japan).

2.6.2 Anti-parasite activity of lectin

Anti-parasite activity of *Sm*LTL on ciliate *Philasterides dicentrarchi* was assayed similar to the experimental scheme of Ganapati (Nakamura et al, 2012). *P. dicentrarchi* generates a severe systemic infection invading internal organs such as brain, gills, liver and intestine that generally results in the death of the host. The ciliates were isolated and cultured in the laboratory under the conditions described by Zhang (Zhang et al, 2013). 800 μL ciliates suspension were transferred to tubes containing different concentration of *Sm*LTL (200 μg/mL, 100 μg/mL, 50 μg/mL, 25 μg/mL, 12.5 μg/mL), and incubated at 18 ℃. Negative control without lectin was used for comparison. Number of live and dead nematodes from each lectin concentration set was counted at different time intervals (3 h, 6 h, 12 h, 24 h and 48 h) under the microscope. Percentage mortality of nematodes was calculated frome the average of triplicate experiments for each concentration.

2.6.3 Binding test with parasite

Interaction of *Sm*LTL with ciliate *P. dicentrarchiwas* investigated using FITC-*SmLTL* by fluorescent microscopy. The ciliates were isolated and cultured in the laboratory under the conditions described by Zhang (Zhang et al, 2013). Suck up 1 mL ciliates suspension, and incubated with a 4:1 dilution of FITC-*Sm*LTL at room temperature in the dark. As a control, fixed ciliates were incubated with 4:1 dilution of FITC-*Sm*LTL in solution containing 400 mmol/L *D*-mannose. Finally, ciliates were washed carefully with PBS (+) buffer to remove excess lectin and were palced on glass slides for observation under the fluorescence microscope (Nikon E800, Japan).

3 Results

3.1 Tissue distribution of *Sm*LTL

To examine the expression of *Sm*LTL mRNA in turbot gill, spleen, liver, head kidney, kidney, heart, intestine, skin and muscle tissue, qRT-PCR analysis was employed using β-actin as a control. *Sm*LTL mRNA was constitutively expressed in all selected tissues. Further analysis showed that *Sm*LTL expression was significantly higher ($P < 0.05$) in skin, intestine and gill, and poorly expressed in the liver, trunk-kidney, spleen and muscle (Fig. 1). This result indicates that *Sm*LTL showed tissue specific variation in turbot tissues.

3.2 Effects of environment stress on *Sm*LTL expression

3.2.1 *Sm*LTL gene expression in response to infection with the ciliate

In order to estimate if the *Sm*LTL is involved response to ciliate challenge, qRT-PCR analysis was used to determine the relative expression of *Sm*LTL in gill, intestine and skin at 24 h, 48 h, 72 h, 96 h and 120 h after infection with the ciliate *P. dicentrarchi*. As shown in Fig. 2, *Sm*LTL expression was found to be up-regulated in gill, intestine and skin in time dependent

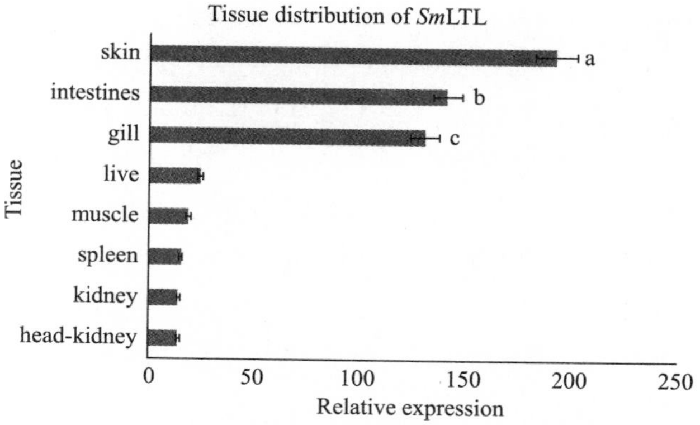

Fig. 1 *Sm*LTL expression in different tissues as measured by qRT-PCR

*Sm*LTL expression levels are shown for liver, spleen, kidney, head-kidney, intestine, muscle, skin and gill. The relative SmLTL expression level as exhibited by $2^{-\Delta\Delta Ct}$ was determined for each group and the values were shown as mean ± SD (n = 3). Means with different letters indicate statistical significance (P < 0.05) between groups

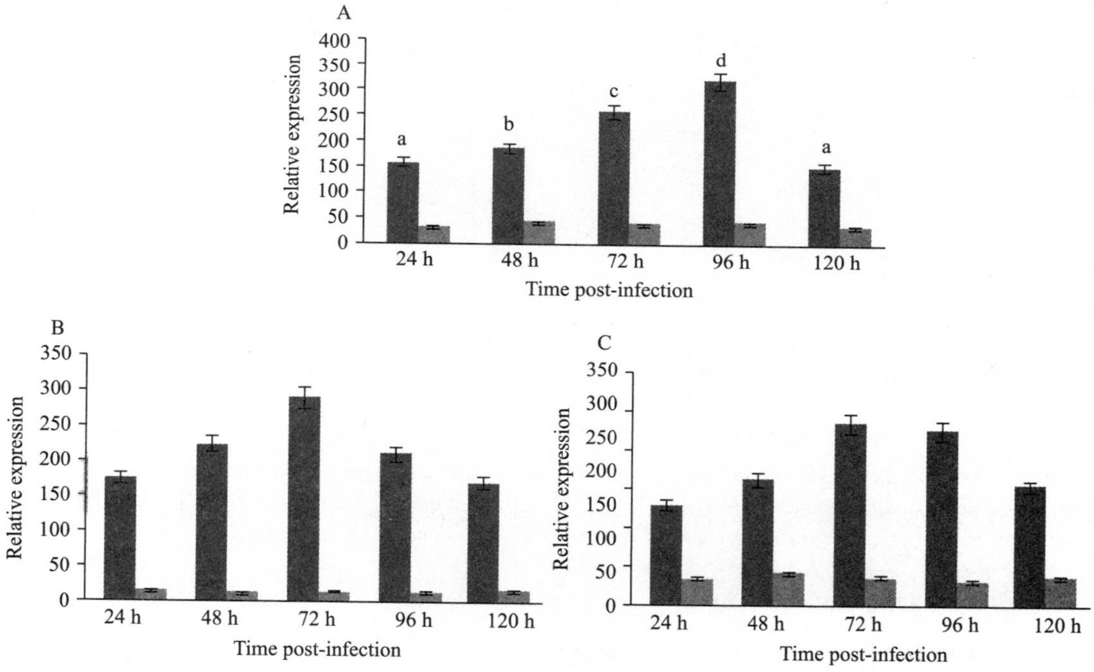

Fig. 2 *Sm*LTL gene expression in turbot gill, intestine and skin at 24～120 h after infection with the ciliate *P. dicentrarchi*

The relative *Sm*LTL expression level as exhibited by $2^{-\Delta\Delta Ct}$ was determined for each group and the values were shown as mean ± SD (n = 3). Means with different letters indicate statistical significance (P < 0.05) between groups

manners, the expression level significantly increased and peaked at 96 h, 72 h and 72 h, respectively (P < 0.05). Althoug it dropped afterward, *Sm*LTL was expressed much higher than in the control groups through 120 h.

3.2.2 *Sm*LTL gene expression in response to temperature change

To investigate whether temperature affects the expression of *Sm*LTL, as suggested by the effect of temperature on the protein expression of *Sm*LTL in previously (Goel et al, 2015), we determined the qRT-PCR analysis of *Sm*LTL in response to higher temperatures. We observed that *Sm*LTL transcription was significantly elevated following exposure to higher temperatures, but expression was decreased at 22 ℃ and 25 ℃ in skin and gill, respectively (Fig. 3). No significant up-regulation was observed in the intestine throughout the experiment.

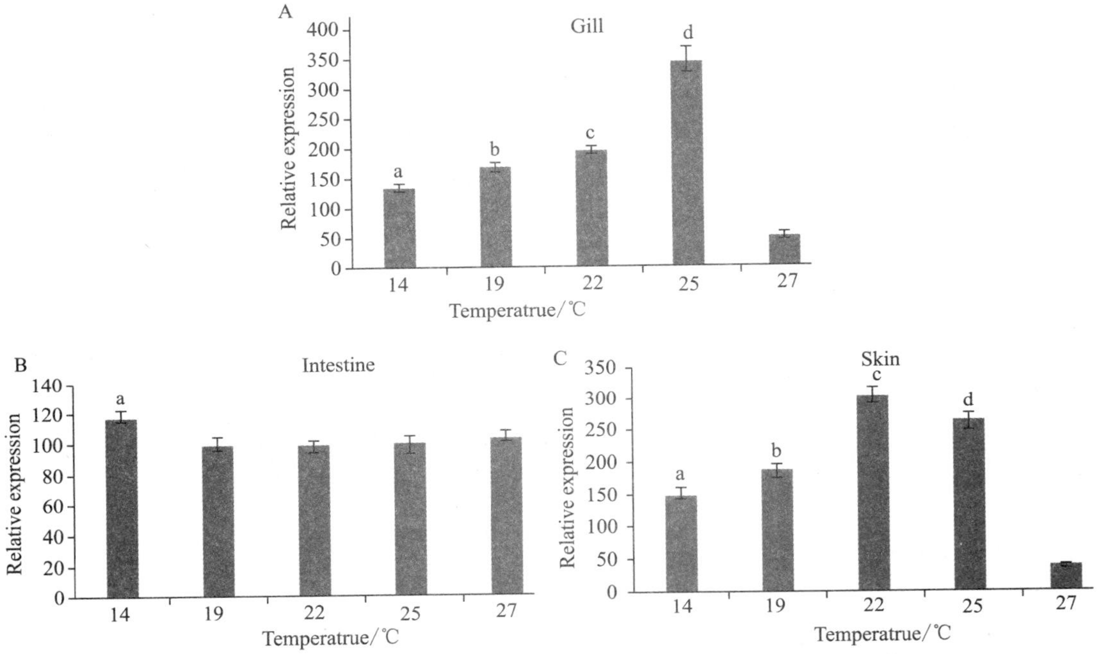

Fig. 3 *Sm*LTL gene expression in turbot gill, intestine and skin in response to temperature change
The relative *Sm*LTL expression level as exhibited by $2^{-\Delta\Delta Ct}$ was determined for each group and the values were shown as mean ± SD (n = 3). Means with different letters indicate statistical significance ($P < 0.05$) between groups

3.2.3 *Sm*LTL gene expression in response to salinity change

As shown in Fig. 4, *Sm*LTL expression was found to be slight increase followed by salinity changed in gill. However, in the skin, *Sm*LTL was down-regulated followed by salinity changed which compared to the control group (30). No significant regulation was observed in the intestine throughout the 20, 30 and 40, while *Sm*LTL transcripts were greatly up-regulated at 10.

3.3 Expression of recombinant protein *Sm*LTL

Recombinant *Sm*LTL was expressed in *Escherichia coli* BL21 (DE3). *Sm*LTL has a deduced molecular mass of 13.47×10^3, and a protein of the appropriate size (including a 4.9×10^3

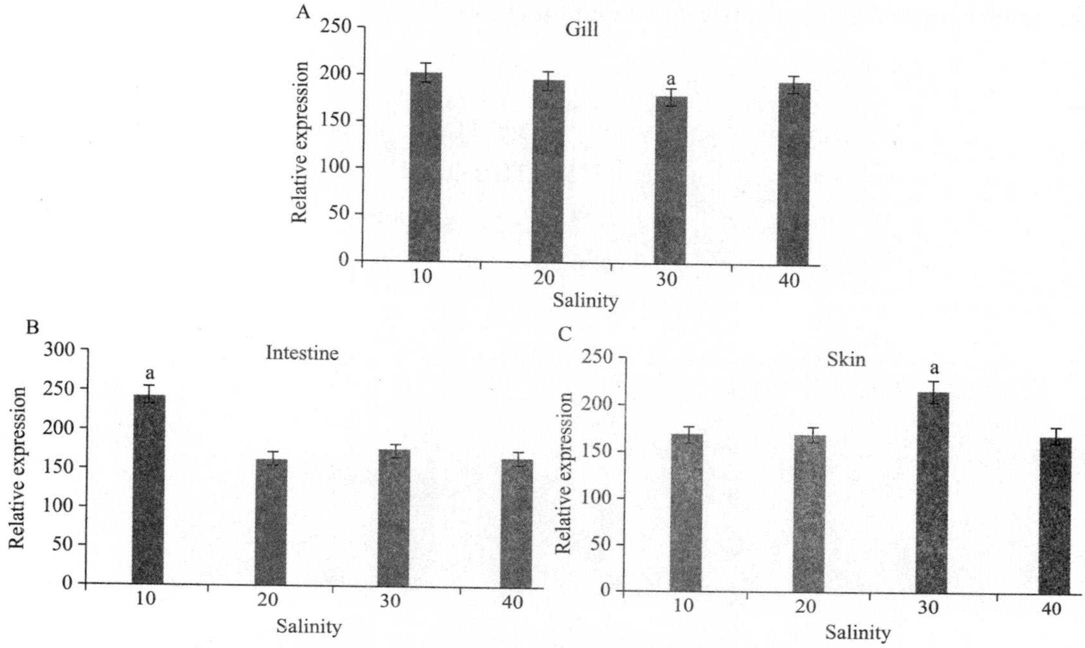

Fig. 4　*Sm*LTL gene expression in turbot gill, intestine and skin in response to salinity change

The relative *Sm*LTL expression level as exhibited by $2^{-\Delta\Delta Ct}$ was determined for each group and the values were shown as mean $\pm$ SD ($n = 3$). Means with different letters indicate statistical significance ($P < 0.05$) between groups

His-tag) was detected by Coomassie Bright Blue staining following separation by SDS-PAGE separation to have an apparent about 13.47×10^3 band (Fig. 5, Lane 1). Western blotting showed the recombinant proteins, indicating they are correctly expressed (Fig. 6, Lane 1).

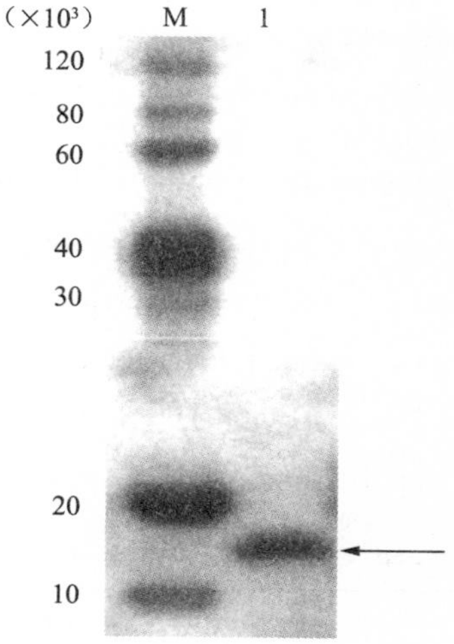

Fig. 5

Lane 1: *Sm*LTL; Lane 2: BSA; M: SDS-PAGE Marker

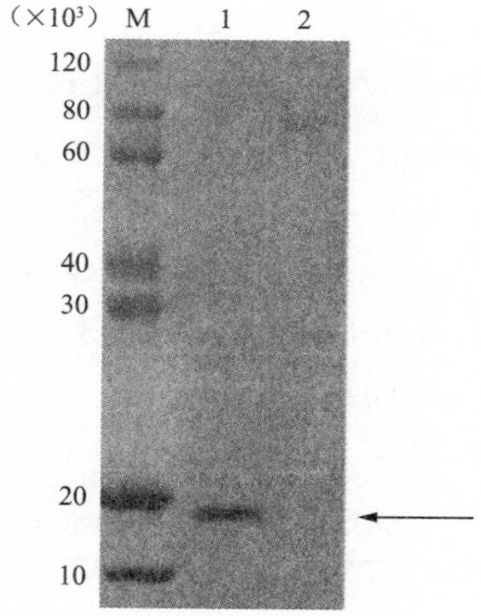

Fig. 6

M: Western Blot Marker; Using Anti-His antibody

Lane 1: recombinant proteins

3.4 Biochemical characterization of recombinant *Sm*LTL

3.4.1 Haemagglutination assay

To examine the haemagglutination activity of *Sm*LTL, the protein was incubated with mouse erythrocytes. The results showed that 4 μg/mL *Sm*LTL induced haemagglutination of erythrocytes in the presence of 10 mmol/L $CaCl_2$ (Fig. 7). The haemagglutination activity was not altered by the addition of either EDTA or $CaCl_2$, indicating that *Sm*LTL induced agglutination in a calcium-independent manner.

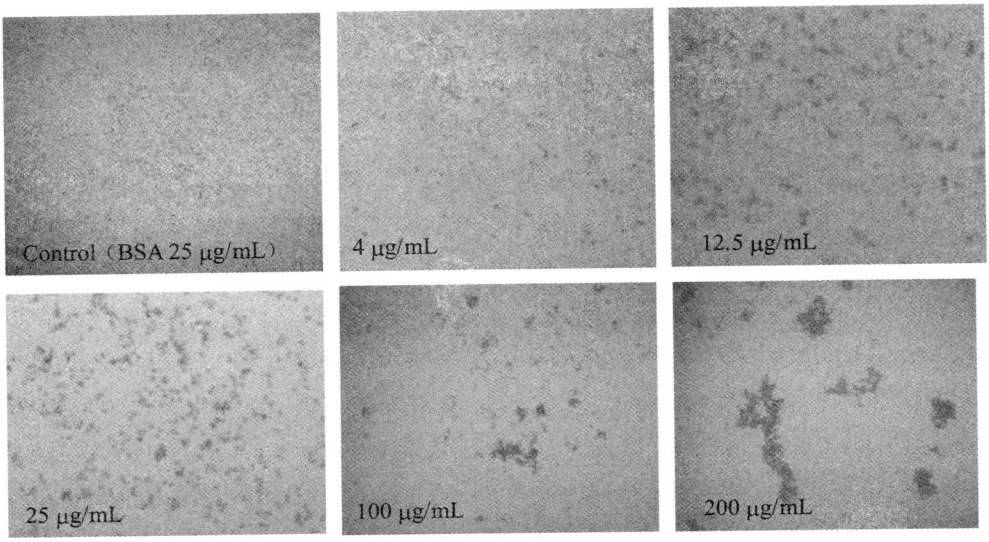

Fig. 7 Agglutination activity of *Sm*LTL was incubated with erythrocytes in the presence of 10 mmol/L $CaCl_2$

Erythrocytes mixed with serial dilutions of *Sm*LTL (4 μg/mL, 12.5 μg/L, 25 μg/mL, 100 μg/mL and 200 μg/mL) at room temperature. Erythrocytes mixed with BSA as control

3.4.2 Sugar binding specificity of *Sm*LTL

Treatment with *D*-mannose and *D*-galactose, but not with *D*-glucose, lactose, sucrose, maltose, *D*-fructose or xylose, disabled the ability of *Sm*LTL to induce haemagglutination of erythrocytes. Carbohydrates were pre-incubated with *Sm*LTL, and the mixtures were then added to mouse erythrocytes. Moreover, *Sm*LTL protein had its own sugar specificity. *D*-mannose and *D*-glucose inhibited agglutination at a concentration of 25 mmol/L and 100 mmol/L respectively, whereas no inhibition was detected with the other sugars up to 200 mmol/L (Table 2).

Table 2 Carbohydrates haemagglutination activity of *Sm*LTL protein

Carbohydrates	Haemagglutination activity
D-glucose	NI[b]
D-xylose	NI[a]

(continue)

Carbohydrates	Haemagglutination activity
D(−)-Fructose	NI[a]
D(−)-Fructose	NI[a]
D-maltose monohydrate	NI[a]
α-lactose	NI[a]
D-galactose	NI[a]
D-mannose	NI[b]

NI[a]: recombinant protein could not inhibit the haemagglutination of erythrocytes under the maximum detectable concentration(> 200 mmol/L). NI[b]: recombinant protein could inhibit the haemagglutination of erythrocytes under the 100 mmol/L detectable concentration

3.4.3 Temperature and pH stability

*Sm*LTL-induced haemagglutination was optimal at pH 7 and at room temperature. The recombinant lectin was stable at neutral pH between 4 ℃ and 45 ℃ (Table 3), and activity was preserved between pH 4 and 8 (Table 4).

Table 3 Temperature affecting the stability of *Sm*LTL protein

Temperature	Haemagglutination activity
4 ℃	NI[b]
10 ℃	NI[b]
25 ℃	NI[b]
37 ℃	NI[b]
45 ℃	NI[b]
60 ℃	NI[a]

NI[a]: recombinant protein could not inhibit the haemagglutination of erythrocytes under the maximum detectable concentration (> 200 mmol/L). NI[b]: recombinant protein could inhibit the haemagglutination of erythrocytes under the 100 mmol/L detectable concentration

Table 4 pH affecting the stability of *Sm*LTL protein

pH	Haemagglutination activity
2	NI[a]
3	NI[a]
4	NI[b]
5	NI[b]
6	NI[b]
7	NI[b]
8	NI[b]

(continue)

pH	Haemagglutination activity
9	NI[a]

NI[a]: recombinant protein could not inhibit the haemagglutination of erythrocytes under the maximum detectable concentration (> 200 mmol/L). NI[b]: recombinant protein could inhibit the haemagglutination of erythrocytes under the 100 mmol/L detectable concentration

3.5 Immunological activity

3.5.1 Bacterial agglutination assay

To examine the agglutination of *Sm*LTL, the protein was incubated with FITC- treated Gram-negative (*E. coli*, *A. hydrophila*, *E. tarda* and *V. anguillarum*) and Gram-positive (*B. mycoides* and *S. aureus*) bacteria. Recombinant *Sm*LTL just showed agglutination activity against *E. tarda* and *V. anguillarum* (Fig. 8). This result indicates that *Sm*LTL protein showed specific binding in bacterial agglutination test.

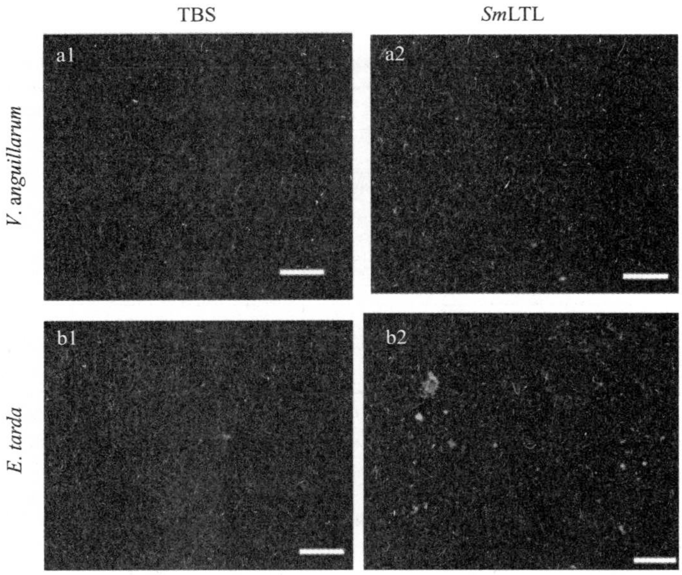

Fig. 8 Agglutination activity of SmLTL incubated with *E. tarda* or *V. anguillarum* overnight (a2 and b2) compared with TBS controls (a1 and b1)

Cells were stained with FITC and visualized using a fluorescence microscope. Bar = 10 μm

3.5.2 Anti-ciliate activity of *Sm*LTL

The anti-ciliate activity of *Sm*LTL against *P. dicentrarchi* was evaluated (Table 5). *Sm*LTL exhibited strong anti-ciliate activity, with over 60% mortality observed after 24 h at a concentration of 100 μg/mL, and all ciliates were killed after 48 h.

Table 5　Anti-ciliate activity of *Sm*LTL

Time / h	Mortality at different concentrations / %					
	200 μg / mL	100 μg / mL	50 μg / mL	25 μg / mL	12.5 μg / mL	Control
3	0	0	0	0	0	0
6	3	1	0	0	0	0
12	45	28	20	14	6	0
24	80	60	52	30	24	10
48	100	100	100	100	100	100

3.5.3　Binding of FITC-*Sm*LTL to ciliates

In order to get an insight to the mode of the *Sm*LTL toxicity effect on *P. dicentrarchi*, we determined the binding of *Sm*LTL after incubating the ciliates in FITC-*Sm*LTL solution. Binding of *Sm*LTL in ciliates was observed by fluorescent microscopy (Fig. 9a). Lectin binding was found on the surface of ciliate body. However, pronounce fluorescence reduced substantially in the presence of 400 mmol / L *D*-mannose (Fig. 9b).

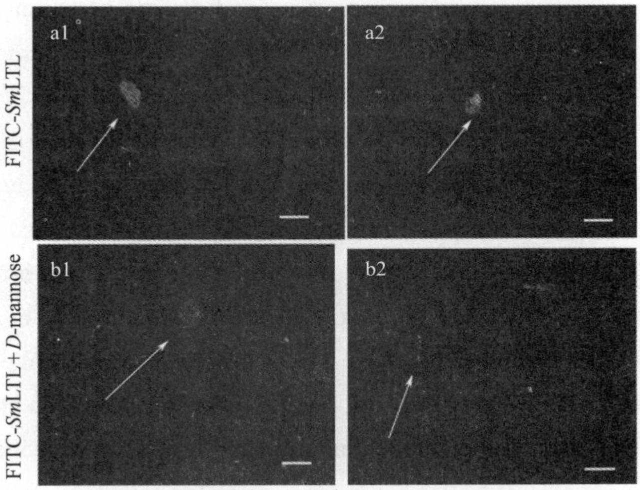

Fig. 9　Incubation of *P. dicentrarchi* with FITC-*Sm*LTL (a1 and a2)

Fluorescence was significantly reduced in the presence of *D*-mannose, which competes for *Sm*LTL binding (b1 and b2). Cells were observed with a fluorescence microscope. Bar = 10 μm

4　Discussion

Due to their primitive evolutionary status among vertebrates and poikilothermic nature, fish exhibit an inefficient acquired immune response. The innate immune system in fish is therefore thought to be of greater importance for combating microbial infection than it is in higher vertebrates. We previously identified lily-type lectin (*Sm*LTL) in the skin mucus of turbot *Scophthalmus maximus* by comparative proteomics (Ma et al, 2013). In the present study, we characterized the immunological functions of this lectin for the first time.

Previous studies demonstrated that different types of lectin exhibit different tissue-specific expression profiles in teleosts (Sun et al, 2016). In our study, *Sm*LTL transcripts were abundant in the skin, intestine and gill, but were less abundantin the liver, head-kidney, spleen and muscle. This expression profile is similar to that of *Cs*LTL-1, which is also significantly higher in the gill, liver, intestine and skin (Arasu et al, 2013), and pufflectin, which is widely expressed in gill, oesophagus and skin (Suzuki et al, 2003). Mucus lectins are produced by club cells that are abundant in the epidermis and epithelium of the upper digestive tract in fish, and secreted into the cutaneous, bronchial and oesophageal mucus (Huang et al, 2011). In teleosts, the gut, skin and gill are the main mucosal surfaces and immune barriers (Goel et al, 2015). Our results suggest that lily-type lectins serve as the first line of defence against microbial infection and therefore play a pivotal role in the mucosal immune system.

Fish live in aquatic environments that support a wide range of potentially infectious microorganisms. The mucosal innate immune system of aquatic vertebrates plays an important role in protecting against microbial infections and changes in environmental conditions. We previously investigated the effect of temperature on the expression of skin mucus proteins, and discovered that *Sm*LTL is influenced by this environmental parameter (Huang et al, 2011). In the present study, we measured *Sm*LTL transcription under various environmental conditions. ① Following exposure to the ciliate *P. dicentrarchi*, expression increased significantly in skin, gill and intestine. Specifically, *Sm*LTL expression in intestine, gill and skin were up-regulated at 24 h after infection (compared with SPS controls), and levels peaked at 72 h in intestine and skin but at 96 h in gills. Expression of lectins in response to bacterial challenge has been studied previously in various fish species, but the response to ciliate parasites has not been reported. The results suggest that the responses inducible and sensitive to parasite infections. ② The effect of temperature on the expression of lectins has been investigated previously (Goel et al, 2015). In the present study, a significant up-regulation was observed with increased temperature, but expression was decreased at 22 ℃ and 25 ℃ in skin and gill, respectively. No significant up-regulation was observed in intestine throughout the experiment. This expression profile was similar to those of other immune-relevant genes in mucus following exposure to high temperatures (Huang et al, 2011). Warmer conditions often favour pathogens, and we presumed that lectin expression would be up-regulated significantly when self-regulation of the body temperature reached the limit of tolerance and invading ciliates were bound. The gill is the site of gaseous exchange, and the skin is covered with mucus that functions as the first mechanical and biochemical barrier to the environment. It was no surprise that these tissues were acutely regulated by environmental stress, as previously observed. ③ *Sm*LTL expression was found to be slight increase followed by salinity changed in gill. However, in the skin, *Sm*LTL expression was down-regulated compared to the control group (salinity 30). In the intestine, *Sm*LTL transcripts were greatly up-regulated at salinity 10. And no significant regulation was observed throughout the salinity 10, 20 and 40 in gill and skin. We hypothesized that such results may be related to less sensitivity of *Sm*LTL to salinity stress stimuli.

Several reports have reported induction of haemagglutination of erythrocytes by lily-type lectins (Tsutsui et al, 2003; Karla et al, 2009; Nakamura et al, 2012; Arasu et al, 2013). In the present study, recombinant *Sm*LTLwas also able to haemagglutinate mouse erythrocytes in the presence of 10 mmol/L $CaCl_2$. Furthermore, haemagglutination was not altered by the addition of either EDTA or $CaCl_2$, indicating that *Sm*LTL-induced agglutination does not require calcium, as previously reported by Tsutsui for pufflectin and Abirami Arasu for *Cs*LTL-1 (Tsutsui et al, 2003; Arasu et al, 2013). Moreover, the haemagglutination activity was inhibited by 25 mmol/L *D*-mannose, but not by other sugars up to the maximum tested concentration of 200 mmol/L. *Sm*LTL possesses the typical lily-type lectin carbohydrate recognition domain (CRD) and mannose-binding motif (Q*x*D*x*N*x*V*x*Y), and it is thermostable like other lectins (Ribeiro et al, 2014; Almeida et al, 2014; Vasconcelos et al, 2015). *Sm*LTL activity was maintained across a wide pH range, with maximum stability between pH 7 and 8, suggesting *Sm*LTL is somewhat sensitive to acidic and alkaline conditions.

Bacterial agglutination activity revealed that recombinant *Sm*LTL could selectively bind to several bacterial species including *E. tarda* and *V. anguillarum*. Contrary to other studies, *Cs*LTL did not exhibit agglutination activity towards Gram-negative (*E. coli*, *A. hydrophila*, *V. vulnificus* and *V. alginolyticus*) or Gram-positive bacteria (*B. mycoides*, *S. pyogenes* and *S. aureus*) (Arasu et al, 2013). Additionally, pufflectin showed no agglutination towards these seven bacterial species (Tsutsui et al, 2003). However, Tsutsui demonstrated that pufflectin-induced agglutination activity is mainly restricted to a few species of the genus *Vibrio*, suggesting pufflectin contributes to the defence mechanism on the skin surface of pufferfish (Tsutsui et al, 2006). In the present study, recombinant *Sm*LTL could selectively bind to several bacterial species including *E. tarda* and *V. anguillarum*, indicating a difference inbinding affinities to different bacterial cells, as reported previously for *Cs*BML1, *Cs*BML2, and *Cs*BML3 from tongue sole (Sun et al, 2016).

P. dicentrarchi is a histophagous ciliate that infects turbot (*Scophthalmus maximus*) in aquaculture. This pathogen causes a severe systemic infection by invading internal organs such as the brain, gill, liver and intestine, usually resulting in the death of the host and heavy economic losses (Paramá et al, 2003). Tsutsui demonstrated that pufflectin can bind to the metazoan parasite *Heterobothrium okamotoi*, suggesting Lily-lectin may play a role in defence against parasites (Tsutsui et al, 2003). However, the influence of *Sm*LTL on ciliates remained unknown. The present study showed that *Sm*LTL exhibited toxicity towards *P. dicentrarchiand* reduced its survival. Similar results have been reported for monocot mannose-binding lectins towards sucking pests and nematodes (Ganapati et al, 2010). In the present study, when turbot were infected with *P. dicentrarchi*, *Sm*LTL mRNA expression levels were increased significantly in the skin, gill and intestine. These results are consistent with the ciliate activity analysis, and indicate for the first time a link between the *in vitro* parasiticidal capacity and the *in vivo* anti-parasitic activity of lectins in teleosts.

In order to gain insight into the mechanism if *Sm*LTL toxicity towards *P. dicentrarchi*, we investigated the binding of FITC-labelled *Sm*LTL to ciliates in solution. The results showed a pronounced fluorescence on the surface of the pathogen surface that was reduced substantially in the presence of 400 mmol/L *D*-mannose, as was observed previously with pufflectin (Tsutsui et al, 2003). Lectin binding therefore appears to inhibit the chemoreception of host signals, which disrupts ciliate binding and invasion.

In summary, we characterized the novel *Sm*LTL lily-type lectin from *Scophthalmus maximus*. Environmental stressors in the form of ciliate invasion and high temperature regulated the expression of *Sm*LTL at the transcription level. Expression was significantly altered in skin, intestine and gill tissue. *Sm*LTL displayed selective binding of various bacterial species including *E. tarda* and *V. anguillarum*. To our knowledge, this is the first study on the anti-ciliate activity of lectins, and the first demonstration of a link between the *in vitro* parasiticidal capacity and the *in vivo* anti-parasitic activity of lectins in teleosts. Together our results suggest that lily-type lectins serve as the first line of defence against microbial infections and play a pivotal role in the innate mucosal immune system. We intend to further investigate the functions of *Sm*LTL in the mucosal immune system through comparative parasitic studies.

Acknowledgements

This work was supported by the earmarked fund for Modern Agro-Industry Technology Research System (CARS-50-G01), fund for Outstanding talents and innovative team of agricultural scientific research, General Financial Grant from the China Postdoctoral Science Foundation (2015M572096), Special Financial Grant from the the China Postdoctoral Science Foundation (2016T90661), Shandong Provincial Natural Science Foundation (ZR2014CP001), the National High Technology Research and Development Program of China (863 Program) (No. 2012AA10A408-8).

References

[1] Alexander J B, Ingram G A. Noncellular nonspecific defence mechanisms of fish[J]. Annu Rev Fish Dis, 1992, 2: 249-279.

[2] Almeida A C, Silva H C, Pereira-Junior F N, et al. Purification and partial characterization of a new mannose/glucose-specific lectin from *Centrolobium tomentosum* Guill. ex Benth seeds exhibiting low toxicity on *Artemia* sp[J]. Int J Indig Med Plants, 2014, 47: 1567-1577.

[3] Balzarini J, Neyts J, Schols D. The mannose-specific plant lectins from Cymbidium hybrid and Epipactis helleborine and the (N-acetylg1ucosamine)-specific plant lectin from *Urtica dioica* are potent and selective inhibitors of human immunodeficiency virus and cytomegalovirus replication *in vitro*[J]. Antiviral Res, 1992, 18(2): 191-207.

[4] Diegane N, Chen Y Y, Lin Y H, et al. The immune response of tilapia *Oreochromis mossambicus* and its susceptibility to *Streptococcus iniae* under stress in low and high temperatures[J]. Fish Shellfish Immunology, 2007, 22(6): 686-694.

[5] Karla D S E, Filipe A, Flávia F D R, et al. Plumieribetin, a fish lectin homologous to mannose-binding B-type lectins, inhibits the collagen-binding $\alpha1\beta1$ integrin[J]. J Biol Chem, 2009, 284(50): 34747-34759.

[6] Arasu A, Kumaresan V, Sathyamoorthi A, et al. Fish lily type lectin-1 contains β-prism architecture: Immunological characterization[J]. Mol Immunol, 2013,56(4): 497-506.

[7] Ganapati G B, Kartika N S, Nagaraja N N. Purification, characterization and molecular cloning of a monocot mannose-binding lectin from *Remusatia vivipara* with nematicidal activity[J]. Glycoconj, 2010, 27(3): 309-320.

[8] Goel C, Barat A, Pande V, et al. Molecular cloning and characterization of mannose binding lectin homologue from snow trout (*Schizothorax richardsonii*)[J]. Protein J, 2015, 34(1): 1-8.

[9] Gomez D, Sunyer J O, Salinas I. The mucosal immune system of fish: The evolution of tolerating commensals while fighting pathogens[J]. Fish Shellfish Immunol, 2013, 35(6): 1729-1 739.

[10] Huang Z H, Ma A J,Wang X A. The immune response of the skin of turbot (*Scophthalmus maximus*) under stress at high temperature[J]. J Fish Dis, 2011, 34(8): 8619-8627.

[11] Kamiya H, Muramoto K, Goto R. Purification and properties of agglutinins from conger eel, *Conger myriaster* (Brevoort), skin mucus[J]. Dev Comp Immunol, 1988, 12(2): 309-318.

[12] Kamiya H, Shimizu Y. Marine biopolymers with cell specificity. Purification and characterization of agglutinins from mucus of windowpane flounder *Lophopsetta maculate*[J]. Biochim Biophys Acta Part II, 1980, 622: 171-178.

[13] Lin J, Yao J H, Zhou X W, et al. Expression and purification of a novel mannose-binding lectin from *Pinellia ternate*[J]. Molecular Biotechnology, 2003, 25(3): 3215-3221.

[14] Ma A J, Huang Z H, Wang X A. Changes in protein composition of epidermal mucus in turbot *Scophthalmus maximus* (L.) under high water temperature[J]. Fish Physilo Biochem, 2013, 39(6): 1411-1418.

[15] Magnadóttir B. Innate immunity of fish (overview)[J]. Fish Shellfish Immunol, 2006, 20: 137-151.

[16] Martins M L, Xu D H, Shoemarker C A, et al. Temperature effects on immune response and hematological parameters of channel catfish *Ictalurus punctatus* vaccinated with live theronts of *Ichthyophthirius multifiliis*[J]. Fish Shellfish Immunology, 2011, 31(6): 774-780.

[17] Muramoto K, Kamiya H. The amino-acid sequence of a lectin from conger eel, *Conger myriaster*, skin mucus[J]. Biochim Biophys Acta, 1992, 1116(2): 129-136.

[18] Muramoto K, Kagawa D, Sato T, et al. Functional and structural characterization of multiple galectins from the skin mucus of conger eel, *Conger myriaster*[J]. Comp Biochem Physiol B, 1999, 123(1): 33-45.

[19] Nakamura O, Watanabe M, Ogawa T. Galectins in the abdominal cavity of the conger eel *Conger myriaster* participate in the cellular encapsulation of parasitic nematodes by host cells[J]. Fish Shellfish Immunol, 2012, 33(4): 780-787.

[20] Okamoto M, Tsutsui S, Tasumi S, et al. Tandem repeat *L*-rhamnose-binding lectin from the skin mucus of ponyfish, *Leiognathus nuchalis*[J]. Biochem Biophys Res Commun, 2005, 333(2): 463-469.

[21] Paramá A, Iglesias R, Álvarez M F, et al. *Philasterides dicentrarchi* (Ciliophora, Scuticociliatida): experimental infection and possible routes of entry in farmed turbot (*Scophthalmus maximus*)[J]. Aquaculture, 2003, 217: 73-80.

[22] Rajan B, Fernandes M O, Caipang M A, et al. Brinchmann, Proteome reference map of the skin mucus of Atlantic cod (*Gadus morhua*) revealing immune competent molecules[J]. Fish Shellfish Immunol, 2011, 31(2): 2224-2231.

[23] Ribeiro A C, Monteiro S V, Carrapic B M, et al. Are vicilins another major class of legume lectins?[J] Molecules, 2014, 19(12): 20350-20373.

[24] Shigeyuki T, Yukie K, Takaya S, et al. A unique epidermal mucus lectin identified from catfish (*Silurus asotus*): first evidence of intelectin in fish skin slime[J]. J Biochem, 2011, 150(5): 5501-5514.

[25] Shiomi K, Uematsu H, Yamanaka H, et al. Purification and characterization of a galactose-binding lectin from the skin mucus of the conger eel *Conger myriaster*[J]. Comp Biochem Physiol B, 1989, 92(2): 255-61.

[26] Sun C, Hu L L, Liu S S, et al. Functional analysis of domain of unknown function (DUF) 1943, DUF1944 and von Willebrand factor type D domain (VWD) in vitellogenin2 in zebrafish[J]. Dev Comp Immunol, 2013, 41(4): 469-476.

[27] Sun Y Y, Liu L, Li J, et al. Three novel B-type mannose-specific lectins of *Cynoglossus semilaevis* possess varied antibacterial activities against Gram-negative and Gram-positive bacteria[J]. Dev Comp Immunol, 2016, 55: 194-202.

[28] Suzuki Y, Tasumi S, Tsutsui S, et al. Molecular diversity of skin mucus lectins in fish[J], Comp Biochem Physiol, Part B, 136(4): 723-730.

[29] Tasumi S, Ohira T, Kawazoe I, et al. Primary structure and characteristics of a lectin from skin mucus of the Japanese eel *Anguilla japonica*[J]. J Biol Chem, 2002, 277(30): 27305-27311.

[30] Tasumi S, Yang W J, Usami T, et al. Characteristics and primary structure of a galectin in the skin mucus of the Japanese eel, *Anguilla japonica*[J]. Dev Comp Immunol, 2004, 28(4): 325-335.

[31] Tsutsui S, Tasumi S, Suetake H, et al. Lectins homologous to those of monocotyledonous plants in the skin mucus and intestine of pufferfish, *Fugu rubripes*[J]. J Biol Chem, 2003, 278(23): 20882-20889.

[32] Tsutsui S, Iwamoto K, Nakamura O, et al. Yeast-binding C-type lectin with opsonic activity from conger eel (*Conger myriaster*) skin mucus[J]. Mol Immunol, 2007, 44(5): 691-702.

[33] Tsutsui S, Okamoto M, Tasumi S, et al. Novel mannose-specific lectins found in torafugu, *Takifugu rubripes*: A review[J]. Comp Biochem Physiol, Part D, 2006, 1(1): 122-127.

[34] Tsutsui S, Hirasawa M, Nakamura A, et al. Common skate (*Raja kenojei*) secretes pentraxin into the cutaneous secretion: the first skin mucus lectin in cartilaginous fish[J]. J Biochem, 2009, 146: 295-306.

[35] Vasconcelos M A D, Alves A C, Carneiroc R F. Purfication and primary structure of a novel mannose-specific lectin from *Centrolobium microchaete* mart seeds[J]. Int J Biol Macromol, 2015, 81: 600-607.

[36] Yang H, Luo T, Li F, et al. Purification and characterization of a calcium-independent lectin (PjLec) from the haemolymph of the shrimp *Penaeus japonicas*[J]. Fish Shellfish Immunol, 2007, 22(1-2): 88-97.

[37] Zhang M, Hu Y, Sun L. Identification and molecular analysis of a novel C-type lectin from *Scophthalmus maximus*[J]. Fish Shellfish Immunol, 2010, 29: 182-188.

[38] Zhang Z, Li B, Wang Y G, et al. Method for preparing and culturing culture medium for supravital culture of seawater Scuticociliatida ciliates[P]. 2013, ZL201110414123. 6.

Cytological studies on induced mitogynogenesis in Japanese flounder *Paralichthys olivaceus* (Temminck et Schlegel)

Jilun Hou[1], Guixing Wang[1], Xiaoyan Zhang[1], Haijin Liu[2]

[1] Beidaihe Central Experiment Station, Chinese Academy of Fishery Sciences, Qinhuangdao 066100, China

[2] Centre for Applied Aquatic Genomics, Chinese Academy of Fishery Sciences, Beijing, 100141, China

Abstract The effect of hydrostatic pressure treatment on the induction of mitogynogenesis in the eggs of Japanese flounder *Paralichthys olivaceus* (Temminck et Schlegel) by using heterospecific sperm were studied. Before the treatment, the eggs were at the metaphase of the first mitosis. The spindle was disassembled by the treatment and then resembled in its pretreatment position, and the chromosomes were rearranged, i.e., the first mitosis was not blocked. During the second mitotic cycle, only a monopolar spindle was assembled in each blastomere and the chromosomes doubled, but cell cleavage was blocked. In the third cycle, mitosis proceeded normally with a bipolar spindle in each blastomere. Flow cytometric analysis of ploidy demonstrated that mitogynogenetic larvae were all diploid. The ultraviolet-irradiated sperm of the red sea bream (*Pagrus major*) was condensed, formed a dense chromatin body, and randomly entered one blastomere.

Keywords Japanese flounder *Paralichthys olivaceus*; mitogynogenesis; cytology; monopolar spindle; hydrostatic pressure

1 Introduction

Gynogenesis refers to the fertilization of eggs with inactivated sperm, and prevents any contribution of the paternal genome to the progeny. As a result, embryonic development proceeds with the inheritance of only the maternal chromosome set(s). In nature, the well-studied examples of spontaneous gynogenetic fish species include the Ginbuna (*Carassius auratus gibelio*; Yamashita et al, 1993) and the Amazon molly (*Poecilia formosa*; Schartl et al, 1995). In fish, due to pre-embryonic events such as insemination, the extrusion of the second polar body and the first mitotic cleavage can be manipulated; the techniques for artificially induced gynogenesis are well established and have been widely utilized. Induced mitogynogenesis is the most convenient and powerful approach to generate homozygous progeny at the first filial generation (Bertotto et al, 2005). It can be achieved by blocking cell division with temperature, pressure, or chemical treatments after ultraviolet (UV)-irradiated sperm inseminated eggs have completed their second meiosis, in which the maternal haploid chromosome set is doubled. In 1981, Streisinger et al published a key study about induced gynogenesis in zebrafish (*Danio rerio*) and described how to produce meiotic diploids, mitotic diploids and homozygous and heterozygous clonal lines of zebrafish (Streisinger et al, 1981). The publication marked the rise of the zebrafish as the leading animal model for research on embryonic development in vertebrates. Mitogynogenesis has been successfully induced in a total of 19 fish species and doubled haploid animals have been produced, including 16 freshwater species and 3 marine species (Komen & Thorgaard, 2007).

There are two distinct hypotheses explaining the mechanism of mitogynogenesis and chromosome doubling. The prevalent hypothesis is that the treatments may disable or disorganize spindles, block the anaphase movement of chromosomes, and form a doubled nucleus without cell division, which is also called "suppression of the first cleavage" (Streisinger et al, 1981; Chourrout, 1984; Ihssen et al, 1990; Kobayashi, 1997; Morelli & Aquacop, 2003; Nam et al, 2001; Sakao et al, 2003). However, after careful study of the behavior of spindles and nuclei in eggs treated with hydrostatic pressure or heat shock around the time of the metaphase of the first cell cycle in the rainbow trout (*Oncorhynchus mykiss*), Zhang & Onozato (2004) found that heat shock or hydrostatic pressure did not inhibit the first cleavage owing to the regeneration of the bipolar spindle but inhibited the second cleavage by forming a monopolar spindle during the second cell cycle, and thus, the chromosome set was doubled. They proposed a new hypothesis in which the destruction of the centrosome by hydrostatic pressure treatment leads to chromosome set doubling. Zhu et al (2006, 2007) adapted indirect immunofluorescence staining to detect cytological changes during mitosis of UV-irradiated homosperm inseminated eggs from the Oliver flounder (another name of the Japanese flounder) treated with cold shock or hydrostatic pressure. The results of both studies supported the theory of Zhang & Onozato.

The objective of this paper was to detect the effects of hydrostatic pressure treatment on the induction of mitogynogenesis in Japanese flounder eggs inseminated using UV-irradiated red sea bream (Pagrus major) sperm by analyzing histological sections.

2 Material and methods

2.1 Animal and gamete collection

All experiments were conducted at Beidaihe Central Experiment Station, Chinese Academy of Fishery Sciences, Qinhuangdao, Heibei Province, China. The fish were reared at 14 ℃ with a 18 h light/6 h dark photoperiod. Sperm from mature red sea bream (3 year old, total length: 35 cm, body weight: 1.4 kg) were collected using a 5 mL plastic syringe by gently pressing the abdomen and avoiding water and urine contamination. Eggs were collected from one fertile, mature Japanese flounder (4 years old, total length: 45 cm, body weight: 2.4 kg). The eggs were stripped from the female and collected in a 1 000 mL glass beaker. Totally, 3 mL of sperm (1.01×10^6 cells per milliliter) and 50 milliliter eggs (1 000 eggs per milliliter) were collected, and were stored at 4 ℃ before use.

2.2 UV irradiation of sperm, insemination and hydrostatic pressure treatment

The UV irradiation of red sea bream sperm was performed according to the method of Yamamoto (1995, 1999). Before irradiation, the sperm were diluted 50 times with Ringer's solution (0.140 8 mol/L NaCl, 0.005 2 mol/L KCl, 0.004 9 mol/L $CaCl_2 \cdot 2H_2O$, 0.001 1 mol/L $MgCl_2 \cdot 6H_2O$, 0.000 8 mol/L $NaH_2PO_4 \cdot 2H_2O$, 0.002 4 mol/L $NaHCO_3$, 0.005 0 mol/L $C_6H_{12}O_6 \cdot H_2O$; pH 6.5～7.2), and then irradiated under UV light at an intensity of 73 mJ/cm^2 (VLX-3W, Cole Parmer Instrument Company, Illinois, USA). Artificial insemination was performed by mixing 1 mL UV-irradiated sperm and 40 mL eggs for 3 min in a 1 000 mL glass beaker, and the mixture was diluted with 1 000 mL filtered sea-water and incubated at 17 ℃ . The mixture was subjected to a pressure of 63.7 MPa for 6 min after 60 min of insemination (Yamamoto, 1995). The eggs were then transferred to 17 ℃ sea-water and incubated after the treatment.

2.3 Cytology

For the cytological observations, after hydrostatic pressure treatment, the eggs were fixed as follows: from insemination to 60 min after insemination (AI) (time just before hydrostatic pressure treatment), the eggs were fixed with the time interval of 3 min in 4.5 mL Bouin's solution. From 68 min AI to 3.5 h AI, the eggs were fixed with the time interval of 2 min. After a fixation of 14 h, the eggs were transferred to 70% ethanol until use.

The histological sections were made according to the method of Sun et al (2005). The membrane of each egg was pricked with a needle at the vegetal pole. Eight to ten eggs at each stage were embedded in a hot agarose solution (1%～3%), and dehydrated in an ethanol series and terpineol. The eggs were then embedded in paraffin, serially sectioned into 8 μm slices, and stained with Harris' hematoxylin and eosin. All slides were observed under an Olympus BX 51 research microscope and photographs were taken with a DP71 CCD (Olympus Corporation, Tokyo, Japan).

2.4 Ploidy determination

The relative content of DNA in 1-day-old hatching larvae were measured to determine their ploidy status by using a flow cytometer (Ploidy Analyzer, Partec GmbH, Münster, Germany). Each larva was placed in a petri dish, 0.5 mL CyStain DNA 1step, including 4′6-deamidino-2-phenylindole (DAPI; Partec GmbH, Münster, Germany) was added, and the larva was rubbed with two tweezers in order to release the cells. Subsequently, added an additional 1.5 mL of CyStain DNA 1step. The cells were then incubated at room temperature for 5 min, filtered through a 30 μm CellTrics disposable filter (Partec GmbH, Münster, Germany), and analyzed using flow cytometry. The normal diploid larvae hatched from normal fertilization of male and female Japanese flounder were used for the 2C diploid value, which were set as the standard, together with the larvae from UV-irradiated sperm inseminated eggs without pressure treatment and normal larvae from mitogynogenetic group. For each group, 10 larvae were analyzed.

3 Results

Before the application of hydrostatic pressure at 60 min AI, most of the eggs (8 of 10) were at the metaphase of the first mitosis. For each egg, the spindle was very conspicuous, the deeply stained chromosomes were arranged in an orderly manner at the centre of the metaphase plate, and the axis of the chromosomes was perpendicular to the spindle axis (Fig. 1a). The spindle was disassembled by hydrostatic pressure immediately after the treatment at 68 min AI and the previously well-arranged chromosomes became disordered (Fig. 1b). At 82 min AI, the spindle resembled in its former position and the chromosomes were rearranged. In this section, a dense chromatin body (DCB) was also observed (Fig. 1c). The egg then entered the normal anaphase (Fig. 1d) and telophase (Fig. 1e) of the first mitosis, which was not blocked.

After the first mitosis, the egg entered the second mitosis. At 118 min AI, the egg was at the prophase, a nucleus was formed, and at only one side of the nucleus, there was an aster that was produced by the centrosome (Fig. 1g). Thirty four minutes later, the nuclear membrane was breakdown and a monopolar spindle was assembled instead of the normal bipolar spindle. The chromosomes were arranged in a line on the microtubules of the monopolar spindle, facing the first cleavage furrow, and the egg entered the metaphase of the second mitosis (Fig. 1h). The monopolar spindle then became bigger and arranged chromosomes were randomLy disturbed on the microtubules, this represents the anaphase and telophase of the second mitosis (Fig. 1i, j). In the second mitosis, the chromosomes were replicated, but cell cleavage was blocked.

At 162 min AI, the egg entered the prophase of the third mitotic cycle, and two asters at opposite ends of the nucleus (Fig. 1l). As the egg developed into metaphase, a bipolar spindle was assembled, the deeply stained chromosomes arranged orderly in the centre of metaphase plate, and the axis of the chromosomes was perpendicular to the spindle axis at 174 min AI (Fig. 1m). The chromosomes were then separated and dragged to the two poles by the spindles, and the cell entered the anaphase and telophase of the third mitosis (Fig. 1n, o).

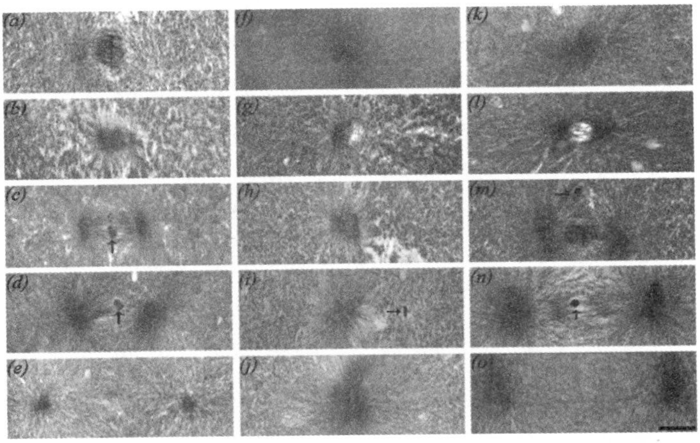

Fig. 1 The developmental process of mitogynogenetic eggs of Japanese flounder (*Paralichthys olivaceus*) from the first mitosis to the third mitosis when incubated at 17 ℃

(a) ~ (e) First mitotic cycle, (f) ~ (j) second mitotic cycle, and (k) ~ (o) third mitotic cycle. (a) Metaphase of the first mitosis (60 min after insemination, AI); (b) 68 min AI, the spindle was disassembled by the hydrostatic pressure treatment; (c) 82 min AI, the spindle resembled in its former position, the arrow indicates the dense chromatin body (DCB); (d) anaphase, the arrow indicates the DCB (94 min AI); (e) telophase (100 min AI); (f) interphase of second mitosis (104 min AI); (g) prophase, one nucleus and one aster (118 min AI); (h) metaphase, monopolar spindle (152 min AI); (i) anaphase, the arrow indicates the dense chromatin body (154 min AI); (j) telophase (156 min AI); (k) interphase of the third mitosis (160 min AI); (l) prophase, one nucleus and two asters (162 min AI); (m) metaphase, bipolar spindle, the arrow indicate the dense chromatin body (174 min AI); (n) anaphase, the arrow indicates the DCB (182 min AI); (o) telophase (190 min AI). Scale bar = 20 μm

The deeply stained DCB, which was formed by the UV-irradiated red sea bream sperm, was observed in the sections (Fig. 1c, d, i, m, n). In the first and third mitotic cycles, the DCB disturbed on the equatorial plate of the bipolar spindle. In the second cycle, the distance between the DCB and the monopolar spindle was half of that between the equatorial plate and one aster in the bipolar spindle (Fig. 1i). The DCB randomLy entered one blastomere. In the section shown in Fig. 2, only one DCB was observed in one of the four blastomeres.

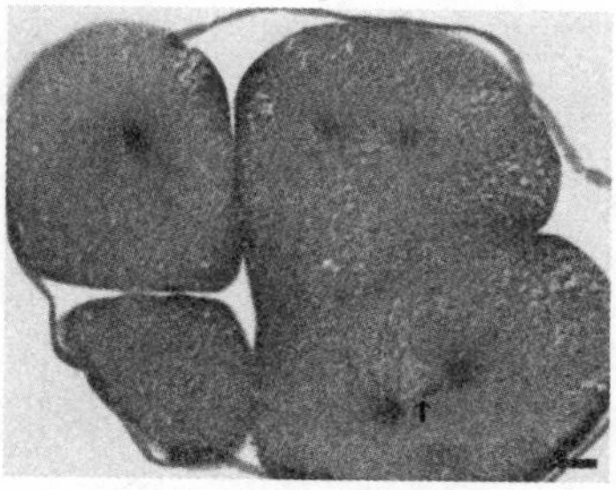

Fig. 2　Four-cell stage

The arrow indicates the only dense chromatin body disrupting the equatorial plate of one blastomere. Scale bar = 20 μm

Seven of 150 eggs at 150 min AI were observed to have three blastomeres— one big blastomere and two small blastomeres (Fig. 3a). In the histological section, the bigger blastomere contained a monopolar spindle (Fig. 3b), but the two smaller blastomeres contained a bipolar spindle (Fig. 3c). The chromosomes arranged on the monopolar spindle were parallel to the first cleavage furrow, while the bipolar spindle's chromosomes were perpendicular to it. The presence of three blastomeres in one egg is one example of the abnormal cell events that occur in eggs following hydrostatic treatment.

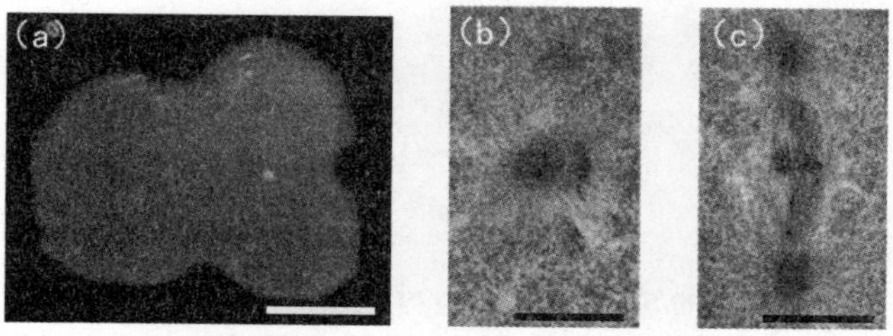

Fig. 3　Histological and morphological observation of an egg with three blastomeres

(a) An embryo with three blastomeres at 150 min after insemination; (b) monopolar spindle in one blastomere; (c) bipolar spindle in the other blastomere. Scale bar in (a) indicates 160 μm. Scale bar in (b) and (c) indicates 20 μm

Some other abnormal cell events were also observed. At 126 min AI, a tripolar spindle was clearly assembled in one blastomere of two (Fig. 4). The areas of the three asters were 47.29 μm^2, 98.24 μm^2, and 144.56 μm^2. The sum of the two small asters' areas (145.53 μm^2) was approximately equal to the area of one large aster. Another abnormal cell event was observed in an egg at 116 min AI wherein two nuclei were formed in one blastomere. These two nuclei were approximately the same size, and both had one aster each located beside it (Fig. 5).

The results of flow cytometric analysis indicated that the relative DNA content of the diploid control were 51.0 ± 1.9 (mean ± SD), while for the haploid it were 27.0 ± 2.5, which were almost half that of the diploid control. The relative DNA content of the mitogynogenetic diploid were 50.0 ± 1.7, which were equal to that of the diploid control (Fig. 6).

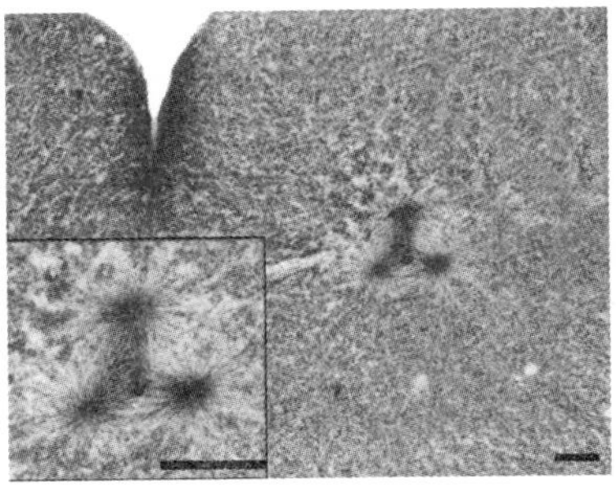

Fig. 4 At 126 min after insemination, a tripolar spindle was well assembled in one blastomere of two

Scale bar = 20 μm

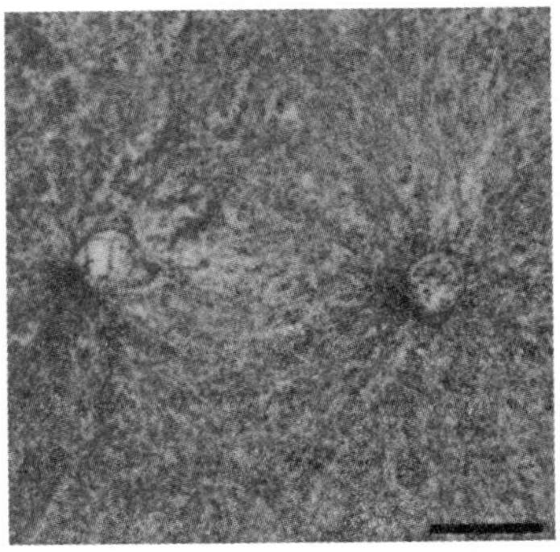

Fig. 5 At 116 min after insemination, two nuclei formed in only one blastomere

Scale bar = 20 μm

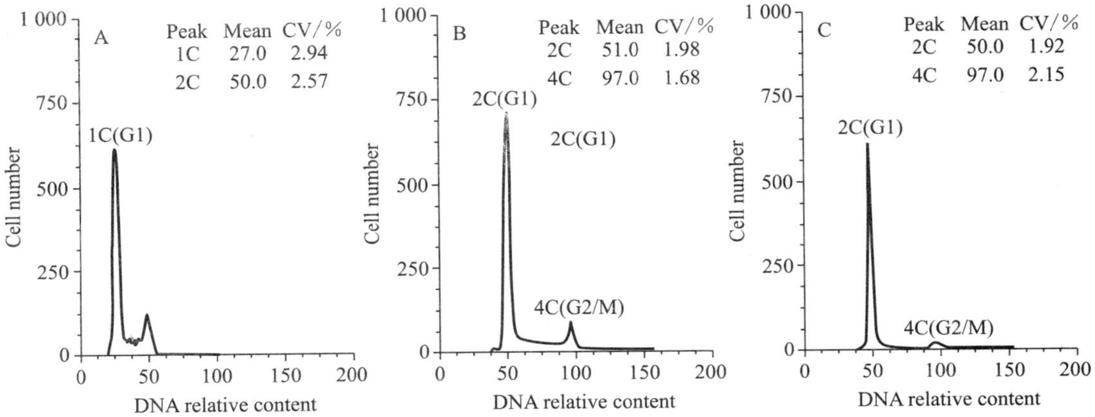

Fig. 6 Histograms of ploidy analysis

(a) Haploid; (b) diploid from normal fertilization; (c) diploid from mitogynogenesis

4 Discussion

Most of the eggs were treated by hydrostatic pressure during the metaphase of the first mitosis; the bipolar spindle of each egg was well assembled and then destroyed by the treatment. Fourteen minutes later, the bipolar spindle reassembled and the first mitosis normally continued. In the second mitotic cycle, a monopolar spindle assembled in each blastomere instead of a bipolar spindle. Mitosis proceeded with a monopolar spindle in each blastomere, and, as a result, the chromosomes were doubled but the cell cleavage was blocked. Subsequently, the third cycle proceeded normally. Flow cytometric analysis indicated that the relative DNA content of mitogynogenetic larva was equal to that of the diploid control, and the relative DNA content of larvae developed from eggs inseminated with UV-irradiated red sea bream sperm without hydrostatic pressure treatment was almost half of the diploid control. These observations demonstrated the effectiveness of UV-irradiation and hydrostatic pressure treatment.

In the studies of rainbow trout (Zhang & Onozato, 2004) and olive flounder (Zhu et al, 2006, 2007), pooled eggs of several females were used. Although eggs from only one female were used in this study, the results we obtained are similar to those observed by Zhang & Onozato (2004), and Zhu et al (2006, 2007). This indicated that the maternal effect to the embryo development process after hydrostatic pressure treatment could be eliminated. Zhang & Onozato (2004) proposed a new model for chromosome set doubling. The spindle and the daughter centrioles are dissembled by hydrostatic treatment. The remaining two single-mother centrioles and pericentriolar matrices (PCMs) regenerate a bipolar spindle and the first cleavage progresses normally. A centriole distributed to a daughter blastomere generates a whole centrosome with a newly formed procentriole, and, as a result, a monopolar spindle is formed, which leads to the failure of the anaphase disjunction of duplicated chromosomes of the second cleavage, eventually resulting in chromosome set doubling (Zhang & Onozato, 2004).

The animal cell's centrosome contains a pair of centrioles surrounded by a PCM. At the time of DNA synthesis, each centriole duplicates to produce a mother and daughter centriole pair, which forms two centrosomes to organize the mitotic spindle poles (Loncarek et al, 2007). The number of centriole pairs determines the number of spindle poles. In sea urchin zygotes, when mitosis is prolonged by artificial methods, the two spindle poles do not further subdivide even when mitosis is prolonged to 20 times its normal duration. Ultrastructural analysis of such tetrapolar spindles reveals that each pole contains only one centriole. After the cell divides into four, each of the half centrosomes assembles a daughter centriole, and then becomes a normal centrosome with full reproductive capacity. However, they do not undergo centriole splitting or centrosome disjunction, and each cell assembles a monopolar spindle at the next mitosis. In effect, the two mitotic centrosomes with normal reproductive capacity subdivide into four centrosomes with half the normal reproductive capacity (Mazia et al, 1960; Sluder & Begg, 1985; Hinchcliffe et al, 1998; Hinchcliffe & Sluder, 2001).

The formation of three blastomeres in one egg can also be interpreted using the model of Zhang & Onozato (2004). As they develop asynchronously, the daughter centrioles in one egg may be mature, immature, or in a transition state from the immature to the mature state when hydrostatic pressure is applied. After the application of hydrostatic pressure, the mature centriole is not destroyed but the immature centriole is, resulting in one of the centrosomes containing one centriole and the other forming a pair. After the first mitosis, the blastomere that receives the centrosome containing only one centriole assembles a monopolar spindle instead of a bipolar spindle, resulting in a diploid blastomere. Another blastomere that has a complete centrosome undergoes normal mitosis. Those embryos may stop developing at some stage or develop into haploid-diploid mosaics.

The probable cause of the tripolar spindle is that the centrosomes are not destroyed by hydrostatic pressure and the first mitosis proceeds normally. At the second mitotic cycle, the centrosome in one blastomere is duplicated at the interphase and assembles into two integrated centrosomes. An unexpected event can result in the separation of two centrioles from one centrosome. The one centrosome and two separated centrioles then assemble a tripolar spindle (Fig. 4). A tripolar spindle is rarely observed in the finfish and it is only observed in spontaneous gynogenetic fish species such as the Ginbuna (*Carassius auratus gibelio*; Yamashita et al, 1993). A tripolar spindle is formed at the first meiosis and the three sets of chromosomes present in triploid Ginbuna oocytes are separated in three directions; this ensures the retention of the original ploidy during gynogenesis. In our study the tripolar spindle was formed after hydrostatic pressure treatment, but it was not spontaneously formed. Chromosomes were not uniformly distributed on the tripolar spindle and resulted in chromosomal malsegregation.

References

[1] Bertotto D, Cepollaro F, Libertini A, et al. Production of clonal founders in the European sea bass, *Dicentrarchus labrax* L., by mitotic gynogenesis[J]. Aquaculture, 2005, 246: 115-124.

[2] Chourrout D. Pressure-induced retention of second polar body and suppression of first cleavage in rainbow trout: production of all-triploids, all-tetraploids, and heterozygous and homozygous diploid gynogenetics[J]. Aquaculture, 1984, 36: 111-126.

[3] Hinchcliff E H, Sluder G. "It takes two to tango": understanding how centrosome duplication is regulated throughout the cell cycle[J]. Genes & Development, 2001, 15: 1167-1181.

[4] Hinchcliffe E H, Cassels G O, Rieder CL, et al. The coordination of centrosome reproduction with nuclear events during the cell cycle in the sea urchin zygote[J]. Journal of Cell Biology, 1998, 140: 1417-1426.

[5] Ihssen P E, McKay L R, McMillan I, et al. Ploidy manipulation and gynogenesis in fishes: cytogenetic and fisheries applications[J]. Transactions of American Fisheries Society, 1990, 119: 698-717.

[6] Kobayashi T. Survival and cytological observations on early development of normal, hybrid, and gynogenetic embryos of amago salmon[J]. Fisheries Science, 1997, 63: 33-36.

[7] Komen H, Thorgaard G H. Androgenesis, gynogenesis and the production of clones in fishes: A review[J]. Aquaculture, 2007, 269: 150-173.

[8] Loncarek J, Sluder G, Khodjakov A. Centriole biogenesis: a tale of two pathways[J]. Nature Cell Biology, 2007, 9: 736-738.

[9] Mazia D, Harris P, Bibring T. The multiplicity of the mitotic centers and the time-course of their duplication and separation[J]. Journal of Biophysical and Biochemical Cytology, 1960, 7: 1-20.

[10] Morelli M, Aquacop. Effects of heat-shock on cell division and microtubule organization in zygotes of the shrimp *Penaeus indicus* (Crustacea, Decapoda) observed with confocal microscopy[J]. Aquaculture, 2003,

216: 39-53.

[11] Nam Y K, Choi G C, Park D J, et al. Survival and growth of induced tetraploid mud loach[J]. Aquaculture International, 2001, 9: 61-71.

[12] Sakao S, Fujimoto T, Tanaka M, et al. Aberrant and arrested embryos from masu salmon eggs treated for tetraploidization by inhibition of the first cleavage[J]. Nippon Suisan Gakkaishi, 2003, 69: 738-748 (in Japanese).

[13] Schartl M, Wilde B, Schlupp I, et al. Evolutionary origin of a parthenoform, the amazon molly *Poecilia formosa*, on the basis of a molecular genealogy[J]. Evolution, 1995, 49: 827-835.

[14] Sluder G, Begg D A. Experimental analysis of the reproduction of spindle poles[J]. Journal of Cell Science, 1985, 76: 35-51.

[15] Streisinger G, Walker C, Dower N, et al. Production of clones of homozygous diploid zebra fish (*Brachydanio rerio*)[J]. Nature, 1981, 291: 293-296.

[16] Sun W, You F, Zhang P J, et al. Study on insemination biology of turbot (*Psetta maxima*)[J]. Marine Sciences, 2005, 29: 75-80 (in Chinese).

[17] Yamamoto E. Studies on sex-manipulation and production of cloned population in hirame flounder, *Paralichthys olivaceus* (Temminck et Schlegel)[J]. Bulletin of Tottori Prefecture Fisheries Experiment Satiation, 1995, 34: 1-145 (in Japanese).

[18] Yamamoto E. Studies on sex-manipulation and production of cloned population in hirame flounder, *Paralichthys olivaceus* (Temminck et Schlegel)[J]. Aquaculture, 1999, 173: 235-246.

[19] Yamashita M, Jiang J, Onozato H, et al. A tripolar spindle formed at meiosis-I assures the retention of the original ploidy in the gynogenetic triploid crucian carp ginbuna, *Carassius auratus langsdorfii*[J]. Development, Growth & Differentiation, 1993, 35: 631-636.

[20] Zhang X L, Onozato H. Hydrostatic pressure treatment during the first mitosis does not suppress the first cleavage but the second one[J]. Aquaculture, 2004, 240: 101-113.

[21] Zhu X P, You F, Zhang P J, et al. Effects of hydrostatic pressure on microtubule organization and cell cycle in gynogenetically activated eggs of olive flounder (*Paralichthys olivaceus*)[J]. Theriogenology, 2007, 68: 873-881.

[22] Zhu X P, You F, Zhang P J, et al. Effects of cold shock on microtubule organization and cell cycle in gynogenetically activated eggs of olive flounder (*Paralichthys olivaceus*)[J]. Marine Biotechnology, 2006, 8: 312-328.

Cold-shock induced androgenesis without egg irradiation and subsequent production of doubled haploids and a clonal line in Japanese flounder, *Paralichthys olivaceus*

Jilun Hou, Guixing Wang, Xiaoyan Zhang,
Zhaohui Sun, Haijin Liu, Yufen Wang

Beidaihe Central Experiment Station, Chinese Academy of Fishery Sciences, Qinhuangdao 066100, China

Abstract Androgenesis is a useful manipulation to fix male-specific genetic traits as well as to restore the genotypes of cryopreserved sperm, and it has been induced by the genetic inactivation of the egg nucleus with gammol / La-, X- or UV-ray irradiation before fertilization. Recently, a technique for cold-shock induced androgenesis was developed in a freshwater species, the loach, *Misgurnus anguillicaudatus*, in which viable androgenetic diploids and doubled haploids (DHs) were successfully produced by the fertilization of diploid sperm and chromosome doubling by heat-shock treatment, respectively. This technique was immol / Lediately applied for cloning in the zebrafish, Danio rerio. Here, we reported the first successful induction of androgenetic development by means of cold-shock treatment (at 0 ℃ for 15 min) just after the fertilization (within 10 s) of eggs in a marine aquaculture fish species, the Japanese flounder, *Paralichthys olivaceus*. Then, androgenetic embryos thus generated were subjected to hydrostatic pressure treatment (650 kg/cm^2, 6 min) after incubation at 17 ℃ so as to produce DHs by chromosome doubling. The yield rate of putative DHs, which was estimated as the frequency of diploid larvae at the first feeding stage in relation to the total eggs used, ranged from 0.81% to 1.79%. The complete homozygosity of these putative DHs was genetically verified using 36 microsatellite markers that covered 24 linkage groups of Japanese flounder. Thus, an androgenetic clonal line was produced with the sperm of a mito-gynogenetic DH male by using cold-shock and subsequent hydrostatic pressure-shock treatments.

Keywords chromosome manipulation; clone; doubled haploid; inbred line; heterosis; gynogenesis

1 Introduction

Japanese flounder, *Paralichthys olivaceus*, is an economically important fish species widely cultured along the coasts of China, Japan and Korea. Females of this species grow much faster than males, and all-female populations have therefore been established by using artificially induced gynogenesis (Yamamoto, 1999). Individuals hatched via induced gynogenesis only have the maternal genome, with the paternal genome being entirely eliminated. However, in some cases, males exhibit better performances in specific traits than females; thus, it is important to develop technology for fixing these traits both quickly and effectively.

Androgenesis is a mode of development in which only the paternal genome genetically contributes to the progeny, without the contribution of maternal chromosomes (Komen and Thorgaard, 2007). Inducing androgenesis can lead to male traits being effectively fixed. The traditional method to induce androgenesis in fish requires the irradiation of unfertilized eggs to genetically inactivate the maternally derived genome. This irradiation has been carried out by using gamma rays, X-rays or UV-rays (Arai, 2001). For all irradiation procedures, specific equipment is needed to ensure the safety of the operators, especially when using gamma- and X-ray irradiation; in addition, protection solutions are also needed to maintain the fertilization potential of the eggs. The most widely used protection solutions are water (Myers et al, 1995), synthetic ovarian fluid (Bongers et al, 1994), natural ovarian fluid (Lin and Dabrowski, 1998; Rothbard et al, 1999), Ringer's solution (David and Pandian, 2006), seminal plasma (Corley-Smith et al, 1996; Fujimoto et al, 2007) or Hank's solution (Yasui et al, 2010). The requirement of specific equipment and protection solutions makes the traditional method of induced androgenesis inconvenient for routine use (Arai, 2001).

In the loach, *Misgurnus anguillicaudatus*, a new method to induce androgenesis was established (Morishima et al, 2011). Activated eggs were immol/Lediately cold-shocked at a temperature of 0 ℃ to 3 ℃ for 60 min, which resulted in more than 30% of the hatched larvae being haploid androgenotes by means of eliminating the maternally derived nucleus. The haploids were non-viable owing to the expression of an abnormality widely referred as haploid syndrome. Thus, viable diploid androgenotes were successfully induced by using the cold-shock method together with diploid sperm from the neo-tetraploid male (Hou et al, 2013). Doubled haploid (DH) androgenetic loach progenies were also produced by the cold-shock method from haploid sperm of diploid male, complemented with a heat-shock at the metaphase of the first mitosis, which doubled the chromosomes (Hou et al, 2014). The cold-shock method to induce haploid androgenesis was also proved to be effective in zebrafish, *Danio rerio*, and the first androgenetic clonal zebrafish line was established using a combination of cold-shock androgenesis and heat-shock chromosome doubling treatments (Hou et al, 2015).

Here, we established an androgenetic induction method using cold-shock of just-activated eggs of Japanese flounder. We optimized the cold-shock treatment duration, after which androgenetic DHs were induced by a combination of cold- and hydrostatic pressure-shock treatments. Finally, an androgenetic clonal line was established by cold-shock induction followed by chromosome doubling with hydrostatic pressure in eggs fertilized with the sperm of a mito-gynogenetic DH male, which had been established previously (Liu et al, 2012a). We determined ploidy status based on cellular DNA content by flow cytometry, and then genetically verified the exclusive transmission of paternal genome and complete homozygosities of DHs as well as the genetic identity among clonal progenies by microsatellite genotyping.

2 Materials and methods

2.1 Ethics

This study was performed in accordance with the Guide for Care and Use of Laboratory Animals of the Chinese Association for Laboratory Animal Sciences (No. 2011-2).

2.2 Fish and gamete collection

Mature male and female Japanese flounders were reared at the Beidaihe Central Experimental Station, Chinese Academy of Fishery Sciences, at 14 ℃ water temperature with an 18 h light/6 h dark photoperiod. Eggs were manually stripped and collected from the females using a 1000 mL glass beaker. Sperm was collected using a 5 mL plastic syringe, with gentle pressure on the abdomen to facilitate collection. At least 2 mL of sperm was collected from each male. The collected eggs and sperm were store at 4 ℃ in dark before use.

2.3 Optimization of the cold-shock treatment duration

Before use, sperm samples were diluted 50 times with Ringer's solution. Diluted sperm samples were added to the eggs at a ratio of 1:40, mixed well, and activated with a small amount of filtered 17 ℃ sea-water. Within the first 10 s after activation, activated eggs were transferred to a 0 ℃ water-bath, and subjected to cold shock for 15 min, 30 min, 45 min or 60 min. During the cold-shock treatment, water temperature was constantly maintained at precisely 0.5 ℃ . After the cold-shock treatment, the eggs of each group were transferred to a 1 000 mL glass beaker that contained filtered 17 ℃ sea-water and maintained in an incubator at constant temperature of 17 ℃ until hatching. The control group was composed of just-fertilized eggs that were directly transferred to the incubator at a constant temperature of 17 ℃ .

2.4 Chromosome doubling by hydrostatic pressure

Just-activated eggs were first transferred to a 0 ℃ water-bath for a cold-shock of 15 min, followed by a 60 min incubation at 17 ℃ in a water-bath. The eggs were then exposed to a 650 kg/cm^2 hydrostatic pressure for 6 min, following the protocol of Yamamoto (1999). After the treatment, eggs were returned to incubator with 17 ℃ water-bath until hatching.

2.5 Ploidy, all-male inheritance and homozygosity

The ploidy of all hatched abnormal larvae at 69 h after activation was analyzed by flow cytometry (PA-II, Partec GmbH, Münster, Germany), following the procedure of Fujimoto et al (2007). Each larva was first digested by using 85 μL of solution A (CyStain DNA 2 step, Cod. 05-5005, Partec GmbH) for 15 min. Then, 15 μL of the digested solution was stained by using 500 μL of solution B (CyStain DNA 2 step, Cod. 05-5005, Partec GmbH), and analyzed by flow cytometry. The remaining 70 μL of digested solution was used for DNA extraction.

To determine all-male inheritance (exclusive paternity), eight haploid larvae and five diploid larvae of control group, together with female and male parents, were genotyped using two Japanese flounder microsatellite markers: *Poli1359TUF* (Castaño-Sánchez et al, 2010) and *HLJYP38* (Liu et al, 2013). Homozygosity of eight diploid (putative DH) progeny, hatched from the cold-shock and hydrostatic pressure treatment groups, was genetically confirmed by analyzing 36 microsatellite markers that covered 24 linkage groups of the Japanese flounder (Castaño-Sánchez et al, 2010; Liu et al, 2013). DNA was extracted using TIANamp Marine Animals DNA Kit (DP324-02, Tangent). PCR was performed in a 15 μL reaction solution (Liu et al, 2012a) under the following conditions: one cycle of initial denaturation for 3 min at 94 ℃ , 25 cycles of denaturation for 30 s at 93 ℃ , annealing for 30 s at 62 ℃ , extension for 30 s at 72 ℃ , and one cycle of final extension for 10 min at 72 ℃ . Electrophoresis of PCR products together with molecular weight markers were performed on 6% denaturing polyacrylamide gel, and after electrophoresis, the gel was stained with silver nitrate (Liao et al, 2007).

2.6 Fertilization, hatch, abnormal and haploid rates

The fertilization rate was calculated as the frequency of gastrula embryos relative to the total eggs used, measured 24 h after activation. The hatching rate was calculated as the proportion of hatched larvae relative to the total eggs used. The abnormality rate was calculated as the rate of externally abnormal larvae relative to the total hatched larvae. The haploid rate was calculated as the rate of haploid larvae relative to the total eggs used. The normal rate of larvae that hatched from eggs subjected to cold-shock and hydrostatic pressure treatments was calculated as the proportion of normal larvae relative to the hatched larvae. Survival rate at first feeding was the frequency of surviving larvae at this stage in relation to the hatched larvae. Finally, the diploid rate was the rate of diploid larvae relative to the total eggs used.

2.7 Production of androgenetic clonal line

To produce the androgenetic clonal line, eggs of normal diploid females were fertilized with the sperm of a mito-gynogenetic DH male, which was produced using the procedure described by Yamamoto (1999) and Liu et al (2012a). Androgenesis was immediately induced by the cold-shock treatment described above, followed by the hydrostatic pressure treatment, which doubles the chromosome. Ploidy determination and microsatellite genotyping to verify clonal nature were conducted as mentioned above.

2.8 Statistical analysis

All experiments were performed at least with three biological replicates. Data used to optimize the cold-shock duration were statistically analyzed by one-way analysis of variance (ANOVA) followed by Student-Newsman-Keuls multiple comparisons (significance at $P < 0.05$), using R software.

3 Results

3.1 Optimization of cold-shock duration

The fertilization, hatch, abnormal and haploid rates of the groups that were subjected to different cold-shock duration treatments are shown together with the rates of the control group in Fig. 1. The fertilization rate in the control group (65.88% ± 6.77%) was not significantly different to the 30, 45, and 60 min duration cold-shocked groups ($P > 0.05$), but the difference was significant ($P < 0.05$) when compared with the 15 min group (46.27% ± 14.92%). However, fertilization rates among the cold-shock treatment groups were not significantly different ($P > 0.05$). The control group had the highest hatching rate (60.82% ± 0.48%). Based on comparisons among the cold-shock treatment groups, the hatching rate of the 15 min group (32.64% ± 1.11%) was significantly different from 30 min group (46.76% ± 5.79%). Considering the abnormal rate, the 15 min group (46.78% ± 2.40%) was highest, significantly different from the other groups ($P < 0.05$).

Ploidy analysis showed that most abnormal larvae of the control group were diploid, only one was hyper-2N (Table 1), and none of them were haploid. Within the cold-shock treatment

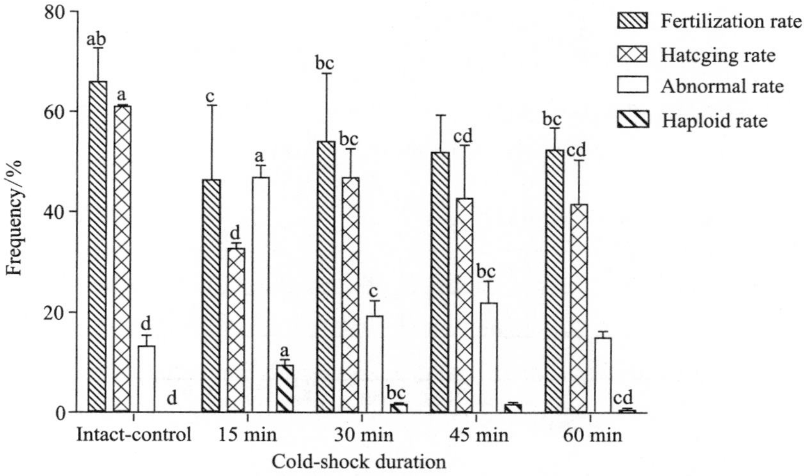

Fig. 1　Fertilization rate, hatching rate, abnormal rate and haploid rate of just-fertilized eggs of the Japanese flounder, *Paralichthys olivaceus*, cold-shocked at 0 ℃ for 15 min, 30 min, 45 min and 60 min

Letters above columns indicate significant differences as determined by one-way ANOVA and Student-Newsman-Keuls multiple comparisons ($P < 0.05$)

groups, haploid (Fig. 2A), together with triploid and hyper-2N larvae were detected. Diploid larvae were also detected in the groups subjected to 15 min and 30 min cold-shock treatments (Fig. 2B). The haploid rate was significantly higher in the 15 min cold-shock treatment group (9.12%±1.38%) compared with the other groups (Fig. 1).

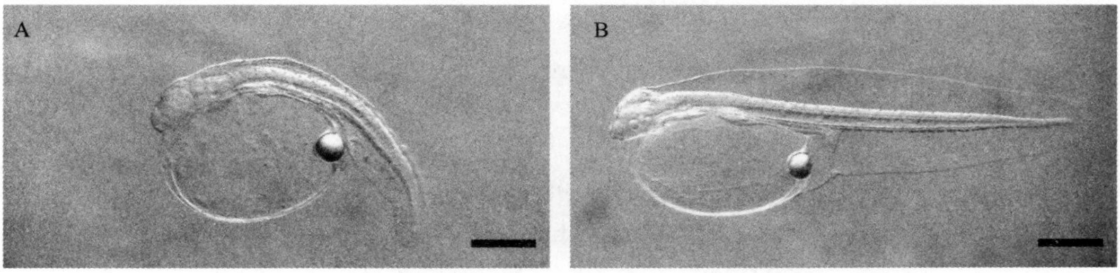

Fig. 2　The two different types of Japanese flounder, *Paralichthys olivaceus*, larvae hatched from just-fertilized eggs subjected to a 0 ℃ cold-shock treatment for 15 min

A: abnormal haploid larva; B: normal diploid larva with similar appearance that the control diploid larva. The ploidy of these larvae was analyzed by flow cytometry. Scale bar denotes 0.5 mm

We verified the all-paternal inheritance of the haploid larvae that hatched from the cold-shock treatments by means of two microsatellite markers: *Poli1359TUF* and *HLJYP38*. In the diploid progeny of control group, one maternally derived allele and one paternally derived allele were detected. Only paternally derived alleles were observed in the haploid progeny (Table 2).

Table 1 Ploidy status of abnormal larvae hatched from different cold-shock durations (0 ℃) in Japanese flounder, *Paralichthys olivaceus*

Treatment	No. of larvae	Ploidy status				
		1N	2N	3N	Hyper-2N	Mosaic
Control	32	0	31	0	1[a]	0
0 ℃ , 15 min	70	43	2	12	13[b]	0
0 ℃ , 30 min	47	9	3	18	15[c]	2[d]
0 ℃ , 45 min	46	7	0	28	10[e]	1[f]
0 ℃ , 60 min	29	2	0	20	7[g]	0

a: 2.6N; b: 2.2N: 4; 2.4N: 2; 2.6N: 3; 2.8N: 4; c: 2.2N: 3; 2.6N: 9; 2.8N: 3; d: 1N-4.6N; 1.5N-2N; e: 2.2N: 3; 2.4N: 2; 2.6N: 3; 2.8N: 2; f: 2.8N-3N; g: 2.2N: 1; 2.4N: 1; 2.6N: 1; 2.8N: 3

Table 2 Microsatellite genotype at two loci in haploid progenies from the cold-shock treatment (0 ℃ , 15 min), and diploid progenies from intact control in Japanese flounder, *Paralichthys olivaceus*

Locus	Female	Male	Haploid progenies from cold-shock treatment	Diploid progenies from intact control
Poli1359TUF	194/194	210/210	210: 8	194/210: 5
HLJYP38	190/190	184/184	184: 8	190/184: 5

Our results indicated that cold-shock treatment at 0 ℃ just after activation of eggs induced haploid androgenesis without any egg irradiation in Japanese flounder, and the 15 min was the optimum parameter for cold-shock duration.

3.2 Chromosome doubling by hydrostatic pressure treatment

Haploid larvae hatched from cold-shock treatments were non-viable owing to their abnormal development. In order to produce viable androgenetic DHs in Japanese flounder, the haploid chromosome set of cold-shocked eggs was duplicated by inducing endomitosis with hydrostatic pressure treatment at the early cleavage stage. Five batches of eggs were treated. The fertilization rate varied from 32.30% to 73.99%, showing high variability between groups (Table 3). At the hatching stage, each batch had normal hatched larvae, but the variability of normality rate was also high between groups. Flow cytometry revealed the presence of surviving diploid larvae. The diploid rate was very low, ranging from 0.81% to 1.79%.

These diploid larvae were designated as the putative androgenetic DHs, and the homozygosity was determined by means of microsatellite markers. Based on this analysis, no heterozygous locus was detected in any of the eight diploid larvae analyzed (Table 4). Therefore, the diploid progenies hatched from cold-shock androgenesis and hydrostatic pressure treatment were confirmed to be androgenetic DHs.

Table 3　Rates of fertility, hatch, normal larvae, survival of feeding larvae and putative diploid larvae developed from eggs subjected to cold-shock and hydrostatics pressure treatment in Japanese flounder, *Paralichthys olivaceus*

Treatment batch	No. of egg used	Fertilization/%	Hatching/%	Normal/%	Survival/%	Diploid/%
1	359	32.30	23.96	56.98	37.21	1.11
2	735	39.32	25.85	39.47	22.63	1.08
3	1107	58.63	23.22	42.80	33.07	0.81
4	223	73.99	50.67	28.31	24.78	1.79
5	1232	56.41	30.60	33.69	18.83	1.22

Table 4　Microsatellite genotyping of putative androgenetic doubled haploid in Japanese flounder, *Paralichthys olivaceus*

Locus (Linkage group)	Male parent	Diploid progenies	Locus (Linkage group)	Male parent	Diploid progenies
BDHYP39 (1)	266/266	266/266: 8	BDHYP468 (12)	264/282	264/264: 3 282/282: 5
BDHYP239 (2)	156/156	156/156: 8	Poli609TUF (13)	192/202	192/192: 6 202/202: 2
Poli1167TUF (3)	174/190	174/174: 4 190/190: 4	Poli1446TUF (13)	204/228	204/204: 6 228/228: 2
Poli1831TUF (3)	98/132	98/98: 5 132/132: 3	BDHYP216 (14)	180/192	180/180: 7 192/192: 1
Poli1361TUF (4)	180/200	180/180: 3 200/200: 5	BDHYP27 (14)	208/222	208/208: 4 222/222: 4
Poli222TUF (4)	171/177	171/171: 5 177/177: 3	BDHYP116 (15)	236/236	236/236: 8
Poli1072TUF (4)	218/236	218/218: 3 236/236: 5	Poli1937TUF (16)	198/216	198/198: 3 216/216: 5
Poli641TUF (5)	192/228	192/192: 6 228/228: 2	Poli954TUF (16)	112/178	112/112: 3 178/178: 5
Poli2022TUF (5)	154/164	154/154: 6 164/164: 2	Poli1235TUF (17)	116/128	116/116: 3 128/128: 5
Poli445TUF (6)	182/206	182/182: 4 206/206: 4	Poli230TUF (18)	114/140	114/114: 5 140/140: 3
Poli2046TUF (7)	154/186	154/154: 3 186/186: 5	BDHYP160 (18)	186/204	186/186: 2 204/204: 6
Poli803TUF (7)	154/192	154/154: 4 192/192: 4	BDHYP170 (19)	208/226	208/208: 4 226/226: 4
Poli194TUF (8)	156/164	156/156: 4 164/164: 4	BDHYP427 (20)	242/242	242/242: 8

（continue）

Locus (Linkage group)	Male parent	Diploid progenies	Locus (Linkage group)	Male parent	Diploid progenies
Poli106TUF (8)	118/128	118/118: 5 128/128: 3	BDHYP271 (21)	150/172	150/150: 3 172/172: 5
Poli1450TUF (8)	132/156	132/132: 5 156/156: 3	Poli2TUF (22)	136/136	136/136: 8
BDHYP455 (9)	182/182	182/182: 8	Poli1817TUF (23)	168/182	168/168: 4 182/182: 4
BDHYP345 (10)	248/248	248/248: 8	Poli1836TUF (24)	246/258	246/246: 6 258/258: 2
BDHYP58 (11)	224/224	224/224: 8	Poli1953TUF (24)	172/200	172/172: 5 200/200: 3

3.3 Production of androgenetic clonal line

To produce an androgenetic clonal line, the eggs of females were activated with the sperm of a previously produced mito-gynogenetic DH male; then, cold-shock androgenesis was induced and eggs were subjected to the hydrostatic pressure treatment for duplication of haploid chromosome set. The fertilization rate, calculated from three batches of induction, was 43.14 %± 8.32% (total eggs used: 579～914). The hatch rate was 29.54% ± 9.65%, and the diploid rate was 0.52% ± 0.33%. Due to the low induction rate of the clonal line, we used 400 mL of eggs for large-scale induction. Six months after hatch, there were 45 diploid survivors.

From these diploid survivors, we randomLy selected 21 individuals to confirm their complete homozygosity and genetic identity, i.e., clonality. Based on analysis using a total of 65 microsatellite markers, a single survivor showed heterozygosity at five loci, but exhibited homozygosity at the other 60 loci. The other 20 progeny from the same male parent were completely homozygous at all the 65 loci analyzed. Moreover, at each locus, all progenies examined had alleles identical to those of the male parent, and no difference was detected among siblings. The above-mentioned microsatellite genotypes clearly indicated that these 20 progenies were genetically identical members of a clone.

4 Discussion

Here, we reported the successful production of an androgenetic clonal line without egg irradiation in Japanese flounder. This is the first successful example of cold-shock induced androgenesis in a marine fish species. Our method entirely eliminates egg irradiation before fertilization as a method of genetic inactivation of the maternally derived genome. Irradiation has been considered the only effective method for genetic inactivation of eggs in artificial induced androgenesis, and has been commonly used in teleosts (Komen and Thorgaard, 2007). However, very recently, cold-shock induced androgenesis was first reported by Morishima et al (2011), and was then shown to be as an effective method in loach (Hou et al, 2013) and zebrafish (Hou et al, 2015). In our study, the yield rate of androgenetic haploids was 9.12% ± 1.38% under the optimum cold-shock duration, and 0.81% to 1.79% for androgenetic diploids. As induced androgenesis using egg irradiation has not been reported in Japanese flounder, it is impossible to compare the effectiveness of the induction associated with cold-shock and irradiation. However, such yield rates are lower than that reported for the loach (10.43% ± 1.69%)(Hou et al, 2014).

In fish, low yield rates of androgenetic DHs and the great variance of yield rates between species could be explained by several factors. For example, some salmonids and cyprinids have a recent evolutionary history of genome duplication, being more tolerant to the diploidization of haploid genome; thus, the probability of their obtaining higher yield rates is high (Komen and Thorgaard, 2007). The genetic load could also be another explanation. It is plausible that the unmasking, and subsequent expression of homozygous deleterious genes due to homozygosity causes the decreased viability and reproducibility of resultant embryos. However, it is difficult to accurately estimate the genetic load, and some studies have provided conflicting evidence. Similar results were obtained in rainbow trout (Babiak et al, 2002; Patton et al, 2007) and common carp (Bongers et al, 1994; Tanck et al, 2001). When inbred homozygous clonal sperm was compared with that from outbred sources, the yield rates of androgenetic fry were very similar. In our study, the yield rate of androgenetic DHs was lower when the sperm from a mito-gynogenetic DH male was used. Other possible reasons may include egg quality, side effects of the treatments and asynchronous embryonic development, among others. In Japanese flounder, multifactorial processes maybe associated with the low yield rate rather than a single major cause.

A cross between genetically different doubled haploid female and male could produce a heterozygous clone (Streisinger et al, 1981). Heterozygous clones are considered free from deleterious recessive genes, and such clonal lines often show heterosis when compared to homozygous clones in terms of growth, reproduction, and other physiological processes (Komen and Thorgaard, 2007). In Japanese flounder, heterozygous clones have been already produced and reported (Yamamoto, 1999; Liu et al, 2012b). Recently, we produced several heterozygous clonal lines, of which only one exhibited heterosis in growth traits. This clonal line grew 62.48% faster than the controls (unpublished data); thus, this line has great potential for commercial application in aquaculture. However, the male parent of this heterozygous clone line is a single DH (the female parent is from the homozygous clonal line). Thus, it would be impossible to produce sufficient fertilized eggs to fulfill the current aquaculture demand using sperm from a single male. Therefore, the population of the DH male should be increased by producing a homozygous clonal line from the male. The androgenetic induction method we present here could satisfy such demand; based on this method, we could successfully produce an androgenetic clonal line using sperm from the DH male. By crossing male and female homozygous clones, it is possible to produce a quantity of heterozygous clonal fishes that is sufficient to satisfy aquaculture demand.

5 Conclusion

In conclusion, we successfully induced androgenetic DHs and an androgenetic clonal line without egg irradiation, by means of cold- and hydrostatic pressure-shock treatments in Japanese flounder. This is the first production of androgenetic DHs and clonal Japanese flounder. Our results suggest broad applicability of the cold-shock induced androgenesis method in commercially important marine fish species.

References

[1] Arai K. Genetic improvement of aquaculture finfish species by chromosome manipulation techniques in Japan[J]. Aquaculture, 2001, 197: 205-228.

[2] Babiak I, Dobosz S, Kuzminski H, et al. Failure of interspecies androgenesis in salmonids[J]. Journal of Fish Biology, 2002, 61: 432-447.

[3] Bongers A B J, in't Veld E P C, Abo-Hashema K, et al. Androgenesis in common carp (*Cyprinus carpio* L.) using UV irradiation in a synthetic ovarian fluid and heat shocks[J]. Aquaculture, 1994, 122: 119-132.

[4] Castaño-Sánchez C, Fuji K, Ozaki A, et al. A second generation genetic linkage map of Japanese flounder (*Paralichthys olivaceus*)[J]. BMC Genomics, 2010, 11: 554.

[5] Corley-Smith G E, Lim C J, Brandhorst BP. Production of androgenetic zebrafish (*Danio rerio*)[J]. Genetics, 1996, 142: 1 265-1 276.

[6] David C J, Pandian T J. Cadaveric sperm induces intergeneric androgenesis in the fish, *Hemigrammus caudovittatus*[J]. Theriogenology, 2006, 65: 1048-1070.

[7] Fujimoto T, Sakao S, Yamaha E, et al. Evaluation of different doses of UV irradiation to loach eggs for genetic inactivation of the maternal genome[J]. Journal of Experimental Zoology, 2007,307: 449-462.

[8] Hou J, Fujimoto T, Saito T, et al. 2015. Generation of clonal zebrafish line by androgenesis without egg irradiation[J]. Scientific Reports, 2015, 5: 13346-13360.

[9] Hou J, Fujimoto T, Yamaha E, et al. 2013. Production of androgenetic diploid loach by cold-shock of eggs fertilized with diploid sperm[J]. Theriogenology, 2013, 80: 125-130.

[10] Hou J, Saito T, Fujimoto T, et al. Androgenetic doubled haploids induced without irradiation of eggs in loach (*Misgurnus anguillicaudatus*)[J]. Aquaculture, 2014, 420-421: S57-S63.

[11] Komen H, Thorgaard G H. Androgenesis, gynogenesis and the production of clones in fishes: A review[J].

Aquaculture, 2007, 269: 150-173.

[12] Liao X, Shao C W, Tian Y S, et al. 2007. Polymorphic dinucleotide microsatellites in tongue sole (*Cynoglossus semilaevis*) [J]. Molecular Ecology Notes, 2007, 7: 1147-1149.

[13] Lin F, Dabrowski K. Androgenesis and homozygous gynogenesis in muskellunge (*Esox masquinongy*): evaluation using flow cytometry [J]. Molecular Reproduction and Development, 1998, 49: 10-18.

[14] Liu Y, Wang G, Liu, Y, et al. Genetic verification of doubled haploid Japanese flounder, *Paralichthys olivaceus* by genotyping telomeric microsatellite loci [J]. Aquaculture, 2012, 324-325: 60-63.

[15] Liu Y, Wang G, Liu, Y, et al. Production and verification of heterozygous clones in Japanese flounder, *Paralichthys olivaceus* by microsatellite marker [J]. African Journal of Biotechnology, 2012b, 10: 17088-17094.

[16] Liu Y, Liu Y, Liu Y, et al. Constructing a genetic linkage map and mapping quantitative trait loci for skeletal traits in Japanese flounder [J]. Biologia, 2013, 68: 1221-1228.

[17] Morishima K, Fujimoto T, Sato M, et al. Cold-shock eliminates female nucleus in fertilized eggs to induce androgenesis in the loach (*Misgurnus anguillicaudatus*), a teleost fish [J]. BMC Biotechnology, 2011, 11: 116.

[18] Myers J M, Penman D J, Basavaraju Y, et al. Induction of diploid androgenetic and mitotic gynogenetic Nile tilapia (*Oreochromis niloticus* L.) [J]. Theoretical and Applied Genetics, 1995, 90: 205-210.

[19] Patton S J, Kane S L, Wheeler P A, et al. 2007. Maternal and paternal influence on early embryonic survival of androgenetic rainbow trout (*Oncorhynchus mykiss*): Implications for measuring egg quality [J]. Aquaculture, 2007, 263: 26-34.

[20] R Core Team. R: A language and environment for statistical computing [J]. R Foundation for Statistical Computing, 2012, Vienna, Austria.

[21] Rothbard S, Rubinshtein I, David L, et al. Ploidy manipulations aimed to produce androgenetic Japanese ornamental (koi) carp, *Cyprinus carpio* [J]. Israel Journal of Aquaculture-Bamidgeh, 1999, 51: 26-39.

[22] Streisinger G, Walker C, Dower N, et al. 1981. Production of clones of homozygous diploid zebra fish (*Brachydanio rerio*) [J]. Nature, 1981, 291: 293-296.

[23] Tanck M W T, Palstra A P, deWeerd M V, et al. 2001. Segregation of microsatellite alleles and residual heterozygosity at single loci in homozygous androgenetic common carp (*Cyprinus carpio* L.) [J]. Genome, 2001, 44: 743-751.

[24] Yamamoto E. Studies on sex-manipulation and production of cloned populations in hirame, *Paralichthys olivaceus* (Temminck et Schlegel) [J]. Aquaculture, 1999, 173: 235-246.

[25] Yasui G S, Fujimoto T, Arai K. Restoration of the loach, *Misgurnus anguillicaudatus*, from cryopreserved diploid sperm and induced androgenesis [J]. Aquaculture, 2010, 308: S140-S144.

Chromatin immunoprecipitation sequencing technology reveals global regulatory roles of low cell density quorum-sensing regulator AphA in pathogen *Vibrio alginolyticus*

GU Dan[1], LIU Huan[4], YANG Zhen[1], ZHANG Yuanxing[1,2,3], WANG Qiyao[1,2,3]

[1] State Key Laboratory of Bioreactor Engineering, East China University of Science and Technology Shanghai 200237

[2] Shanghai Engineering Research Center of Maricultured Animal Vaccines Shanghai 200237

[3] Shanghai Collaborative Innovation Center for Biomanufacturing Technology Shanghai 200237

[4] School of Food and Biological Engineering Shanxi University of Science and Technology Xi'an 710021

Abstract Quorum-sensing (QS) is an important regulatory system in virulence expression and environmental adaptation in bacteria. The master QS regulators (MQSR) LuxR and AphA reciprocally control QS gene expression in vibrios. However, the molecular basis for the regulatory functions of AphA remains undefined. Here, we characterized its regulatory roles in *Vibrio alginolyticus*, an important zoonotic pathogen causing diseases in marine animals, as well as in humans. AphA is involved in the motility ability, biofilm formation, and *in vivo* survival of *V. alginolyticus*. Specifically, AphA is expressed at low cell density growth phases. In addition, AphA negatively regulates the expression of the main virulence factor, alkaline serine protease (Asp) through LuxR. Chromatin immunoprecipitation (ChIP) followed by high-throughput DNA sequencing (ChIP-seq) detected 49 enriched loci harboring AphA-binding peaks across the *V. alginolyticus* genome. An AphA specific binding motif was identified and further confirmed by electrophoretic mobility shift assay (EMSA) and mutagenesis analysis. A quantitative real time-polymerase chain reaction (qRT-PCR) assay further validated the regulation of AphA on these genes. Among its regulon, AphA binds directly to the *aphA* promoter and negatively regulates its own expression. Moreover, AphA directly regulates genes encoding for adenylate cyclase, anti-σ^D, FabR, and a small RNA CsrB, revealing versatile regulatory roles of AphA in its physiology and virulence. Furthermore, our data indicated that AphA modulates motility through the coordinated function of LuxR and CsrB. Collectively, this work contributes to better understanding of the regulatory roles of AphA in QS and non-QS genes.

Keywords AphA, ChIP-seq, binding motif, quorum-sensing, *Vibrio alginolyticus*

1 Instruction

Bacteria are highly social organisms capable of sophisticated cooperative behaviors mediated by the Quorum-sensing (QS) system, a cell-density dependent signaling circuit (Miller et al, 2001). In *Vibrio harveyi*, there are three well-defined QS signaling molecules, known as autoinduers (AIs), including acylated homoserine lactones (HAI-1), furanosyl borate diester (AI-2), and CAI-1 (Ng et al, 2009). The membrane sensors LuxN, LuxQ, and CqsS can detect their cognate AIs and transduce the signals required to trigger gene expression through phosphorelay cascades (Ng et al, 2009). At low cell densities (LCDs; low AI concentration), phosphates are directed by kinases in upstream QS cascades to the pivotal regulator LuxO. This leads to repression of the master QS regulator (MQSR), LuxR, by multiple small regulatory RNAs (*qrr1–qrr5*) (Ng et al, 2009). In contrast, at high cell densities (HCD; high AI concentration), the phosphate circuit runs in the opposite direction, resulting in the dephosphorylation and inactivation of LuxO, which de-represses the expression of LuxR (Ng et al, 2009). MQSR then binds to and regulates the expression of many (～150) target genes, leading to diverse phenotypes, including bioluminescence, biofilm formation, and motility (van Kessel et al, 2013). *V. harveyi* and *V. cholerae* also deploy other regulators, such as AphA (Rutherford et al, 2011), and various qrr-mediated regulatory mechanisms (Feng et al, 2015) to optimize the output of QS regulation, thus demonstrating exquisite and sophisticated regulatory architectures in vibrio QS systems.

AphA was first identified as an activator of *tcpP* and *tcpH* expression and for virulence gene expression in *V. cholerae* (Skorupski et al, 1999). The crystal structure of AphA reveals that it belongs to the winged helix DNA-binding transcriptional factor superfamily (De Silva et al, 2005). AphA cooperates with AphB, a LysR family transcriptional factor, as well as with CRP (cAMP receptor protein) on the promoter region of *tcpPH*, to regulate virulence expression in *V. cholerae* (Kovacikova et al, 1999; Kovacikova et al, 2001; Kovacikova et al, 2004). Interestingly, the *aphA* gene is situated between *ppc* (phosphoenolpyruvate carboxylase) and *cysE* (serine acetyltransferase) in the vibrios chromosome, a region containing genes involved in normal metabolic functions, suggesting that this gene is encoded on the ancestral vibrio chromosome and not associated with a pathogenicity island (Skorupski et al, 1999). In *V. cholerae*, AphA is demonstrated to be involved in the acetoin biosynthesis and motility/biofilm formation, which are not associated to the vibrio pathogenicity island (VPI) (Kovacikova et al, 2005). The gene cluster *cpsQ-mfpABC*, which contributes to biofilm development in *V. parahaemolyticus*, is also regulated by AphA (zhou et al, 2013).

One interesting aspect of the AphA regulatory network is that its expression is regulated by

QS and it responds to cell density (Kovacikova et al, 2002). At LCD, AphA levels are relatively high and control the expression of ~167 to 296 genes, whereas at HCD, LuxR binds to the promoter of AphA and represses its expression (van Kessel et al, 2013). This results in the regulation of ~625 genes in *V. harveyi*, rending an overlap of 77 AphA and LuxR co-regulated genes (Rutherford et al, 2011; van Kessel et al, 2013). The reciprocal control of QS by the MQSR AphA and LuxR at LCD and HCD, respectively, optimizes the QS output in terms of temporal and strength, and contributes to an exquisite and fine-tuned QS regulatory architecture. It will be intriguing to decipher the regulon and the molecular basis of the regulation of AphA across chromosomes. This will contribute to the understanding of how conserved regulatory systems coordinate the expression of QS and non-QS genes, as well as how the core-genome cross-talks with the pan-genome via AphA.

V. alginolyticus is a zoonotic pathogen that infects a wide range of sea animals and causes intra- and extra-intestinal diseases in humans (Austin 2010). The major virulence factors of *V. alginolyticus* are extracellular proteases, motility, siderophore-dependent iron uptake systems, biofilm, type III secretion systems (T3SS), and type VI secretion system (T6SS) (Wang et al, 2007; Rui et al, 2008; Rui et al, 2009; Cao et al, 2010; Zhao et al, 2011; Sheng et al, 2013). These factors and some other virulence-related genes are closely regulated by the QS system in *V. alginolyticus*. In our recent investigation, AphA was identified and demonstrated to be involved in the regulation of LuxR expression (Gu et al, 2016). In this study, we further explored other genes regulated by AphA via ChIP-seq, and the electrophoretic mobility shift assays (EMSA) and qRT-PCR experiments were used to verify these results. The AphA consensus sequence was identified by using the Multiple EM for Motif Elicitation (MEME) (Bailey et al, 2009). Overall, this work not only pinpoints the binding motif and the direct targets of AphA *in vivo*, but also reveals that AphA plays important roles in the regulation of QS and other non-QS genes at LCD.

2 Materials and methods

2.1 Bacterial strains, plasmids and growth conditions

V. alginolyticus strains and derivatives were normally grown at 30 ℃ in Luria-Bertani (LB) medium supplemented with 3% NaCl (LBS). *Escherichia coli* strains were grown at 37 ℃ in LB medium. When appropriate, the medium was supplemented with ampicillin (100 μg/mL), chloramphenicol (25 μg/mL), kanamycin (100 μg/mL), Isopropyl β-*D*-1-thiogalactopyranoside (1 mmol/mL), or *L*-arabinose (0.04%, wt/vol).

2.2 Construction of *V. alginolyticus aphA-flag* fusion and *csrB* deletion mutant in the genome

The construction of mutant strains followed a previous strategy developed with the suicide plasmid pDM4 (Sheng et al, 2013). To construct the *aphA-flag* fusion, PCR was performed to amplify sequences upstream and downstream of the termination codon of *aphA* (TAA), and the *flag* DNA fragment was inserted in the front of TAA. The PCR products were cloned into the *Xba*I site of pDM4 by isothermal assembly, as previously described (Gibson et al, 2009). The recombinant plasmid was conjugated into a wild-type EPGS with selection for ampicillin and chloramphenicol resistance. Then, the correct colony was screened for sucrose (12%) sensitivity, which typically indicates a double crossover event and thus, the occurrence of gene replacement. Additionally, the *aphA-flag* fusion strain was confirmed by PCR and sequencing. The construction of a *csrB*-deletion mutant followed the same strategy.

2.3 Construction of the promoter-reporter plasmids

The plasmid pDM8 was used to construct the promoter-*luxAB* fusion of *aphA*. The *aphA* promoter regions were amplified by PCR and then cloned into the *Sma*I site in the pDM8 by isothermal assembly (Gibson et al, 2009). The recombinant plasmid was introduced into the wild type and Δ*aphA* by conjugation. The recombinant plasmids were confirmed by DNA sequencing.

2.4 qRT-PCR

Equal amounts of RNA (1 μg) were used to generate complementary DNA (Toyobo, Tsuruga, Japan) using random primers. Three independent experiments were performed, each in triplicate, with specific primer pairs on an Applied Biosystems 7500 Real Time System (Applied Biosystems, Foster City, CA, USA), and transcript levels were normalized to 16*S* RNA in each sample by the ΔΔCT method.

2.5 Western blot assay

For the immunoblotting assay, supernatants and bacterial cell pellets were harvested at the same OD600. Then, 15 μL of each sample was loaded onto a 12% denaturing polyacrylamide gel; proteins were resolved by electrophoresis and finally transferred to a PVDF membrane (Millipore, Bedford, MA, USA). The membranes were blocked with a 10% skim milk powder solution, incubated with a 1∶2 000 dilution of the Asp-, LuxR- (GL Peptide Ltd., Shanghai, China) or Flag- (Sigma-Aldrich, St. Louis, MO, USA) specific antiserum, and then incubated with a 1∶2 000 dilution of the Horseradish peroxidase-conjugated goat anti-rabbit and anti-mouse IgG

(Santa Cruz Biotechnology, Santa Cruz, CA, USA). Finally, the blots were visualized with an ECL reagent (Thermo Fisher Scientific Inc., Waltham, MA USA).

2.6 Motility and biofilm assay

The motility and biofilm assay were performed as previous described (Sheng et al, 2013). The overnight cultured wild type (WT), Δ*aphA*, and $aphA^+$ strains were diluted to OD_{600} 1.0, and then spotted to the LBS containing 0.3% (swimming) and 1.5% (swarming) agar. The cultures (50 μL) were diluted to 5 mL LBS medium in glass tubes and incubated at 30 ℃ without shaking for 48 h. Total biofilm was measured by 2% crystal violet staining. The experiments were performed at least three times and one representative result was shown.

2.7 Extracellular protease activity assay (HPA)

Extracellular protease (ECP) activity was determined using hide powder azure (HPA) digestion as previously described (Sheng et al, 2013). Briefly, strains were grown in LBS medium at 30 ℃ for 9 h. The cell density was measured at 600 nm. The bacterial cultures were centrifuged and then the supernatants were filtered through 0.22 μm filters (Millipore). After that, 1 mL filtered supernatants was mixed with 1 mL phosphate buffered saline (PBS, pH 7.2) and 10 mg of HPA (Sigma-Aldrich). The mixture was incubated with shaking at 37 ℃ for 2 h. After stopping the reaction by adding 10% trichloracetic acid, total protease activity was measured at 600 nm. ECP activity was normalized by dividing total activity with OD_{600} for each strain.

2.8 EMSA

Purified 6X His-tagged AphA was incubated with different cy5-labeled DNA probes in 20 μL of the loading buffer (10 mmol/L NaCl, 0.1 mmol/L DTT, 0.1 mmol/L EDTA, 10 mmol/L Tris, pH 7.4). After incubation at 25 ℃ for 30 min, the samples were resolved by 6% polyacrylamide gel electrophoresis in 0.5 × Tris/Boric Acid/EDTA (TBE) buffer on ice at 100 V for 120 min. The gels were then scanned with a FLA 9500 (GE healthcare, Uppsala, Sweden).

2.9 ChIP-seq analysis

The pBAD33Flag::AphA and pBAD33Flag plasmids, encoding AphA-Flag and Flag tag alone, respectively, were transferred to the Δ*aphA* strain for ChIP assays as previously described (Gu et al, 2016). Overnight cultures of each strain in LBS medium were diluted (1:100) in 50 mL fresh LBS medium with 0.04% *L*-arabinose. After 2 h of growth with shaking, the protein-DNA complexes in the bacteria cells were cross-linked *in vivo* with 1% formaldehyde at room temperature for 10 min. Cross-linking was stopped by addition of 125 mmol/L glycine.

Bacteria were then washed twice with cold PBS and resuspended in 5 mL SDS lysis buffer (50 mmol/L HEPES-KOH pH 7.5, 0.1% sodium deoxycholate, 150 mmol/L NaCl, 0.1% SDS, 1 mmol/L EDTA pH 8, 1% Triton X-100, protease inhibitors). Then the bacteria were sonicated, and the DNA was fragmented to 100～500 bp at 200 W. Insoluble cellular debris was removed by centrifugation and the supernatant used as the input sample in the following IP experiments. Both Input and IP samples were washed with 50 μL protein G beads for 1 h. Then, IP samples were incubated overnight with 50 μL of Flag-beads (Sigma-Aldrich). After incubation, the beads were washed twice with a low salt wash buffer (10 mmol/L Tris-HCl pH 8, 150 mmol/L NaCl, 0.1% SDS, 1 mmol/L EDTA pH 8, 1% Triton X-100), twice with 1 mL of a high salt wash buffer (10 mmol/L Tris-HCl pH 8, 500 mmol/L NaCl, 0.1% SDS, 1 mmol/L EDTA pH 8, 1% Triton X-100), twice with 1 mL of LiCl wash buffer (10 mmol/L Tris-HCl pH 8, 250 mmol/L LiCl, 1 mmol/L EDTA pH 8, 0.5% Triton X-100, 0.5% sodium deoxycholate), and twice with 1 mL of TE buffer. The beads were resuspended in 200 μL of elution buffer (50 mmol/L Tris HCl pH 8, 10 mmol/L EDTA, 1% SDS), incubated for 2 h at 65 ℃, and centrifuged at 5 000 g for 1 min. The supernatants containing the immunoprecipitated DNA were collected and 8 μL of 5 mol/L NaCl was added to all of the tubes (IPs and Inputs). The tubes were incubated at 65 ℃ overnight to reverse the DNA-protein crosslinks. Then, 1 μL of RNase A was added, and the solutions were incubated for 30 min at 37 ℃. Four microliters of 0.5 mol/L EDTA, 8 μL of 1 mol/L Tris-HCl, and 1 μL of proteinase K were added to each tube and incubated at 45 ℃ for 2 h. The DNA was purified using phenol-chloroform.

DNA fragments were used for library construction with the VAHTS™ Turbo DNA Library Prep Kit (Vazyme, Nanjing, China). The number of reads per microliter of each library was quantitated by quantitative real-time-PCR against a standard curve and sequenced with a MiSeq sequencer (Illumina, San Diego, CA, USA). ChIP-seq reads were mapped to the *V. alginolyticus* EPGS genome using Bowtie 2 (Feng et al, 2012). The enriched peaks were identified using MACS software (Feng et al, 2012) followed by MEME analysis to generate the AphA-binding motif (Philip et al, 2011). The KEGG pathway analysis was performed with the Kobas 2.0 to illustrate the above-enriched gene function (Xie et al, 2011).

2.10 Competitive index (CI) assay

CI of WT, Δ*aphA*, and *aphA*$^+$ was performed using the WT strain harboring a *lacZ* gene behind the glmS locus. The infection dose was 10^5 CFU/fish; seven zebra fish were used and sacrificed at 24 h post infection, then plated to the LBS containing X-Gal and Amp. The plates were incubated overnight at 30 ℃. The WT strain was differentiated from the mutant strain based on the production of blue and white colonies (blue for WT strains). The ratio of WT counts to the

Δ*aphA* or *aphA*$^+$ counts were used to determine the CI.

All animal experiments presented in this study were approved by the Animal Care Committee of the East China University of Science and Technology (2006272). The Experimental Animal Care and Use Guidelines from Ministry of Science and Technology of China (MOST-2011-02) were strictly adhered.

3 Results

3.1 Cell density-dependent expression of AphA in *V. alginolyticus*

To investigate the function of AphA and its relationship to the QS system and virulence in *V. alginolyticus*, we cloned and identified *aphA* from the genome of *V. alginolyticus* EPGS. In our recent investigation, AphA was identified and determined to bind directly to the luxR promoter and repress the expression of LuxR in *V. alginolyticus* (Gu et al, 2016). Here, we set out to investigate AphA production throughout growth phases by assaying the relative transcript and protein production profiles of AphA at different stages. Transcriptional levels of *aphA* were decreased as the cell grew and the densities increased, as determined with a transcriptional fusion construct of P_{aphA}-*luxAB* (Fig. 1A). The transcriptional activity of P_{aphA} was highest at the early exponential growth stage; however, its transcriptional activity reduced dramatically when the cells were approaching late exponential or stationary growth phases. Western blot with a strain

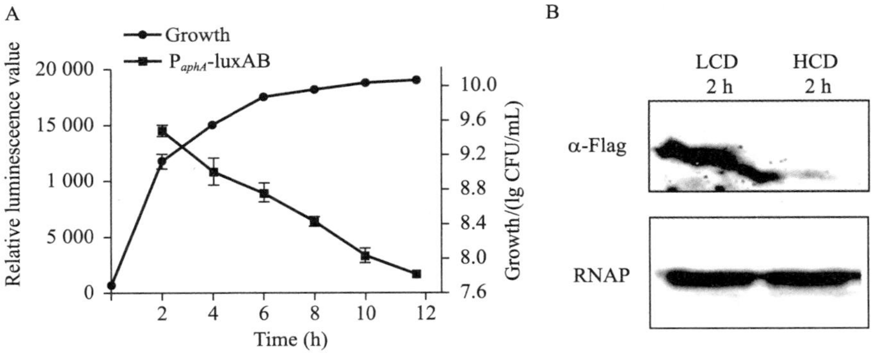

Fig. 1 The low cell density-dependent expression of *aphA* in *V. alginolyticus*

A. Growth of *V. alginolyticus* and transcriptional levels of *aphA* at different growth phases were determined. The cultures were sampled at various time points and plate-counted after serial dilutions with fresh LBS medium. To assay the transcriptional activity of *aphA*, a pDM8 derivative plasmid carrying the *aphA* promoter region (P_{aphA}) fused to promoterless *luxAB* (pDM8-P_{aphA}-*luxAB*) was conjugated into a WT strain and the fluorescence was assayed. B. Western-blotting analysis of AphA expression. Flag-tagged AphA was expressed from a WT strain carrying *aphA-flag* translational fusion at the *aphA* locus (AphA-flag/EPGS) and blotted against a Flag-specific antibody to determine the expression levels of AphA in WT cells. RNAP was used as a loading control of the cell lysates carrying a translational fusion of *aphA-flag* in the *aphA* locus indicated that AphA was expressed at LCDs (Fig. 1B). Taken together, these data indicate that the expression of AphA is cell density-dependent, and the production of AphA is highest at LCDs.

3.2 Essential roles of AphA in motility, biofilm and *in vivo* survival of zebra fish

Two flagella systems have been identified in *V. alginolyticus*. A single polar flagellum serves to swim and numerous lateral flagella serve to swarm (Tian et al, 2008). Both the swarming and swimming abilities were significantly reduced in the Δ*aphA* strain; these abilities were restored to WT in the complement strain (Fig. 2A). Furthermore, the biofilm formation was significantly decreased in the Δ*aphA* strain as compared to the WT strain, and the complement strain restored biofilm formation (Fig. 2B). To further test how AphA affects *in vivo* growth of the bacterium, WT and Δ*aphA* were mixed in a 1:1 ratio and inoculated into zebra fish. A profound competitive defect was observed in *in vivo* growth (～50-fold) of the Δ*aphA* mutant strain in comparison to that of the WT strain (Fig. 2C); no competitive defect was observed for either WT or Δ*aphA* strains in LBS cultures. These results suggest that AphA plays important roles in the regulation of motility, biofilm, and *in vivo* survival in zebra fish. The roles in controlling motility and biofilm by AphA have also been observed in *V. parahaemolyticus* and *V. cholerae* (Wang et al, 2013; Srivastava et al, 2011; Yang et al, 2010; Lee et al, 2008), suggesting an AphA-dependent common regulatory network in these aspects in vibrios.

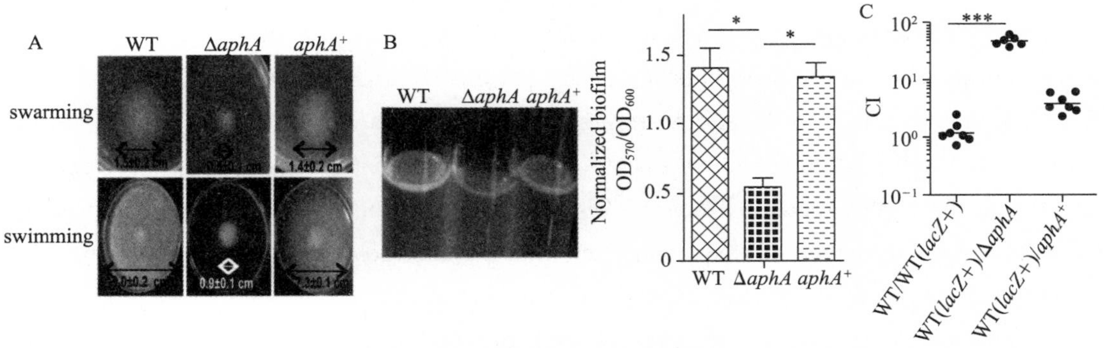

Fig. 2 Essential roles of AphA in motility, biofilm formation, and in vivo survival of *V. alginolyticus*

A. Motility assays of WT, Δ*aphA*, and $aphA^+$ complement strain. Swarming motility assays were performed on LBS plates with 1.5% agar, whereas swimming motility assays with 0.3% agar. The diameter of the colonies was shown for the motility assays. The data were presented as mean ± SD. (n = 3). B. Biofilm formation in glass tubes after 48 h was quantitatively assayed with violet crystal staining of the biofilm cells. The colorimetric assay of the released violet crystal from the biofilm cells was measured at OD_{570} and normalized with OD_{600} value of the cultures. C. CI assays for the Δ*aphA* strain against WT ($lacZ^+$), the WT strain carrying *lacZ* following the glmS locus. $^*P < 0.05$ and $^{***}P < 0.001$, Student's t-test

3.3 AphA represses exotoxin asp expression

Alkaline serine protease (Asp) is the main extracellular toxin (exotoxin) of *V. alginolyticus*, and it is closely regulated by QS. Here, we examined whether AphA could regulate the expression of Asp. A notable increase in Asp activity was displayed in the Δ*aphA* strain, whereas the *aphA* complement strain decreased to WT levels (Fig. 3A). Western blot analysis using an Asp-specific antibody also demonstrated an increased level of Asp production in the supernatant pellets of this mutant strain. The complement strain restored extracellular Asp production to WT levels (Fig. 3B).

We further tested the AphA regulation of exotoxin Asp with an *L*-arabinose-inducible AphA strain. The results showed that low concentrations of *L*-arabinose, which generated low levels of AphA, led to Asp activities that were similar to the sample without *L*-arabinose or the strain harboring pBAD33-empty plasmids (Fig. 3C). In contrast, high AphA levels, induced by high *L*-arabinose concentrations, strongly repressed Asp production (Figs. 3C and 3D). Finally, qRT-PCR analysis also confirmed a significant repression of *asp* transcription in the AphA-overexpressing strain (Fig. 3E). Taken together, our data suggest that AphA represses the expression of Asp.

3.4 AphA regulates Asp expression via LuxR

We had previously reported that the MQSR, LuxR, positively regulates the expression of Asp (Rui et al, 2008; Rui et al, 2009), and that AphA represses LuxR expression by binding directly to the *luxR* promoter region (Gu et al, 2016). Here, we tested whether AphA regulates the expression of Asp in a LuxR-dependent manner. For this reason, we constructed a Δ*aphA*Δ*luxR* double mutant strain. The HPA digestion assay and western blot analysis showed that the Δ*aphA*Δ*luxR* strain exhibited no expression of Asp, equal to that of the Δ*luxR* strain (Fig. 3A), suggesting that AphA regulates Asp expression via LuxR. Then, we detected the expression levels of LuxR induced by different concentration of *L*-arabinose in the *L*-arabinose-driven AphA overexpression strain. The results showed that a low concentration of *L*-arabinose produced no or low levels of AphA, whereas it increased the expression of LuxR (Fig. 4A). Increased levels of AphA led to a significant reduction in LuxR expression (Fig. 4A). Additionally, qRT-PCR analysis also confirmed a significant repression of *luxR* transcription in the AphA overexpression strain (Fig. 4B). Taken together, our results demonstrated that AphA regulates the expression of Asp via LuxR in *V. alginolyticus*.

3.5 Genome-wide screen of AphA-binding regions by ChIP-seq

As shown above, AphA is the master regulator of QS and virulence. We further investigated

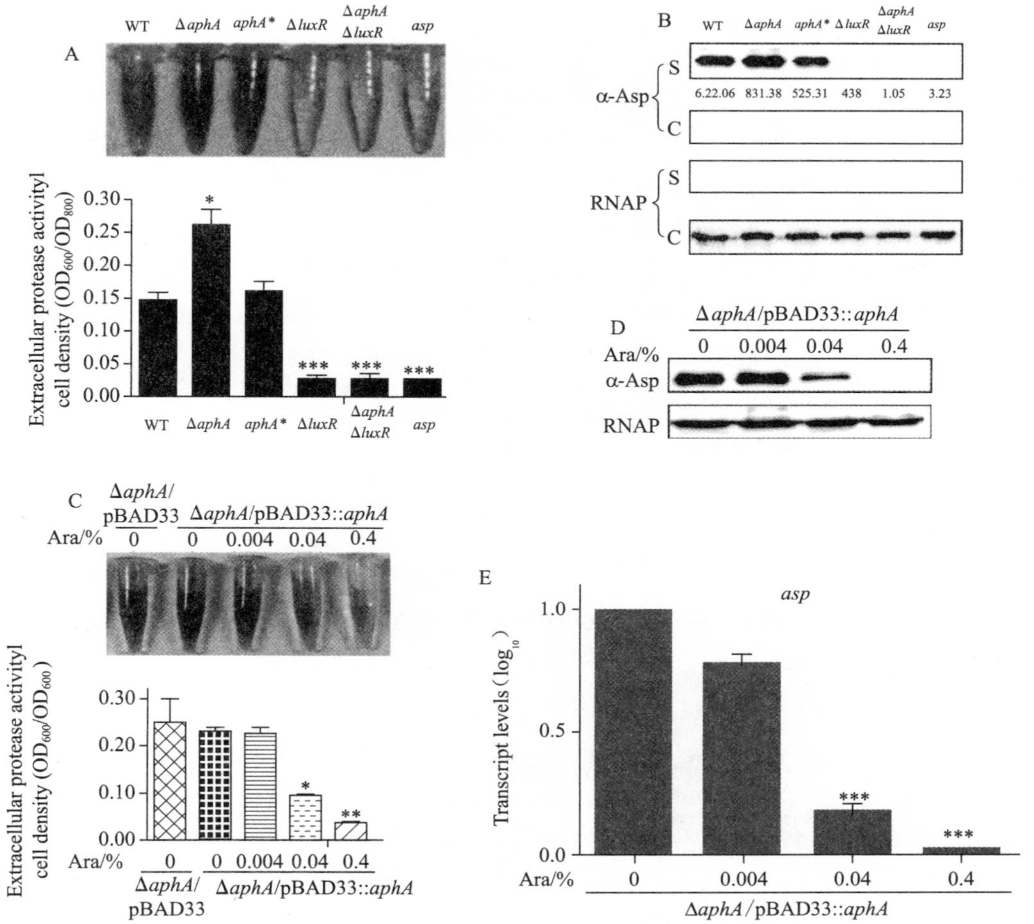

Fig. 3 AphA negatively regulates Asp expression

A. HPA digestion assays for the extracellular Asp activity in different strains. After 9 h, culture supernatants were used to measure the protease activities of each strain, which were normalized by cell density.
B. Western blotting analysis with concentrated supernatants (S) and cellular pellets (C) were performed using an Asp-specific antiserum. RNAP was used as a loading control of the blots. C–E. HPA digestion assays (C) and Western blotting analysis (D) for the extracellular Asp activity, and qRT-PCR analysis (E) of asp expression in Δ*aphA*/pBAD33::*aphA* with different concentrations of *L*-arabinose (Ara). The results are shown as mean ± SD. ($n = 3$). $^{*}P < 0.05$, $^{**}P < 0.01$, $^{***}P < 0.001$, student's *t*-test compared to corresponding samples from WT or the inducible strain without addition of arabinose

all the possible AphA-binding loci on the chromosomes of *V. alginolyticus* using ChIP-seq experiments with a FLAG-tagged AphA, which was overexpressed from plasmid pBAD33 in the Δ*aphA* strain. As a control, first we verified that FLAG-AphA behaves similarly to WT AphA *in vivo* (Fig. 5A) by assaying their activity in the repression of Asp production and by measuring FLAG-AphA occupancy at the luxR promoter region that contains a known AphA binding site in *V. alginolyticus* (Gu et al, 2016). Our ChIP-qPCR analysis revealed a ~5.5-fold enrichment of

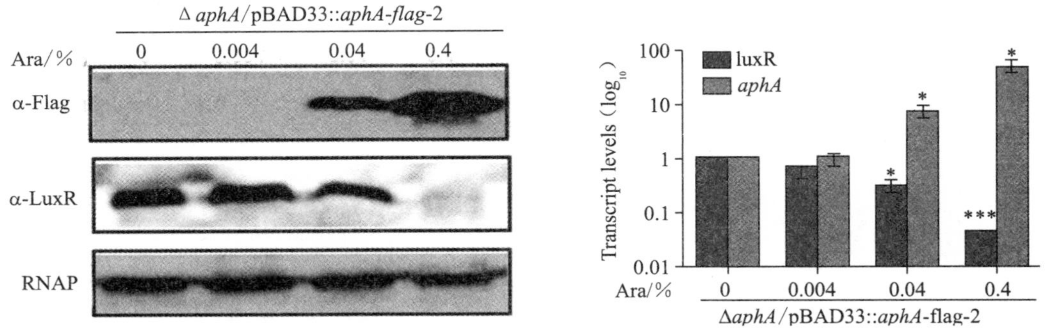

Fig. 4 AphA regulates Asp expression via LuxR

A. Western blotting assay of LuxR expression with various concentrations of AphA (induced with *L*-arabinose). Bacterial cells of Δ*aphA* harboring plasmid pBAD33::*aphA-flag*-2 were cultured in LBS medium for 9 h, harvested, and blotted with specific anti-Flag and anti-LuxR antibodies. B. qRT-PCR analysis of the transcriptional level of *aphA* and *luxR* in Δ*aph*A/pBAD33::*aphA-flag*-2 with different concentrations of *L*-arabinose. The bacterium was cultured in LBS for 9 h, and mRNA transcripts of *aphA* and *luxR* were detected by qRT-PCR. The 16*S* rDNA was selected as a control. The results were shown as mean ± SD (n = 3). $^{*}P < 0.05$, $^{**}P < 0.01$, $^{***}P < 0.001$, Student's *t*-test as compared to that without addition of *L*-arabinose

AphA binding at the promoters of luxR, but not at gyrB, a promoter that is not bound by AphA (Fig. 5A).

Sequence reads were obtained from three independent ChIP-seq assays and mapped to the *V. alginolyticus* genome. We identified 49 enriched loci harboring AphA binding peaks, which were enriched by > 4-fold (P < 0.001) than those in control samples from WT/pBAD33-FLAG strain (WT strain only expressing the FLAG tag). These 49 loci are located across the genome, specifically in intergenic regions (51.02%) and within coding regions (48.98%), suggesting that AphA is a global transcriptional regulator in *V. alginolyticus* (Fig. 5B). Using the MEME suite, a 19-bp AphA consensus sequence (TATTCGN$_7$GCTTAT) was identified (Fig. 5C). This sequence matches with the motif in the luxR promoter region, which was revealed by the DNase I footprinting assay (Gu et al, 2016), and shows high similarity to the binding sites for AphA previously identified in other vibrios (Yang et al, 2010). The count matrix was provided to describe the alignment of AphA binding sites, and represents the relative frequency of each base at a different position in the consensus sequence (Fig. 5C).

In the enriched list, the luxR promoter region was identified to be a binding substrate of AphA by 4.1-fold (P < 0.001), validating the approach, and as reported in other vibrios (van Kessel et al, 2013). Furthermore, the genes corresponding to 4 loci (*aphA*, Exoribonuclease R, DNA-directed RNA polymerase beta' subunit, and 5′-3′ Exonuclease) in the enriched list have also been shown to be regulated by AphA in *V. harveyi* (Rutherford et al, 2011; van Kessel et al, 2013), further validating our ChIP-seq methodology. The other 44 genes were, for the first time, identified to be directly regulated by AphA. Several transcriptional regulator genes (n = 8), and

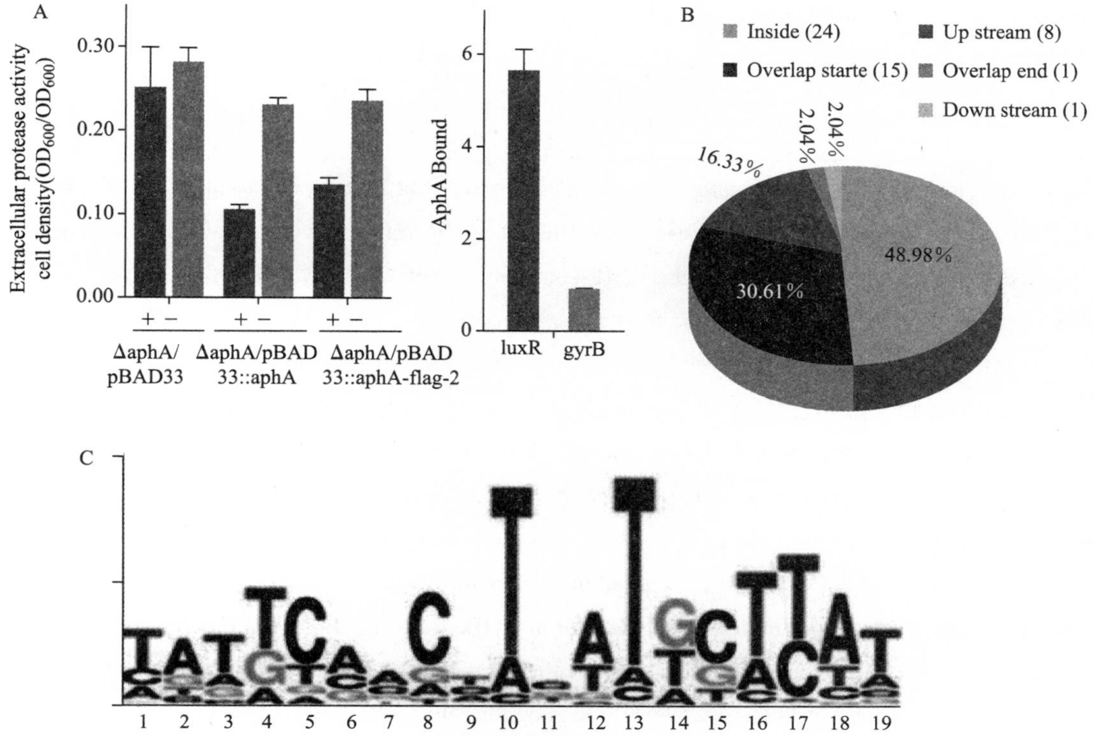

Count Matrix

	11	33	11	9	5	28	26	7	15	10	6	33	5	10	2	14	2	38	12
	13	7	5	0	38	15	16	39	18	3	22	3	4	0	24	6	22	7	7
	3	11	10	18	5	13	10	9	5	0	16	6	1	29	14	1	1	3	6
T	32	8	33	32	11	3	7	4	21	46	15	17	49	20	9	38	34	11	34

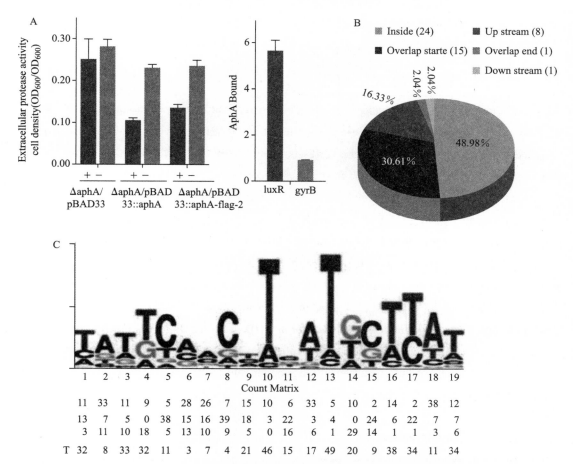

Fig. 5　ChIP-seq reveals *in vivo* binding sites of AphA in the *V. alginolyticus* chromosomes

A. Behaviors of AphA and Flag tagged AphA (AphA-Flag) expressed with a pBAD33 plasmid with (+) or without (−) addition of *L*-arabinose on the production of Asp activities (left panel) or on the binding of the *luxR* promoter region in *V. alginolyticus* using ChIP-qPCR analysis (right panel). The DNA region in *gyrB*, which does not bind to AphA, was used as a negative control. B. Pie chart of 49 enriched AphA-binding peaks indicating the location of binding sites in various genes. C. The most significant motif derived from a ChIP-seq binding sequence generated by the MEME tool (Bailey et al, 2009; Philip et al, 2011). The height of each letter represents the relative frequency of each base at a different position in the consensus sequence. A position frequency matrix describes the alignment of AphA sites, and denotes the frequency of each nucleotide at each position

genes including sigma D, anti-sigma D, type I/IIA topoisomerases, RNAP beta subunit, a putative translation factor related to gene transcription/translation processes, and FabR (involved in the unsaturated fatty acids biosynthesis) were enriched in the ChIP-seq peaks. In the list were also the genes related to adenylate cyclase (related to cAMP generation) and F0F1-type ATP synthase, which is involved in virulence regulation networks in vibrios (Kovacikova et al, 2001; Lee et al, 2008). In addition, AphA seemed to directly bind to a small RNA gene that belongs to the CsrB/RsmB family. Finally, many metabolism-associated genes seemed to also be regulated by

AphA. Collectively, these newly identified target genes strongly suggest that AphA is a MQSR in multiple virulence and metabolic associated pathways at LCDs.

3.6 AphA represses its own expression

In *V. cholerae*, *V. harveyi* and *V. parahaemolyticus*, AphA has been established to repress its own expression (Rutherford et al, 2011; Sun et al, 2012; Lin et al, 2007). Among the 49 enriched loci that were identified in the ChIP-seq experiment, *aphA* (VEPGS-04106) represented one of the highest peaks (14.62-fold) in the *V. alginolyticus* genome bound by AphA (Fig. 6A). As expected, the AphA protein bound directly to the *aphA* promoter region in a concentration-dependent manner in the EMSA in the presence of high concentrations (10-fold) of non-specific poly(dI:dC) competitor (Fig. 6B). Blast analysis showed that *aphA* promoter contains the specific binding site of AphA protein and the absence of this binding site abolished the capacity of AphA to bind to its own promoter (Figs. 6A and 6B). Given this direct interaction, we next attempted to determine whether *aphA* expression is regulated by AphA *in vivo*. To this end, we compared the transcriptional levels of *aphA* in WT and Δ*aphA* with the transcriptional fusion construct of P_{aphA}-*luxAB* at 2 h of culture. The luminescence in the Δ*aphA* mutant was significantly higher than

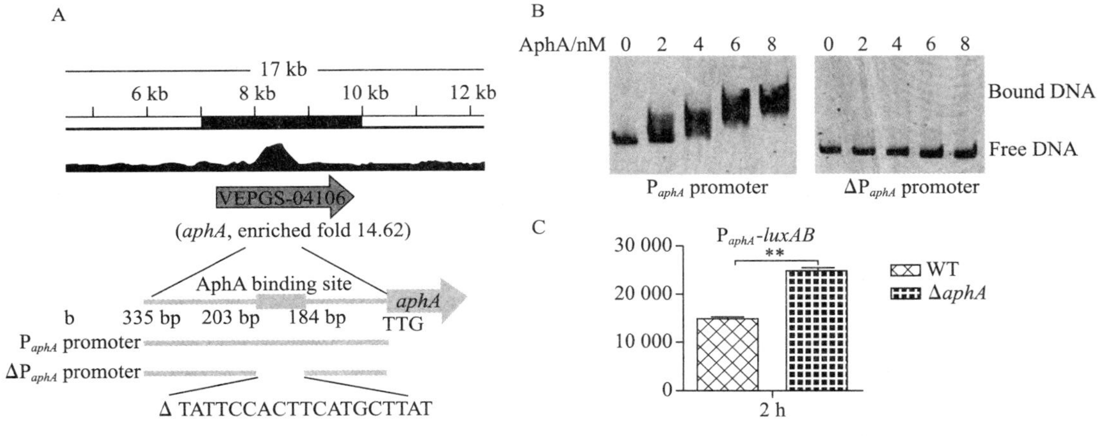

Fig. 6 AphA binds directly to its own promoter for autorepression

A. AphA binds to its own promoter region as illustrated with the peak (enriched by 14.6-fold) identified by the ChIP-seq experiment. The AphA binding site and the probes (335 bp P_{aphA} promoter and 316 bp ΔP_{aphA} promoter with deletion of AphA binding site ranging 184～203 bp relative to start site TTG) used for EMSA experiments in B, are illustrated. B. EMSA of the *aphA* promoter region (left panel) or the *aphA* promoter region with deletion of the specific AphA-binding box (ΔP_{aphA}) (right panel) with purified AphA. The amounts of AphA protein (nmol/L) used were as indicated and 20 ng of each Cy5-labeled probe was added to the EMSA reactions. The specificity of the shifts was verified by adding a 10-fold excess of non-specific competitor DNA (poly(dI:dC)) to the EMSA reactions. C. P_{aphA}-*luxAB* transcriptional analysis in WT and Δ*aphA* strains. The WT and Δ*aphA* strains carrying the P_{aphA}-*luxAB* reporter plasmid were cultured in LBS for 2 h and assayed for luminescence. The results are shown as mean ± SD (n = 3). $^{**}P < 0.01$, Student's t-test as compared to the WT strain harboring P_{aphA}-*luxAB* that in the WT (Fig. 6C). These results showed that AphA binds directly to its own promoter to negatively regulate the expression of AphA at LCDs in *V. alginolyticus*, which represents another evidence that AphA auto-regulates itself.

3.7 Multiple roles of AphA in adenylate cyclase, anti-σ^D, FabR, and CsrB expression

We further analyzed the ChIP-seq data by choosing genes of interest in the list (VEPGS-04385 for adenylate cyclase, VEPGS-04316 for anti-σ^D, VEPGS-04332 for FabR, and small RNA CsrB). In *V. harveyi*, the AphA protein can bind directly to the promoter region of *qrr4* (Rutherford *et al*, 2011). An alignment of the AphA-protein binding site in these four promoters and *qrr4* of *V. harveyi* is shown in Fig. 7A, and these binding sites are similar to the AphA logo identified by ChIP-seq. Next, we used EMSA to test the binding of AphA to these four promoters *in vitro*. AphA bound efficiently to all four probes in a concentration-dependent manner in the presence of high concentrations (10-fold excess of the probe DNA) of a non-specific competitor, whereas the negative control (*gyrB*) remained unbound even at the highest protein concentrations (Fig. 7B). Then, qRT-PCR was performed in the WT, Δ*aphA*, and *aphA*$^+$ complemented strains. The expression of adenylate cyclase, anti-σ^D, FabR and a putative small RNA CsrB (see below) were significantly increased 3- to 5-fold in Δ*aphA* as compared to WT (Fig. 7C); furthermore,

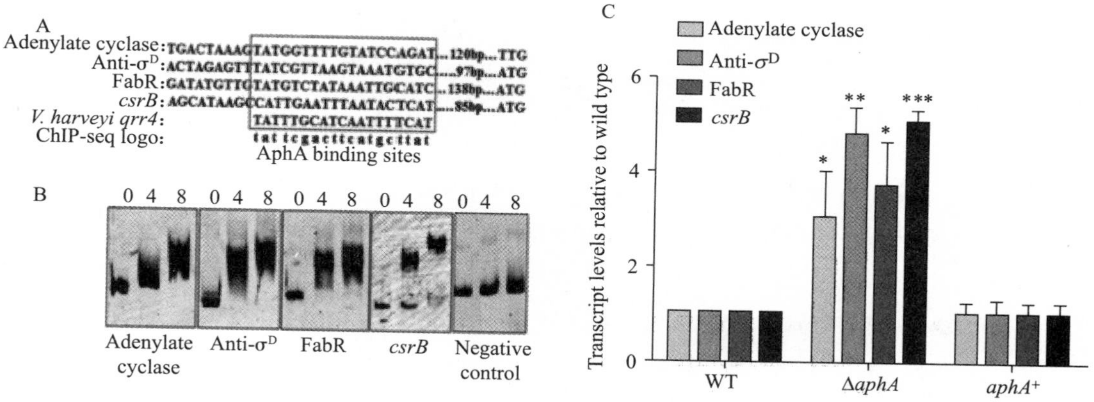

Fig. 7　Validation of ChIP-seq results *in vitro* and *in vivo*

A. Alignment of the selected promoter region of genes containing an AphA-binding box (revealed by ChIP-seq analysis) and as established previously (Rutherford et al, 2011). The conserved consensus motif was boxed and their positions relative to the start codon were shown. B. AphA specifically binds to the selected target regions, as revealed by EMSA. The promoter regions of adenylate cyclase, anti-σ^D, FabR, and *csrB* were chosen and the PCR products containing the indicated fragment were added to the reaction mixtures with different concentrations of AphA protein and a 10-fold excess of non-specific competitor (poly(dI:dC)). The negative control (gyrB gene fragment) showed no binding to AphA. C. qRT-PCR analysis of the transcripts of selected genes. The WT, Δ*aphA*, and, *aphA*$^+$ strains were cultured in LBS for 2 h, and mRNA transcripts were detected by qRT-PCR. The 16*S* rDNA was selected as a control. The results were shown as mean ± SD (n = 3) $^*P < 0.05$, $^{**}P < 0.01$, $^{***}P < 0.001$, student's *t*-test as compared to their corresponding WT results the complemented strain was restored to WT levels. Taken together, these results confirmed the ChIP-seq data and suggested multiple novel roles of AphA in various cellular processes of *V. alginolyticus*.

3.8 AphA regulates motility through CsrB and LuxR

We were intrigued that sRNA CsrB was bound directly and negatively regulated by AphA in the bacteria (Fig. 7). Bioinformatics analyses indicated that the CsrB gene of *V. alginolyticus* shared 98%, 95%, 87%, and 84% identities to the homologous gene present in *V. harveyi*, *V. parahaemolyticus*, *V. cholerae*, and *V. fischeri*, respectively (Fig. 8A). CsrB had been identified by Northern blot analysis in *V. cholerae* and *V. fischeri* (Lenz et al, 2005; Kulkarni et al, 2006) with functions similar to the CsrB families described in *E. coli* (acting redundantly to titrate the activity of the global regulator protein CsrA) (Romeo et al, 1993; Pernestig et al, 2001). The predicted secondary structures of CsrB (Zuker 2003) are shown in Fig. 8B; specifically, the CsrB has 13 AGGA motifs and 10 ARGGA motifs (R stands for G, C and T) in loop regions of stem-loops or in single-stranded regions, which are the hallmark sites of the interaction with CsrA in *E. coli* (Liu et al, 1997; Weilbacher et al, 2003). In *V. cholerae* and *V. fischeri*, the CsrB has 18 and 21 AGGA/ARGGA motifs, respectively (Lenz, 2005; Kulkarni et al, 2006).

Further analysis showed that CsrB positively regulates motility in *V. alginolyticus*, as deletion of *csrB* in the bacterium significantly down-regulated both swarming and swimming abilities and complement *csrB* restored motility in the Δ*csrB* strain (Fig. 8C). Overexpression of CsrB in the Δ*aphA* mutant strain (Δ*aphA* $csrB^{+}$) dramatically enhanced the motility of Δ*aphA* (Fig. 8C). Because AphA controls motility capacities (Figs. 2A and 8C), and AphA can negatively regulate the expression of CsrB, we inferred that AphA might regulate motility through CsrB and other regulator(s). LuxR negatively regulates motility in *V. alginolyticus* (Fig. 8C) (Rui et al, 2008) and AphA directly represses the expression of LuxR (Gu et al, 2016). Therefore, we further explored how LuxR and CsrB coordinate the motility capacities with a ΔluxRΔcrsB double mutant strain. The Δ*luxR*Δ*crsB* strain partially restored the motility capacities of Δ*crsB* to WT level but was weaker than that of the Δ*luxR* strain. These data suggested that both *LuxR* and *CsrB* modulate motility, and LuxR might exert stronger inhibition on the motility as compared to the relatively weaker activation by CsrB. We further investigated the transcription of luxR and csrB in different strains at HCD (10 h LBS culture) using qRT-PCR. The results indicated that AphA negatively regulated both LuxR and CsrB, but LuxR and CsrB appeared to not regulate each other (Fig. 8D). Taken together, these results demonstrated that AphA can regulate motility through CsrB and LuxR.

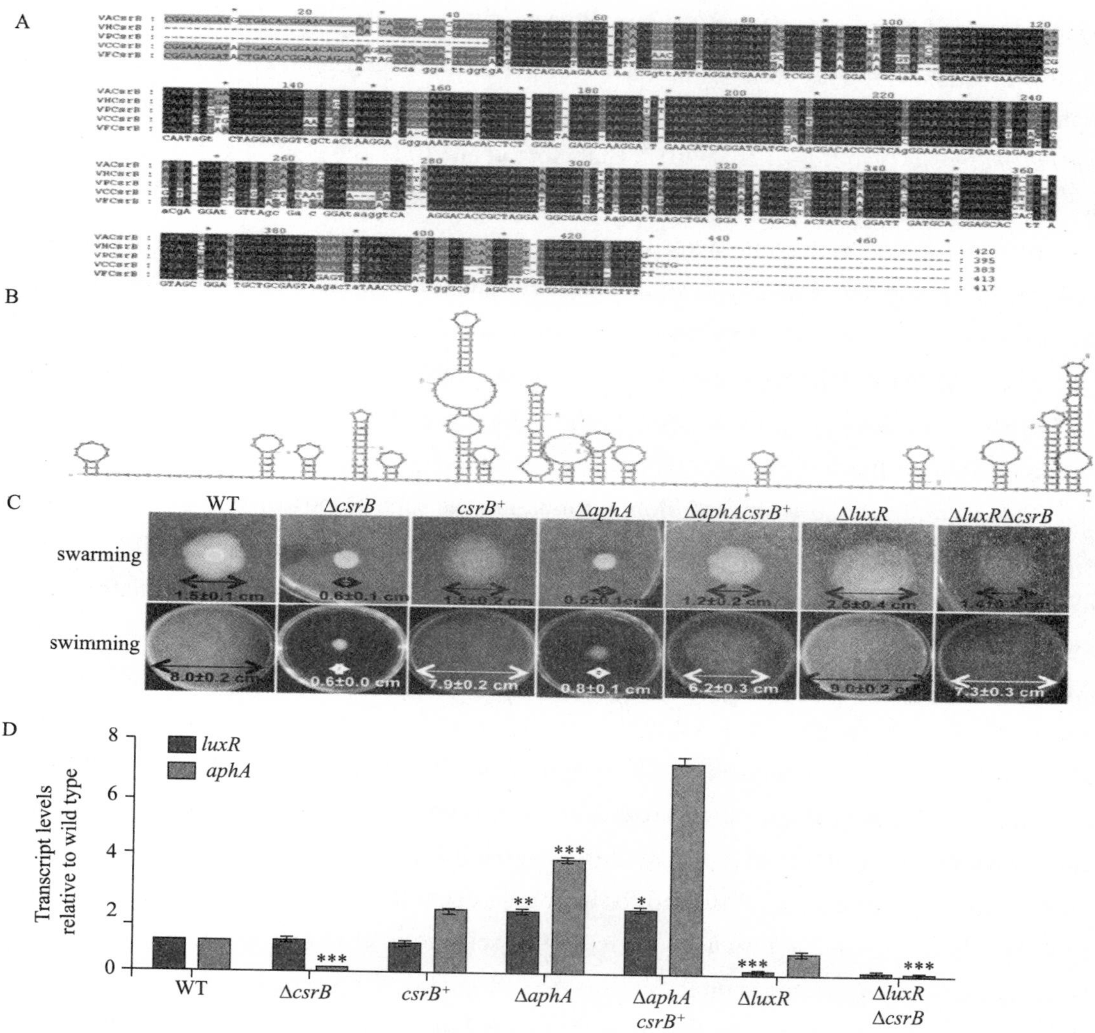

Fig. 8　An AphA-controlled sRNA *csrB* gene is involved in the mobility ability of *V. alginolyticus*

A. Alignment of DNA sequences encoding the *csrB* sRNAs in *V. alginolyticus* (VA), *V. cholerae* (VC), *V. fischeri* (VF), *V. parahaemolyticus* (VP), and *V. harveyi* (VH). B. Secondary structure of CsrB in *V. alginolyticus*. Predicted secondary structure for CsrB obtained using MFOLD (Zuker 2003) in V. alginolyticus showing multiple AGGA/ARGGA motifs in the loop regions or in the single-stranded regions. C. Motility assays of WT, various mutant strains, and strains over-expressing *csrB* ($csrB^+$). Swarming motility assays were performed on LBS plates with 1.5% agar, whereas swimming motility assays with 0.3% agar. D. qRT-PCR analysis of *luxR* and *csrB* transcripts. The strains were cultured in LBS for 10 h, and the mRNA transcripts were detected by qRT-PCR. The 16*S* rDNA was selected as a control. The results were shown as mean ± SD ($n = 3$) $^{*}P < 0.05$, $^{**}P < 0.01$, $^{***}P < 0.001$, student's *t*-test as compared to their corresponding WT results

4　Discussion

QS systems are intricately regulated as the global regulatory networks of gene expression in bacteria. In *V. alginolyticus*, a *V. harveyi*-like QS signaling system has been characterized, which involved pivotal LuxO and LuxR regulators (Wang et al, 2007; Rui et al, 2008; Rui et al, 2009; Cao et al, 2010; Cao et al, 2011; Zhao et al, 2011; Sheng et al, 2013; Tian et al, 2008). AphA has been reported as another master regulator of QS in *V. harveyi*, *V. cholerae*, and *V. parahaemolyticus* (van Kessel et al, 2013; Rutheford et al, 2011; Sun et al, 2012). AphA is present at LCDs, and regulates the expression of ～167 genes in *V. harveyi* (van Kessel et al, 2013). The molecular mechanisms of QS systems on the virulence and their dependence on AphA remain undefined in *V. alginolyticus*. We have recently found that LuxR and AphA can bind directly to the *luxR* promoter, and repressed the expression of LuxR in *V. alginolyticus* (Gu et al, 2016). Here, we characterized the *aphA* gene of *V. alginolyticus* and identified the direct targets of AphA by ChIP-seq. An AphA-dependent binding motif was disclosed that revealed novel regulation targets of AphA in the bacterium.

In *V. alginolyticus*, motility and biofilm formation has been proposed as important virulence-associated factors that could be regulated by QS (Sheng *et al*, 2013; Tian et al, 2008; Liu et al, 2011). Here we reported that AphA, another master regulator of QS, could also modulate motility, biofilm formation, and in vivo survival in zebra fish in the bacterium (Fig. 2). In *V. cholerae*, AphA enhances biofilm formation by activating expression of biofilm regulator VpsT, and the c-di-GMP and QS are also involved in the regulation of biofilm by controlling the expression of VpsT and AphA (Srivastava et al, 2011; Yang et al, 2010). AphA might be also involved in biofilm formation through the similar regulatory network in *V. alginolyticus*. Furthermore, AphA could regulate the extracellular protein Asp through LuxR (Figs. 3 and 4). AphA as a global regulator protein has been well studied for its role in controlling virulence gene expression in *V. cholerae V. harveyi* and *V. parahaemolyticus* (Kovacikova et al, 2001; van Kessel et al, 2013; Wnag et al, 2013). AphA binds directly to the *qrr4* promoter to repress its transcription; the same mechanism of repression is expected for *qrr2* and *qrr3* in *V. harveyi* (van Kessel et al, 2013). At LCDs, the Qrr sRNAs are constitutively produced and activate *aphA* expression. This activation maybe due to Qrr sRNAs bound to the target mRNA, which induces a conformation in the *aphA* 5′-UTR conducive to ribosome binding in *V. harveyi* (Rutherford et al, 2011). Then, AphA directly binds to its own promoter and auto-represses its transcription in both *V. cholerae* and *V. harveyi* (Rutherford et al, 2011). LuxR can also directly bind to *aphA* promoter in *V. cholerae*, *V. harveyi* and *V. parahaemolyticus* (Rutherford et al, 2011; Kovacikova et al, 2002; Zhang et al, 2012). AphA can cooperate with AphB to regulate the expression of *V. cholerae* virulence factors, toxin-co-regulated pilus (TCP) and cholerae toxin (CT) (Kovacikova et al, 2004). The research on AphA-coordinated AphB binding to specific gene promoters in *V. alginolyticus* is underway in the lab, which will further illuminate the complex regulatory networks and roles of AphA in vibrios.

Given that AphA is a transcription factor, it will be intriguing to examine how it controls the expression of additional genes in *V. alginolyticus*. We performed the ChIP-seq analysis to identify all potential *in vivo* binding sites of AphA in *V. alginolyticus*. Our analysis demonstrated that AphA is produced at LCD (Fig. 1B); therefore, ChIP-seq samples were collected after a 2h incubation period, and 49 AphA-binding peaks were detected. Previous microarray analyses established that ~300 genes are regulated by AphA at LCDs in *V. harveyi*, a subset of which was also controlled by LuxR (Rutherford et al, 2011). Another research also showed that AphA regulates 167 genes, LuxR regulates 625 genes, and they co-regulate 77 genes in *V. harveyi* (van Kessel et al, 2013). Blast indicated that among the 49 detected peaks, 4 loci overlapped in the ChIP-seq data of *V. alginolyticus* and microarray analyses of *V. harveyi*, including VEPGS-04037 (exoribonuclease R), VEPGS-04106 (AphA), VEPGS-04317 (DNA-directed RNA polymerase beta' subunit) and VEPGS-01513 (exonuclease). Because AphA can directly bind to its own promoter, which has been demonstrated in other vibrios (Rutherford et al, 2011; Sun et al, 2012; Lin et al, 2007), the overlapped list of genes validated, to some extent, our ChIP-seq analysis pipeline. Additionally, it suggested the versatile roles of AphA in different bacterial context. Work in different bacterium will help elucidate its functions on QS and non-QS regulations.

In this study, we defined a consensus AphA-binding sequence (TATTCGN$_7$GCTTAT), which is consistent with the motif identified by DNase I footprinting assays of the *luxR* promoter in *V. parahaemolyticus*, *V. cholerae* and *V. alginolyticus* (Rutherford et al, 2011; Gu et al, 2016; Sun et al, 2012). In another study the authors collected nine known or predicted direct targets of AphA from different vibrios, and generated an AphA 20bp consensus as ATATGCAN$_6$TGCATAT that contained imperfect inverted repeats of ATATGCA with a 6nt centered spacer (Sun et al, 2012). Our ChIP-seq generated consensus sequence was similar but not completely identical to this consensus, which may also reflect that the binding sites of AphA vary in different bacteria or the preferential binding activities differs in the bacteria. The research in *V. cholera* showed that the binding site of AphA to *vpsT* promoter was a 7-nt centered spacer, while the binding sites of AphA to promoters of *tcpPH* and *alsRD* were a 6-nt centered spacer (Yang et al, 2010). The above-mentioned discrepancies in the length of the consensus and spacer sequence maybe due to the difference in modulators (e.g. AphB) in terms of modulator species and cooperative strength with AphA that facilitates AphA binding to specific promoter regions. In *V. harveyi* and *V. parahaemolyticus*, AphA can directly bind to the *qrr2-4* promoters (Rutherford et al, 2011; Sun et al, 2012), but these targets were missing in our ChIP-seq data. However, the biochemical and bioinformatic analyses also showed that AphA could directly regulate the expression of putative *qrr4* in *V. alginolyticus*. Generally, ChIP-seq analysis may lose some specific targets because of the detection limits of ChIP with specific antibodies.

According to the results of ChIP-seq, we used the Kobas 2.0 (Xie et al, 2011) to analyze the possible pathways of genes containing an AphA-binding motif (Fig. 9). Among the major AphA targets are several transcription factors such as *aphA*, which inhibits its own expression (Fig. 9). VEPGS-04385 is predicted as an adenylate cyclase, a protein that transforms ATP to cAMP and GMP to cGMP (Fig. 9). In *V. cholerae*, the cAMP-CRP complex binds to sites in the *tcpPH* promoter that overlap with AphA and AphB binding sites, thus competing with the capacity of these two activators to bind (Kovacikova et al, 2001). In *V. vulnificus*, the cAMP-CRP complex could directly bind to the promoter regions of rpoS and result in the repression of *rpoS*, hence modulating numerous cellular processes (Lee et al, 2008). ChIP-seq data also showed that AphA could regulate thiamine metabolism, valine biosynthesis and shikimate pathway (Fig. 9), indicating the profound roles of AphA in a variety of metabolic pathways. The results also indicated that AphA could regulate the GlnL/GlnG two-complement system involved in the nitrogen metabolism regulation (Fig. 9). In addition, we also found that AphA could regulate the anti-σ^D, taking part in the regulation of σ^D and F0F1-type ATP synthase expression.

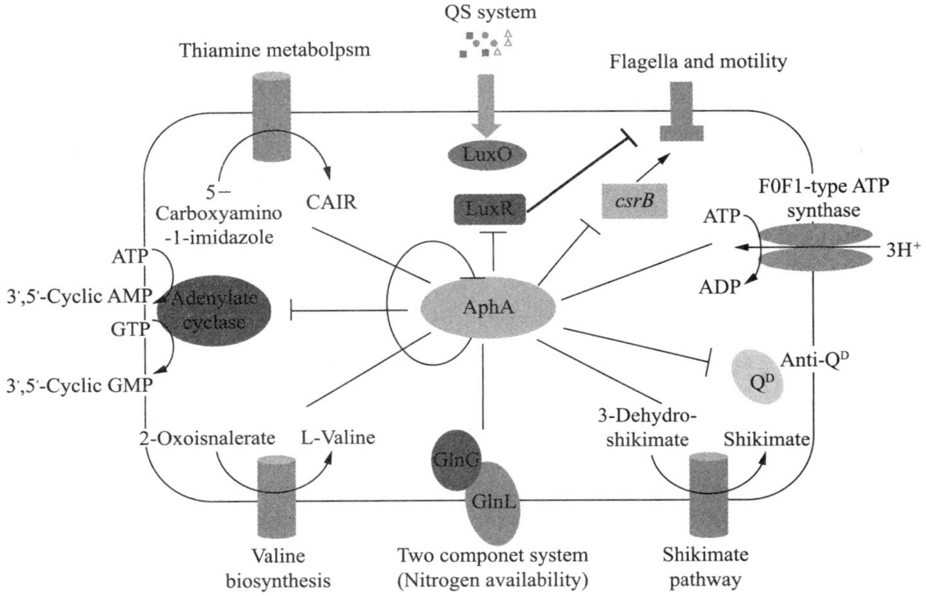

Fig. 9 Schematic summarization of the regulatory roles of AphA in *V. alginolyticus*

The KEGG pathway analysis was performed with the Kobas 2.0 (Xie et al, 2011). The various pathways and their respective cellular locations, as well as the regulatory roles of AphA, are illustrated with arrows (activation) or bar-ended lines (repression) and are discussed in the text. The inferred stronger inhibition of motilities by LuxR outcompetes CsrB-mediated activation of motilities and is depicted with a thicker bar-ended line. See the related text for detailed explanation

A small RNA belonging to the CsrB/RsmB family in vibrios (Lenz et al, 2005; Kulkarni et al, 2006), was also determined to be regulated by AphA (Fig. 9). AphA binding to the corresponding binding sites was confirmed *in vitro* and *in vivo* by EMSA and qRT-PCR (Fig. 7).

Moreover, our data also showed that AphA could modulate motility via LuxR and CsrB, probably in a mutual independent manner (Fig. 8). In *V. cholerae*, the VarS/VarA system could regulate the expression of CsrB and then control the QS through CsrA, a global regulatory protein (Lenz et al, 2005). The VarS/VarA-CsrA/B/C/D system is involved in the control of HapA expression and biofilm production through HapR in *V. cholerae* (Jiang et al, 2010; Tsou et al, 2011). In this study, we found that CsrB could be involved in the regulation of the motility in *V. alginolyticus*, and predicted the targets of CsrB by sRNA Target (Cao et al, 2009). A total of 23 flagellar associated genes were predicted as targets of CsrB, including the structural genes and regulatory genes of both lateral and polar flagella. CsrB could directly influence these genes expression to modulate motility. We assumed that AphA controls motility through both LuxR- and CsrB- mediated independent pathways with the following observations: i) LuxR and CsrB appear to be mutual-independent as deletion in *luxR* didn't affect *csrB* expression, and vice versa (Fig. 8D), ii) both LuxR and CsrB control motility through repression and activation on the motility associated pathway, respectively (Fig. 8C) (17), iii) AphA tightly regulates motility (Fig. 8C) and controls the expression of LuxR and CsrB at LCD (Figs. 7B and 7C) (Gu et al, 2016), and iv) Δ*luxR*Δ*csrB* double mutant shows faster motility than that of Δ*csrB* mutant, which suggests that the stronger de-repression of motilities by LuxR deletion outcompetes the repression rendered by CsrB abrogation, and thus promotes motility in the bacterium (Fig. 8C). Taken together, these data suggested that LuxR plays as a stronger regulator than CsrB does in terms of motility capacities, and both is controlled by AphA in *V. alginolyticus* (Fig. 9). The molecular mechanisms on how CsrB expression is controlled by AphA and mediates motility, possibly independent of LuxR, needs to be further elucidated in the light of spatial-temporal regulation as well as involvement of other factors in *V. alginolyticus*.

In summary, the results presented in this work provide an improved understanding of the AphA-LuxR centered complex regulatory network of vibrios QS. AphA can regulate the extracellular protein Asp through LuxR. Additionally, AphA can bind directly to the *luxR* and *aphA* promoter regions to then repress the expression of LuxR and AphA. AphA also plays essential roles in biofilm formation, motility, and survival in zebra fish. AphA motility modulation might be dependent on the action of CsrB and LuxR. Our ChIP-seq results also provided useful information that would allow us to further characterize other non-QS targets and processes regulated by AphA in the future.

References

[1] Austin B. Vibrios as causal agents of zoonoses[J]. Veterinary Microbiology, 2010, 140:310-317.

[2] Bailey T L, Boden M, Buske F A, et al. MEME SUITE: tools for motif discovery and searching[J]. Nucleic Acids Research, 2009, 37:W202-W208.

[3] Cao X D, Wang Q Y, Liu Q, et al. *Vibrio alginolyticus* MviN is a *luxO*-regulated protein and affects cytotoxicity towards epithelioma papulosum cyprini (EPC) cells[J]. Journal of Microbiology and Biotechnology, 2010, 20:271-280.

[4] Cao X D, Wang Q Y, Liu Q, et al. Identification of a *luxO*-regulated extracellular protein Pep and its roles in motility in *Vibrio alginolyticus*[J]. Microbial Pathogenesis, 2011, 50:123-131.

[5] Cao Y, Zhao Y L, Cha L, et al. sRNATarget: a web server for prediction of bacterial sRNA targets[J]. Bioinformation, 2009, 3:364-366.

[6] De Silva R S, Kovacikova G, Lin W, et al. Crystal structure of the virulence gene activator AphA from *Vibrio cholerae* reveals it is a novel member of the winged helix transcription factor superfamily[J]. Journal of Biological Chemistry, 2005, 280:13779-13783.

[7] Derrick H L, Melissa B M I, Zhu J, et al. CsrA and three redundant small RNAs regulate quorum sensing in *Vibrio cholerae*[J]. Molecular Microbiology, 2005, 58:1186-1203.

[8] Feng J X, Liu T, Qin B, et al. Identifying ChIP-seq enrichment using MACS[J]. Nature Protocols, 2012, 7:1728-1740.

[9] Feng L, Rutherford S T, Papenfort K, et al. A Qrr noncoding RNA deploys four different regulatory mechanisms to optimize quorum-sensing dynamics[J]. Cell, 2015, 160:228-240.

[10] Gibson D G, Young L, Chuang R Y, et al. Enzymatic assembly of DNA molecules up to several hundred kilobases[J]. Nature Methods, 2009, 6:343-345.

[11] Gu D, Guo M, Yang M J, et al. A σ^E-mediated temperature gauge controls a switch from LuxR-mediated virulence gene expression to thermal stress adaptation in *Vibrio alginolyticus*[J]. PLoS Pathogen, 2016, 12:e1005645.

[12] Guzman L M, Belin D, Carson M J, et al. Tight regulation, modulation, and high-level expression by vectors containing the arabinose PBAD promoter[J]. Journal of Bacteriology, 1995, 177:4121-4130.

[13] Jang J Y, Jung K T, Yoo C K, et al. Regulation of hemagglutinin/protease expression by the VarS/VarA-CsrA/B/C/D system in *Vibrio cholerae*[J]. Microbial Pathogenesis, 2010, 48:245-250.

[14] Kovacikova G, Lin W, Skorupski K. *Vibrio cholerae* AphA uses a novel mechanism for virulence gene activation that involves interaction with the LysR-type regulator AphB at the *tcpPH* promoter[J]. Molecular Microbiology, 2005, 53:129-142.

[15] Kovacikova G, Lin W, Skorupski K. Dual regulation of genes involved in acetoin biosynthesis and motility/biofilm formation by the virulence activator AphA and the acetate-responsive LysR-type regulator AlsR in *Vibrio cholerae*[J]. Molecular Microbiology, 2005, 57:420-433.

[16] Kovacikova G, Skorupski K. A *Vibrio cholerae* LysR homolog, AphB, cooperates with AphA at *the tcpPH* promoter to activate expression of the ToxR virulence cascade[J]. Journal of Bacteriology, 1999, 181:4250-4256.

[17] Kovacikova G, Skorupski K. Overlapping binding sites for the virulence gene regulators AphA, AphB and cAMP-CRP at the *Vibrio cholerae tcpPH* promoter[J]. Molecular Microbiology, 2001, 41:393-407.

[18] Kovacikova G, Skorupski K. Regulation of virulence gene expression in *Vibrio cholerae* by quorum sensing: HapR functions at the *aphA* promoter[J]. Molecular Microbiology, 2002, 44:1135-1147.

[19] Lee H J, Park S J, Choi S H, Lee K H. *Vibrio vulnificus rpoS* expression is repressed by direct binding of cAMP-cAMP receptor protein complex to its two promoter regions[J]. Journal of Biological Chemistry, 2008, 283:30438-30450.

[20] Lin W, Kovacikova G, Skorupski K. The quorum sensing regulator HapR downregulates the expression of the virulence gene transcription factor AphA in *Vibrio cholerae* by antagonizing Lrp- and VpsR-mediated activation[J]. Molecular Microbiology, 2007, 64:953-967.

[21] Liu H, Wang Q Y, Liu Q, et al. Roles of Hfq in the stress adaptation and virulence in fish pathogen *Vibrio alginolyticus* and its potential application as a target for live attenuated vaccine[J]. Applied Microbiology and Biotechnology, 2011, 91:353-364.

[22] Liu M Y, Gui G, Wei B, et al. The RNA molecule CsrB binds to the global regulatory protein CsrA and antagonizes its activity in *Escherichia coli*[J]. Journal of Biological Chemistry, 1997, 272:17502-17510.

[23] Miller M B, Bassler B L. Quorum sensing in bacteria[J]. Annual Review of Microbiology, 2001, 55:165-199.

[24] Ng W L, Bassler B L. Bacterial quorum-sensing network architectures[J]. Annual Review of Genetics, 2009, 43:197-222.

[25] Pernestig A K, Melefors O, Georgellis D. Identification of UvrY as the cognate response regulator for the BarA sensor kinase in *Escherichia coli*[J]. Journal of Biological Chemistry, 2001, 276:225-231.

[26] Philip M, Timothy L B. MEME-ChIP: motif analysis of large DNA datasets[J]. Bioinformatics, 2011, 2712:1696-1697.

[27] Prajna R K, Cui X H, Joshua W W, et al. Prediction of CsrA-regulating small RNAs in bacteria and their experimental verification in *Vibrio fischeri*[J]. Nucleic Acids Research, 2006, 34:3361-3369.

[28] Romeo T, Gong M, Liu M Y, et al. Identification and molecular characterization of *csrA*, a pleiotropic gene from *Escherichia coli* that affects glycogen biosynthesis, gluconeogenesis, cell size, and surface properties[J]. Journal of Bacteriology, 1993, 175:4744-4755.

[29] Rui H P, Liu Q, Ma Y, et al. Roles of LuxR in regulating extracellular alkaline serine protease A, extracellular polysaccharide and mobility of *Vibrio alginolyticus*[J]. FEMS Microbiology Letter, 2008, 285:155-162.

[30] Rui H P, Liu Q, Wang Q Y, et al. Role of alkaline serine protease, Asp, in *Vibrio alginolyticus* virulence and regulation of its expression by LuxO-LuxR regulatory system[J]. Journal of Microbiology and Biotechnology, 2009, 19:431-438.

[31] Rutherford S T, van Kessel J C, Shao Y, et al. AphA and LuxR/HapR reciprocally control quorum sensing in Vibrios[J]. Genes and Development, 2011, 25:397-408.

[32] Sheng L L, Lv Y Z, Liu Q, et al. Connecting type VI secretion, quorum sensing, and c-di-GMP production in fish pathogen *Vibrio alginolyticus* through phosphatase PppA[J]. Veterinary Microbiology, 2013, 162:652-662.

[33] Skorupski K, Taylor R K. A new level in the *Vibrio cholerae* ToxR virulence cascade: AphA is required for transcriptional activation of the *tcpPH* operon[J]. Molecular Microbiology, 1999, 31:763-771.

[34] Srivastava D, Harris R C, Waters CM. Integration of cyclic di-GMP and quorum sensing in the control of *vpsT* and *aphA* in *Vibrio cholera*[J]. Journal of Bacteriology, 2011, 173:6331-6341.

[35] Sun F J, Zhang Y Q, Wang L, et al. Molecular characterization of direct target genes and *cis*-acting consensus recognized by quorum-sensing regulator AphA in *Vibrio parahaemolyticus*[J]. PLoS ONE, 2012, 7:e44210.

[36] Tian Y, Wang Q Y, Liu Q, et al. Involvement of LuxS in the regulation of motility and flagella biogenesis in *Vibrio alginolyticus*[J]. Bioscience Biotechnology and Biochemistry, 2008, 72:1063-1071.

[37] Tsou A M, Liu Z, Cai T, Zhu J. The VarS/VarA two-component system modulates the activity of the *Vibrio cholerae* quorum-sensing transcriptional regulator HapR[J]. Microbiology, 2011, 157:1620-1628.

[38] Van Kessel J C, Rutherford S T, Shao Y, et al. Individual and combined roles of the master regulators AphA and LuxR in control of the *Vibrio harveyi* quorum-sensing regulon[J]. Journal of Bacteriology, 2013, 195:436-443.

[39] Van Kessel J C, Ulrich L E, Zhulin I B, et al. Analysis of activator and repressor functions reveals the requirements for transcriptional control by LuxR, the master regulator of quorum sensing in *Vibrio harveyi*[J]. mBio, 2013, 4:e00378-13-e00378-13.

[40] Wang L, Ling Y, Jiang H J, et al. AphA is required for biofilm formation, motility, and virulence in pandemic *Vibrio parahaemolyticus*[J]. International Jouranl Food Microbiology, 2013, 160:245-251.

[41] Wang Q Y, Liu Q, Ma Y, et al. LuxO controls extracellular protease, haemolytic activities and siderophore production in fish pathogen *Vibrio alginolyticus*[J]. Journal of Applied Microbiology, 2007, 103:1525-1534.

[42] Wang S Y, Lauritz J, Jass J, et al. A ToxR homolog from *Vibrio anguillarum* serotype O1 regulates its own production, bile resistance, and biofilm formation[J]. Journal of Bacteriology, 2003, 184:1630-1639.

[43] Weilbacher T, Suzuki K, Dubey A K, et al. A novel sRNA component of the carbon storage regulatory system of *Escherichia coli*[J]. Molecular Microbiology, 2003, 48:657-670.

[44] Xie C, Mao X Z, Huang J J, et al. KOBAS 2.0: a web server for annotation and identification of enriched pathways and diseases[J]. Nucleic Acids Research, 2011, 39:W316-W322.

[45] Yang M H, Frey E M, Liu Z, et al. The virulence transcriptional activator AphA enhances biofilm formation by *Vibrio cholerae* by activating expression of the biofilm regulator VpsT[J]. Infection and Immunity, 2010, 78:697-703.

[46] Zhang Y Q, Qiu Y F, Tan Y F, et al. Transcriptional regulation of *opaR*, *qrr2–4* and *aphA* by the master quorum-sensing regulator OpaR in *Vibrio parahaemolyticus*[J]. PLoS ONE, 2012, 7:e34622.

[47] Zhao Z, Zhang L P, Ren C H, et al. Autophagy is induced by the type III secretion system of *Vibrio alginolyticus* in several mammalian cell lines[J]. Archives of Microbiology, 2011, 193:53-61.

[48] Zhou D S, Yan X J, Qu F, et al. Quorum sensing modulates transcription of *cpsQ-mfpABC* and *mfpABC* in *Vibrio parahaemolyticus*[J]. International Journal of Food Microbiology, 2013, 166:458-463.

[49] Zuker M. Mfold web server for nucleic acid folding and hybridization prediction[J]. Nucleic Acids Research, 2003, 31:3406-3415.

Effects of stocking density on antioxidative status, metabolism and immune response in juvenile turbot (*Scophthalmus maximus*) reared in a land-based recirculating aquaculture system

Baoliang Liu[1], Rui Jia[2], Cen Han[3], Bin Huang[1], Ji-Lin Lei[1]

[1] Yellow Sea Fisheries Research Institute, Chinese Academy of Fishery Science, Qingdao 266071, China

[2] Wuxi Fisheries College, Nanjing Agricultural University, Wuxi 214081, China

[3] College of Fisheries and Life Science, Dalian Ocean University, Dalian 116023, China

Abstract This study was designed to evaluate the physiological and immune responses of juvenile turbot to stocking density in a land-based recirculating aquaculture system (RAS). Turbot (average weight 185.4 g) were reared for 120 days under three stocking densities: low density (LD, 9.3 kg/m^2～26.1 kg/m^2, initial to final density), medium density (MD, 13.6 kg/m^2～38.2 kg/m^2) and high density (HD, 19.1 kg/m^2～52.3 kg/m^2). Fish were sampled at day 0, 40, 80 and 120 to obtain growth parameters and liver tissues. No significant difference was detected in growth, biochemical parameters and genes expression among the three densities until at the final sampling (day 120). At the end of this trial, fish reared in HD group showed lower specific growth rate (SGR) and mean weight than those reared in LD and MD groups. Similarly, oxidative stress and metabolism analyses represented that antioxidants (superoxide dismutase (SOD), catalase (CAT), glutathione (GSH) and metabolic enzymes (glycerol-3-phosphate dehydrogenase (G3PDH) and glucose-6-phosphate dehydrogenase (G6PDH) clearly reduced in the liver of turbot reared in HD group. The gene expression data showed that glutathione S-transferase (GST), cytochrome P450 1A (CYP1A), heat shock protein 70 (HSP70) and metallothionein (MT) mRNA levels were significantly up-regulated, and lysozyme (LYS) and hepcidin (HAMP) mRNA levels were significantly down-regulated in HD group on day 120. Overall, our results indicate that overly high stocking density might block the activities of metabolic and antioxidant enzymes, and cause physiological stress and immunosuppression in turbot.

Keywords *Scophthalmus maximus*; stocking density; antioxidative enzymes; metabolism; gene expression

1 Introduction

Stocking density is considered one of the most important variables in intensive aquaculture influencing growth, welfare and health of farmed fish (Castillo-Vargasmachuca et al, 2012; Ellis et al., 2002; Ni et al, 2014). In commercial aquaculture, operation at higher stocking densities can reduce production costs, but high stocking density is a stressor, which induces chronic stress associated with deterioration in water quality, adverse social interactions or over-crowding, resulting in negative physiological and biochemical changes (Bolasina et al, 2006; Montero et al, 1999). A number of studies on cultured flatfish in which the effects of stocking density on growth performance is investigated have reported a growth reduction in higher-density treatments (Bolasina et al, 2006; Costas et al, 2008; Irwin et al, 1999). Thus, for a cost-effective production, it is necessary to find a balance between the maximum profit and the minimum incidence of physiological disorders and growth inhibition (Björnsson et al, 2012; Herrera et al, 2009).

Oxidative stress occurs when there is an imbalance between the activity of antioxidants and the generation of reactive oxygen species (ROS) (Lushchak, 2011). The antioxidants include antioxidant enzymes such as superoxide dismutase (SOD), glutathione peroxidase (GPx) and catalase (CAT), and nonenzymatic antioxidant components such as glutathione (GSH) (Jia et al, 2013; Sayeed et al, 2003). In aquaculture, high density is a source of stress increasing the production of ROS (Braun et al, 2010; Vijayan et al, 1990). These ROS could damage cellular components and autoxidation function (Sayeed et al, 2003), result in oxidative injury and increase the incidence of diseases and deaths (Herrera et al, 2009; Iwama et al, 2011).

Like higher vertebrates, the immune system of teleost fish is composed of non-specific and specific immune system (Zapata et al, 2006). Non-specific defences are the most important defense mechanisms in fish. Many reports have shown the links of stress, depression of immune system and disease (Magnadóttir, 2006). The effects of crowding stress on the immune parameters of few species of freshwater and marine fish reared in indoor tanks have been documented (Li et al, 2006; Montero et al, 1999; Sadhu et al, 2014). High stocking density is a stressor that activates stress responses in fish, which caused alteration of immune-related enzymes, proteins or genes (Ni et al, 2014; Salas-Leiton et al, 2010; Vargas-Chacoff et al, 2014). Costas et al (2013) reported that high stocking density reduced plasma lysozyme (LYS), alternative complement pathway (ACP) and peroxidase activities, suggesting some degree of immunosuppression in *Solea senegalensis*. Similar results were also reported in *Sparus aurata*, *Oncorhynchus mykiss* and *Oreochromis niloticus* reared in high density (Montero et al., 1999; Telli et al, 2014; Yarahmadi et al, 2014). Moreover, high density affected expression of genes involved in physiological stress such as cytochrome P450 1A (CYP1A), heat shock proteins HSP70 and HSP90, and innate immune system such as interleukin 1β (IL-1β), g-LYS and hepcidin (HAMP) (Ni et al, 2014; Salas-Leiton et al, 2010; Tapia-Paniagua et al, 2014). However, effects of stocking density on the immune functions of turbot reared in recirculating aquaculture system (RAS) is scarcely understood

Turbot (*Scophthalmus maximus*) is a suitable species for aquaculture in Europe and China because of its considerable commercial value. This species has been intensively cultured in land-based farms including recirculation and flow-through systems (Irwin et al, 1999; Li et al, 2013). Initial ongrowing results indicated that this species could be adequately cultivated at high stocking densities (Aksungur et al, 2007; Baer et al, 2011). Nevertheless, there is little analysis regarding physiological and molecular responses caused by stocking density. Therefore, the purpose of this study aimed to evaluate the effects of stocking density on the antioxidative status, metabolism and expression of immune-related genes of turbot reared in a land-based RAS.

2 Materials and methods

2.1 Experimental design and animals

The study took place in the farm of Shandong Oriental Ocean Sci-Tech Co., Ltd. (Shandong, China), and was operated in recirculating system supplied with high-quality well water, stabilized at 17 ℃ ~ 19 ℃. The RAS consisted of ten 30 m^3 tanks connected to a biofilter and a mechanical filter (Fig. 1). The oxygen content and water level (0.5 m ~ 0.55 m in depth) in the fish tanks were monitored via the RAS computer. Make up water (16 ℃ ~ 18 ℃) was pumped from a 20 m depth from the Laizhou Bay of China, mechanically filtered by two sand filters (5 μm filtration) and UV sterilized before entering the RAS units, and the water volume was ≤ 12% of the total system volume per day. The water residence time in the fish tanks was 1 hour. The photoperiod was maintained at 12hour light/12hour dark using fluorescent light banks. Temperature, dissolved oxygen (DO), pH and salinity in each tank varied slightly throughout the day, but in all cases remained at (18 ± 1.5) ℃, (8 ± 1) mg/L, 7.9 ± 0.3, and (27.3 ± 3) g/L, respectively. Total ammonium nitrogen (TAN), nitrite and orthophosphates (PO_4-P) never exceeded 0.3 mg/L, 0.25 mg/L and 7.5 μmol/L, respectively.

The turbot were obtained from this farm and reared in the RAS for 15 d to acclimatize to the experimental environment. A total of 20 400 fish (average individual weight 185.4 g) were reared for 120 d under three densities: low density (LD) with 1500 fish per tank (9.3 kg/m^2 ± 0.11 kg/m^2 at initial density), medium density (MD) with 2 200 fish per tank (13.6 kg/m^2 ± 0.21 kg/m^2 at initial density), and high density (HD) with 3 100 fish per tank (19.1 kg/m^2 ± 0.38 kg/m^2 at initial density). Each density was tested in triplicate. No differences in weight and coefficient of variation (CV) for initial weight were found among the three densities. The turbot were fed a commercial-pellet diet (Ningbo Tech-Bank Co., LTD., Zhejiang, China), which contained 52% crude protein, 12% crude lipids, 16% crude ash, 3% crude fiber, 12% water, 5% Ca, 0.5% P, 2.3% lysine and 3.8% sodium chloride. The fish were fed two times per day (08:00 and 20:00), and the daily feed rations (approximately 1.2% of fish weight/day) were adjusted based on observation of the feeding behavior and weight of fish. Fish were not fed 24 hours before sampling (Arends et al, 1999).

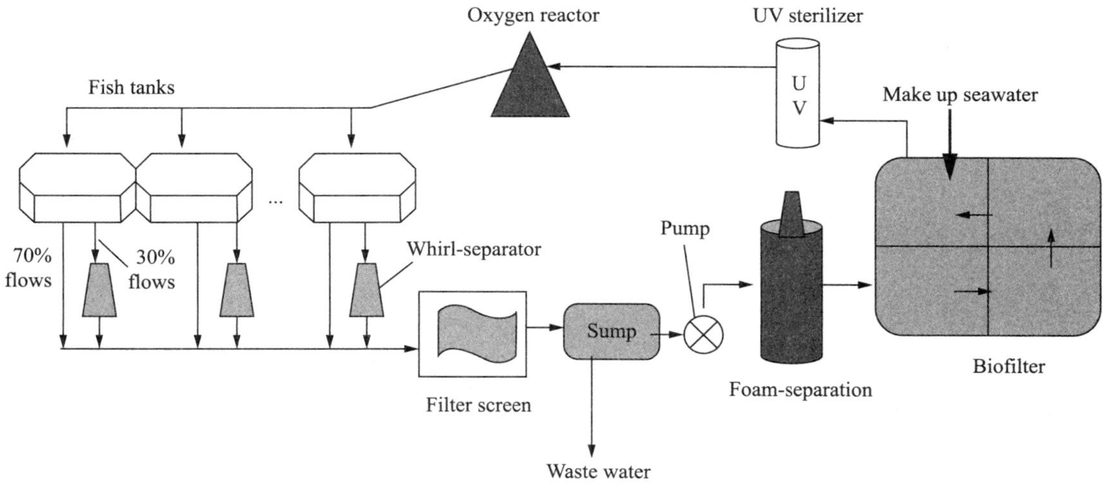

Fig. 1 Schematic diagram of the experimental land-based recirculating aquaculture system
Water flows from rearing tanks → whirl-separator (30% flows) → filter screen → sump → pump → foam-separation → biofilter → UV sterilizer → oxygen reactor → rearing tanks

2.2 Sampling

Growth was monitored by measuring weight and total length at 0, 40, 80 and 120 days after the commencement of the assay. Total sample size represented 15% of the whole population in per tank (Garcia et al, 2013).

To perform the biochemical parameters determinations and genes expression analyses, 20 individuals of each tank were randomly netted at Day 0, 40, 80 and 120, and immediately anesthetized in 0.05% tricaine methane sulfonate (MS-222, Sigma Diagnostics INS, MO, USA). Liver was rapidly dissected, frozen in liquid nitrogen, and stored at −80 ℃ until use.

2.3 Growth parameters

Specific growth rate (SGR) was calculated from individual weights recorded on Day 0, 40, 80 and 120. SGR/(%/d) = 100 × (ln final weight − ln initial weight)/number of days. The corresponding stocking density (kg biomass/m^2) = number of individuals × average weight/surface.

2.4 Biochemical parameter assays

Oxidative stress parameters, including SOD, GPx, CAT, GSH and malondialdehyde (MDA), in the liver tissues were measured using commercially available kits (Jiancheng Institute of Biotechnology, Nanjing, China).

The activities of liver metabolic enzymes fructose-1,6-bisphosphatase (FBPase, EC 3.1.3.11), glycerol-3-phosphate dehydrogenase (G3PDH, EC 1.1.1.8), hexokinase (HK, EC 2.7.1.11) and glucose-6-phosphate dehydrogenase (G6PDH, EC 1.1.1.49) were determined using a Microplate Reader (Multiscan Go, Thermo Scientific, USA), using Software Scan 3.2 to Multiscango. Enzyme reaction rates were determined by the increase or decrease in absorbance of NAD(P)H at 340 nm. The reactions were started by addition of homogenates (15 μL) at a pre-established protein concentration, omitting the substrate in control wells (final volume 275～295 μL) and allowing the reactions to proceed at 37 ℃ for pre-established times (Vargas-Chacoff et al, 2014). Protein concentration of liver homogenate was determined by the Bradford method, using bovine serum albumin as a standard (Bradford, 1976).

2.5 Total RNA isolation

The total RNA was isolated from liver tissues using a fast pure RNA kit (Dalian Takara, China) according to the manufacturer's instruction. The amount of RNA was measured using GeneQuant 1300 (GE Healthcare Biosciences, Piscataway, NJ) and its quality was checked on an agarose gel.

2.6 Quantification of genes expression levels

Total RNA (2 μg) from each sample was reverse-transcribed using PrimeScript RT reagent Kit with gDNA Eraser (Takara, Dalian, China) according to the manufacturer's instruction. The real-time quantitative PCR (qRT-PCR) was performed with an ABI PRISM 7500 Detection System (Applied Biosystems, USA). Each reaction mixture (20 μL) contained 4 μL of cDNA template (ten-fold diluted), 10 μL of SYBR® Premix Ex TaqTM (Perfect Real Time) (Takara Bio., China), 0.4 μL of ROXII, 0.4 μL of each primer (10 μmol/L) and 4.8 μL of ddH_2O.

The primers used for amplification and genes expression analyses are presented in Table 1. The following PCR profile was set to run for one cycle at 95 ℃ for 10 s, 40 cycles at 95 ℃ for 5 s and at 60 ℃ for 34 s. The melting curve was used to ensure that a single product was amplified and to check for the absence of primer-dimer artifacts. All qRT-PCRs were performed at least in triplicate. The results were normalized to β-actin and relative gene quantification was performed using the $2^{-\Delta\Delta CT}$ method (Livak and Schmittgen, 2001).

Table 1 Primer utilized for gene expression analysis

Genes	Primer sequence (5′-3′)	Amplicon size / bp	GenBank accession No.	Reference
β-actin	F: TGAACCCCAAAGCCAACAGG R: AGAGGCATACAGGGACAGCAC	107	EU686692.1	(Wang et al, 2008)
CYP1A	F: ATCGCTCTCCTCTTCTCTCT R: TTAGAGGTGCAGTGTGGAAT	115	AJ310694.1	(Hermann et al, 2016)
GST	F: GGGTTCGCATCGCTTTT R: GGCCTGGTCTCGTCTATGTACT	196	DQ848966.1	(Reiser et al, 2011)
HSP 70	F: CTGTCCCTGGGTATTGAGAC R: GAACACCACGAGGAGCA	220	EF191027.1	(Reiser et al, 2011)
HSP 90	F: CCGCCTACCTCGTTGC R: TAGCCGATGAACTGCGAGT	229	EU099575.1	(Reiser et al, 2011)
MT	F: TGCTCCAAGAGTGGAACCTG R: CGCATGTCTTCCCTTTGCAC	148	EF406132.1	This study
TNF-α	F: GGGCTGGTACAACACCATCTATC R: TTCAATTAGTGCCACGACAAAGA	165	FJ654645.1	(Muñoz-Atienza et al, 2014)
IL-1β	F: ACCAGACCTTCAGCATCCAGCGT R: TTCAGTGCCCCATTCCACCTTCCA	81	AJ295836.2	(Hermann et al, 2016)
LYS	F: CTCTCAACGTTCCCACTGGTTCTA R: GGGGTCATGAAGTGTCTGTAGAT	191	AJ250732.1	(Muñoz-Atienza et al, 2014)
HAMP	F: CACTCGTGCTCGCCTTTGTTT R: TGTCATTGCTCCCTGCCTCTT	204	AY994075.1	This study

2.7 Statistical analysis

The statistical analysis was performed using SPSS (version 18.0) software. Data was presented as mean ± standard error of the mean (SEM) and analyzed by one-way analysis of variance (ANOVA) after checking for normality. Logarithmic transformations of the data were performed when necessary to fulfill the conditions of the ANOVA, but the data was shown in their decimal values for clarity. Post-hoc comparisons were performed using Tukey's test; significance was taken at $P < 0.05$.

3 Results

3.1 Growth performance

During the course of the experiment, no disease outbroke or other signs of disease were observed. Survival rate was extremely high in all treatments (survival rate > 99.1%) with no significant differences between treatments. The established initial stocking densities 9.3 ± 0.11, (13.6 ± 0.21) kg/m^2 and (19.1 ± 0.32) kg/m^2 increased up to (26.1 ± 0.16) kg/m^2, (38.2 ± 0.18) kg/m^{2}and (52.3 ± 0.19) kg/m^2 at the end of the experiment, respectively. There were no significant difference in growth performance among the three groups until at the final weighing (Day 120) when the mean weight and SGR in the HD group were significantly lower than in MD and LD groups ($P < 0.05$; Table 2).

Table 2 The effects of different stocking densities on mean weight and SGR of turbot for 0, 40, 80 and 120 days

Time	Group	Stoking density /(kg/m^2)	Mean weight	SGR/(%/d)
Day 0	LD	9.3 ± 0.11	186.0 ± 0.97	—
	MD	13.6 ± 0.21	185.2 ± 0.82	—
	HD	19.1 ± 0.32	185.0 ± 1.57	—
Day 40	LD	14.1 ± 0.26	281.3 ± 1.12	1.03 ± 0.02
	MD	20.5 ± 0.21	279.6 ± 1.87	1.03 ± 0.01
	HD	28.9 ± 0.38	279.8 ± 1.92	1.03 ± 0.02
Day 80	LD	20.3 ± 0.14	406.6 ± 2.55	0.92 ± 0.02
	MD	29.5 ± 0.17	402.6 ± 1.30	0.91 ± 0.01
	HD	41.4 ± 0.26	400.5 ± 4.22	0.90 ± 0.03
Day 120	LD	26.1 ± 0.16	540.4 ± 4.52^a	0.71 ± 0.02^a
	MD	38.2 ± 0.18	534.4 ± 2.25^a	0.71 ± 0.01^a
	HD	52.3 ± 0.19	514.2 ± 3.37^b	0.61 ± 0.02^b

The values are means ± SEM, n = 20 fish per tank. Different letter denote significant differences between densities within sampling day. LD, low density; MD, medium density; HD, high density

3.2 Antioxidant activities and lipid peroxidation in liver

At the end of trial, the activities of SOD and CAT apparently decreased in HD and MD groups compared with LD group ($P < 0.05$; Fig. 2A and 2B). Similarly, the GSH content in HD group was obviously lower than in LD group ($P < 0.05$; Fig. 2D). However, the levels of GPx and MDA had no difference between the different densities during the 120 days culture (Fig. 2C and 2E).

3.3 Metabolic enzymes activities in liver

The liver G3PDH and G6PDH activities were highest in the LD group and lowest in the HD group on Day 120 ($P < 0.05$; Fig. 3A and 3D), while the FBPase and HK activities did not show

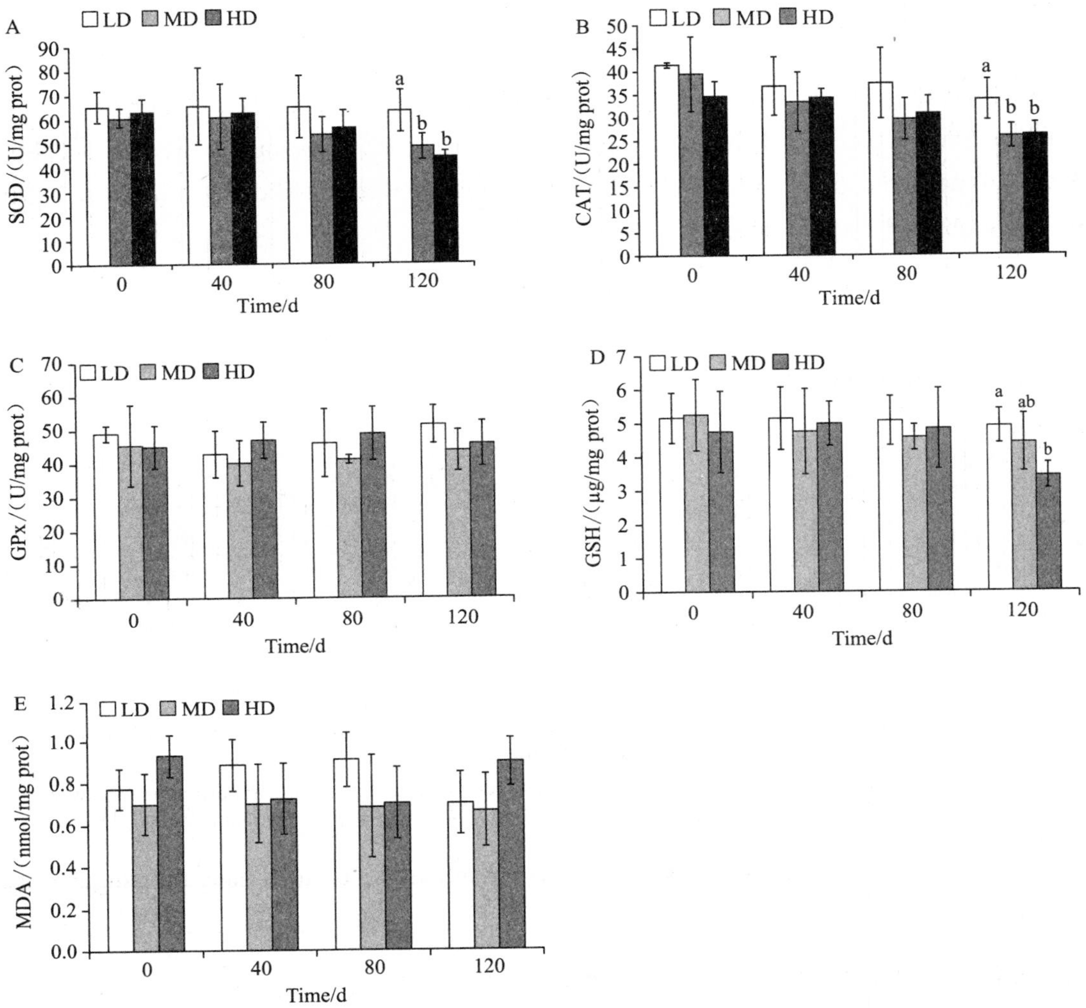

Fig. 2 The effects of different stocking densities on antioxidant capacity and lipid peroxidation in the liver of juvenile turbot

The values are means ± SEM, n = 20 fish per tank. Different letter denote significant differences between densities within sampling day. LD, low density; MD, medium density; HD, high density

any differences among the three groups (Fig. 3B and 3C).

3.4 Stress-related genes expression in liver

No variations in these stress-related genes were observed in the first two periods (days 0 to 40, and days 40 to 80), while significant differences among the three density groups were observed at the final sampling. Compared with the LD and MD groups, fish reared in the HD group displayed significantly lower values in glutathione-s-transferase (GST) mRNA level and higher values in CYP1A, HSP70 and metallothionein (MT) mRNA levels ($P < 0.05$; Fig. 4A,

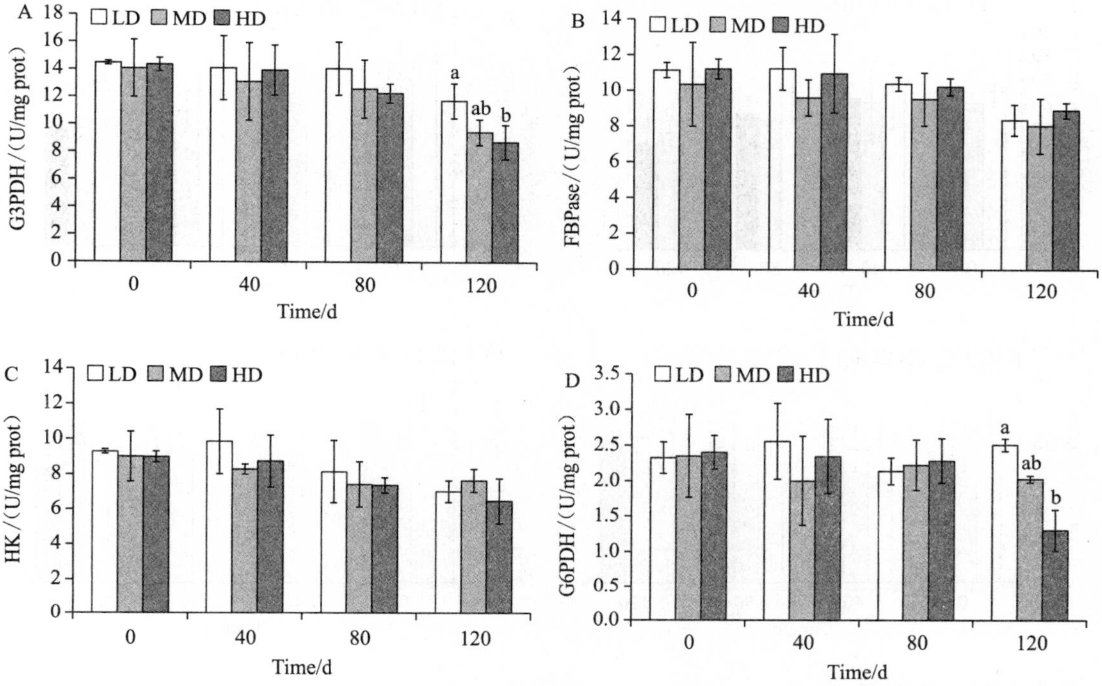

Fig. 3　The effects of different stocking densities on metabolic enzymes activities in the liver of juvenile turbot

The values are means ± SEM, n = 20 fish per tank. Different letter denote significant differences between densities within sampling day. LD, low density; MD, medium density; HD, high density

4B, 4C and 4E). But the HSP90 mRNA level was not influenced by increasing stocking density (Fig. 4D).

3.5　Immune-related genes expression in liver

Liver HAMP and LYS genes expressions were lower in the HD group relative to the LD group only in the fourth sampling ($P < 0.05$; Fig. 5A and 5B), whereas the tumor necrosis factor-α (TNF-α) and IL-1β genes expressions did not show any differences among the three groups during the trial (Fig. 5C and 5D).

4　Discussion

The stocking density is an important factor for the cultivation of fish. Increase in the stocking density is an option that combines the maximum water utilization with a higher fish production. However, overly high stocking density can lead to stress, negatively affecting the growth, survival, immunologic response and behavior (Andrade et al, 2015; Montero et al, 1999;

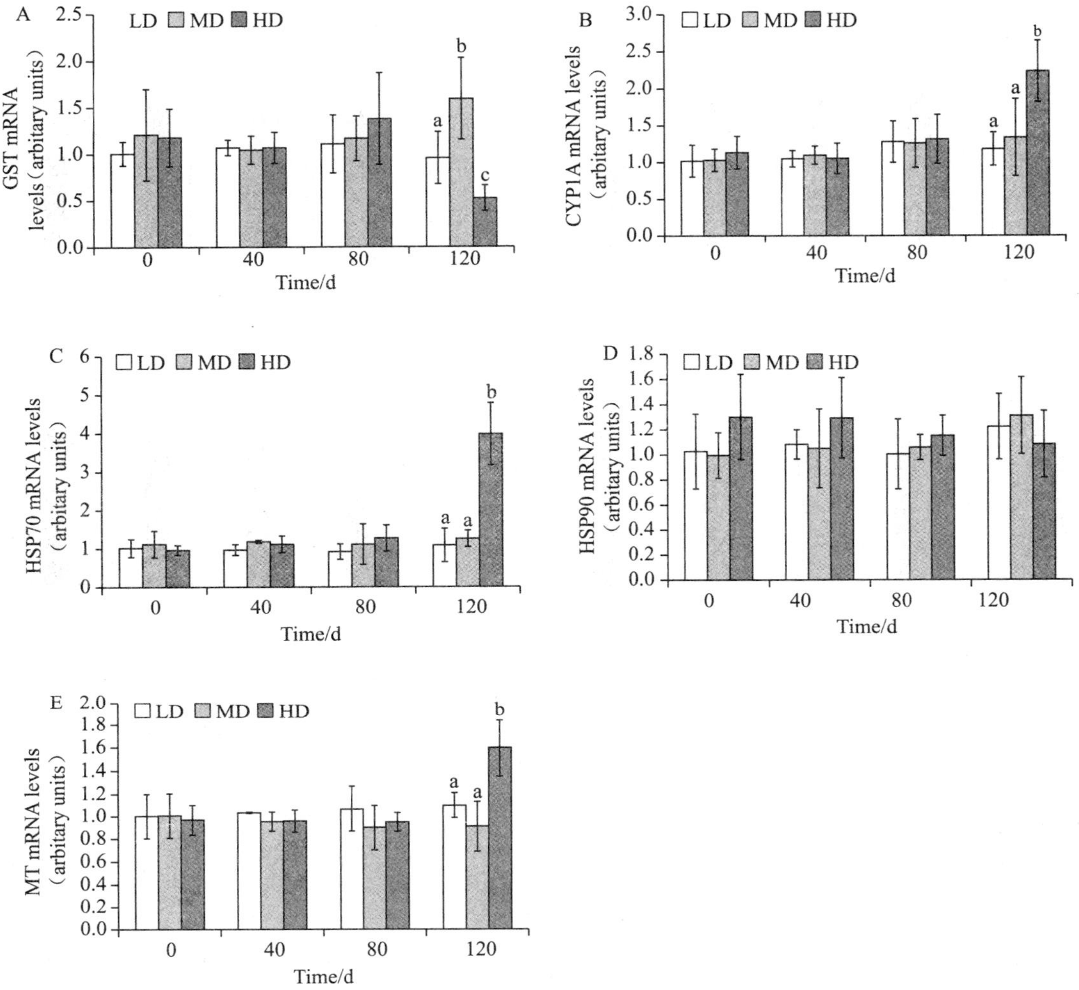

Fig. 4 The effects of different stocking densities on expression of stress-related genes in the liver of juvenile turbot

The values are means ± SEM, n = 20 fish per tank. Different letter denote significant differences between densities within sampling day. LD, low density; MD, medium density; HD, high density

Vargas-Chacoff et al, 2014). The results of this study showed that mean weight and SGR of turbot at the HD were significantly lower than at MD and LD at the end of the trial, which indicated that growth performance was negatively influenced when the density was above a certain value. Similar results also were obtained in previous studies with different fish species (Andrade et al, 2015; Garcia et al, 2013; Sánchez et al, 2013).

Oxidative stress is particularly relevant in aquaculture, since the oxidative damage of fish tissues is directly associated with animal welfare and the quality of the final product (Halver and Hardy, 2002; Senso et al, 2007). Stress caused by high stocking density can enhance the

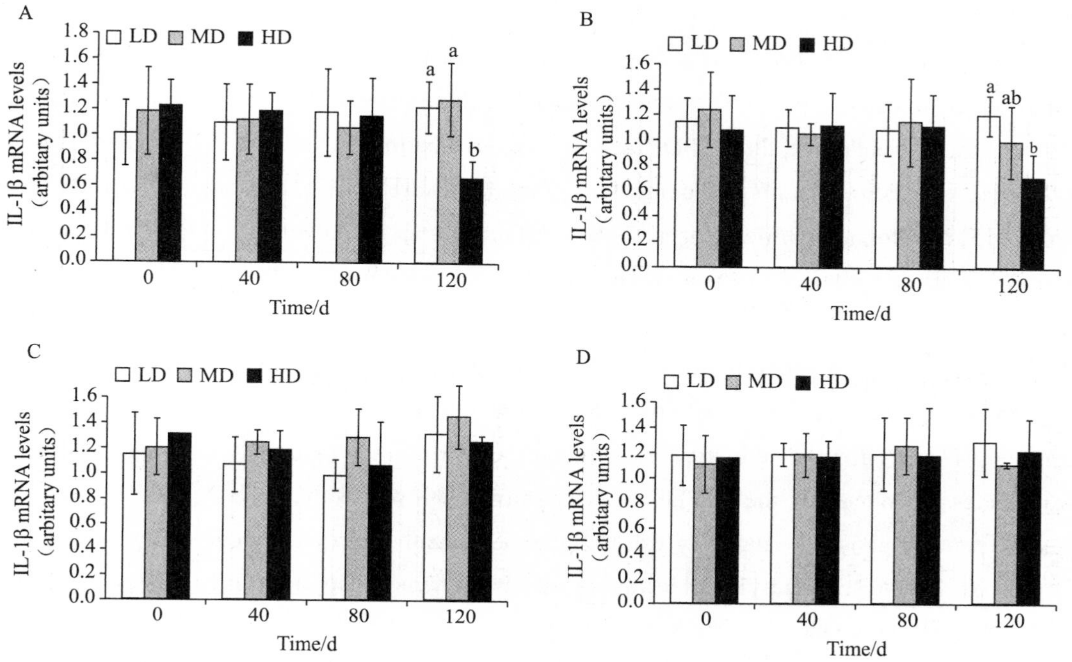

Fig. 5 The effects of different stocking densities on expression of immune-related genes in the liver of juvenile turbot

The values are means ± SEM, *n* = 20 fish per tank. Different letter denote significant differences between densities within sampling day. LD, low density; MD, medium density; HD, high density

intracellular formation of ROS (Ahmad et al, 2000). The ROS is able to attack antioxidant defense system, leading to the loss of antioxidant components (SOD, CAT, GPx and GSH) (Apel and Hirt, 2004; Bopp et al, 2008). In the present study, hepatic SOD, CAT and GSH levels were remarkably depressed in fish held at HD group on day 120. The depression was considered to be a response to the continuous stresses of stocking density, and might reflect the limited abilities for antioxidant systems in turbot to wholly remove these harmful superoxide radicals, finally leading to oxidative damage (Andrade et al, 2015; Braun et al, 2010; Costas et al, 2013). Lipid peroxidation is considered a valuable biomarker of oxidative stress and usually is reflected by level of MDA, a decomposition product of lipid hydroperoxides (Draper and Hadley, 1989). Some studies have showed that stress such as high stocking density in fish disturb the balance between the production of ROS and the antioxidant systems, inducing lipid peroxidation (Andrade et al, 2015; Sahin et al, 2014). However, results from this study did not show differences in MDA formation after 120 days. This finding suggested that high stocking density (up to 52.3 kg/cm^2 ± 0.19 kg/m^2) might not induce lipid peroxidation in turbot, in line with that previously observed in African Catfish (*Clarias gariepinus*) (Wang et al, 2013).

It is well established that high stocking density produces stress responses in fish, which affects different metabolic enzymes related to lipid, carbohydrate and protein metabolism (Menezes et al, 2015). G3PDH is an interesting enzyme in lipid metabolism, which catalyzes the oxidation of glycerol-3-phosphate into dihydroxyacetone phosphate (Van Weerd and Komen, 1998). In the current work, the G3PDH activity displayed a lowest value in HD group at the end of the trial, suggesting that high stocking density might disturb lipid metabolism in liver. Menezes et al (2015) also found significant decrease for G3PDH activity in silver catfish (*Rhamdia quelen*) rearing at high density. G6PDH is a crucial metabolic enzyme involved in the pentose phosphate pathway. The primary products of pentose phosphate pathway are NADPH and ribose 5-phosphate, which are essential for reductive biosynthesis and nucleic acid synthesis (Boulekbache, 1981). Previous study reported that G6PDH activity was stimulated under high stocking density, supplying NADH for the reductive biosynthesis of fatty acids (Vargas-Chacoff et al, 2014). Nevertheless, in our study the G6PDH activity showed a lowest value in HD group on day 120, thus the pentose phosphate pathway was depressed. In rainbow trout, high stocking density also suppressed the G6PDH activity (Aksakal et al, 2011), which supported our suggestion. Some reports showed increases in liver HK and FBPase activities at high density (Menezes et al, 2015; Vargas-Chacoff et al, 2014; Vijayan et al, 1990; Vijayan et al, 1997), but in this study both the enzymes did not show significant differences between treatments.

Physiological and cellular stress response is facilitated by the action of various stress proteins including GST, CYP1A, HSPs and MT. GST is an important enzyme for the biotransformation of xenobiotics and degradation of endogenous compounds (Oshima, 1995). It is involved in the oxidative stress pathway and catalyzes conjugation GSH with nucleophile xenobiotics or ROS-impaired nuclear component (O'Brien et al, 2000). In the present work, the lowest mRNA levels of GST was observed in HD group on day 120, implying that intense oxidative stress conditions might suppress the GST synthesis due to depletion of GSH (Andersson and Förlin, 1992). Like GST, CYP1A is a marker of metabolism of xenobiotics and indicator of the presence of contaminants (Collier et al, 1995). It also was used to evaluate the physiological pressures and immune response in fish. Gornati et al (2004) and Ni et al (2014) reported that high density caused an increasing expression of CYP 1A gene in Acipenser schrenckii and Dicentrarchus labrax. Similar expression of CYP1A gene was observed in our study confirming that overly high rearing density was a physiological stressor.

HSPs, mainly acting as molecular chaperones, are involved in protein folding, assembly, degradation, and intracellular localization. In fish, the two main HSP families are HSP70 and HSP90, which are induced by various stressors including chemical stress, heat stress and crowding, and participated in the immune response (Kim et al, 2013; Viant et al, 2003; Zhang et al, 2015). Our experimental results showed that the mRNA levels of HSP70, but not of HSP90, increased in HD group on day 120, which were consistent with previous studies in Gadus morhua and Acipenser schrenckii (Caipang et al, 2008; Gornati et al, 2004). The inducible HSP70 is important for allowing cells to cope with acute stressor insults, especially those affecting protein machinery (Boone and Vijayan, 2002). MTs are a group of nonenzymatic proteins with a high binding capacity for metals (Hamilton and Mehrle, 1986). One function of MT is to regulate redox homeostasis and protects against superoxide and hydroxyl radical (Nzengue et al, 2012; Wu et al, 2015). Moreover, the expression of its transcript, generally present at basal levels, is increased after physiological stress (Gagné et al, 2013; Ricketts et al, 2015). Our results demonstrated that MT mRNA level was obviously up-regulated in HD group at the end of trial. The up-regulation was also observed in Dicentrarchus labrax reared at high density, indicating that high stocking density resulted in a stress response and the MT could be useful to monitor fish stress conditions (Gornati et al, 2004).

There is considerable evidence that long-term exposure to stressors like high stocking density can cause immune system disorder, increasing susceptibility to infectious diseases in fish (Costas et al, 2013; Ni et al, 2014). Several immune parameters have been successfully employed to evaluate the immune state of fish subjected to various stresses. LYS and HAMP are molecules that participate in the innate immune response by acting as a first barrier against a bacterial infection (Douglas et al, 2003; Jiménez-Cantizano et al, 2008). It has been reported that mRNA levels of both genes were highly reduced by a high stocking density in Solea senegalensis (Salas-Leiton et al, 2010; Tapia-Paniagua et al, 2014). In line with previous study, LYS and HAMP mRNA levels clearly decreased in turbot held at high density on day 120 in our study, indicating significant immunosuppression. IL-1β and TNF-1α are the two crucial pro-inflammatory cytokines that can lead to the activation of the inflammatory response by regulating the expression of other cytokines (Dan et al, 2013). It has been reported that both IL-1β and TNF-1α were significantly up-regulated following crowding stress (Caipang et al, 2008; Kang et al, 2011). However, our experiment did not show any differences among three stocking densities in the expressions of IL-1β and TNF-1α genes during the whole culture period. This indicated that an inflammatory response did not take place in turbot when subjected to the high density (up to $52.3\ kg/m^2 \pm 0.19\ kg/m^2$).

5 Conclusion

In conclusion, our results indicated that juvenile turbot (180 g ～ 510 g) can be reared in commercial-scale RAS to high density. However, when the stocking density reached up to $52.3\ kg/m^2 \pm 0.19\ kg/m^2$, responses associated to a crowding stress could negatively modify antioxidative status, metabolism and expression of immune-related genes in liver of juvenile turbot. The fish showed no signs of disease but displayed symptoms of immunosuppression. Thus, husbandry procedures in commercial aquaculture should consider these effects while addressing better management strategies. Moreover, this work would provide a reference for choosing an available stocking density of turbot in a commercial land-based RAS.

5.1 Acknowledgments

This work was supported by the National Natural Science Foundation of China (No. 31402315 and 31240012), the Modern Agriculture Industry System Construction of Special Funds (CARS-50-G10), Key R & D program of Jiangsu Province (BE2015328) and the Key Laboratory of Mariculture & Stock Enhancement in North China's Sea, Ministry of Agriculture, P.R. China.

References

[1] Ahmad I, Hamid T, Fatima M, et al. Induction of hepatic antioxidants in freshwater catfish (*Channa punctatus* Bloch) is a biomarker of paper mill effluent exposure[J]. Biochimica et Biophysica Acta (BBA)-General Subjects, 2000, 1523: 37-48.

[2] Aksakal E, Ekinci D, Erdoğan O, et al. Increasing stocking density causes inhibition of metabolic-antioxidant enzymes and elevates mRNA levels of heat shock protein 70 in rainbow trout[J]. Livestock Science, 2011, 141: 69-75.

[3] Aksungur N, Aksungur M, Akbulut B, et al. Effects of stocking density on growth performance, survival and food conversion ratio of turbot (*Psetta maxima*) in the net cages on the southeastern coast of the Black Sea[J]. Turkish J Fish Aquat Sci, 2007, 7: 147-152.

[4] Andersson, T, Förlin, L. Regulation of the cytochrome P450 enzyme system in fish[J]. Aquat Toxicol, 1992, 24: 1-19.

[5] Andrade T, Afonso A, Pérez-Jiménez A, et al. Evaluation of different stocking densities in a Senegalese sole (*Solea senegalensis*) farm: Implications for growth, humoral immune parameters and oxidative status[J]. Aquaculture, 2015, 438: 6-11.

[6] Apel K, Hirt H. Reactive oxygen species: metabolism, oxidative stress, and signal transduction[J]. Annu Rev Plant Biol, 2004, 55: 373-399.

[7] Arends R, Mancera J, Munoz J, et al. The stress response of the gilthead sea bream (*Sparus aurata* L.) to air exposure and confinement[J]. J Endocrinol, 1999, 163: 149-157.

[8] Baer A, Schulz C, Traulsen I, et al. Analysing the growth of turbot (*Psetta maxima*) in a commercial recirculation system with the use of three different growth models[J]. Aquacult Int, 2011, 19: 497-511.

[9] Björnsson B, Steinarsson A, Oddgeirsson M, et al. Optimal stocking density of juvenile Atlantic cod (*Gadus morhua* L.) reared in a land-based farm[J]. Aquaculture, 2012, 356: 342-350.

[10] Bolasina S, Tagawa M, Yamashita Y, et al. Effect of stocking density on growth, digestive enzyme activity and cortisol level in larvae and juveniles of Japanese flounder, *Paralichthys olivaceus*[J]. Aquaculture, 2006, 259: 432-443.

[11] Boone A N, Vijayan M M. Constitutive heat shock protein 70 (HSP70) expression in rainbow trout hepatocytes: effect of heat shock and heavy metal exposure[J]. Comp Biochem Phys C Toxicol Pharmacol, 2002, 132: 223-233.

[12] Bopp S K, Abicht H K, Knauer K. Copper-induced oxidative stress in rainbow trout gill cells[J]. Aquat Toxicol, 2008, 86: 197-204.

[13] Boulekbache, H. Energy metabolism in fish development[J]. Am Zool, 1981, 21: 377-389.

[14] Bradford M M. A rapid and sensitive method for the quantitation of microgram quantities of protein utilizing the principle of protein-dye binding[J]. Anal Biochem, 1976, 72: 248-254.

[15] Braun N, de Lima R L, Baldisserotto B, et al. Growth, biochemical and physiological responses of *Salminus brasiliensis* with different stocking densities and handling[J]. Aquaculture, 2010, 301: 22-30.

[16] Caipang C M A, Brinchmann M F, Berg I. Changes in selected stress and immune-related genes in Atlantic cod, *Gadus morhua*, following overcrowding[J]. Aquacult Res, 2008, 39: 1533-1540.

[17] Castillo-Vargasmachuca S, Ponce-Palafox J, García-Ulloa M, et al. Effect of stocking density on growth performance and yield of subadult Pacific Red Snapper cultured in floating sea cages[J]. N Am J Aquacult, 2012, 74: 413-418.

[18] Collier T K, Anulacion B F, Stein J E, et al. A field evaluation of cytochrome P4501A as a biomarker of contaminant exposure in three species of flatfish. Environ[J]. Toxicol Chem, 1995, 14: 143-152.

[19] Costas B, Aragão C, Dias J. Interactive effects of a high-quality protein diet and high stocking density on the stress response and some innate immune parameters of Senegalese sole *Solea senegalensis*[J]. Fish Physiol Biochem, 2013, 39: 1141-1151.

[20] Costas B, Aragão C, Mancera J M, et al. High stocking density induces crowding stress and affects amino acid metabolism in Senegalese sole *Solea senegalensis* (Kaup 1858) juveniles[J]. Aquacult Res, 2008, 39: 1-9.

[21] Dan X M, Zhang T W, Li Y W, et al. Immune responses and immune-related gene expression profile in orange-spotted grouper after immunization with *Cryptocaryon irritans* vaccine[J]. Fish Shellfish Immunol, 2013, 34: 885-891.

[22] Douglas S E., Gallant J W, Liebscher R S, et al. Identification and expression analysis of hepcidin-like antimicrobial peptides in bony fish[J]. Dev Comp Immunol, 2003, 27: 589-601.

[23] Draper H, Hadley M. Malondialdehyde determination as index of lipid peroxidation[J]. Methods Enzymol, 1989, 186: 421-431.

[24] Ellis T, North B, Scott A, et al. The relationships between stocking density and welfare in farmed rainbow trout[J]. J Fish Biol, 2002, 61: 493-531.

[25] Gagné F, Smyth S, André C, et al. Stress-related gene expression changes in rainbow trout hepatocytes exposed to various municipal wastewater treatment influents and effluents[J]. Environ Sci Pollut R, 2013,

20: 1706-1718.

[26] Garcia F, Romera D M, Gozi K S, et al. Stocking density of Nile tilapia in cages placed in a hydroelectric reservoir[J]. Aquaculture, 2013, 410: 51-56.

[27] Gornati R, Papis E, Rimoldi S, et al. Rearing density influences the expression of stress-related genes in sea bass (*Dicentrarchus labrax,* L.)[J]. Gene, 2004, 341: 111-118.

[28] Halver J E, Hardy R W. Fish nutrition[M].San Diego, CA: Academic Press, 2002.

[29] Hamilton S J, Mehrle P M. Metallothionein in fish: review of its importance in assessing stress from metal contaminants[J]. Trans Am Fish Soc, 1986, 115: 596-609.

[30] Hermann B, Reusch T, Hanel R, 2016. Effects of dietary purified rapeseed protein concentrate on hepatic gene expression in juvenile turbot (*Psetta maxima*)[J]. Aquacult Nutr, 2016, 22: 170-180

[31] Herrera M, Vargas-Chacoff L, Hachero I, et al. Physiological responses of juvenile wedge sole *Dicologoglossa cuneata* (Moreau) to high stocking density[J]. Aquacult Res, 2009, 40: 790-797.

[32] Irwin S, O'halloran J, FitzGerald R. Stocking density, growth and growth variation in juvenile turbot, *Scophthalmus maximus* (Rafinesque)[J]. Aquaculture, 1999, 178: 77-88.

[33] Iwama G K, Pickering A, Sumpter J. Fish stress and health in aquaculture[M]. Londom: Cambridge University Press, 2011.

[34] Jia R, Cao L, Du J, et al. The protective effect of silymarin on the carbon tetrachloride (CCl4)-induced liver injury in common carp (*Cyprinus carpio*)[J]. In Vitro Cell Dev An, 2013, 49: 155-161.

[35] Jiménez-Cantizano R M, Infante C, Martin-Antonio B, et al. Molecular characterization, phylogeny, and expression of c-type and g-type lysozymes in brill (*Scophthalmus rhombus*)[J]. Fish Shellfish Immunol, 2008, 25: 57-65.

[36] Kang S Y, Ko Y H., Moon Y S, et al. Effects of the combined stress induced by stocking density and feed restriction on hematological and cytokine parameters as stress indicators in laying hens[J]. Asian-Australas J Anim Sci, 2011, 24: 414-420.

[37] Kim J H, Dahms H U, Han K N. Biomonitoring of the river pufferfish, *Takifugu obscurus* in aquaculture at different rearing densities using stress-related genes[J]. Aquacult Res, 2013, 44: 1835-1846.

[38] Li X, Liu Y, Blancheton J P. Effect of stocking density on performances of juvenile turbot (*Scophthalmus maximus*) in recirculating aquaculture systems[J]. Chin J Oceanol Limnol, 2013, 31: 514-522.

[39] Li Y, Li J, Wang Q. The effects of dissolved oxygen concentration and stocking density on growth and non-specific immunity factors in Chinese shrimp, *Fenneropenaeus chinensis*[J]. Aquaculture, 2006, 256: 608-616.

[40] Livak K J, Schmittgen T D. Analysis of relative gene expression data using real-time quantitative PCR and the $2^{-\Delta\Delta C_T}$ method[J]. Methods, 2001, 25: 402-408.

[41] Lushchak V I. Environmentally induced oxidative stress in aquatic animals[J]. Aquat Toxicol, 2011, 101: 13-30.

[42] Magnadóttir B. Innate immunity of fish (overview)[J]. Fish Shellfish Immunol, 2006, 20: 137-151.

[43] Menezes C, Ruiz-Jarabo I, Martos-Sitcha J A, et al. The influence of stocking density and food deprivation

in silver catfish (*Rhamdia quelen*): A metabolic and endocrine approach[J]. Aquaculture, 2015, 435: 257-264.

[44] Montero D, Izquierdo M, Tort L, et al. High stocking density produces crowding stress altering some physiological and biochemical parameters in gilthead seabream, *Sparus aurata*, juveniles[J]. Fish Physiol Biochem, 1999, 20: 53-60.

[45] Muñoz-Atienza E, Araújo C, Magadán S, et al. *In vitro* and *in vivo* evaluation of lactic acid bacteria of aquatic origin as probiotics for turbot (*Scophthalmus maximus* L.) farming[J]. Fish Shellfish Immunol, 2014, 41: 570-580.

[46] Ni M, Wen H, Li J. The physiological performance and immune responses of juvenile Amur sturgeon (*Acipenser schrenckii*) to stocking density and hypoxia stress[J]. Fish Shellfish Immunol, 2014, 36: 325-335.

[47] Nzengue Y, Steiman R, Rachidi W, et al. Oxidative stress induced by cadmium in the C6 cell line: role of copper and zinc[J]. Biol Trace Elem Res, 2012, 146: 410-419.

[48] O'Brien M, Kruh G D, Tew K D. The influence of coordinate overexpression of glutathione phase II detoxification gene products on drug resistance[J]. J Pharmacol Exp Ther, 2000, 294: 480-487.

[49] Oshima Y. Chemical and enzymatic transformation of paralytic shellfish toxins in marine organisms[J]. Harmful Marine Algal Blooms, 1995: 475-480.

[50] Reiser S, Wuertz S, Schroeder J, et al. Risks of seawater ozonation in recirculation aquaculture: Effects of oxidative stress on animal welfare of juvenile turbot (*Psetta maxima*, L.)[J]. Aquat Toxicol, 2011, 105: 508-517.

[51] Ricketts C D, Bates W R, Reid S D. The effects of acute waterborne exposure to sub-lethal concentrations of molybdenum on the stress response in rainbow trout (*Oncorhynchus mykiss*)[J]. PloS one, 2015, 10: e0115334.

[52] Sánchez: P, Ambrosio P P, Flos R. Stocking density affects Senegalese sole (*Solea senegalensis*, Kaup) growth independently of size dispersion, evaluated using an individual photo-identification technique[J]. Aquacult Res, 2013, 44: 231-241.

[53] Sadhu N, Sharma S K, Joseph S, et al. Chronic stress due to high stocking density in open sea cage farming induces variation in biochemical and immunological functions in Asian seabass (*Lates calcarifer*, Bloch)[J]. Fish Physiol Biochem, 2014, 40: 1105-1113.

[54] Sahin K, Yazlak H, Orhan C, et al. The effect of lycopene on antioxidant status in rainbow trout (*Oncorhynchus mykiss*) reared under high stocking density[J]. Aquaculture, 2014, 418: 132-138.

[55] Salas-Leiton E, Anguis V, Martín-Antonio B, et al. Effects of stocking density and feed ration on growth and gene expression in the Senegalese sole (*Solea senegalensis*): potential effects on the immune response[J]. Fish Shellfish Immunol, 2010, 28: 296-302.

[56] Sayeed I, Parvez S, Pandey S, et al. Oxidative stress biomarkers of exposure to deltamethrin in freshwater fish, *Channa punctatus* Bloch[J]. Ecotoxicol Environ Saf, 2013, 56: 295-301.

[57] Senso L, Suárez M, Ruiz-Cara T, et al. On the possible effects of harvesting season and chilled storage on

the fatty acid profile of the fillet of farmed gilthead sea bream (*Sparus aurata*) [J]. Food Chem, 2007, 101: 298-307.

[58] Tapia-Paniagua S, Vidal S, Lobo C, et al. The treatment with the probiotic Shewanella putrefaciens Pdp11 of specimens of *Solea senegalensis* exposed to high stocking densities to enhance their resistance to disease [J]. Fish Shellfish Immunol, 2014, 41: 209-221.

[59] Telli G S, Ranzani-Paiva M J T, de Carla Dias D, et al. Dietary administration of Bacillus subtilis on hematology and non-specific immunity of Nile tilapia *Oreochromis niloticus* raised at different stocking densities [J]. Fish Shellfish Immunol, 2014, 39: 305-311.

[60] Van Weerd J, Komen J. The effects of chronic stress on growth in fish: a critical appraisal [J]. Comp Biochem Physiol A Mol Integr Physiol, 1998, 120: 107-112.

[61] Vargas-Chacoff L, Martínez D, Oyarzún R, et al. Combined effects of high stocking density and *Piscirickettsia salmonis* treatment on the immune system, metabolism and osmoregulatory responses of the Sub-Antarctic Notothenioid fish *Eleginops maclovinus* [J]. Fish Shellfish Immunol, 2014, 40: 424-434.

[62] Viant M, Werner I, Rosenblum, E et al. Correlation between heat-shock protein induction and reduced metabolic condition in juvenile steelhead trout (*Oncorhynchus mykiss*) chronically exposed to elevated temperature [J]. Fish Physiol. Biochem, 2003, 29: 159-171.

[63] Vijayan M, Ballantyne J, Leatherland J. High stocking density alters the energy metabolism of brook charr, *Salvelinus fontinalis* [J]. Aquaculture, 1990, 88: 371-381.

[64] Vijayan M M, Pereira C, Grau E G, et al. Metabolic responses associated with confinement stress in tilapia: the role of cortisol [J]. Comp Biochem Phys C Pharmacol Toxicol Endocrinol, 1997, 116: 89-95.

[65] Wang C, Zhang X H, Jia A, et al. Identification of immune-related genes from kidney and spleen of turbot, *Psetta maxima* (L.), by suppression subtractive hybridization following challenge with *Vibrio harveyi* [J]. J Fish Dis, 2008, 31: 505-514.

[66] Wang X, Dai W, Xu M, et al. Effects of stocking density on growth, nonspecific immune response, and antioxidant status in African catfish (*Clarias gariepinus*) [J]. The Israeli Journal of Aquaculture, 2013, 65: 1-6.

[67] Wu S M, Liu J H, Shu L H, et al. Anti-oxidative responses of zebrafish (*Danio rerio*) gill, liver and brain tissue upon acute cold shock [J]. Comp Biochem Physiol A Mol Integr Physiol, 2015, 187: 202-213.

[68] Yarahmadi P, Miandare H K, Hoseinifar S H, et al. The effects of stocking density on hemato-immunological and serum biochemical parameters of rainbow trout (*Oncorhynchus mykiss*) [J]. Aquacult Int, 2014, 23: 55-63.

[69] Zapata A, Diez B, Cejalvo T, et al. Ontogeny of the immune system of fish [J]. Fish Shellfish Immunol, 2006, 20: 126-136.

[70] Zhang C N, Li X F, Tian H Y, et al. Effects of fructooligosaccharide on immune response, antioxidant capability and HSP70 and HSP90 expressions of blunt snout bream (*Megalobrama amblycephala*) under high ammonia stress [J]. Fish Physiol Biochem, 2015, 41: 203-217.

Dietary arachidonic acid differentially regulates the gonadal steroidogenesis in the marine teleost, tongue sole (*Cynoglossus semilaevis*), depending on fish gender and maturation stage

Houguo Xu[1], Lin Cao[1], Yuanqin Zhang[1],
Ronald B. Johnson[2], Yuliang Wei[1], Keke Zheng[1], Mengqing Liang[1,3]

[1]Yellow Sea Fisheries Research Institute, Chinese Academy of Fishery Sciences, 106 Nanjing Road, Qingdao 266071, China
[2]Environmental and Fisheries Science Division, Northwest Fisheries Science Center, National Marine Fisheries Service, 2725 Montlake Blvd E., Seattle 98112, USA
[3]Laboratory for Marine Fisheries Science and Food Production Processes, Qingdao National Laboratory for Marine Science and Technology, 1 Wenhai Road, Qingdao 266237, China

Abstract A 3-month feeding trial with tongue sole *Cynoglossus semilaevis* broodstock was conducted before and during the spawning season to investigate the effects of dietary arachidonic acid (ARA) on the production of sex steroid hormones and gonadal gene expression of key proteins in steroidogenesis. Three isonitrogenous and isolipidic experimental diets were formulated to contain different ARA levels: the control diet without ARA supplementation (C, 0.58% ARA of total fatty acids (TFA)) and two diets with low (5.14% of TFA, ARA-L) or high ARA (15.44% of TFA, ARA-H) supplementation. The diets were randomLy assigned to 9 tanks of 3-year-old tongue sole (10 females and 15 males in each tank). Fish were reared in a flowing seawater system and fed to apparent satiation twice daily. At the end of the feeding trial, tissue samples from mature females (MF, with spontaneous ovulation), immature females (IMF, early vitellogenesis), and mature males (Mm, expressing milt) were collected to assay the production of sex steroid hormones, gonadal gene expression of sex steroid-synthesizing proteins, as well as the fatty acid profiles of gonad, liver and muscle lipids. Results showed that ARA supplementation significantly reduced the estradiol production in females, but stimulated the testosterone production in males. ARA supplementation significantly reduced the mRNA expression of aromatase in ovaries but significantly increased the gene expression of 3β-hydroxysteroid dehydrogenase (3β-HSD) in testes. In mature ovaries, diet ARA-L significantly reduced the gene expression of follicle stimulating hormone receptor (FSHR), 3β-HSD, and 17β-HSD; however, in immature ovaries, it significantly increased the gene expression of FSHR, steroidogenic acute regulatory protein (StAR), cholesterol side-chain cleavage enzyme (P450ssc), 3β-HSD, and 17β-HSD. In all gonads, 17α-hydroxylase (P450c17) responded to dietary ARA differently from other sex steroid-synthesizing proteins. ARA was preferentially accumulated in tongue sole gonad lipids. ARA concentrations were highest in gonad, liver and muscle lipids of Mm fish and lowest in MF fish. Compared to female tongue sole, males had higher DHA concentrations in gonad lipids, but lower concentrations in liver lipids. In conclusion, results suggest dietary ARA regulates sex steroid hormone synthesis in tongue sole broodstock, and accumulates in gonad lipids, depending on both fish gender and maturation stage. Dietary ARA supplementation appears more important for male fish than for female fish, and more important for immature females than for mature females.

Keywords *Cynoglossus semilaevis*; diet; 20:4n-6; sex steroid hormone synthesis

1 Introduction

Lipid and fatty acid compositions of broodstock diets have been identified as major dietary factors that determine successful reproduction and survival of offspring in fish (Izquierdo et al, 2001). The important roles of n-3 long chain-polyunsaturated fatty acids (LC-PUFA) in fish reproductive processes have been demonstrated in a number of studies (Harel et al, 1994; Cerdá et al, 1995; Abi-Ayad et al, 1997; Navas et al, 1997; Almansa et al, 2001; Watanabe and Vassallo-Agius, 2003; Rodríguez-Barreto et al, 2012; Beirão et al, 2015; Butts et al, 2015; Luo et al, 2015); however, the importance of n-6 LC-PUFA, primarily arachidonic acid (ARA), has been relatively neglected. In the past decade, the functions of ARA in fish reproduction have been gaining more and more attention and it has been shown that moderate levels of ARA in broodstock diets exert significant positive effects on the spawning performance, egg quality, and offspring quality of several fish species including European sea bass *Dicentrarchus labrax* (Bruce et al, 1999), Japanese flounder *Paralichthys olivaceus* (Furuita et al, 2003), Atlantic halibut *Hippoglossus hippoglossus* (Mazorra et al, 2003), and rice field eel *Monopterus albus* (Zhou et al, 2011). The analysis of fatty acid profiles in gonads and gametes, as well as the dynamics of fatty acids during fish maturation and embryogenesis, additionally suggests the importance of ARA in fish reproduction (Grigorakis et al, 2002; Salze et al, 2005; Rodríguez-Barreto et al, 2012; Johnson, 2012; Støttrup et al, 2013; Hauville et al, 2015).

To date, however, little information has been available on the mechanisms involved in the effects of dietary ARA on fish reproductive processes. The limited work suggests that dietary ARA modulates the oocyte maturation, ovulation, and the binding of sex steroids in plasma (Van Der Kraak and Biddiscombe, 1999; Sorbera et al, 2001; Patiño et al, 2003). Very few studies have been conducted to investigate the effects of dietary ARA on the synthesis of sex steroid hormones (Norambuena et al, 2013a), which is a very important process in fish reproduction. In mammals and birds, ARA has been shown to influence gonadal steroidogenesis (Johnson et al, 1991; Lopez-Ruiz et al, 1992). Some *in vitro* studies have demonstrated the modulation of testosterone synthesis by ARA, but the effects of ARA on estradiol synthesis is still lacking. Moreover, the *in vitro* studies have a limited ability to elucidate the real *in vivo* reproductive processes. Therefore, the present study was aimed to investigate the modulation of fish sex steroid hormone synthesis by dietary ARA through a broodstock feeding trial. At the same time, this study evaluates the effects of dietary ARA on the gene expression of key proteins involved in the key processes of sex steroid synthesis, i.e., the endocrine response to gonadotrophins, the delivery of cholesterol substrate, and biosynthetic reactions. These physiological processes play key roles in sex steroid synthesis, gonad maturation, and spawning performance; however, little information is available regarding their modulation by dietary nutrients.

This study was conducted on tongue sole (*Cynoglossus semilaevis*), an important aquaculture species in China. The demand for tongue sole has increased in the past decade; however, the unpredictable and variable reproductive performance of the species was a limiting factor to the successful mass production of juveniles. In recent years, our laboratory has investigated the variation in reproductive performance of this fish by altering dietary lipids (Xu et al, 2015) and fatty acid profile, including the dietary n-3/n-6 fatty acid ratio (Liang et al, 2014). As a following-up study, the present study is aimed at investigating the ability of dietary ARA, the primary n-6 LC-PUFA in fish gonads, to regulate sex steroid synthesis in tongue sole, as well as at elucidating the involved physiological processes. Along with evaluating changes in fatty acid profiles of gonad and other tissue lipids in response to experimental diets, this study provides needed information on the role of dietary fatty acids in tongue sole reproductive success.

2 Materials and methods

2.1 Experimental diets

Three isonitrogenous and isolipidic (55.5% crude protein and 13.0% crude lipid) experimental diets were formulated to contain different levels of ARA (Table 1). The control diet (C) was formulated using fish meal, casein, and wheat meal as protein sources, and soy lecithin, olive oil, and tristearin as lipid sources. An ARA enriched oil (ARA concentration, 41% of total fatty acids (TFA); in the form of triglyceride; Jiangsu Tiankai Biotechnology Co., Ltd., Nanjing, China) was supplemented to the control diet, replacing tristearin, to formulate two diets with

Table 1 Formulation and proximate composition of the experiment diets (g/kg dry matter)

Ingredient	C	ARA-L	ARA-H
Fish meal	600.0	600.0	600.0
Casein	120.0	120.0	120.0
Wheat meal	156.2	156.2	156.2
Vitamin premix[1]	10.0	10.0	10.0
Mineral premix[2]	10.0	10.0	10.0
Monocalcium phosphate	10.0	10.0	10.0
Choline chloride	5.0	5.0	5.0
L-ascorbyl-2-polyphosphate	5.0	5.0	5.0
Mold inhibitor[3]	1.0	1.0	1.0
Ethoxyquin	0.5	0.5	0.5
Soy lecithin	15.0	15.0	15.0
Olive oil	10.0	10.0	10.0
n-3 LC-PUFA enriched oil[4]	13.0	13.0	13.0
ARA enriched oil[5]	0.0	14.3	44.3
Tristearin[6]	44.3	30.0	0.0
Proximate composition			
Crude protein	558.1	556.4	559.0
Crude lipid	128.7	129.8	130.1
Ash	127.9	128.5	129.3

[1]Vitamin premix (mg or g/kg diet): thiamin 25 mg; riboflavin, 45 mg; pyridoxine HCl, 20 mg; vitamin B_{12}, 0.1 mg; vitamin K_3, 10 mg; inositol, 800 mg; pantothenic acid, 60 mg; niacin, 200 mg; folic acid, 20 mg; biotin, 1.2 mg; retinol acetate, 32 mg; cholecalciferol, 5 mg; alpha-tocopherol, 120 mg; wheat middling, 13.67 g

[2]Mineral premix (mg or g/kg diet): $MgSO_4 \cdot 7H_2O$, 1 200 mg; $CuSO_4 \cdot 5H_2O$, 10 mg; $ZnSO_4 \cdot H_2O$, 50 mg; $FeSO_4 \cdot H_2O$, 80 mg; $MnSO_4 \cdot H_2O$, 45 mg; $CoCl_2 \cdot 6H_2O$ (1%), 50 mg; $NaSeSO_3 \cdot 5H_2O$ (1%), 20 mg; $Ca(IO_3)_2 \cdot 6H_2O$ (1%), 60 mg; zoelite, 13.485 g

[3]Mold inhibitor: contained 50% calcium propionic acid and 50% fumaric acid

[4]n-3 LC-PUFA enriched oil: containing 37% DHA and 21% EPA (of total fatty acids); in the form of triglyceride; Hebei Haiyuan Health Biological Science and Technology Co., Ltd., Changzhou, China

[5]ARA enriched oil: containing 41% ARA (of total fatty acids); in the form of triglyceride; Jiangsu Tiankai Biotechnology Co., Ltd., Nanjing, China

[6]Tristearin: HUDONG Daily Chemicals Co., Ltd., Jiaxing, China

low (ARA-L, with 1.43% (dry matter) ARA enriched oil) or high (ARA-H, with 4.43% ARA enriched oil) ARA content. A constant level of n-3 LC-PUFA enriched oil (containing 37% DHA and 21% EPA (of TFA); in the form of triglyceride; Hebei Haiyuan Health Biological Science and Technology Co, Ltd., Changzhou, China) was supplemented to all the diets to meet the n-3 LC-PUFA requirement. The diets were made, packed and stored following the common procedures in our laboratory (Xu et al, 2016). The fatty acid compositions of the experimental diets are presented in Table 2. The ARA content in C, ARA-L, and ARA-H was 0.58%, 5.14% and 15.44% of TFA, respectively.

Table 2　Fatty acid compositions of the experimental diets (% total fatty acids)

Fatty acid	C	ARA-L	ARA-H
C14:0	3.5	3.5	3.1
C16:0	24.8	22.0	17.4
C18:0	25.1	19.2	6.8
ΣSFA	53.4	44.6	27.3
C16:1$n-7$	2.4	2.7	2.7
C18:1$n-9$	10.9	12.3	14.7
C20:1$n-9$	0.5	0.5	0.6
ΣMUFA	13.8	15.5	18.0
C18:2$n-6$	5.2	6.0	7.9
C20:4$n-6$	0.6	5.1	15.4
Σ$n-6$ PUFA	5.8	11.1	23.4
C18:3$n-3$	0.8	0.7	0.8
C20:5$n-3$	3.8	4.3	4.1
C22:6$n-3$	8.8	8.7	8.9
Σ$n-3$ PUFA	13.3	13.7	13.8
Σ$n-3$/Σ$n-6$	2.3	1.2	0.6

SFA: saturated fatty acid; MUFA: monounsaturated fatty acid; PUFA: polyunsaturated fatty acid

2.2 Experimental fish and feeding procedure

Three-year-old tongue sole *Cynoglossus semilaevis* broodstock, which have been reared with formulated feeds from the early juvenile stage, were used in the present study. Prior to the start of the feeding trial, experimental fish were reared in concrete tanks (25 m^3) and fed the control diet for 7 d to acclimate to the experimental conditions. At the onset of the feeding trial, experimental fish were distributed into 9 polyethylene tanks (diameter: 230 cm, height: 100 cm) and each diet was randomLy assigned to triplicate tanks. Each tank had 25 fish (10 females and 15 males) and the tanks were supplied with flowing filtered seawater at a rate of 50 L/min. Fish were hand-fed to apparent satiation twice daily. The feeding trial lasted for 90 d, from July to October, which is the natural spawning season of tongue sole. Fish were reared under the natural photoperiod and ambient temperature of Haiyang, Shandong, China (N36°41′, E121°07′). During the experiment, the water temperature ranged from 22 ℃～26 ℃ ; salinity, 30～31; pH, 7.4～8.5; and dissolved oxygen, 5 mg/L～7 mg/L. The ponds were cleaned daily by siphoning out residual feed and feces.

2.3 Sampling

At the end of the feeding trial, serum, liver, muscle, and gonad samples from 3 mature females (MF), 2 immature females (IMF), and 3 mature males (MM) per tank were collected. The maturity of the females was confirmed when spontaneous ovulation was observed. Since it was difficult to precisely determine the maturation stages of immature females from visual observation, 4 immature female fish from each tank, with similar estimated maturation stage, were collected and sacrificed. Microscopic examination of oocyte morphology was then used to precisely determine the stage of ovarian maturation. Microscopic examination showed that the majority of immature ovaries were in the stage of early vitellogenesis. Ovary samples from 2 fish per tank, which showed the most typical characteristics of early vitellogenesis, i.e., the centralized appearance of spherical, eosinophilic, and vitellogenic yolk granules/globules in oocyte cells, were collected to represent IMF fish. The maturity of male fish was confirmed by the release of milt when handled. Since very few males did not release milt when handled, no immature testis samples were collected.

After being anesthetized with eugenol (1:10 000), blood was drawn from all fish via the caudal vein to collect serum samples. Fish were then dissected to collect liver, muscle and gonad samples. All samples were frozen with liquid nitrogen immediately, and then stored at －80 ℃ prior to analysis.

All sampling protocols, as well as fish rearing practices, were reviewed and approved by the animal care and use committee of the Yellow Sea Fisheries Research Institute.

2.4 Proximate composition analysis and fatty acids analysis

The proximate composition analyses of experimental diets were performed in accordance with the standard methods of AOAC (AOAC, 2000). Moisture content of feed samples was determined by drying to a constant weight at 105 ℃. Protein content was estimated by measuring nitrogen content via the Kjeldahl method and multiplying by a factor of 6.25. Lipid content was determined via Soxhlet extraction using petroleum ether as the solvent. Ash content was determined by combustion at 550 ℃.

The fatty acid compositions of diet and fish tissue lipids were analyzed via a gas chromatograph, using a flame ionization detector (FID). Fatty acids in freeze-dried samples were esterified first with KOH-methanol and then with HCL-methanol, on 72 ℃ water bath. Fatty acid methyl esters were extracted with hexane and then separated via gas chromatography (HP6890,

Agilent Technologies Inc., Santa Clara, California, USA) with a fused silica capillary column (007-CW, Hewlett Packard, Palo Alto, CA, USA). The column temperature was programmed to rise from 150 ℃ up to 200 ℃ at a rate of 15 ℃/min, and then from 200 ℃ to 250 ℃ at a rate of 2 ℃/min. Both the injector and detector temperatures were 250 ℃ . Results are expressed as the percentage of each fatty acid with respect to total fatty acids.

2.5 Quantitative real-time polymerase chain reaction (qRT-PCR) analysis and the analysis of estradiol and testosterone in serum

Total RNA in gonads was extracted using RNAiso Plus (TaKaRa Biotechnology (Dalian) Co., Ltd., Dalian, China) and reverse transcribed with PrimeScript™ RT reagent Kit with gDNA Eraser (Perfect Real Time) (TaKaRa Biotechnology (Dalian) Co., Ltd., Dalian, China) according to the user manual.

Specific primers for key proteins in steroidogenesis and the reference gene β-actin were designed based on the sequences available in the GenBank database and synthesized by Sangon Biotech (Shanghai, China) (Table 3). The real-time PCR was carried out with SYBR Green Real-time PCR Master Mix (TaKaRa Biotechnology (Dalian) Co., Ltd., Dalian, China) in a quantitative thermal cycler (Mastercycler eprealplex, Eppendorf, German). The detailed program was similar to Xu et al (2014). The mRNA expression levels were studied by qRT-PCR method: $2^{-\Delta\Delta C_T}$ (Livak and Schmittgen, 2001).

Serum estradiol and testosterone levels were determined using an electrochemiluminescence method with a Roche COBAS-6000 automatic electrochemiluminescence immunoassay analyzer located in the affiliated hospital of Qingdao University (Qingdao, China).

Table 3　Sequences of the primers used in this work

Primer	Sequence (5′–3′)
FSHR-F	AAGATCAAGGGAAAACGCTA
FSHR-R	CTCAGATGGTTGGAGGAAAG
StAR-F	ACCTCGTGGGTGACCATCGTGT
StAR-R	AGGACGGCTGGACCACTGAAAT
P450ssc-F	TTCTGTGCTGTATGGCGAAC
P450ssc-R	CTTTTGACCCAATCCGTCTC
P450c17-F	GCCCACTCGCTCCCTACATACT
P450c17-R	GTCTTTCCCATCTCGGGTCAG
3β-HSD-F	CACCACTGGGTAAGCACTATC
3β-HSD-R	AGGTTATCGCAAACAGCATT

（continue）

Primer	Sequence (5′–3′)
17β-HSD-F	AATGTGCAGGCTCTAACTGCTTC
17β-HSD-R	AGGTTCCTCATGGTGGCGTA
Aromatase-F	TGCGATTTCAGCCCGT
Aromatase-R	TGCGACCCGTGTTCAGA
β-actin-F	GCTGTGCTGTCCCTGTA
β-actin-R	GAGTAGCCACGCTCTGTC

2.6 Statistical methods

All data were subjected to one-way analysis of variance (ANOVA) in SPSS 16.0 for Windows. All percentage data were arcsine transformed prior to analysis. Significant differences between the means were detected by Tukey's multiple range test. The level of significance was chosen at $P < 0.05$ and the results are presented as means ± standard error.

3 Results

3.1 Concentrations of estradiol and testosterone

In both MF and IMF fish, serum estradiol concentrations were significantly ($P < 0.05$) lower in groups ARA-L and ARA-H than in group C (Fig. 1A and 1B). However, in Mm fish, the serum testosterone concentration of group ARA-H was significantly ($P < 0.05$) higher than that of group C, with no significant difference observed either between groups ARA-H and ARA-L or between groups C and ARA-L (Fig. 1C).

3.2 Gonadal mRNA expressions of sex steroid-synthesizing proteins

In immature ovaries, the mRNA expressions of follicle stimulating hormone receptor (FSHR), steroidogenic acute regulatory protein (StAR), cholesterol side-chain cleavage enzyme (P450ssc), and 3β-hydroxysteroid dehydrogenase (3β-HSD) were significantly higher ($P < 0.01$) in group ARA-L than in groups C and ARA-H, but no significant differences were observed between groups C and ARA-H (Fig. 2A). In contrast, the mRNA expressions of 17α-hydroxylase (P450c17) and aromatase were significantly lower ($P < 0.01$) in groups ARA-L and ARA-H compared to group C. The mRNA expression of 17β-HSD in group ARA-L was significantly higher ($P < 0.05$) than that in group ARA-H while no significant difference ($P > 0.05$) was observed either between groups C and ARA-L or between groups C and ARA-H.

In mature ovaries, groups ARA-L and ARA-H showed significantly lower mRNA expressions of FSHR ($P < 0.01$) and aromatase ($P < 0.05$) than group C (Fig. 2B). The mRNA

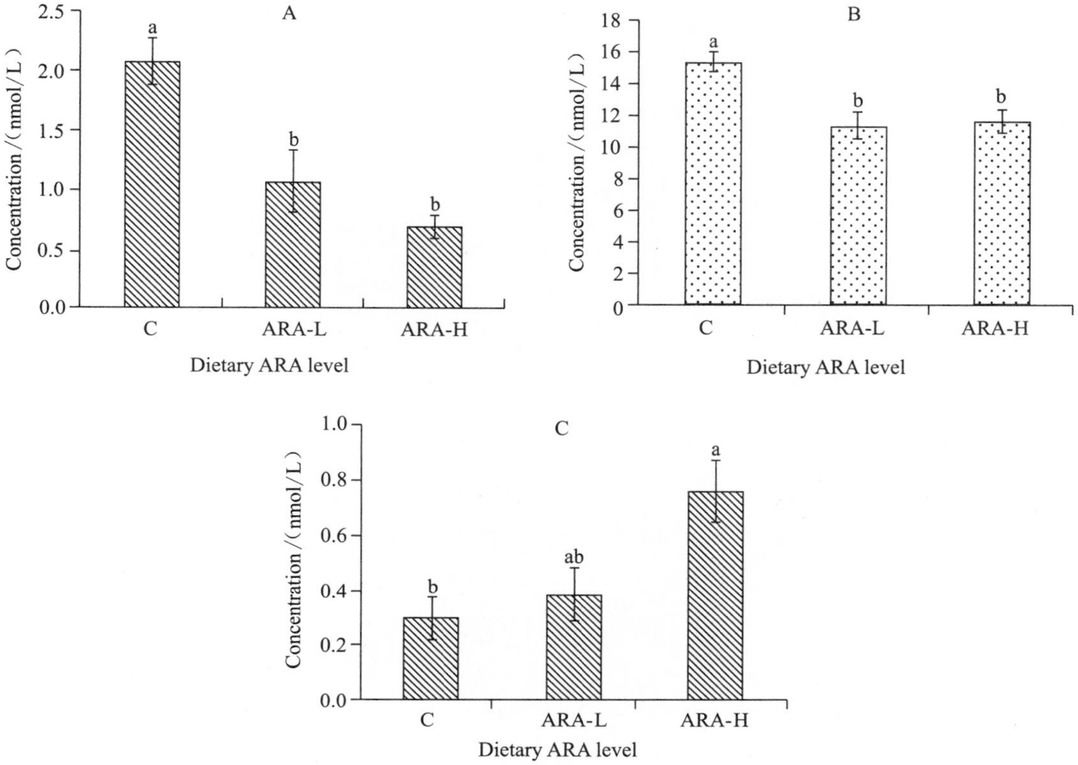

Fig. 1　Effects of dietary ARA on the concentrations of sex steroid hormones in serum of tongue sole broodstock

A, estradiol in immature females; B, estradiol in mature females; C, testosterone in mature males. Values (means ± standard error, $n = 3$) in bars that do not have the same letter are significantly different ($P < 0.05$)

expressions of 3β-HSD ($P < 0.05$) and 17β-HSD ($P < 0.01$) in group ARA-L were significantly lower than that of group C, but no significant differences were observed between groups C and ARA-H. No significant differences in the gene expression of StAR, P450ssc or P450c17 were observed between the groups ($P > 0.05$).

In mature testes, group ARA-H showed significantly higher ($P < 0.01$) mRNA expression of P450ssc than groups ARA-L and C while group ARA-L showed significantly lower ($P < 0.01$) mRNA expression of P450c17 than groups C and ARA-H (Fig. 2C). The mRNA expression of 3β-HSD significantly ($P < 0.01$) increased with increasing dietary ARA. No significant differences in StAR or 17β-HSD expression were observed between the groups ($P > 0.05$).

3.3　Tissue fatty acid profiles

The fatty acid profiles of gonad, liver, and muscle lipids differed in response to the experimental diets, and are presented in Table 4, 5 and 6, respectively. The ARA concentrations in all fish tissue lipids significantly increased with increasing dietary ARA ($P < 0.01$).

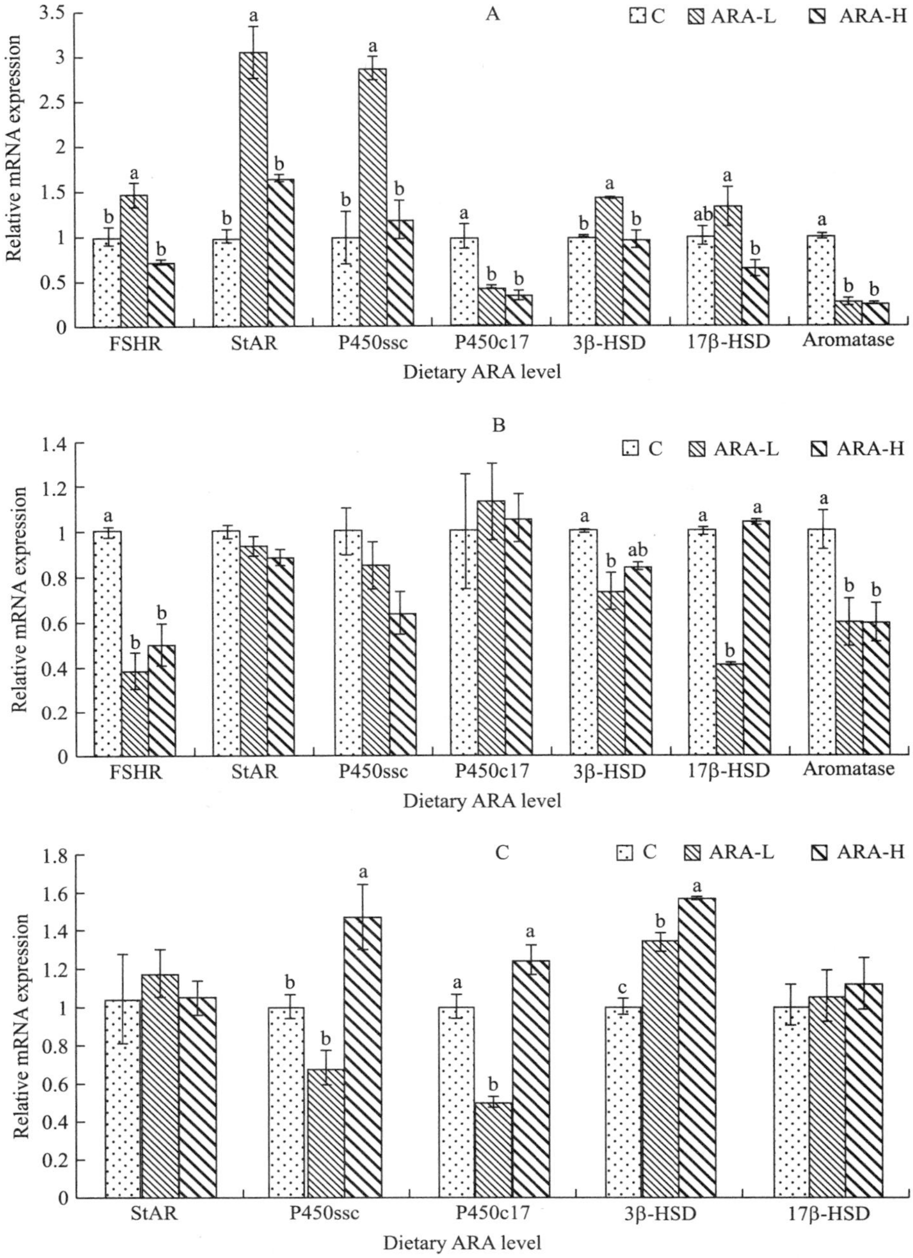

Fig. 2　Effects of dietary ARA on the mRNA expressions of sex steroid-synthesizing proteins in gonads of tongue sole broodstock

A, immature ovaries; B, mature ovaries; C, mature testes. Values (means ± standard error, $n = 3$) in bars that do not have the same letter are significantly different ($P < 0.05$)

In IMF fish, the concentrations of C16:0, total saturated fatty acids, C16:1n-7, total monounsaturated fatty acids, and C18:2n-6 in ovary lipids decreased significantly with increasing

Table 4 Gonad fatty acid compositions of tongue sole fed the experimental diets (% total fatty acids, means ± standard error, $n = 3$)

Fatty acid	Immature ovary			Mature ovary			Mature testis		
	C	ARA-L	ARA-H	C	ARA-L	ARA-H	C	ARA-L	ARA-H
C14:0	1.7 ± 0.2^{a}	1.9 ± 0.1^{a}	1.1 ± 0.1^{b}	2.1 ± 0.3	2.0 ± 0.2	1.8 ± 0.0	1.7 ± 0.2	1.7 ± 0.2	1.6 ± 0.2
C16:0	14.7 ± 0.3^{a}	12.5 ± 0.7^{b}	11.8 ± 0.4^{b}	16.0 ± 0.6	15.8 ± 0.2	14.9 ± 0.2	19.7 ± 0.9	20.2 ± 1.0	21.1 ± 0.3
C18:0	6.3 ± 0.2	5.9 ± 0.6	7.0 ± 0.2	7.9 ± 0.5	7.1 ± 0.3	6.0 ± 0.5	8.00 ± 0.4	7.6 ± 0.5	9.0 ± 0.4
ΣSFA	22.8 ± 0.6^{a}	20.3 ± 1.0^{ab}	19.9 ± 0.3^{b}	26.6 ± 1.0^{a}	24.9 ± 0.4^{ab}	22.7 ± 0.6^{b}	29.3 ± 0.8	29.5 ± 0.9	31.7 ± 0.1
C16:1$n-7$	2.7 ± 0.3^{a}	2.8 ± 0.3^{a}	1.6 ± 0.1^{b}	3.4 ± 0.2	3.4 ± 0.1	3.2 ± 0.2	2.3 ± 0.2	2.0 ± 0.2	1.8 ± 0.2
C18:1$n-7$	3.5 ± 0.2	2.8 ± 0.2	3.1 ± 0.3	2.9 ± 0.1	2.4 ± 0.1	2.7 ± 0.2	3.4 ± 0.3	3.4 ± 0.2	3.2 ± 0.1
C18:1$n-9$	14.6 ± 0.9	14.4 ± 0.6	11.9 ± 0.8	16.9 ± 0.2	16.5 ± 0.5	16.2 ± 0.1	9.4 ± 0.2	8.5 ± 0.6	8.6 ± 0.6
C20:1$n-9$	1.2 ± 0.3	1.3 ± 0.3	1.4 ± 0.0	1.6 ± 0.4	2.3 ± 0.2	1.9 ± 0.1	0.6 ± 0.1	0.7 ± 0.1	0.7 ± 0.1
ΣMUFA	22.0 ± 1.1^{a}	21.3 ± 1.0^{ab}	18.0 ± 0.7^{b}	24.7 ± 0.2	24.6 ± 0.4	24.0 ± 0.1	15.6 ± 0.8	14.5 ± 0.9	14.2 ± 0.8
C18:2$n-6$	7.5 ± 0.4^{a}	7.5 ± 0.2^{a}	5.7 ± 0.5^{b}	7.8 ± 0.3	7.5 ± 0.1	7.4 ± 0.1	6.1 ± 0.3	5.7 ± 0.4	5.1 ± 0.1
C20:4$n-6$	1.8 ± 0.2^{c}	5.4 ± 0.6^{b}	11.2 ± 0.5^{a}	1.3 ± 0.1^{c}	3.5 ± 0.5^{b}	5.6 ± 0.3^{a}	2.8 ± 0.1^{c}	7.4 ± 0.2^{b}	12.2 ± 1.1^{a}
Σ$n-6$ PUFA	9.3 ± 0.2^{c}	12.9 ± 0.5^{b}	16.9 ± 0.5^{a}	9.1 ± 0.3^{b}	11.0 ± 0.6^{b}	13.1 ± 0.3^{a}	8.9 ± 0.3^{c}	13.1 ± 0.5^{b}	17.3 ± 1.0^{a}
C18:3$n-3$	0.7 ± 0.1	0.7 ± 0.1	0.3 ± 0.1	0.8 ± 0.1	0.8 ± 0.0	0.7 ± 0.0	0.4 ± 0.0	0.4 ± 0.0	0.3 ± 0.0
C20:5$n-3$	3.1 ± 0.1^{a}	2.7 ± 0.0^{b}	1.9 ± 0.1^{c}	3.7 ± 0.1^{a}	2.8 ± 0.1^{b}	2.7 ± 0.1^{b}	2.8 ± 0.3^{a}	2.2 ± 0.3^{ab}	1.6 ± 0.2^{b}
C22:5$n-3$	3.4 ± 0.3	3.9 ± 0.2	3.9 ± 0.3	2.9 ± 0.2	3.5 ± 0.3	3.0 ± 0.3	6.4 ± 0.3^{a}	4.0 ± 0.4^{ab}	3.0 ± 0.1^{b}
C22:6$n-3$	19.6 ± 0.9	20.8 ± 1.5	19.2 ± 1.6	17.3 ± 0.8	16.7 ± 0.5	16.1 ± 0.2	24.0 ± 0.2^{a}	22.0 ± 0.1^{b}	18.7 ± 0.3^{c}
Σ$n-3$ PUFA	26.7 ± 0.9	28.1 ± 1.1	25.3 ± 1.7	24.7 ± 0.9	23.7 ± 0.6	22.5 ± 0.4	33.4 ± 0.7^{a}	28.6 ± 0.5^{b}	23.5 ± 0.6^{c}
Σ$n-3$/Σ$n-6$	2.9 ± 0.2^{a}	2.2 ± 0.2^{b}	1.5 ± 0.1^{c}	2.7 ± 0.2^{a}	2.2 ± 0.2^{ab}	1.7 ± 0.1^{b}	3.9 ± 0.1^{a}	2.2 ± 0.1^{b}	1.4 ± 0.1^{c}
Total lipid	16.6 ± 0.6	18.3 ± 0.8	17.6 ± 0.5	17.8 ± 0.3	17.8 ± 0.3	18.3 ± 0.4	15.4 ± 0.9	16.4 ± 0.8	15.9 ± 0.5

Values in the same row with different superscript letters are significantly different ($P < 0.05$). Total lipid is expressed as % dry matter. SFA: saturated fatty acid; MUFA: monounsaturated fatty acid; PUFA: polyunsaturated fatty acid

Table 5 Liver fatty acid compositions of tongue sole fed the experimental diets (% total fatty acids, means ± standard error, n = 3)

Fatty acid	Immature female			Mature female			Mature male		
	C	ARA-L	ARA-H	C	ARA-L	ARA-H	C	ARA-L	ARA-H
C14:0	1.2 ± 0.1	1.3 ± 0.1	1.1 ± 0.1	1.2 ± 0.1	1.0 ± 0.1	1.3 ± 0.1	1.8 ± 0.1	1.7 ± 0.0	1.5 ± 0.1
C16:0	20.1 ± 0.8	18.2 ± 0.6	19.1 ± 0.6	18.0 ± 1.0	18.3 ± 0.4	19.1 ± 0.9	21.4 ± 0.6	20.7 ± 0.7	20.7 ± 0.6
C18:0	16.4 ± 0.9	17.0 ± 0.3	16.0 ± 0.7	16.7 ± 1.2	15.9 ± 1.0	15.1 ± 0.7	11.6 ± 1.0	12.2 ± 0.8	13.2 ± 0.4
ΣSFA	37.7 ± 0.3	36.5 ± 0.7	36.2 ± 0.1	35.8 ± 0.4	35.3 ± 0.6	35.5 ± 0.3	34.8 ± 1.2	35.5 ± 1.7	35.4 ± 1.0
C16:1$n-7$	1.3 ± 0.1	1.3 ± 0.1	1.1 ± 0.1	1.3 ± 0.0	1.2 ± 0.0	1.4 ± 0.1	2.5 ± 0.3	2.0 ± 0.2	1.9 ± 0.1
C18:1$n-7$	2.4 ± 0.2	2.3 ± 0.2	2.2 ± 0.2	3.2 ± 0.3	3.2 ± 0.2	3.3 ± 0.2	2.6 ± 0.0	2.4 ± 0.2	2.2 ± 0.1
C18:1$n-9$	10.9 ± 0.4	10.9 ± 0.5	9.6 ± 0.2	12.5 ± 0.8	13.2 ± 0.6	13.3 ± 0.8	10.9 ± 0.2	9.7 ± 0.3	10.0 ± 0.7
C20:1$n-9$	0.5 ± 0.1	0.5 ± 0.1	0.5 ± 0.0	0.6 ± 0.1	0.8 ± 0.1	0.8 ± 0.1	0.8 ± 0.1	0.8 ± 0.1	0.7 ± 0.1
ΣMUFA	15.1 ± 0.5	15.1 ± 0.8	13.2 ± 0.3	17.6 ± 1.0	18.5 ± 0.5	18.8 ± 1.0	16.8 ± 0.4	14.8 ± 0.2	14.8 ± 0.9
C18:2$n-6$	6.6 ± 0.1[a]	4.8 ± 0.1[b]	4.1 ± 0.3[b]	4.8 ± 0.4[a]	3.9 ± 0.1[ab]	3.5 ± 0.2[b]	7.2 ± 0.3[a]	5.6 ± 0.4[b]	4.8 ± 0.1[b]
C20:4$n-6$	3.8 ± 0.1[c]	9.2 ± 0.4[b]	13.1 ± 1.2[a]	3.9 ± 0.3[c]	6.0 ± 0.2[b]	8.2 ± 0.4[a]	3.1 ± 0.1[c]	8.0 ± 0.6[b]	12.7 ± 0.3[a]
Σ$n-6$ PUFA	10.3 ± 0.1[b]	14.0 ± 0.5[ab]	17.2 ± 1.5[a]	8.7 ± 0.2[b]	9.9 ± 0.1[b]	11.7 ± 0.6[a]	10.4 ± 0.2[c]	13.6 ± 0.7[b]	17.5 ± 0.3[a]
C18:3$n-3$	0.3 ± 0.0	0.3 ± 0.0	0.2 ± 0.1	0.3 ± 0.0	0.3 ± 0.0	0.3 ± 0.1	0.4 ± 0.0[a]	0.3 ± 0.0[ab]	0.3 ± 0.0[b]
C20:5$n-3$	3.4 ± 0.3[a]	2.8 ± 0.1[ab]	1.9 ± 0.2[b]	4.4 ± 0.3[a]	2.8 ± 0.1[b]	2.4 ± 0.3[b]	5.1 ± 0.1[a]	3.7 ± 0.2[b]	2.5 ± 0.1[c]
C22:5$n-3$	1.7 ± 0.1	1.8 ± 0.0	1.6 ± 0.0	1.8 ± 0.1	1.9 ± 0.2	1.6 ± 0.1	2.2 ± 0.1	1.9 ± 0.1	2.1 ± 0.1
C22:6$n-3$	21.3 ± 0.2	19.1 ± 0.9	20.1 ± 1.3	23.0 ± 0.7	20.7 ± 1.1	19.7 ± 1.7	19.1 ± 0.6	17.1 ± 1.2	17.8 ± 0.5
Σ$n-3$ PUFA	26.8 ± 0.4	23.9 ± 0.7	23.7 ± 1.2	29.5 ± 0.5	25.7 ± 1.0	23.9 ± 2.0	26.9 ± 0.6[a]	23.0 ± 1.2[b]	22.1 ± 0.6[b]
Σ$n-3$/Σ$n-6$	2.6 ± 0.1[a]	1.7 ± 0.1[b]	1.4 ± 0.2[b]	3.4 ± 0.2[a]	2.6 ± 0.1[ab]	2.1 ± 0.3[b]	2.6 ± 0.1[a]	1.7 ± 0.1[b]	1.3 ± 0.0[c]
Total lipid	11.6 ± 0.6	10.8 ± 0.6	11.0 ± 0.7	11.8 ± 0.6	12.2 ± 0.5	12.2 ± 0.7	11.8 ± 0.6	11.6 ± 0.9	11.6 ± 0.7

Values in the same row with different superscript letters are significantly different ($P < 0.05$). Total lipid is expressed as % dry matter. SFA: saturated fatty acid; MUFA: monounsaturated fatty acid; PUFA: polyunsaturated fatty acid

Table 6　Muscle fatty acid compositions of tongue sole fed the experimental diets (% total fatty acids, means ± standard error, n = 3)

Fatty acid	Immature female			Mature female			Mature male		
	C	ARA-L	ARA-H	C	ARA-L	ARA-H	C	ARA-L	ARA-H
C14:0	3.1 ± 0.4	3.3 ± 0.2	2. ± 0.1	3.1 ± 0.3	3.5 ± 0.3	2.8 ± 0.4	3.1 ± 0.14	2.7 ± 0.3	2.8 ± 0.1
C16:0	22.6 ± 0.2^{a}	22.3 ± 0.8^{ab}	20.5 ± 0.2^{b}	21.4 ± 0.5	21.9 ± 0.2	22.4 ± 0.4	21.7 ± 0.8	21.0 ± 0.6	22.4 ± 1.0
C18:0	6.6 ± 0.5	6.5 ± 0.9	6.0 ± 0.2	6.7 ± 0.2	5.9 ± 0.7	7.5 ± 0.8	6.8 ± 1.0	7.2 ± 0.4	7.4 ± 0.3
ΣSFA	32.3 ± 0.3	32.0 ± 1.1	29.4 ± 0.2	31.2 ± 0.5	31.2 ± 0.4	32.7 ± 0.8	31.7 ± 1.6	30.9 ± 0.8	32.7 ± 1.4
C16:1$n-7$	5.1 ± 0.4	5.3 ± 0.4	4.7 ± 0.2	4.9 ± 0.4	5.2 ± 0.5	4.1 ± 0.8	5.4 ± 0.4	4.3 ± 0.6	4.5 ± 0.2
C18:1$n-7$	2.9 ± 0.2	3.0 ± 0.1	2.7 ± 0.1	2.9 ± 0.2	3.0 ± 0.1	2.6 ± 0.3	2.8 ± 0.3	2.8 ± 0.1	2.8 ± 0.2
C18:1$n-9$	14.4 ± 1.1	15.5 ± 0.5	13.8 ± 0.2	15.2 ± 0.2^{a}	14.5 ± 0.4^{a}	11.4 ± 0.7^{b}	14.8 ± 0.4	13.9 ± 0.6	13.6 ± 0.7
C20:1$n-9$	1.4 ± 0.2	1.2 ± 0.1	1.3 ± 0.1	1.4 ± 0.1	1.5 ± 0.1	1.6 ± 0.4	1.1 ± 0.0	1.4 ± 0.2	1.4 ± 0.1
ΣMUFA	20.9 ± 1.4	22.0 ± 1.0	19.8 ± 0.4	21.6 ± 0.6	21.2 ± 1.0	17.1 ± 1.7	21.3 ± 0.8	19.7 ± 1.4	19.2 ± 1.5
C18:2$n-6$	9.2 ± 0.2^{a}	8.5 ± 0.1^{ab}	7.4 ± 0.4^{b}	8.9 ± 0.5	8.3 ± 0.2	8.3 ± 0.3	7.1 ± 0.3	6.9 ± 0.4	6.7 ± 0.3
C20:4$n-6$	2.2 ± 0.3^{c}	4.4 ± 0.2^{b}	6.1 ± 0.6^{a}	1.7 ± 0.2^{b}	2.7 ± 0.5^{b}	6.5 ± 0.7^{a}	2.5 ± 0.6^{c}	4.9 ± 0.2^{b}	7.9 ± 0.2^{a}
Σ$n-6$ PUFA	11.4 ± 0.2	12.9 ± 0.1	13.5 ± 0.9	10.6 ± 0.3^{b}	11.0 ± 0.4^{b}	14.8 ± 0.5^{a}	9.5 ± 0.4^{c}	11.8 ± 0.6^{b}	14.6 ± 0.1^{a}
C18:3$n-3$	0.9 ± 0.1	1.0 ± 0.1	0.8 ± 0.0	1.2 ± 0.1	1.0 ± 0.1	0.9 ± 0.1	0.8 ± 0.1	0.6 ± 0.1	0.8 ± 0.0
C20:5$n-3$	4.4 ± 0.3	4.7 ± 0.1	3.7 ± 0.2	4.9 ± 0.5	4.5 ± 0.3	4.9 ± 0.2	6.0 ± 0.3^{a}	5.0 ± 0.1^{ab}	4.4 ± 0.2^{b}
C22:5n-3	4.2 ± 0.1	4.3 ± 0.2	4.1 ± 0.4	3.8 ± 0.4	4.4 ± 0.3	3.9 ± 0.2	3.4 ± 0.2	3.6 ± 0.1	3.5 ± 0.3
C22:6n-3	19.6 ± 1.9	17.4 ± 1.2	17.4 ± 1.6	18.7 ± 1.4	17.5 ± 0.9	17.8 ± 1.9	18.6 ± 1.7	18.9 ± 0.8	17.4 ± 1.7
∑n-3 PUFA	29.1 ± 1.7	27.3 ± 1.2	25.9 ± 1.7	28.5 ± 1.5	27.4 ± 0.9	27.4 ± 2.3	28.5 ± 1.4	28.2 ± 0.9	25.2 ± 2.0
∑n-3/∑n-6	2.6 ± 0.1	2.1 ± 0.1	1.9 ± 0.2	2.7 ± 0.1^{a}	2.5 ± 0.0^{a}	1.9 ± 0.1^{b}	3.0 ± 0.1^{a}	2.4 ± 0.1^{b}	1.7 ± 0.1^{c}
Total lipid	6.3 ± 0.3	5.8 ± 0.7	5.4 ± 0.4	6.5 ± 0.2	6.3 ± 0.3	6.8 ± 0.0	6.6 ± 0.7	5.5 ± 0.6	5.7 ± 0.7

Values in the same row with different superscript letters are significantly different ($P < 0.05$). Total lipid is expressed as % dry matter. SFA: saturated fatty acid; MUFA: monounsaturated fatty acid; PUFA: polyunsaturated fatty acid

dietary ARA. In contrast, in MF fish, only total saturated fatty acids decreased significantly ($P < 0.05$). In Mm fish, the concentrations of DPA (C22:5n-3) and DHA in testes lipids significantly decreased ($P < 0.01$) with increasing dietary ARA. Compared with IMF fish, MF fish showed lower ARA contents in ovary lipids. Compared with MF fish, Mm fish had higher concentrations of ARA, DPA and DHA in gonad lipids, but lower concentrations of EPA.

The C18:2n-6 concentrations in livers lipids of all types of tongue sole significantly decreased with increasing dietary ARA. It was also observed that the C18:3n-3 concentrations in Mm liver lipids significantly ($P < 0.05$) decreased with increasing dietary ARA. The average ARA concentration in MF liver lipids was lower than that of IMF and Mm fish, and the average DHA concentrations in liver lipids of male fish was lower than that of female fish.

In muscle lipids, the concentrations of C16:0 in IMF fish ($P < 0.05$) and C18:1n-9 ($P < 0.01$) in MF fish significantly decreased with increasing dietary ARA. The average ARA concentration in MF muscle lipids was lower than that of Mm fish. The DHA concentrations remained constant in muscle lipids of all types of tongue sole.

4 Discussion

The present study showed that dietary ARA (15.44% of TFA in 130 g lipid / kg dry diet) increased the testosterone production in male tongue sole. This is in agreement with the previous study on Senegalese sole *Solea senegalensis*, which showed that 6% (of TFA) dietary ARA increased the production of 11-ketotestosterone and testosterone (Norambuena et al, 2013a). Previous in vitro results on goldfish *Carassius auratus* also showed that ARA, and its metabolite prostaglandin E2, stimulated the gonadotropin-induced testosterone production in both testes and ovarian follicles (Van Der Kraak and Chang, 1990; Wade and Van Der Kraak, 1993; Wade et al, 1994; Mercure and Van Der Kraak, 1995, 1996).

In contrast to the effects of dietary ARA on testosterone production in male fish, estradiol production in female fish was reduced by dietary ARA supplementation in the present study. The previous study on Senegalese sole also showed increases in dietary ARA had different effects on testosterone production in males and estradiol production in females; however, estradiol production in females did not change with changes in dietary ARA (Norambuena et al, 2013a). To date, little information has been available regarding the effects of ARA on estradiol production in female fish. However, several studies indicate ARA is metabolized differently in male and female fish and suggest that ARA may be more important for the reproductive success of males (Norambuena et al, 2012, 2013b).

In this study, the gene expression of aromatase, a key protein in the biosynthesis of estrogens, in both immature and mature ovaries was reduced by dietary ARA supplementation. This may explain the reduction of estradiol production in females receiving the supplemented diets. Despite the similar reductions in aromatase gene expression in both immature and mature ovaries, gene expressions of other sex steroid-synthesizing proteins responded differently between immature and mature ovaries to dietary ARA supplementation. In mature ovaries, the ARA-L feed (5.14% of TFA) reduced the gene expression of 3β-HSD and 17β-HSD, and both the ARA-L and ARA-H feeds reduced the gene expression of FSHR. However, in immature ovaries, it seemed that dietary ARA had the opposite effect on gene expression. In IMF fish, the ARA-L feed stimulated the gonadal gene expression of FSHR, StAR, P450ssc, 3β-HSD, and 17β-HSD. These discrepancies might be related to the feedback regulation of ARA effects by the sex steroid hormones and/or gonadotropin (Lee et al, 2001; Sohn et al, 2001; Yamaguchi et al, 2003, 2006). Mercure and Van Der Kraak (1996) reported that at low dosages of human chorionic gonadotropin, ARA potentiated the steroidogenic actions in ovarian follicles of goldfish while with maximal gonadotropin stimulation, ARA attenuated the steroid production. Feedback regulation of ARA effects by the sex steroid hormones may also partly explain the observed different responses to dietary ARA between female and male tongue sole. It has been reported that estradiol, but not testosterone, stimulates gonadotropin II concentrations in marine fish (Lee et al, 1999). The high gonadotropin levels in mature ovaries, through feedback regulation, may reduce the potential effects of supplemental ARA on steroidogenesis in mature females.

Regarding the responses of gene expression of sex steroid-synthesizing proteins to dietary ARA in mature testes, it is noteworthy that the most effective dietary ARA level was the highest tested, ARA-H (15.44% of TFA). This is different than observed in immature ovaries, where the lower ARA supplemented feed, ARA-L (5.14% of TFA) was the most effective for most proteins. This indicates that, compared to female tongue sole, male tongue sole might require higher dietary ARA supplementation to support steroidogenic processes and testis maturation. Another noteworthy observation is that, for the sex steroid-synthesizing proteins analyzed in testes, only the gene expression of 3β-HSD was positively correlated with testosterone production. This indicates that 3β-HSD may play a more important role than other enzymes in testosterone synthesis. Moreover, as a protein mediating the rate-limiting step of steroid synthesis, the delivery of cholesterol substrate, StAR was not regulated by dietary ARA in tongue sole testes. StAR was regulated by dietary ARA only in immature ovaries. In terrestrial animals, it has been observed that in the steroid synthesis ovarian cells and Leydig cells used cholesterol substrate supplied in different ways (Hu et al, 2010). This phenomenon may also exist in fish and explain the discrepancies in StAR function between male and female fish.

Similar to the observed differences in steroid synthesis response to dietary ARA between male and female fish, and mature and immature fish, in this study, the accumulation of ARA in gonadal lipids also varied with fish gender and stage of maturation. ARA concentrations in testes lipids were higher than that in ovary lipids. This could explain why dietary ARA had positive effects on testosterone production, but negative effects on estradiol production in this study. In ovary lipids, ARA concentrations were much higher in immature fish than in mature fish. Although in both immature and mature fish, estradiol production was reduced by increases in dietary ARA, this discrepancy in ARA content between immature and mature ovary lipids may be reflected by the differential effects of dietary ARA on gene expression of steroidogenic proteins such as FSHR, StAR, 3β-HSD, and 17β-HSD between immature and mature ovaries, i.e., gene expression of these proteins were enhanced in immature ovaries but reduced in mature ones by moderate dietary ARA. Similar to this study, higher ARA concentrations have previously been observed in immature vs. mature ovary lipids of fiddler crab *Uca tangeri* (Mourente et al, 1994) and coho salmon *Orcorhynchus kisutch* (Johnson, 2012).

Although feedback regulation of fatty acid deposition by steroid hormones has also been observed in mammals (McNamara et al, 2009; Marks et al, 2013), it appears that the specific ARA accumulation in gonad lipids of fish confirms the high biological activity for this LC-PUFA in reproductive processes (Bell and Sargent, 2003; Emata et al, 2003; Johnson, 2009). The higher ARA concentrations in gonad lipids of wild fish vs. farmed fish is suspected to be at least partially responsible for the higher reproductive performance of wild fish (Grigorakis et al, 2002; Salze et al, 2005; Rodríguez-Barreto et al, 2012; Støttrup et al, 2013; Hauville et al, 2015). In the present study, as well as in other previous studies, gonad lipids typically have higher ARA concentrations than other tissue lipids, such as those from muscle and adipose tissue. This indicates that ARA is preferentially accumulated in gonad lipids and may provide the material base for regulating maturation functions in the gonads.

With respect to the gonadal accumulation of other fatty acids, EPA concentrations were low in all types of gonad lipids, indicating EPA probably is less important than ARA for tongue sole reproduction. In contrast, the EPA concentrations in muscle lipids were higher and relatively less sensitive to dietary changes, indicating EPA might have more important functions in muscle tissue.

Compared with the EPA concentrations, however, DHA concentrations were much higher in tongue sole gonad lipids in this study. This indicates that DHA might additionally have important roles in tongue sole reproduction. Moreover, the DHA accumulation in gonad lipids also varied with fish gender and maturation stage, although not as that much as ARA. DHA concentrations were highest in mature testes lipids, followed by immature ovary lipids, and lowest in mature ovary lipids. This indicates DHA could be more important for mature testes and immature ovaries than for mature ovaries. However, the situation was opposite in liver lipids. DHA concentrations in liver lipids ranked as follows: Mm < IMF < MF. This is suggestive of active DHA transport from the liver to the gonads where DHA may have important roles in maturation and gametogenesis. Considering the species-specific importance of DHA in fish reproduction (Yanes-Roca et al, 2009; Beirão et al, 2015), the effects of DHA on the reproductive performances of tongue sole, as well as the involved mechanisms, warrant further studies.

In conclusion, results of this study suggest that dietary ARA enhanced the testosterone production in mature tongue sole testes, but reduced the estradiol production in both immature and mature ovaries. Dietary ARA also differentially regulated the gene expression of sex steroid-synthesizing proteins, depending on fish gender and maturation stage. The ARA accumulations in gonad lipids corresponded well with the ARA effects on steroid synthesis. Dietary ARA seemed more important for male fish than for female fish, and more important for immature females than for mature females. The amount of ARA integrated into commercial diets for tongue sole broodstock should be considered carefully according to fish gender and maturation stage in order to enhance the sex steroid production and the consequent reproductive performances.

Acknowledgments

This work was supported by National Natural Science Foundation of China (31402309), Class General Financial Grant from the China Postdoctoral Science Foundation (2014M551991), the Grant from Modern Agro-industry Technology Research System (CARS-50-G08, Government of China), and Special Financial Grant from the China Postdoctoral Science Foundation (2015T80763).

References

[1] Abi-Ayad A, Mélard C, Kestemont P. Effects of n–3 fatty acids in Eurasian perch broodstock diet on egg fatty acid composition and larvae stress resistance[J]. Aquac Int, 1997, 5: 161-168.

[2] Almansa E, Martín M V, Cejas J R, et al. Lipid and fatty acid composition of female gilthead seabream during their reproductive cycle: effects of a diet lacking n−3 HUFA[J]. J Fish Biol, 2001, 59: 267-286.

[3] Association of Official Analytical Chemists (AOAC). Official Methods of Analysis of AOAC International [M]. 17th ed. Arlington, VA:AOAC, 2000.

[4] Beirão J, Soares F, Pousão-Ferreira P, et al. The effect of enriched diets on *Solea senegalensis* sperm quality[J]. Aquaculture, 2015, 435: 187-194.

[5] Bell J G, Sargent J R. Arachidonic acid in aquaculture feeds: current status and future opportunities[J]. Aquaculture, 2003, 218(s1-4): 491-499.

[6] Bruce M P, Oyen F, Bell J G, et al. Development of broodstock diets for the European sea bass (*Dicentrarchus labrax*) with special emphasis on the importance of n−3 and n−6 HUFA to reproductive performance[J].

Aquaculture,1999, 177: 85-98.

[7] Butts I A E, Baeza R, Støttrup J G, et al. Impact of dietary fatty acids on muscle composition, liver lipids, milt composition and sperm performance in European eel[J]. Comp Biochem Physiol A Mol Integr Physiol, 2015,183: 87-96.

[8] Cerdá J, Zanuy S, Carrillo M, et al. Short- and long-term dietary effects on female sea bass (*Dicentrarchus labrax*): seasonal changes in plasma profiles of lipids and sex steroids in relation to reproduction[J]. Comp Biochem Physiol C Pharmacol Toxicol Endocrinol, 1995, 111: 83-91.

[9] Emata A C, Ogata H Y, Garibay, E S, et al. Advanced broodstock diets for the mangrove red snapper and a potential importance of arachidonic acid in eggs and fry[J]. Fish Physiol Biochem, 2003, 28(1): 489-491.

[10] Furuita H, Yamamoto T, Shima T, et al. Effect of arachidonic acid levels in broodstock diet on larval and egg quality of Japanese flounder *Paralichthys olivaceus*[J]. Aquaculture, 2003, 220: 725-735.

[11] Grigorakis K, Alexis M A, Taylor K D. Comparison of wild and cultured gilthead seabream (*Sparus aurata*): composition, appearance and seasonal variations[J]. Int J Food Sci Tech, 2002, 37: 477-484.

[12] Harel M, Tandler A, Kissil G W. The kinetics of nutrient incorporation into body tissues of gilthead seabream (*Sparus aurata*) females and the subsequent effects on eggs composition and egg quality[J]. Br J Nutr, 1994, 72: 45-48.

[13] Hauville M R, Rhody N R, Resley M J, et al. Comparative study of lipids and fatty acids in the liver, muscle, and eggs of wild and captive common snook broodstock[J]. Aquaculture, 2015, 446: 227-235.

[14] Izquierdo M S, Fernández-Palacios H, Tacon A G J. Effect of broodstock nutrition on reproductive performance of fish[J]. Aquaculture, 2001, 197(s1-4): 25-42.

[15] Hu J, Zhang Z H, Shen W J, et al. Cellular cholesterol delivery, intracellular processing and utilization for biosynthesis of steroid hormones[J]. Nutr Metab, 2010, 7(1): 1-25.

[16] Johnson A L, Tilly J L, Levorse J M. Possible role for arachidonic acid in the control of steroidogenesis in hen theca[J]. Biol Reprod, 1991, 44: 338-344.

[17] Johnson R B. Lipid deposition in oocytes of teleost fish during secondary oocyte growth[J]. Rev Fish Sci, 2009, 17: 78-89.

[18] Johnson R B. Examining the transport of dietary lipids to egg and muscle tissue in cultured coho salmon *Oncorhynchus kisutch* during secondary oocyte growth[D]. Moscow: University of Idaho, 2012: 184.

[19] Lee Y H, Du J L, Yen F P, et al. Regulation of plasma gonadotropin II secretion by sex steroids,aroamtase inhibitors, and antiestrogens in the protandrous black porgy, *Acanthopagrus schlegeli* bleeker[J]. Comp Biochem Physiol B Biochem Mol Biol, 2001, 129(3): 399-406.

[20] Lee Y H, Du J L, Yueh W S, et al. 17β-estradiol, but not testosterone stimulates gonadotropin II concentrations in the protandrous black porgy, *Acanthopagrus schlegeli* bleeker[J]. Fish Physiol Biochem, 1999, 21(4): 345-351.

[21] Liang M Q, Lu Q K, Qian C, et al. Effects of dietary n-3 to n-6 fatty acid ratios on spawning performance and larval quality in tongue sole[J]. Aquacult Nutr, 2014, 20: 79-89.

[22] Livak K J, Schmittgen T D. Analysis of relative gene expression data using real-time quantitative PCR and

the $2^{-\Delta\Delta CT}$ method[J]. Methods, 2001: 25, 402-408.

[23] Lopez-Ruiz M P, Choi M S K, Rose M P, et al. Direct effect of arachidonic acid on protein kinase C and LH-stimulated steroidogenesis in rat Leydig cells: evidence for tonic inhibitory control of steroidogenesis by protein kinase C[J]. Endocrinology, 1992, 130: 1122-1130.

[24] Luo L, Ai L, Li T, et al. The impact of dietary DHA/EPA ratio on spawning performance, egg and offspring quality in Siberian sturgeon (*Acipenser baeri*)[J]. Aquaculture, 2015, 437: 140-145.

[25] Marks K A, Kitson A P, Stark K D. Hepatic and plasma sex differences in saturated and monounsaturated fatty acids are associated with differences in expression of elongase 6, but not stearoyl-CoA desaturase in Sprague-Dawley rats[J]. Genes Nutr, 2013, 8(3): 317-327.

[26] Mazorra, C, Bruce, M, Bell, J G, et al. Dietary lipid enhancement of broodstock reproductive performance and egg and larval quality in Atlantic halibut (*Hippoglossus hippoglossus*)[J]. Aquaculture, 2003, 227(s 1-4): 21-33.

[27] McNamara R K, Able J, Jandacek R, et al. Gender differences in rat erythrocyte and brain docosahexaenoic acid composition: Role of ovarian hormones and dietary omega-3 fatty acid composition[J]. Psychoneuroendocrinology, 2008, 34(4): 532-539.

[28] Mercure F, Van Der Kraak G. Inhibition of gonadotropin-stimulated ovarian steroid production by polyunsaturated fatty acids in teleost fish[J]. Lipids, 1995, 30(30): 547-554.

[29] Mercure F, Van Der Kraak G. Mechanisms of action of free arachidonic acid on ovarian steroid production in the goldfish[J]. Gen Comp Endocr, 1996, 102(1): 130-140.

[30] Mourente G, Medina A, González S, et al. Changes in lipid class and fatty acid contents in the ovary and midgut gland of the female fiddler crab *Uca tangeri* (Decapoda, Ocypodiadae) during maturation[J]. Mar Biol, 1994, 121(1): 187-197.

[31] Navas J M, Bruce M, Thrush M, et al. The impact of seasonal alteration in the lipid composition of broodstock diets on egg quality in the European sea bass[J]. J Fish Biol, 1997, 51: 760-773.

[32] Norambuena F, Estévez A, Carazo I, et al. Self-selection of diets with different contents of arachidonic acid by Senegalese sole (*Solea senegalensis*) broodstock[J]. Aquaculture, 2012, 364-365(1): 198-205.

[33] Norambuena F, Estévez A, Mañanós E, et al. Effects of graded levels of arachidonic acid on the reproductive physiology of Senegalese sole (*Solea senegalensis*): fatty acid composition, prostaglandins and steroid levels in the blood of broodstock bred in captivity[J]. Gen Comp Endocr, 2013a, 191(9): 92-101.

[34] Norambuena F, Morais S, Estévez A, et al. Dietary modulation of arachidonic acid metabolism in Senegalese sole (*Solea senegalensis*) broodstock reared in captivity[J]. Aquaculture, 2013b, 372(1): 80-88.

[35] Patiño R, Yoshizaki G, Bolamba D, et al. Role of arachidonic acid and protein kinase C during maturation-inducing hormone-dependent meiotic resumption and ovulation in ovarian follicles of Atlantic croaker [J]. Biol Reprod, 2003, 68(2): 516-523.

[36] Rodríguez-Barreto D, Jerez S, Cejas J R, et al. Comparative study of lipid and fatty acid composition in different tissues of wild and cultured female broodstock of greater amberjack (*Seriola dumerili*)[J].

Aquaculture, 2012, s360-361(665): 1-9.

[37] Salze G, Tocher D R, Roy W J, et al. Egg quality determinants in cod (*Gadus morhua* L.): egg performance and lipids in eggs from farmed and wild broodstock[J]. Aqua Res, 2002. 36: 1488-1499.

[38] Sohn Y C, Kobayashi M, Aida K. Regulation of gonadotropin β subunit gene expression by testosterone and gonadotropin-releasing hormones in the goldfish, *Carassius auratus*[J]. Comp Biochem Physiol B Biochem[J]. Mol Biol, 2001, 129(2-3): 419-426.

[39] Sorbera L A, Asturiano J F, Carrillo M, et al. Effects of polyunsaturated fatty acids and prostaglandins on oocyte maturation in a marine teleost, the European sea bass (*Dicentrarchus labrax*)[J]. Biol Reprod, 2001, 64(1): 382-389.

[40] Støttrup J G, Jacobsen C, Tomkiewicz J, et al. 2013. Modification of essential fatty acid composition in broodstock of cultured European eel *Anguilla anguilla* L[J]. Aquacult Nutr, 2013, 19(2): 172-185.

[41] Van Der Kraak G, Biddiscombe S. Polyunsaturated fatty acids modulate the properties of the sex steroid binding protein in goldfish[J]. Fish Physiol Biochem, 1999. 20(2): 115-123.

[42] Van Der Kraak G, Chang J P. Arachidonic acid stimulates steroidogenesis in goldfish preovulatory ovarian follicles[J]. Gen Comp Endocr, 1990, 77(2): 221-228.

[43] Wade M G, Van Der Kraak G. Arachidonic acid and prostaglandin E2 stimulate testosterone production by goldfish testis *in vitro*[J]. Gen Comp Endocr, 1993, 90(1): 109-118.

[44] Wade M G, Van Der Kraak G, Gerrits, M F, et al. Release and steroidogenic actions of polyunsaturated fatty acids in the goldfish testis[J]. Biol Reprod, 1994, 51(1): 131-139.

[45] Watanabe T, Vassallo-Agius R. Broodstock nutrition research on marine finfish in Japan[J]. Aquaculture, 2003, 227: 35-61.

[46] Xu H, Dong X, Ai Q, et al. Regulation of tissue LC-PUFA contents, Δ6 fatty acyl desaturase (FADS2) gene expression and the methylation of the putative FADS2 gene promoter by different dietary fatty acid profiles in Japanese seabass (*Lateolabrax japonicus*)[J]. PLoS ONE, 2014, 9(1): e87726.

[47] Xu H, Lv Q, Liang M, et al. Effects of different dietary lipid sources on spawning performance, egg and larval quality, and egg fatty acid composition in tongue sole *Cynoglossus semilaevis*[J]. Isr J Aquacult Bamid, 2015, 64: 1091.

[48] Xu H, Mu Y, Zhang Y, et al. Graded levels of fish protein hydrolysate in high plant diets for turbot (*Scophthalmus maximus*): effects on growth performance and lipid accumulation[J]. Aquaculture, 2016, 454: 140-147.

[49] Yamaguchi S, Gen K, Okuzawa K, et al. Regulation of gonadotropin subunit genes expression by 11-ketotestosterone during early spermatogenesis in male red seabream, *Pagrus major*[J]. Fish Physiol Biochem, 2003, 28(1): 111-112.

[50] Yamaguchi S, Gen K, Okuzawa K, et al. Influence of estradiol-17β, testosterone, and 11-ketotestosterone on testicular development, serum steroid hormone, and gonadotropin secretion in male red sea bream *Pagrus major*[J]. Fisheries Sci, 2006, 72(4): 835-845.

[51] Yanes-Roca C, Rhody N, Nystrom M, et al. Effects of fatty acid composition and spawning season patterns

on egg quality and larval survival in common snook (*Centropomus undecimalis*)[J]. Aquaculture, 2009, 287(3-4): 335-340.

[52] Zhou Q B, Wu H D, Zhu C S, et al. Effects of dietary lipids on tissue fatty acids profile, growth and reproductive performance of female rice field eel (*Monopterus albus*)[J]. Fish Physiol. Biochem, 2011, 37(37): 433-445.

鲆、鲽鱼类室内水泥池养殖易发疾病及其防治

郑春波　王世党　于诗群(山东省威海市文登区水产技术推广站)
张开富(山东省青岛市中仁药业有限公司)

摘　要　本文介绍了鲆、鲽鱼类在室内无公害养殖过程中,所经常发生的主要疾病及其防治方法。主要是病毒性疾病中的弹状病毒病、淋巴囊肿病,细菌性疾病中的腹水病、链球菌病、滑走细菌病、细菌性败血病,寄生虫病中的盾纤类纤毛虫、白点病、车轮虫病、鱼波豆虫病,营养性疾病中的黑化病,从病原、症状、流行、及防治进行了详细介绍。

关键词　鲆、鲽鱼;室内无公害养殖;疾病的防治

随着水产养殖业的迅速发展,目前对于水产养殖动物的病害防治及水产动物无公害养殖已越来越引起人们的重视。对病害防治的发展趋势是从以化学药物防治为主,逐渐转向以生物制剂和免疫接种的方面、提高养殖对象的免疫机能、选育抗病能力强的养殖品种、采取生态防治病害等进行综合防治为主,使养殖产品成为绿色产品。

鲆、鲽鱼类室内养殖,具有管理方便、产量稳定等优点,近年来在我们威海发展迅速,成为渔民发家致富的好途径,但随着养殖规模、数量的不断增大,由于受养殖技术及养殖条件的限制,时常出现许多问题,主要表现为病害频发、养殖成活率低等,给部分养殖户造成了很大的损失。因此对鲆、鲽类室内养殖疾病的防治,采取“以防为主、防重于治,防、治结合”的原则,坚持“全面预防,积极治疗”的方针,这样才能或减少养殖的疾病,降低因疾病所造成的损失。

1　病毒性疾病

1.1　弹状病毒病

病原:由弹性病毒(HRV)引起。

症状:病鱼体表及鳍充血和出血,腹部膨胀,内有腹水,肌肉、鳍基部可见点状出血,生殖腺淤血,脑出血等。

流行:发病季节为早春和冬季,水温 10 ℃时为发病高峰期,水温升到 16 ℃以上,疾病自然减轻或停止发生。稚、幼、成鱼均可感染此病。

防治：鱼卵经 40 mg/L～50 mg/L（含有效碘 10%）的有机碘消毒 15 min～20 min，再移入孵化池；孵化用水经紫外线照射或臭氧消毒，可有效预防仔稚鱼患此病。治疗将养殖水温升至 18 ℃以上，可逐渐自愈。

1.2　淋巴囊肿病

病原：淋巴囊肿病毒（LCV），形态呈正 20 面体，大小 200 nm～250 nm，具囊膜，双 DNA 病毒。

症状：病鱼的皮肤、鳍及眼球等处出现小泡状的肿胀物，这是一种慢性皮肤瘤。

流行：10 月至次年的 5 月，0 龄～1 龄鱼易发病，但发病率低，但感染率可达 80% 以上，死亡率可达 30%。

防治：保持良好水质，加强饲养管理。治疗可采用投喂抗生素药饵，如 1 kg 饲料添加氟苯尼考 2 g～3 g，连喂 7 日～10 日。

2　细菌性疾病

2.1　腹水病

病原：由爱德华氏菌（*Edwardsiella*）感染引起，革兰氏阴性，短杆菌，大小（0.5 μm～1 μm）×（1 μm～3 μm），具有运动性。

症状：病鱼腹部膨胀，肛门扩张发红，有的鱼肠道从肛门脱出；解剖发现有出血性腹水，肝脏、肾脏发生脓肿性出血或肿大贫血，也有的出现肝脏局部坏死和出血。

流行：7 月～8 月 20 ℃以上的高水温期为发病高峰期。夏季水温越高，发病期越长，秋冬季的危害也越大，高水温长期持续，则发病危险性更大。从仔鱼到成鱼都可能患此病。

防治：养殖过程中要保持大的换水量，保持养殖环境清洁；要投喂洁净的配合饵料，并严格控制投饵量，使鱼处在 70% 的饱食状态；降低饲养密度，减少鱼的应激性刺激；将病鱼迅速除去。治疗可投喂抗生素药饵，如 1 kg 饲料添加氟苯尼考 2 g～3 g，连喂 7 日～10 日。

2.2　链球菌病

病原：β 溶血性链球菌属（*Streptococus* sp.）一种，革兰氏阳性的二链或连锁状的球菌，大小 0.7 μm～1.4 μm，无运动性。

症状：病鱼摇摇晃晃，体色变黑；眼球白浊、充血、突出；头部和上下颚发红，鳃盖内发红，鳍出血；解剖可发现腹腔内有积水，肠管发炎，肝脏出血。

流行：主要发生于高水温期。

防治：口服四环素类药物疗效。但最好的预防措施是降低放养密度，避免过量投饵，清除残饵，改善水质。

2.3 滑走细菌病

病原：海岸曲桡杆菌(*Flexibacter maritimus*)和曲桡杆菌属(*Flexibacter* sp.)的一种，革兰氏阴性菌，弯曲而长的杆菌，0.5 μm×(2 μm～30 μm)，可滑行。

症状：病鱼各鳍糜烂、缺损，游泳缓慢，食欲下降；病症恶化时，头部腹面发红、充血，皮肤溃疡。内脏一般无异常。

流行：春秋两季水温为15 ℃～20 ℃时易发生。对3 cm～10 cm的幼鱼危害最大，可引起大量死亡。

防治：掌握合适的饲养密度，提高换水率，投喂优质的饵料，饲养过程中，要防止鱼体受伤，尽量不使鱼的体表黏液脱落，若发现病鱼，早发现、早治疗，捞出病鱼，并与其他养殖池隔离。治疗时可用磺胺类或盐酸土霉素50 mg/L～100 mg/L药浴1 h～2 h，每日一次，连续3日～4日，必要时可口服药饵。

2.4 细菌性败血病

病原：变异微球菌(*Micrococcus uarians*)，革兰氏阳性菌，圆形或椭圆形，大小0.5 μm～1.8 μm，不运动。

症状：病鱼在池中焦躁不安，常游到水表面，翻身又跌入水中，如此反复，不久沉入池底而死亡，病鱼食欲下降，皮下大面积弥漫性出血，体色发黑，腹部膨大；各鳍基部、口腔、鳃裂充血出血。解剖镜检，可见肝脏土黄色，脾暗红，消化道积水，心肝脾细胞大量坏死，鳃丝发红呈贫血状。

流行：危害各期鱼体，死亡率在10%～40%不等。

防治：对养殖用水，用饵进行严格杀菌消毒，及时隔离病鱼。治疗时可投喂每1 kg饲料添加3 g盐酸环丙沙星的药饵，连续投喂5日～7日，或用20 mg/L的盐酸环丙沙星进行药浴。

3 寄生虫病

3.1 盾纤类纤毛虫

病原：一般为喙突拟舟虫(*Paralembus rostratus*)，是盾纤类纤毛虫(*Scudcociliad ciliate*)的一种，为兼性寄生纤毛虫。虫体呈瓜子形，大小为20 μm～50 μm，内质不透明，常有多个食物胞，虫体的前半部稍向内侧弯曲，体披纤毛，后端有一根较长的尾毛，大核位于身体的中央，小核1个，以横二分裂方式进行繁殖。

病症：病鱼摄食量减少，在养殖池中分布散乱，且常不安地狂动、狂游，体色发黑或发红，眼球突出，鳃盖及吻端溃疡或出血，体表及鳍的基部糜烂，鳃丝苍白。镜检溃烂处可发现葵花籽状活泼游动全身被覆纤毛的虫体。

流行：常在高温季节发生，寄生于鱼的体表、鳃，严重的进入肌肉、脑颅内，可引起 0 龄～1 龄鱼短期内大量死亡。

防治：定期清扫池底和消毒，及时清除养殖池底污物，投饵后马上大排水，保持池底清洁卫生，每 10 日～15 日对鱼池用甲醛或含碘消毒剂消毒 1 次防治。治疗可用淡水浸泡 10 min～20 min，并结合移池，可将体外纤毛虫全部清除；采用 0.2 mL/L～0.25 mL/L 浓度的福尔马林浸泡 1 h～2 h 后，大排水；采用 30% 的双氧水 0.1 mL/L～0.15 mL/L 药浴 1 h～2 h 后，大排水；采用硫酸铜 0.5 mg/L～0.7 mg/L 与硫酸亚铁 0.2 mg/L～0.3 mg/L 合用药浴 1 h～2 h，大排水。

3.2　白点病（小瓜虫病）

病原：刺激隐核虫（*Cryptocaryon irritans*），为寄生纤毛虫类。虫体卵圆形或球形，体被均匀一致的纤毛，大小为 40 μm～300 μm。

症状：病鱼发病初期表现为摄食量减少、焦躁不安、起水环游；中期基本不摄食，头尾上翘，呼吸急促，肛门处可见白色粘性粪便；后期出现死亡，体表有“白点状”病斑，且基部糜烂，解剖时肠道充血，并有大量白色黏液。

流行：常在高温季节发生，虫体寄生于体表或鳃，可引起 1 龄～2 龄鱼短期内大量死亡。

防治：减小放养密度，加大换水量，对池水用漂白粉、高锰酸钾消毒。治疗可用市售双氧水 0.1 mL/L～0.15 mL/L 药浴 1 h～2 h，每日 2 次，连用 4～5 日；采用 0.025 mL/L 的福尔马林浸泡 24 h，也有较好的效果；用淡水浸浴 10 min～15 min；采用 0.3 mg/L 醋酸铜全池泼洒，每日一次，连用 3 次～4 次。

3.3　车轮虫病

病原：车轮虫属（*Trichodina*）的一种，为寄生纤毛虫类。虫体卵圆形，大小为 20 μm～100 μm。

症状：车轮虫主要寄生在鱼体表皮肤及鳃上，引起体表及鳃黏液分泌增多，食欲减退，呼吸困难，行为呆滞，且体色发黑；解剖观察，肝脏肿大、褪色，其他内脏器官无明显异常。

流行：该病发生后可引起病鱼慢性死亡，并且会引发细菌的感染，主要危害幼鱼。

防治：提高换水量，保持水质及饲养环境清洁，及时清除池底污物。治疗可用 0.2 mL/L～0.25 mL/L 的福尔马林浸泡 2 h～3 h，每日一次，3 日～5 日为一个疗程。亦可用 30% 的双氧水 0.1 mL/L～0.15 mL/L 药浴 1 h～2 h，大排水。

3.4　淀粉卵涡鞭虫病

病原：淀粉卵涡鞭虫属（*Amyloodinium*）的一种，为寄生鞭毛虫类，呈球形、卵形或梨形，大小 20 μm ～150 μm。

症状：病鱼的体表和鳃、鳍，肉眼可见有许多小白点；患病鱼鳃盖开闭不规则或不闭合，呼吸频率加快，口不能闭，鳃呈灰白色；病鱼浮于水面，游动迟缓，有时在四周的物体上摩擦身体；鱼体逐渐瘦弱，呼吸困难，最后衰竭而死亡。

流行：该病主要发生在 7 月 ～ 9 月高温季节，淀粉卵涡鞭虫寄生在幼鱼和成鱼的体表皮和鳃丝上，大量寄生时引起成鱼呼吸困难，而导致连续死亡。

防治：提高换水率，改善水质，减少淀粉卵涡鞭虫的附着机会。治疗方法是将鱼体在淡水中浸洗 5 min ～ 10 min，第二日重复一次；用硫酸铜 0.5 mg/L ～ 1 mg/L 全池泼洒，第二天换水 1/3 ～ 1/2，每 3 日 ～ 4 日一次，连续 3 次 ～ 4 次；用福尔马林 0.025 mL/L ～ 0.03 mL/L 全池泼洒，每 3 天 ～ 4 日一次，连续 3 次 ～ 4 次。

3.5 鱼波豆虫病

病原：鱼波豆虫(*Ichthyobodo* sp.)，为寄生鞭毛虫类，虫体侧面观卵形、梨形或近似圆形，偏于侧面的一边有一鞭毛沟，由此长出 2 根鞭毛。

症状：虫体寄生在鱼体表面、皮肤、鳃、鳍上，引起鱼皮肤及鳃分泌黏液增多，形成白絮状膜。随着病情的发展，寄生处充血、发炎、糜烂；重者个体发黑，鳞囊内积水，竖鳞，行动迟缓，常沉于水底，尾部略上翘，肌肉痉挛，呈扭曲状，肾、肝、脾肿大。

流行：该虫繁殖适宜水温为 12 ℃ ～ 20 ℃，在此水温范围内可引起 2 cm ～ 5 cm 的鲆、鲽鱼类幼鱼大量死亡。

防治：福尔马林 0.25 mL/L 浸浴鱼体 0.5 h ～ 1 h，能起到良好的杀虫效果。

4 营养性疾病

4.1 黑化病(体色变黑)

症状：病鱼体色变黑，鱼体消瘦、行动缓慢、食欲不振、视觉能力下降、生长缓慢、存活率低。解剖观察，消化道空，肝脏颜色浅淡，经检查无寄生虫和细菌性疾病。

病因：长期缺乏维生素、DHA、各种微量元素；饲料中的脂肪酸，氧化生成过氧化物，鱼体连续摄食后引起肝脏代谢障碍。

流行：一般全长 5 cm 以上的幼鱼，四季都可发生。

防治：加强营养，及时补充投喂维生素等。投喂配合饵料并且其蛋白质(必需氨基酸)、脂肪、碳水化合物、矿物质、维生素等营养成分的配比科学合理，各种添加剂用量适宜，完全能满足鲆、鲽鱼类的生长、生理需求的优质饲料。要根据不同的品种、不同的规格而合理配制相应的饵料配方，以增强鱼的体质、免疫力和抗病能力、促进其生长发育。

参考文献

[1] 蔡良候．无公害海水养殖综合技术[M]．北京：中国农业出版社，2002：33-41；90-98.

[2] 张晓刚，张胜宇，金殿凯，等. 无公害水产养殖病害的防治技术要点[J]. 齐鲁渔业，2005(2)：26-28.

[3] 韩茂森，张晓凌，宋立勋，等. 关于健康养殖(Healthy Culture)有关问题的探讨[J]. 现代渔业信息，2002，17（3）：25-27.

[4] 周丽，官庆礼. 海水鱼虾蟹贝病害防治技术[M]. 青岛：青岛海洋大学出版社，1998：32-47.

[5] 陈辉，尹文林，李秀梅，等. 水生动物病害防治技术(上册)[M]. 北京：中国农业出版社，2004：70-74；90-95.

[6] 王印庚，张正，秦蕾，等. 养殖大菱鲆主要疾病及防治技术[J]. 海洋水产研究，2004，25(6)：61-68.

[7] 窦海鸽，刘彦，张悦，等. 大菱鲆病毒性疾病的综合防治对策[J]. 渔业现代化 2004(6)：29-30.

[8] 郑春波，王世党，于诗群，等. 大菱鲆牙鲆盾纤毛虫病的防治技术[J]. 中国水产，2003(6)：51.

[9] 周进. 大菱鲆主要疾病及其防治[J]. 水产养殖，2003(1)：31-32.

[10] 金玉先，王国玲，于洪彬，等. 牙鲆大菱鲆常见疾病的防治[J]. 科学养鱼 2005(1)：50.

[11] 高永刚，郭旭强，王大建，等. 牙鲆网箱养殖技术及病害防治探讨[J]. 齐鲁渔业，2006(9)：9-10.

我国水产品质量安全监管技术新动向

林洪　李梦哲　米娜莎

中国海洋大学 食品科学与工程学院,山东青岛 266003

摘　要　近年来,我国水产品消费量不断增加,水产品供应链愈加复杂,一些新技术在水产品生产中应用的同时伴随着新的风险产生,非传统食品安全问题也日益凸显。由此,水产品质量安全监管面临着更大的挑战。目前,我国的水产品安全性虽然已经达到可控的状态,质量检测与控制技术也获得了长足进步,但新兴问题的出现促使新型水产品质量安全监管技术的发展。本文依据近年来国内外相关研究报道,对水产品风险监测、原料溯源、危害控制、产品召回以及突发事件的应急评估5方面的新型监管技术进行综述,为水产品监管部门提供技术支撑,并为确保水产品质量安全提供参考。

关键词　风险监测;溯源;危害控制;召回;应急评估

1　引言

近些年来,我国水产品消费量不断增长,同时随着贸易全球化的发展,水产品的供应链也更加复杂[1]。据统计,2014年1～11月我国水产品来进料加工贸易出口量达到106.3万吨,同比增长0.9%[2]。水产企业的全球化、跨境电商的兴起和发展[3],都使得我国对水产品安全的监管面临更大的挑战。同时,新技术诸如辐照技术[4]、新型生物制剂[5]以及转基因技术[6]等的应用在为消费者提供更多可选择的优质产品的同时,也对我国水产品质量安全的监管提出了更高的要求。除此之外,日益突出的非传统食品安全问题也促使我国加强对水产品质量安全的监管[7],为消费者提供更安全、更放心的水产品。本文将从水产品产业链的质量与安全监测、原料追溯、危害因子控制、问题水产品召回以及突发水产品安全事件应急评估5个方面进行技术方法的论述,为水产品质量与安全的监管提供技术支撑。

2　风险监测:发展实时快检技术,突破非定向筛查监测

根据我国《食品安全风险监测管理规定》,食品安全风险监测是通过系统和持续地收集食源性疾病、食品污染以及食品中有害因素的监测数据及相关信息,并进行综合分析和及时通报的活动[8]。然而在实施监测计划的过程中存在一些急需改进的技术问题。

随着新兴技术的革新，更多便捷实用的监测或检测装置可应用在水产品全产业链中。水产品中富含蛋白质、脂质等营养物质，易在贮运、销售过程中因控温变化而使其品质劣化甚至腐败变质[9]。目前，以时间为基础的传统食品标签制度虽然可以提供指示食品腐败程度的保质期，但其受限于反映食品在实际贮运过程中所经历的温度历史，难以精确指示食品的新鲜程度。因此，对于生食或生鲜水产品而言，更为准确而直观地判断水产品新鲜度的方法的确立是十分必要的。比如基于酶反应的时间－温度指示器（TTIs）可指示冷冻水产品贮运过程中的温度及品质变化历程，可较准确地反映水产品的新鲜度[10]。另外，生鲜水产品因其新鲜味美、营养价值高而广受消费者喜爱，但其货架期较短，对其进行监管检测的周期会影响其食用质量。因此，快速检测技术因其操作简便而快速、经济又准确的优点[11]能较好地满足生鲜水产品初筛检测的要求，这也是目前生鲜水产品检测技术的主要发展趋势[12]。免疫层析技术是广泛认可且适用的快检方法[13]，由于水产品种类繁多、成分复杂、影响因素较多而易造成假阳性结果。目前正在对现有技术进行改进，不断优化胶体金免疫层析法以满足水产品的快检要求。例如改进前处理过程、消除样品的基质效应[14]、基底材质的优化[15]，采用辅助的读数设备实现检测结果的半定量和定量[16]以及建立水产品多残留检测的方法[17]等，实现水产品快速检测技术的逐步完善和发展。

水产品的风险监测不仅需要排除已知有害因素的风险，更要及时发现并排除未知风险因子的危害。目前我国对于水产品中化学危害物的检测主要以定向检测为主，即已经明确了要检测的目标物，然后按照标准进行检测。然而，现有的监测方法不能实现对未知的潜在危害物的发现和检测，不能有效地预防未知潜在危害物。2008年的“三聚氰胺”事件足以引起我们对于未知危害物监测的重视。吴永宁[18]在针对“十三五”国家重点研究计划在食品安全部署的工作中介绍，食品化学危害物检测从定向检测向非定向筛查转变，发展食品污染物和恐怖毒物人群健康危害的生物标志物监测技术。这一监测重点的转移就需要相应技术方法的支持，如色谱[19]、质谱[20]、核磁共振[21]等大型精密设备走进水产品企业，及时、有效地监测潜在危害物是否存在于水产品中，为安全水产品的供给提供保障。

3　原料溯源：革新溯源预警技术，集成安全信息大数据

水产品原料是水产品生产和消费的源头，因此对其进行质量与安全的控制是极其重要的。近些年来，水产品原料的安全性事件时有发生，以低值虹鳟冒充高值三文鱼，使用不适宜食用的油鱼来冒充银鳕鱼，将养殖大黄鱼粉饰冒充野生大黄鱼，将非阳澄湖的大闸蟹标注为原产地阳澄湖，将淡水鱼糜冒充海水鱼糜。这些现象的出现不仅影响了水产品的声誉，有些甚至危害消费者的健康。由此，水产原料物种及真伪鉴别技术成为了水产品质量与安全控制中的新兴研究热点。

水产品原料溯源技术是建立在消费者与生产商、加工商间的重要桥梁。目前可应用于水产品监管的追溯技术研究较为成熟，主要分为4类，包括物理追溯、化学追溯、生物追溯

以及感官追溯。其中物理追溯是目前应用较广的技术方法,包括条码、射频识别技术(Rapid frequency identification, RFID)等[22]。以蟹苗种为例,上海已经采用RFID和二维码对其进行唯一识别,并结合GPS和GIS技术,建立了基于Web的中华绒螯蟹蟹种精细养殖物联网智慧服务系统,构建了中华绒螯蟹蟹种质量的全程动态追溯体系[23]。对于消费者而言,感官追溯是最直观的鉴别方法,可以结合水产品的外观、大小、形态、色泽、质地等综合因素[24],对水产品来源地进行追溯。而以同位素指纹为标志的化学追溯[25,26]和以不同物种DNA为特征的生物溯源[27,28]都可为产地溯源提供科学的身份鉴定信息。然而目前溯源预警系统仍较分散,地域性特征较明显,尚未集成涵盖广域水产品的风险预警的大数据系统。因此,将溯源预警从分散趋于统一,整合全国乃至全球水产品资源追溯信息是目前水产品溯源的主要发展方向。例如,建立水产品的生物信息数据库,其中包含水产品的特征信息如红外光谱图谱等,这样可有效解决信息不对称问题,便于消费者对水产品进行鉴真辨伪、追根溯源。

原料溯源可以追溯水产品产业链信息,利于水产品的质量安全监管。另外,一些食源性疾病的爆发更需要通过技术手段对病源进行追溯以防止影响扩大。近些年来,微生物导致的食源性疾病广受关注[29],然而常用的微生物检测方法仅限于指示已知微生物的种类和数量。微生物的全基因组测序技术成熟后,不仅能快速检测特定、已知的微生物,也可以将未知的微生物一次性筛选出来。这也是"十三五"规划中的既定目标,"微生物诊断溯源由传统技术向下一代全基因测序转移"[18],这不仅能够快速诊断由微生物引起的食源性疾病并进行溯源,而且对新型抗体的识别和制备技术的发展都大有裨益。

4 危害控制:以脆弱性评估引导控制,消减水产品产中产后危害

水产品的质量安全在对危害因子进行监测、溯源的同时,更需要主动地控制,从而减少水产品安全事件的发生,包括对水产品原料、水产品产中及产后的危害因素的控制与消减工作。以冷冻牡蛎为例,近几年,我国出口日本的冷冻牡蛎产品被多次通报检出腹泻性贝毒(DSP),然而调查结果显示,其生物检测方法的结果为假阳性。这是因为牡蛎产品中富含多不饱和脂肪酸酯,在低温贮藏情况下水解酶依然存有活性而产生游离脂肪酸,生物检测法的结果显示其具有较强的毒性而呈假阳性[30]。对现有传统检测方法存在问题的认识促进更为科学的检测方法的出现和替代,比如质谱、液相等仪器检测的引进;同时也科学地指导了水产品贮运条件的合理控制,确保水产品安全。比如,−25 ℃贮存温度条件下,贮存时间长达12个月的牡蛎产品中游离脂肪酸含量也未达到临界值,一直处于安全范围内[30]。这样就可以通过已有的生产技术成果指导安全控制措施,从而确保水产品的安全。

另外,关于水产品安全的控制理念也随近几年安全事件的发生特征而转变。"十三五"规划中提出"食品安全控制理念从HACCP向脆弱性评估为基础全程控制转变"[18]。HAC-

CP[31]即危害分析和关键控制点，能够确保食品在消费的生产、加工、制造和食用等过程中的安全，在危害识别、评价和控制方面是一种科学、合理和系统的方法，这对加工贮藏过程中病原微生物的防控[32]，渔药、重金属以及环境污染物等典型危害因子的消减及脱除等[33]都有重要意义。然而，食品掺假、替代等风险的存在就急需脆弱性评估（BRC评估）[34]的实施，这一评估方法旨在水产品的生产、贮运、销售以及消费者使用过程中对任何水产原料、辅料掺假、替代风险进行严格分析，从而避免此类现象出现。在水产品监管中该评估方式主要涉及食物链脆弱性评估、产品真实性溯源、产地污染以及水产品安全控制等技术，将最新的检测仪器和方法应用于水产品全产业链，同时促进水产品现场和在线监管所需的新型技术。

5　产品召回：水产品监管信息化，杜绝问题水产品危害扩散

根据食品召回制度的规定，由食品生产者自己主动、或者经国家有关部门责令，对已经上市销售的不符合食品安全标准的食品，由生产者公开回收并采取相应措施，及时消除或减少食品安全危害的制度。

由于消费者获得水产品质量信息的成本高，供求双方对水产品质量信息掌握的不对称性[35]，导致市场自身对水产品质量的调节失效。食品安全事件发生后，其召回方式的有效性决定着对消费者的影响范围[36]。美国泰森食品在遭到消费者针对产品异味和食用产品后有轻微病状的投诉之后召回了相关产品。美国农业部食品安全检验局（FSIS）于2015年11月16日发出通告，要求泰森食品公司召回的鸡翅大概有52 486磅，这些鸡翅有异味，涉嫌掺假。目前官方尚未收到消费者食用这些产品后导致不良反应的确认报告[37]。1998年12月美国沙莉集团召回了近3 500万磅的熟食和热狗，但因召回不及时，导致爆发了由李斯特菌引起的全球性疾病，事件导致21人死亡、100余人住院[38]。由此可见，有效、及时地召回不符合生产标准的产品才能保证食品的质量和安全。这几个案例给予了水产品召回很好的借鉴。

目前美国农业部食品安全与检验局向公众告知采取产品召回措施主要有两种方式：一是新闻发布；二是召回告知报告。我国国家食品药品监督管理总局发布的《食品召回管理办法》的第三章第十七条也对不安全食品召回方式做了相应要求，规定食品召回公告要通过省级食品药品监督管理部门网站和省级主要媒体发布，并应当与国家食品药品监督管理总局网站链接[38]。然而，水产企业多而分散，还有一些小型企业未被纳入监管系统，这就给不合格水产品及时召回带来了困难。因此，健全水产品追溯系统，使水产品在整个流通环节中"来源可查、去向可追、责任可究"，既能快速锁定风险源头，又能快速启动产品召回，提高召回效率，这也是水产品监管信息化[39]的主要要求。水产品"身份证明"不仅可以追溯原料产地及产品信息，也可以完整记录水产品在整个产业链中的流通情况，在水产品货架期内完整保留各项记录信息，为确保水产品质量安全提供保障。

6 应急评估:发展传统评估替代技术,加速应急评估进程

一些突发的水产品安全事件或者可能会影响到水产品安全却缺乏相应评估手段的事件,极易引起消费者的恐慌并有可能会威胁消费者的健康[40]。如2005年11月13日,中石油吉林石化公司双苯厂发生爆炸事故,事故区域排出的污水致使污染物硝基苯和苯对松花江水及江中的生物造成污染,造成该地区渔业生产人员的极大困惑,不知江中的水产品是否还能符合食品安全标准。由此可见,有些水产品安全事件的发生不局限于常规检测项目不合格等引发,对于该类事件需采用相应的应急评估方法预测其风险,进而提出科学的防控手段、消除生产者和消费者的恐慌。

传统的对化学物的安全性评价主要基于动物实验进行,包括急性毒性实验、亚慢性和慢性毒性实验、致突变实验、致癌实验、生殖发育毒性实验以及免疫毒性实验等[41],通过实验结果来推测化学物对人体造成的危害。但是通过上述动物实验评估化学物的安全性周期长、通量低[42],不能满足应急事件发生后短时间内评估危害物的危害程度及特点的要求。因此,寻找有效且快速地评估化学品的毒理学替代方法成为了毒理学方法学研究领域的热点。吴永宁[18]在"十三五"规划中指出"以动物实验为基础的传统评估技术向人为基础的新型评估技术的转移"。目前实现这项转移的关键技术领域是"突破人源性细胞体外替代毒性实验"。并且美国国家研究委员会(NRC)"21世纪的毒性测试:愿景与策略"这一极具里程碑意义的研究报告的发表[43],预示着毒理学实验方法将发生革命性的改变。而基于人体细胞的靶向性测试和毒性通路测试将取代以实验动物为中心的传统毒性评价方法成为研究热点[42]。除此之外,一些新型生物制剂在应用之前也需通过对其安全性进行相应评价,分子生物学技术如PCR[44]、全基因组测序分析[45]等方式都可初步推测一些新型微生物的致病因子、毒性情况。这些新技术的引进,不仅可以较为准确地评估新兴风险因子的毒性情况,还可以加速评估进程,快速、及时地向消费者报告危害程度,可有效规避水产品安全事件的发生。

7 结语

我国水产品质量安全监管技术日益完善,并可充分利用新兴技术严格防控水产品从产前、产中到产后各个环节中可能发生的危害。为避免水产品安全问题的出现,实时监测水产品全产业链中各风险因素的指标,以期及时发现并阻止安全事件的发生成为趋势。与此同时,积极控制并消减水产品产前、产中和产后各个环节中的危害因子,从而实现水产原料可溯源、危害可控制这一目标。当水产品安全事件爆发后,一方面能通过科学、有效的手段及时召回,缩小危害影响范围。另一方面,对于突发的水产品安全事件通过快速有效的应急评估手段给予消费者科学、健康的饮食指导。

参考文献

[1] 孙琛，沈媛. 基于流通渠道视角的我国水产品质量安全问题研究[J]. 食品工业科技，2014，35(13)：275–279.

[2] 陈述平，吴晨. 2014 年全国水产品进出口贸易形势及 2015 年展望[J]. 科学养鱼，2015(3)：4–5.

[3] 鲁将，靳晓泓，高洁，等. 水产品电商的品牌建设研究——以梁子湖螃蟹为例[J]. 安徽农业科学，2015，43(20)：315–316.

[4] 张宁波，乔宇，廖涛，等. 辐照技术在水产品安全控制中的应用[J]. 湖北农业科学，2011，50(22)：4537–4540.

[5] 胥亚夫，林洪，张茜，等. 冷藏牙鲆中主要腐败菌卵黄抗体的抑菌性能[J]. 食品科学，2010，31(23)：109–113.

[6] 柳淑芳，杜永芳，李振，等. 海洋生物技术的研究进展与转基因水产品安全性评价[J]. 中国食品学报，2006，6(1)：383–389.

[7] 崔秀华，严莉莉. 我国非传统食品安全问题研究[J]. 安徽农业科学，2014，42(1)：221–222.

[8] 李宁，杨大进，郭云昌，等. 我国食品安全风险监测制度与落实现状分析[J]. 中国食品学报，2011，11(3)：5–8.

[9] 姜兴为，杨宪时. 水产品腐败菌与保鲜技术研究进展[J]. 湖南农业科学，2010(15)：100–103.

[10] 葛蕾，李振兴，林洪. 基于酶促褐变原理的时间温度指示系统研究[C]. 第二届中国食品与农产品质量安全检测技术国际论坛暨展览会，北京，2013.

[11] 唐益，周俊. 食品微生物检测中快速检测法的应用探究[J]. 化学工程与装备，2015(10)：224–225.

[12] 曹立民，林洪，隋建新. 水产品中化学危害物筛选检测技术的研究应用[J]. 食品工业科技，2009，30(1)：337–340.

[13] 吴刚，姜瞻梅，霍贵成，等. 胶体金免疫层析技术在食品检测中的应用[J]. 食品工业科技，2007，28(12)：216–218.

[14] 李文玲，庄松娟，林洪，等. 海参呋喃唑酮代谢物胶体金免疫层析检测中的基质效应及消除方法[J]. 中国渔业质量与标准，2015，5(6)：35–42.

[15] Du S，Lin H，Sui J，et al. Nano-gold capillary immunochromatographic assay for parvalbumin [J]. Anal Bioanal Chem，2014，406(26)：6 637–6 646.

[16] 万宇平. 快速检测技术在食品安全监管中的应用及发展新方向[J]. 北京工商大学学报(自然科学版)，2011，29(4)：1–5.

[17] 王媛，于慧娟，钱蓓蕾，等. 胶体金免疫层析法在水产品快速检测中的应用和发展方向[J]. 农产品质量与安全，2013(1)：41–43.

[18] 吴永宁. 我国食品安全科学研究现状及“十三五”发展方向[J]. 农产品质量与安全，2015(6)：3–6.

[19] 李秀勇. 色谱法在食品分析中的应用研究[D]. 兰州：兰州大学，2008.

[20] 周凝，刘宝林，王欣. 核磁共振技术在食品分析检测中的应用[J]. 食品工业科技，2011，32(1)：325–329.

[21] 夏俊，凌培亮，虞丽娟，等. 水产品全产业链物联网追溯体系研究与实践[J]. 上海海洋大学学报，

2015，24(2)：303–313.

[22] 虞丽娟，杨劲松，凌培亮，等. 基于物联网智慧服务的中华绒螯蟹蟹种质量动态追溯系统研究[J]. 水产学报，2013，37(8)：1262–1269.

[23] 赵峰，周德庆，丛楠，等. 水产品的真伪鉴别与评价技术研究进展[J]. 中国农业科技导报，2015，17(6)：111–117.

[24] 郭波莉，魏益民，潘家荣. 同位素指纹分析技术在食品产地溯源中的应用进展[J]. 农业工程学报，2007(3)：284–289.

[25] 赵燕，吕军，杨曙明. 稳定同位素技术在农产品溯源领域的研究进展与应用[J]. 农产品质量与安全，2015(6)：35–40.

[26] 李新光. 基于DNA条形码的鱼片（肉）真伪鉴别技术研究[D]. 上海：上海海洋大学，2013.

[27] 李敏，凌超，邬琦沁，等 .DNA条形码技术在食品溯源中的应用[J]. 乳业科学与技术，2014. 37(5)：25–30.

[28] 郭旦怀，崔文娟，郭云昌，等 . 基于大数据的食源性疾病事件探测与风险评估[J]. 系统工程理论与实践，2015，35（10）：2523–2530.

[29] 沈志刚 . 牡蛎产品腹泻性贝毒检测假阳性结果的产生原因及其控制措施[D]. 北京：中国农业科学院，2009.

[30] 郝丹. 基于HACCP监测的冷冻水产品冷链物流研究[J]. 市场周刊（理论研究），2009(2)：82–83.

[31] 张宾，邓尚贵，林慧敏，等. 水产品病原微生物安全控制技术的研究进展[J]. 中国食品卫生杂志，2011，23(6)：581–586.

[32] 林洪，李萌，曹立民 . 我国水产食品安全与质量控制研究现状和发展趋势[J]. 北京工商大学学报（自然科学版），2012，30(1)：1–5.

[33] 楼甜甜，陆柏益，张星联. 脆弱性评价及其在食用农产品质量安全风险评估中的应用[J]. 食品工业科技，2015，36(24)：352–355.

[34] 郭崇义，李宁. 水产食品流通环节的食品安全问题及调控[J]. 食品科学技术学报，2013，31(2)：15–20.

[35] 王菁，刘文. 国外食品召回制度的现状与特点以及对我国的启示[J]. 食品科技，2007(12)：5–8.

[36] Johnston G N. Tyson foods inc. recalls chicken product due to possible adulteration[EB/OL].（2015–11–17）http://www. fsis. usda. gov/wps/portal/fsis/topics/recalls-and-public-health-alerts/recall-case-archive/archive/2015/recall-141-2015-release.

[37] 高秦伟. 美国食品安全监管中的召回方式及其启示[J]. 国家行政学院学报，2010(1)：112–115.

[38] 刘鹏飞，张立涛. 我国食品安全监管信息化应用体系研究[J]. 中国管理信息化，2014，17(6)：30–33.

[39] 张亚平. 突发性化学危害事件现场应急检测技术与仪器设备[J]. 中国卫生检验杂志，2006，16(3)：370–375.

[40] 戴寅，陈君石，李悠慧，等. 食品安全性毒理学评价程序和方法[J]. 医学研究通讯，1997，26(12)：9–10.

[41] 王磊. 多器官人源细胞系共培养模型的建立与初步应用[D]. 上海：第二军医大学，2014.

[42] Council N R. Committee on toxicity testing，assessment of environmental agents，board on environmental

studies, toxicology, institute for laboratory animal research. toxicity testing in the 21st century: a vision and a strategy [Z]. Washington, DC: National Academies Press, 2007.

[43] Guilbaud C, Morris C E, Barakat M, et al. Isolation and identification of *Pseudomonas syringae* facilitated by a PCR targeting the whole *P. syringae* group [J]. FEMS Microbiol Ecol, 2015.

[44] Kang H W, Kim J W, Jung T S, et al. wksl3, a new biocontrol agent for *Salmonella enterica* serovars enteritidis and typhimurium in foods: characterization, application, sequence analysis, and oral acute toxicity study[J]. Appl Environ Microbiol, 2013, 79(6): 1956–1968.

鲆鲽类养殖不同产业链模式的经济效益比较研究

王微波　杨正勇　李佳莹

上海海洋大学经济管理学院,上海 201306

摘　要　产业结构直接影响产业经济效益,而产业结构是产业链长期发展和优化的产物。产业链越长,分工经济越明显,但随之而来的交易环节增多会导致费用增加,分工带来的经济效益的提升有可能被交易费用的增加所抵消。同一产业内不同养殖户对产业链各环节的不同组合会形成不同的产业链模式。对不同产业链模式经济效益的比较研究,不仅有利于养殖户效益最大化,且有利于产业结构的优化,进而促进整个产业的发展。本文以我国鲆鲽类养殖为例,基于2013年不同鲆鲽类养殖模式经济效益的跟踪调查,用秩和检验的方法对"育苗 + 成鱼养殖"模式、"仅育苗"模式及"仅成鱼养殖"模式的成本、收益和资本回报率进行了对比分析。结果显示:① 专业化分工程度越低的产业链模式,其可变成本越低,固定成本中的要素成本越高;② 专业化分工程度越高的产业链模式,单位净收益越高,资本回报率越高。基于此,建议管理部门通过合理整顿市场来提高专业化分工程度,为养殖生产者经济效益提供更好的市场环境,建议养殖生产者加强技术创新和科学管理能力从而提高经济效益。

关键词　鲆鲽;产业链模式;专业化分工;经济效益

2014年,我国的水产养殖产量占到全国水产品总产量的73.49%,已连续多年超过捕捞量。作为我国的水产养殖业、尤其是海水鱼类养殖的重要代表,鲆鲽类养殖业在海水养殖第四次浪潮中脱颖而出并迅速发展起来。随着养殖过程中现代化建设的推进,科技先导型养殖观念深入人心,工厂化养殖及产业集群现象越来越普遍。在现代水产养殖中,旨在提高效益的竞争不仅依靠单个生产环节,而且有赖于产业结构及产业链各环节的合理化。产业结构是产业链持续发展与优化的产物[1,2]。随着产业链的拉长,交易费用会增加,分工经济提升的效益就有可能被交易费用的增加所抵消。因此,养殖户为了最大化利润,会通过理性对比交易费用与内部管理组织费用的高低来选择是进行市场交易还是推进企业内分工,进而决定纳入养殖户内部的产业链环节数量,从而最终决定产业链模式。不同的产业链模式打破了养殖户一体化发展的模式,在追求利润最大化过程中不再仅仅着眼于一体化的全产业链模式,而是选择了更为灵活的专业分工和协作。对比分析不同专业化分工程度的产业链模式的经济效益情况,有利于揭示在当前养殖环境下,不同产业链模式的生产成本收益情况,从而为养殖生产者在专业分工的选择方面提供科学依据。

尽管亚当•斯密等古典经济学家早就有关于分工的论断，但国外20世纪80年代以来水产养殖经济研究的著作多数集中于技术经济效益评估[3-7]，从产业经济角度开展的研究甚少，且仅聚焦于水产养殖产业链的某一环节[8-11]，关于不同产业链模式的经济效益分析的著作就更少。国内这方面研究起源于21世纪初，但可能由于数据获得性较低等原因，研究以定性方法为主，多数选择从宏观层面分析水产养殖产业链的现状[12-23]。纵观国内外文献，用定量方法研究水产养殖产业链模式的相关文献仍较少，尚有较大拓展空间。鉴于此，本文以中国鲆鲽类种业为例，对2013年主产区的两类产业链模式展开跟踪调研，运用非参数秩和检验的分析方法，对比分析不同模式的要素成本、净收益和资本报酬率，剖析两种不同专业化程度的产业链模式的成本、收益及资本投资回报情况，从而评估出在现有环境下，鲆鲽类养殖产业链的最优模式，以期为产业链模式优化研究抛砖引玉。

1　研究对象、方法及数据来源

1.1　研究对象的描述

本研究采用了比较分析法，针对不同专业化分工程度的鲆鲽类养殖产业链模式，分析其要素成本、收益以及资本投资回报率。在调查中发现，鲆鲽类养殖产业链模式可根据专业化分工的程度分为“育苗+成鱼养殖”模式、“仅育苗”模式及“仅成鱼养殖”模式。“育苗+成鱼养殖”模式是指育苗与成鱼养殖在同一经济主体内进行，将交易费用内部化，产生更多养殖户内部的管理组织费用，是专业化分工程度较低的产业链模式；“仅育苗”模式及“仅成鱼养殖”模式指育苗、成鱼养殖两个环节在两个不同经济主体之间进行的产业链模式，专业化分工程度较高，彼此间产生的费用为养殖户之间的交易费用。本研究以“育苗+成鱼养殖”模式、“仅育苗”模式及“仅成鱼养殖”模式为研究对象来评估在当前环境下不同程度专业化分工的产业链模式的经济效益情况。

1.2　研究指标及其计算方法

1.2.1　研究指标

本文研究指标主要有两方面：要素成本和收益。其中净收益等于总收益减去总成本，总收益为该经营主体销售所养殖的产品而获得的总收入。为方便对比，将所有成本及收益转换为单位产量成本与收益。

1.2.2　计算方法

（1）成本收益及资本回报率。总成本包括固定成本和可变成本。其中固定成本为不随产量变动的成本，根据养殖生产者的实际情况，将厂房折旧、设备折旧、租金、维修费用、一般管理人员工人工资归为固定成本。可变成本为随着养殖产量变动的成本，根据实际情况，将苗种（鱼卵）支出、鱼药支出、饲料支出、水电煤支出、临时工工资、其他支出归为可变成本。

资本回报率用于衡量投出资金的使用效率,本文利用该指标来衡量不同产业链模式的经营主体对资金的利用效率情况。

(2) 非参数秩和检验。秩和检验指的是以秩和作为统计量进行假设检验的方法,包括参数秩和检验和非参数秩和检验。由于样本总体分布类型未知,故本文选用非参数秩和检验,即不考虑总体分布类型是否已知,不比较总体参数,只比较总体分布的位置是否相同的统计方法。根据实际情况,在进行三组对比时用 Kruskal-Wallis 秩和检验。

Kruskal-Wallis 秩和检验用来检验多个样本是否来自相同或同分布的总体,先将所有样本看成单一样本,然后按照由小到大的顺序统一编秩,转换成秩样本后再对其进行方差分析。与方差分析的区别在于构造的统计量不是组间平均平方和除以组内平均平方和,而是组间平方和除以全体样本秩方差。需要检验的原假设 H_0: 各组之间不存在差异; 备择假设 H_1: 各组之间存在差异。采用卡方分布检验。

Kruskal-Wallis 秩和统计量 $KW = \frac{12}{n(n+1)}\sum_{i=1}^{k}\left(\frac{R_i}{n_i} - \frac{n+1}{2}\right)^2$

如果样本中存在结值,将上述公式调整为: $KW_c = \frac{KW}{1 - \frac{\sum(\tau_j^3 - \tau_j)}{n^3 - n}}$

式中: 表示 Kruskal-Wallis 秩和统计量, k 表示有 k 组样本, n_i 表示第 i 组样本中的观察数, n 表示所有样本中的观察总数, R_i 表示第 i 组样本中的秩和, μ 表示平均值, σ 表示标准差, τ_j 表示第 j 个结值的个数。KW_c 表示调整后的 Kruskal-Wallis 秩和统计量。

在本研究中,原假设为不同产业链模式之间不存在差异。若原假设为真,则秩将大约均匀分布在样本中,说明不同产业链模式不会导致经济效益差异;若备择假设为真,则说明产业链模式的不同会导致经济效益存在显著差异,此时进而比较两种模式之间的平均值便可说明其中某种模式之优劣。

1.3 数据来源

本文数据来源主要来自两部分:一是资料查阅,包括《中国渔业统计年鉴(2015)》和《中国渔业统计年鉴(1949—1985)》;二是通过调查问卷对相关数据进行采集。此调研数据包括国家鲆鲽类产业技术体系产业经济岗位团队的调研数据和该体系各综合试验站在 2013 年跟踪调研数据。调研区域涵盖了鲆鲽类主产区辽宁、河北、天津、山东、江苏、福建等省市 40 多个区县。根据实地调研观察发现,选择"仅育苗"和"育苗 + 成鱼养殖"养殖模式的总量相对较少,因此抽样调查的户数也较少。共回收有效问卷 89 份,其中"仅育苗"模式 6 份,"育苗 + 成鱼养殖"模式 11 份,"仅成鱼养殖"模式 72 份。

1.4 基本信息描述

从养殖户规模来看,职工总数在 1～116 人,拥有 2 位职工数的最多占 25.45%,职工数在 10 人以下的养殖户占 83.64%，自有劳动力人数在 1～10 人，拥有 1 位自有劳动力的养殖户最多，占 58.13%，说明养殖户以家庭为单位的养殖形式居多；调研的三类模式中养殖户主年龄在 32～68 岁之间；从受教育程度看，43.51% 的养殖者为高中毕业，33.77% 为初中水平，14.29% 的养殖者为小学程度，这说明养殖户的文化水平还普遍较低，但同时也有高学历养殖者。表 1 描述了抽样调查中不同产业链模式产量的基本情况。

表 1　2013 年不同产业链模式养殖概况

产业链模式	样本量	产量（$\times 10^4$）/kg			
		平均值	最大值	最小值	标准差
“仅育苗”模式	6	46.04	100	11.25	33.08
“育苗＋成鱼养殖”模式	11	89.39	309.00	13.75	87.62
“仅成鱼养殖”模式	72	4.79	21.75	0.28	4.09
总计	89	18.02	309.00	0.28	84.99

2　结果及分析

2.1　成本分析

从整体成本结构来看，三种模式的所有要素成本都存在显著差异，并以三种方式呈现：

（1）专业化分工程度低的产业链模式的成本 < 专业化分工程度高的产业链模式的成本。即表现为“育苗＋成鱼养殖”模式的成本 <“仅育苗”模式的成本 <“仅成鱼养殖”模式的成本，或者“育苗＋成鱼养殖”模式的成本 <“仅成鱼养殖”模式的成本 <“仅育苗”模式的成本。

（2）“仅育苗”模式的成本 <“育苗＋成鱼养殖”模式的成本 <“仅成鱼养殖”模式的成本；

（3）专业化分工程度高的产业链模式的成本 < 专业化分工程度低的产业链模式的成本。即表现为“仅育苗”模式的成本 <“仅成鱼养殖”模式的成本 <“育苗＋成鱼养殖”模式的成本，或者“仅成鱼养殖”模式的成本 <“仅育苗”模式的成本 <“育苗＋成鱼养殖”模式的成本。

第一种差异方式表明，专业化分工程度越低的产业链模式成本越低。主要体现在苗种（卵）、鱼药、水电煤和一般管理人员工人工资上，且从表 2 中可发现，该项差异大部分体现在可变成本中。说明育苗环节内部化有助于降低苗种（卵）、鱼药和水电煤，从而降低可变成本，通过降低一般管理人员工人工资来降低固定成本。从每个要素成本上看，专业化分工程度比较低的产业链模式在苗种（卵）上支出降低幅度最大，比“仅成鱼养殖”模式的低 74.7%；从下降的绝对量上看，专业化分工程度低的模式水电煤支出最大，“仅成鱼养殖”模式的平均值为4.36元/千克,而“育苗+成鱼养殖”模式的仅1.55元/千克,两者差额达到2.81元/千克。造成这种专业化分工程度的低模式可变成本低的原因可能是,当前单独的育苗和成鱼养殖环节的技术水平有限,专业化分工程度加深后,技术水平上升带来的技术效率不明显,导致经济组织内部管理费用低于企业间交易费用低。

第二种差异方式表明,专业化分工程度在该项成本上作用并不明显,其主要原因是混养模式起到成本分摊的作用。在以这种方式体现要素成本差异的方式中,“育苗+成鱼养殖”模式的要素成本位于“仅育苗”模式和“仅成鱼养殖”模式之间。在总成本、固定成本、可变成本上均有体现。整体上看,“育苗+成鱼养殖”模式单位总成本为20.61元/千克,略低于“仅育苗”模式和“仅成鱼养殖”模式的平均值23.68元/千克;“育苗+成鱼养殖”模式可变成本为13.76元/千克,略低于“仅育苗”模式和“仅成鱼养殖”模式的平均值17.47元/千克;但是,“育苗+成鱼养殖”模式固定成本为6.85元/千克,要略高于“仅育苗”模式和“仅成鱼养殖”模式的平均值6.21元/千克。在作为可变要素的饲料及临时工工资、作为固定成本的厂房折旧和维修费上,“育苗+成鱼养殖”模式的成本均要高于“仅育苗”模式和“仅成鱼养殖”模式的平均值。

第三种差异方式表明,专业化分工程度越高的产业链模式成本越低。从表2看,主要体现在租金和设备折旧上,以此影响固定成本。说明提高专业化分工程度更有利于固定资产的充分利用,从而降低固定成本。就租金而言,“仅育苗”模式和“仅成鱼养殖”模式的平均值比“育苗+成鱼养殖”模式低1.45元/千克;对于设备折旧来说,“仅育苗”模式和“仅成鱼养殖”模式的平均值比“育苗+成鱼养殖”模式低0.66元/千克。

综上,产业链模式的专业化分工程度对于成本的大致影响在于:

(1)整体上看,产业链模式的专业化分工程度在总成本、固定成本以及可变成本上的影响并不显著。相对来说,“育苗+成鱼养殖”模式起到的成本分摊的作用更明显。成本整体呈现如下趋势:“育苗+成鱼养殖”模式的要素成本位于“仅育苗”模式和“仅成鱼养殖”模式之间;

(2)分开来看,产业链模式的专业化分工程度越低,可变成本中的要素成本越低,固定成本中的要素成本越高。可能原因在于将育苗环节内部化后,有助于在苗种(卵)、鱼药、水电煤上降低可变成本,并通过一般管理人员工资降低固定成本。但同时也需注意到,将育苗环节内部化后,可能增加租金和设备折旧从而使得固定成本升高。

表2 2013年不同产业链模式要素成本秩和检验对比分析

要素成本/(元/千克)	Kruskal-Wallis检验卡方值	平均值			差异呈现方式
		“育苗+成鱼养殖”模式	“仅育苗”模式	“仅成鱼养殖”模式	
苗种(卵)支出	36.30***	0.73	0.94	2.87	①
渔药支出	7.91**	0.27	0.30	0.39	①
饲料支出	30.32***	11.35	0.89	19.36	②

（续表）

要素成本/(元/千克)	Kruskal-Wallis 检验卡方值	平均值			差异呈现方式
		“育苗+成鱼养殖”模式	“仅育苗”模式	“仅成鱼养殖”模式	
水电煤费	11.30***	1.55	3.08	4.36	①
临时工工资	11.67***	0.47	0.02	0.62	②
其他支出	21.06***	0.12	1.85	0.23	①
可变成本	33.77***	13.76	7.07	27.87	②
厂房折旧	8.35**	2.01	0.71	2.86	②
租金	9.56***	1.76	0.04	0.58	③
维修费用	15.30***	0.92	0.04	0.97	②
一般管理人员工人工资	11.53***	1.20	2.93	3.70	①
设备折旧	49.45***	0.95	0.46	0.12	③
固定支出	6.83**	6.85	4.18	8.23	②
总成本支出	33.84***	20.61	11.25	36.10	②

*、**、*** 分别表示在 10%、5%、1%水平下显著

2.2　收益及资本回报率分析

从收益角度看，专业化分工程度越高的产业链模式收益越高，对资金利用程度也越高。从表 3 中发现，专业化分工程度高的“仅育苗”模式与“仅成鱼养殖”模式的单位净收益均为正，分别为 1.61 元/千克和 5.52 元/千克；资金回报率也相对较高，分别为 0.15 和 0.22。而专业化分工程度低的产业链模式的净收益及资金回报率为负，且 Kruskal-Wallis 秩和检验表明不同专业化分工的产业链模式的净收益与资本回报率在 1%水平存在显著差异。净收益以及资本回报率对比情况均体现为：“育苗+成鱼养殖”模式 <“仅育苗”模式 <“仅成鱼养殖”模式。

在综合考虑成本与收益情况下，“仅育苗”模式与“仅成鱼养殖”模式要比“育苗+成鱼养殖”模式的经济效益要高。“仅育苗”模式相比于“育苗+成鱼养殖”模式，成本低约一半，并且在净收益与资本回报率上为正；“仅成鱼养殖”模式相比于“育苗+成鱼养殖”模式，成本更高，净收益与资本报酬率均为正。

表 3　2013 年成本收益秩和检验分析

变量	Kruskal-Wallis 秩和检验卡方值	平均值		
		“仅育苗”模式	“育苗+成鱼养殖”模式	“仅成鱼养殖”模式
总成本支出/(元/千克)	33.84***	11.25	20.61	36.10
净收益/(元/千克)	9.40***	1.61	−1.71	5.52

(续表)

变量	Kruskal-Wallis 秩和检验卡方值	平均值		
		“仅育苗”模式	“育苗+成鱼养殖”模式	“仅成鱼养殖”模式
资本回报率 ROIC	6.16**	0.15	−0.10	0.22

*、**、*** 分别表示在 10%、5%、1%水平下显著

3 结论及建议

3.1 结论

本文用秩和检验的方法分析了 2013 年我国 40 多个地区不同产业链模式下鲆鲽类养殖的经济效益,结论如下:

(1) 我国鲆鲽类养殖产业链模式按专业化分工程度分,有“育苗+成鱼养殖”模式、“仅育苗”模式及“仅成鱼养殖”三种模式。其中“育苗+成鱼养殖”模式专业化分工程度相对较低,“仅育苗”模式及“仅成鱼养殖”模式专业化分工程度较高。三种模式中“仅成鱼养殖”模式占据的比例较高,其余两种模式则相对较低。

(2) 鲆鲽类养殖的经济效益受到产业链模式的显著影响。不同产业链模式下鲆鲽类养殖的成本和收益不同,并最终影响到鲆鲽类养殖的资本回报率。秩和检验的结果表明“育苗+成鱼养殖”模式的整体要素成本位于“仅育苗”模式和“仅成鱼养殖”模式之间,三种模式中可变成本中的要素成本越低,固定成本中的要素成本越高。“育苗+成鱼养殖”模式的收益低于“仅育苗”模式和“仅成鱼养殖”模式。三种模式的资本回报率对比情况是“育苗+成鱼养殖”模式<“仅育苗”模式<“仅成鱼养殖”模式。

(3) 产业链模式对鲆鲽类养殖经济效益的影响因专业化分工程度而异。专业化分工程度更高的“仅育苗”模式及“仅成鱼养殖”模式经济效益明显高于专业化程度较低的“育苗+成鱼养殖”模式。高度的专业化分工有利于育苗和成鱼养殖企业技术提升,进而带来技术效率,从而降低整体要素成本并提高单位净收益。

3.2 对策建议

(1) 整顿产业,促进专业化分工。建议有关部门通过合理整顿产业来提高专业化分工程度,进而加速整个鲆鲽类养殖的产业结构转型。充分考虑种业产业链前端的特殊性,对于亲本质量进行严格控制及技术研发将是整个产业链模式健康发展的基础;有效监督交易过程,防止道德风险及信息不对称带来交易成本的加大,这是产业链模式发展的关键因素;适当宣传产业转型相关知识,稳定养殖生产者投资心理,这是产业链模式发展的最有效推动力。

(2) 创新技术,提高专业化程度。建议养殖生产者加强技术创新和科学管理能力从而

提高经济效益。专业化分工程度高的产业链模式将会是一个趋势，养殖生产者提高养殖的专业化分工程度能更好地改善当前利润空间被压缩的现状。加强养殖过程中的技术创新和科学管理能力，通过提高专业化程度来提升经济效益，不仅可提高养殖户资本回报率，同时也可为整个产业转型和发展夯实基础。

3.3　不足及展望

本研究还存在以下问题需要在将来进一步探讨：① 因条件有限，研究选取的样本较少。本文仅有 89 份有效样本。在条件允许的情况下，增加样本量可能会使本研究的结果更加稳健。② 本文采用的研究指标不够全面。影响养殖户经济效益的因素很多，受条件限制，很多因素，比如组织化程度、品牌建设情况、资源禀赋条件等并未加以考虑，因此本文得到的是一个初步的研究结果，亟待后续更为深入的研究。

参考文献

[1]　林学钦. 传统渔业向现代渔业转化论述[J]. 厦门科技，2003(2)：18–21.

[2]　赵绪福. 农业产业链优化的内涵、途径和原则[J]. 中南民族大学学报(人文社会科学版)，2006(6)：119–121.

[3]　Baliao D D, Franco N M, Agbbayani R F. The economics of retarding milkfish growth for fingerling production in brackishwater ponds[J]. Aquaculture, 1987, 62(3/4): 195–205.

[4]　Santiago C B, Pantastico J B, Baldia S F, et al. Milkfish (*Chanos chanos*) fingerling production in freshwater ponds with the use of natural and artificial feeds[J]. Aquaculture, 1989, 77(4): 307–318.

[5]　Keshavanath P, Gangadhar B, RAMESH T J, et al. Effects of bamboo substrate and supplemental feeding on growth and production of hybrid red tilapia fingerlings (*Oreochromis mossambicus* × *Oreochromis niloticus*)[J]. Aquaculture, 2004, 235(1/4): 303–314.

[6]　Ofor C O. A comparison of the yield and yield economics of three types of semi-intensive grow out systems, in the production of *Heterobranchus longifilis* (Teleostei: Clariidae) (Val. 1840), in Southeast Nigeria[J]. Aquaculture, 2007, 269(1/4): 402–413.

[7]　Bonnieux F, Gloaguen Y, Rainelli P, et al. Potential benefits of biotechnology in aquaculture: The case of growth hormones in French trout farming[J]. Technological Forecasting and Social Change, 1993, 43(3/4): 369–379.

[8]　Sapkota P, Engle C, Heikes D, et al. Cost analysis of a mobile fish nursery system for growing hybrid striped bass fry[J]. Aquacultural Engineering, 2012, 49: 18–22.

[9]　Bui T M, Phuong N T, Nguyen G H, et al. Fry and fingerling transportation in the striped catfish, *Pangasianodon hypophthalmus*, farming sector, Mekong Delta, Vietnam: A pivotal link in the production chain[J]. Aquaculture, 2013, 388/391: 70–75.

[10] Nasr-allah A M, Dickson M W, Al-kenawy D A R, et al. Technical characteristics and economic performance of commercial tilapia hatcheries applying different management systems in Egypt[J].

Aquaculture, 2014, 426/427: 222-230.

[11] 农业部渔业局养殖处. 我国水产种苗生产现状和存在的问题[J]. 渔业致富指南，1998(16): 4-5.

[12] 佟屏亚. 中国种业上市公司的现状与发展[J]. 中国种业，2002(4): 11-14.

[13] 李巍，胡红浪. 分析我国水产苗种产业发展规律预测未来发展趋势[J]. 中国水产，2006(12): 66-70.

[14] Li W, Hu H L. Analysis of development rule and trend of Chinese fingerling industry[J]. China Fisheries, 2006(12): 66-70.

[15] 苏跃中. 福建省水产苗种产业的现状及发展思路[J]. 现代渔业信息，2006(9): 21-23.

[16] 胡红浪. 我国水产养殖种苗现状及发展对策[J]. 科学养鱼，2007(10): 1-3.

[17] 勾维民，刘佳丽. 大连水产苗种产业现状及发展对策[J]. 黑龙江水产，2012(2): 36-38.

[18] 毛汉奇，曾可为，李明光. 打造武汉水产苗种产业链的初步设想[J]. 养殖与饲料，2012(9): 26-28.

[19] 杨正勇，郭鸿鹄，张钰研. 鲆鲽类苗种技术创新与推广对策——基于生产者技术需求调研的思考[J]. 渔业信息与战略，2012(3): 183-188.

[20] 王清印. 我国水产种业现状及发展愿景[J]. 当代水产，2013(11): 50-52.

[21] 相建海. 中国水产种业发展过程回顾、现状与展望[J]. 中国农业科技导报，2013(6): 1-7.

[22] 李巍，贾延民，鲁国延. 我国水产种业产业现状、面临挑战及发展途径[J]. 中国水产，2014(6): 19-23.

[23] 唐曾，韩兴勇. 基于主成份分析法的中国水产苗种产业区域划分及影响因素研究[J]. 中国农学通报，2014(2): 89-94.

中国大菱鲆养殖产业集群的影响因素研究

曹自强 杨正勇 李佳莹 杨明晨

上海海洋大学经济管理学院，上海 201306

摘 要 近年来，随着大菱鲆养殖产业的发展，产业出现养殖成本上涨、收购价格走低、水资源匮乏等问题，而产业集群作为当今一种极有效率的产业经济组织形式，能实现资源的优化配置，提升企业和产业的竞争力。故本文先通过区位熵测算我国大菱鲆养殖主产区的产业集聚度，再结合GEM模型从基础、企业和市场三方面构建产业集群的影响因素指标体系。然后在我国2013年大菱鲆养殖产业主产区的数据基础上，运用灰色关联理论评价各影响因素与产业集聚度的关联程度，表明：基础要素中技术推广水平、海水资源禀赋的关联度较高，市场需求和大型生产者数量的关联度分别在市场要素和企业要素中较高。因此建议：加大养殖技术培训力度，加快研发大菱鲆育种繁殖技术和病害防治技术；推动循环水养殖模式，规范养殖企业标准化管理方式；建立合理的价格策略和加大产品宣传；引进和扶持大型企业，推行合作社或股份公司的建设。

关键字 大菱鲆养殖；产业集群；GEM模型；区位熵；灰色关联理论

1 引言

1992年，我国以稚鱼的形式引进大菱鲆，并在短短几年里，实现苗种规模化的培育，同时创立了“温度大棚＋深井海水”的工厂化养殖模式，此后吸引了大批投资者进入该领域，从事大菱鲆养殖。我国大菱鲆养殖区域最初集中在山东莱州，短时间内迅速扩大到山东全省，然后逐步向全国沿海省地区扩展[1]。2005年，我国大菱鲆养殖产量为5万吨，占全球总量的87.5%，虽然，2006年受上海“多宝鱼风波”事件的影响，养殖产量下降至4万吨以下，但是在社会各界的共同努力下，迅速恢复，产业规模于2007年后平稳发展[2]，到2013年主产区的养殖产量已达到5.63万吨。与此同时，根据国家鲆鲽类产业技术体系产业经济岗位团队的实地调查发现，近年来，大菱鲆养殖产业出现养殖成本上涨、收购价格走低、水资源匮乏等问题，不利于大菱鲆养殖产业的竞争力和可持续发展[3]。

针对上述问题，本文试图从产业经济学相关研究中寻找解决思路，发现产业集群有利于提升企业和产业的竞争力，实现资源的优化配置，降低企业间的交易费用，最终提高产业的经济效率[4]。关于产业集群的研究最早可追溯到19世纪末英国经济学家马歇尔的产业区理论，他认为产业的大量企业的地理集聚可以产生地方化的外部规模经济，即地方化经济[5]。此后，国外学者从经济学和经济地理学等不同视角研究产业集群理论。其中，最具代表性的学者有韦伯[6]、克鲁格曼[7]、迈克尔·波特[8]等。国内学者也在产业集群的内涵、形成原因、竞争优势及测度方法等方面进行了研究[9-12]，但研究对象主要集中在工业、服务业等方面[13-15]，对农业尤其是水产养殖产业集群的研究相对较少。近年来，国内外学者关于水产养殖产业集群的研究主要有 Label、Anh 和 Ha 等关于虾养殖产业集群的研究[16-18]和李艳红、韩振芳等关于鲆鲽类养殖产业的集聚现象的研究[19, 20]。目前，鲆鲽类养殖产业已发展成为我国北方海水养殖业的重要产业，在一定程度上代表了我国水产养殖业的发展方向，尤其是海水养殖业朝着工业化现代渔业发展之方向[21]。国家鲆鲽类产业技术体系的调查表明，大菱鲆养殖产业作为我国鲆鲽类养殖业的代表性产业，已经出现了养殖生产者在某一区域的地理集中现象，并表现出了一定的产业集群特征[22]，但是国内外学者尚未对该产业集群的水平进行测度，也未对影响这种产业集群现象的因素进行深入的分析。故本文将利用区位熵系数测算大菱鲆产业集群的集聚程度，了解大菱鲆养殖主产区的集群现状，并结合 GEM 模型分析该产业集群的影响因素，再运用灰色关联理论评价影响程度。最后，根据结果分析和讨论因素对产业集群的影响，并提出相应的建议，为解决大菱鲆产业所面临的困难提供参考。

2 中国大菱鲆养殖产业集群

国家鲆鲽类产业技术体系调查表明，2013 年我国大菱鲆养殖总面积已达到 610. 74 万平方米，工厂化养殖面积增长到 596. 01 万平方米，占养殖总面积的 97. 6%，养殖产业集中在山东、辽宁、河北、天津、江苏、浙江和福建 7 个沿海省市，其中，山东省和辽宁省分别占 47. 2% 和 38. 6%。依据经合组织（OECD）关于农业产业集群的界定，即农业产业集群是指一组在地理上相互邻近的以生产和加工农产品为对象的企业和互补机构，在农业生产基地周围，由于共性或互补性联系在一起形成的有机整体[23]。那么，在一定区域内，大菱鲆养殖个体、养殖企业、相关机构组织等要素在地理上的集中是否可看作是广义上的农业产业集群？在下文中，笔者根据 2013 年鲆鲽类产业经济体系采集、整理的大菱鲆养殖主产区的产业数据和《中国渔业统计年鉴（2013）》的水产养殖数据，利用区位熵对大菱鲆养殖区域的产业集聚度进行了测度，进而判断各区域的大菱鲆养殖产业是否存在产业集群现象。

区位熵也称为专门化率，是反映某一区域内某一产业部门在其更大的区域范围内的相对集中程度，从而确定该区域的产业集聚程度，识别出这一地区的优势产业[14]。区位熵 LQ_{ij} 的计算公式如下：

$$LQ_{ij} = \frac{q_{ij}/q_j}{q_i/q}$$

其中，j 代表地区，i 代表大菱鲆养殖产业，LQ_{ij} 就是 j 地区的大菱鲆养殖产业在全国的区位熵，q_{ij} 为 j 地区的大菱鲆养殖产业的养殖产量，q_j 为 j 地区所有水产养殖产业的养殖产量，q_i 指在全国范围内大菱鲆产业的养殖产量，q 为全国所有水产养殖产业的养殖产量。计算结果见表 1。

表 1　大菱鲆养殖业区域集聚度

地区(j) / 系数	山东(1)	辽宁(2)	河北(3)	天津(4)	江苏(5)	福建(6)	浙江(7)
LQ_{ij} (X_0)	3. 16	5. 16	3. 95	1. 48	0. 44	0. 24	0. 09

数据来源：大菱鲆养殖主产区的养殖产量数据来源于鲆鲽类产业经济数据库，全国及各养殖主产区的所有水产养殖产业的养殖产量数据来源于《中国渔业统计年鉴》(2013)

一般来说，当 $LQ_{ij}>1$ 时，我们认为 j 地区的大菱鲆养殖产业出现产业集群，系数越大说明集聚程度越高，则该地区的产业集群具有比较优势；当 $LQ_{ij}<1$ 时，系数越接近 0 表示 j 地区的大菱鲆养殖产业的集聚现象越不明显，在全国来说具有劣势[14]。不难发现，我国的山东、辽宁、河北和天津四个地区的大菱鲆养殖产业已出现明显的产业集群现象，浙江、江苏和福建的产业集群现象却不明显。故本文选用存在产业集群的山东、辽宁、河北、天津四个地区，利用 GEM 模型对大菱鲆养殖产业集群的影响因素进行分析。

3　中国大菱鲆产业集群影响因素研究的理论模型、方法与数据

3.1　GEM 模型及指标构建

GEM 模型是 Padmore 和 Gibson 对波特“钻石模型”进行改进后的一种基于区域范围分析产业竞争力的模型[24]。目前，国内利用 GEM 模型对产业集群的影响因素研究主要文献有张晗、吴晨等[25, 26]。GEM 模型将产业集群的影响因素分为三个方面：基础、企业和市场。其中，基础包括资源与设施二因素，主要涉及自然资源禀赋、交通运输条件、科研机构及政府政策等；企业包括“养殖企业和相关辅助产业”与“企业的结构战略和竞争”二因素；市场包括本地市场与外部市场二因素。在 GEM 模型中，各方面的每个因素之间都具有良好的互补作用，如图 1 所示。

本文根据大菱鲆养殖产业的实际情况和集群特征，对影响大菱鲆产业集群的因素及指标进行构建，如表 2 所示。

3.2　数据来源

本文利用农业部鲆鲽类产业经济项目组 2013 年的实地调研数据，其中包括辽宁、天津、河北、山东、江苏、浙江、福建等沿海省份的上千家养殖生产者。另外，本文分析中国大菱鲆养殖产业集群的影响因素情况，还涉及各地区相关指标的基本数据主要来源于《中国统计年

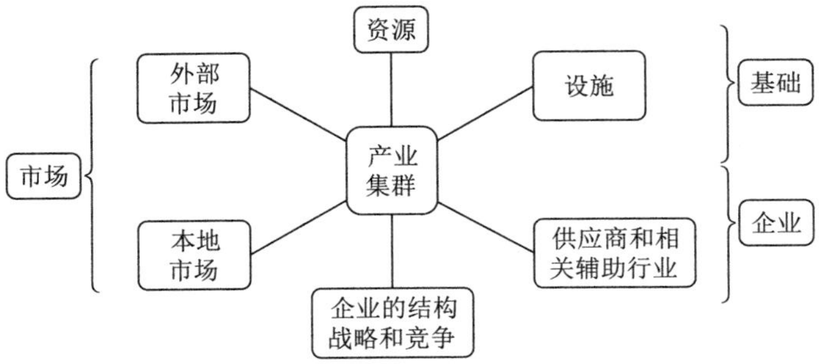

图 1　GEM 模型

表 2　大菱鲆养殖产业集群的影响因素及指标体系

要素	因素	相关因素	衡量指标（X_i）
基础要素	资源	沿海土地资源	近岸及海岸面积（X_1）
		海水资源	海水可养殖面积（X_2）
		劳动力资源	海水养殖的专业从业人数（X_3）
	设施	基础设施	交通、仓储、邮政业的投资额（X_4）
		技术推广水平	水产技术推广人员数量（X_5）
		政府经费支出	水产技术推广经费支出（X_6）
企业要素	企业的结构	大型养殖生产者	生产者数量（X_7）
		养殖生产者	养殖生产者总数量（X_8）
	供应商与辅助行业	苗种企业	苗种企业数量（X_9）
		水产饲料企业	饲料企业数量（X_{10}）
市场要素	本地市场	市场需求	家庭人均水产品类消费支出（X_{11}）
	外部市场	市场份额	区域该产业销量占全国总销量的比（X_{12}）

鉴》、《中国渔业统计年鉴》、《中国海洋统计年鉴》和《中国饲料工业统计年鉴》。

3.3　评价方法

灰色系统理论是一种运用特定方法描述信息不完全的系统并进行预测、决策、控制的崭新的系统理论，由邓聚龙教授于 1982 年提出[27]。此后，谭学瑞、邓聚龙提出灰色关联分析法（GRA），它是根据因素之间发展态势的相似或相异程度来衡量因素间关联程度的方法[28]。目前，国内利用该方法评价因素对产业集群的影响程度的主要文献有白丁和钟祥喜[29, 30]。本文尝试利用该方法，结合 GEM 模型下的指标体系，评价影响因素对大菱鲆产业集群的影响程度，评价过程如下所示。

（1）变量选取及说明。

本文将反映区域产业集聚程度的区位熵作为关键因素，即主行为因素，记作 X_0；影响因

素的衡量指标 $X_1 - X_{12}$ 作为非关键因素，即比较因素。因此，可构建影响因素空间 S_{inu} 表示如下：

$$S_{\text{inu}} = \left\{ X_i(j) \middle| \begin{array}{l} i = 1, 2, 3, \cdots, 12; j = 1, 2, 3, 4; X_0 = (3.16, 5.16, 3.95, 1.48), \cdots \\ X_{12} = (42.32\%, 45.49\%, 3.08\%, 1.16\%) \end{array} \right\}$$

上式中，$X_i(j)$为地区的指标的数值，例如 $X_0(1)$表示山东省的区位熵系数。各地区具体指标的指标值，如表 3 所示。

表 3　2013 年大菱鲆养殖主产区的产业集群及影响因素的相关数据

省份(j) 指标	山东(1)	辽宁(2)	河北(3)	天津(4)
X_0	3. 16	5. 16	3. 95	1. 48
X_1 / 千公顷	1 210. 90	738. 10	278. 80	58. 1
X_2 / 千公顷	358. 21	725. 84	111. 37	18. 49
X_3 / 人	166 395	11 5380	27 022	1 210
X_4 / 亿元	1 657. 20	1 070. 10	1 543. 30	729. 90
X_5 / 人	3 404	3 033	1071	375
X_6 / 万元	13 324. 40	6 125. 00	3 584. 15	1 796. 76
X_7 / 个	123	234	36	10
X_8 / 个	1 669	640	42	10
X_9 / 个	124	4	4	12
X_{10} / 个	14	5	1	4
X_{11} / 元	384. 33	474. 35	177. 91	564. 76
X_{12}	42. 32%	45. 49%	3. 08%	1. 16%

（2）计算灰色关联度的方法。

从上文可以看出，在影响因素空间中，X_i 均为极大值极性，但是各因素量纲不同，且数量级不同。故本文为了解决空间中的序列的不可比问题，第一步，对原始序列进行无量纲化处理，本文选用均值化处理方法，得到新序列 $x_0 - x_{13}$，并构建灰色关联因素空间 S_{GRE}，如下所示。

例如：$x_0(j) = [x_0(1), ..., x_0(4)]$

$$= \left[\frac{x_0(1)}{\text{avg}[x_0(j)]}, \frac{x_0(2)}{\text{avg}[x_0(j)]}, \frac{x_0(3)}{\text{avg}[x_0(j)]}, \frac{x_0(4)}{\text{avg}[x_0(j)]}\right]$$

$$= (0.92, 1.50, 1.15, 0.43)$$

则：$S_{\text{GRE}} = \{x_0(j) \mid i = 0, 1, 2, ..., 12; j = 1, 2, 3, 4; x_0(j) = [x_0(1), ..., x_0(4)], ..., x_{12}(j)\}$

上式中，$x_0(j)$表示区位熵序列，$\text{avg}[x_0(j)]$为平均值。

第二步，令 $x_0(j)$为参考序列，$x_1(j), ..., x_{12}(j)$为比较序列，分别计算比较序列对参考

序列的差异序列 Δ_{0i}(取绝对值)。

例如:$\Delta_{01}(1) = |x_0(1) - x_1(1)| = |2.12 - 0.92| = 1.20; \Delta_{01} = |\Delta_{01}(1), ..., \Delta_{01}(4)|$

则:$\Delta_{0i} = \left\{ \Delta_{0i}(j) \middle| \begin{matrix} i = 1, 2, ..., 12; j = 1, 2, 3, 4; \\ \Delta_{01} = [\Delta_{01}(1), ..., \Delta_{01}(4)], ..., \Delta_{012} \end{matrix} \right\}$

第三步,求 Δ_{0i} 上的两级上、下环境参数:

$$\Delta_{0i}(\max) = \max(i)\max(j)\Delta_{0i}(j)$$

$$\Delta_{0i}(\min) = \min(i)\max(j)\Delta_{0i}(j)$$

上式中,$\Delta_{0i}(\max)$为差异序列 Δ_{0i} 的最大值,$\Delta_{0i}(\min)$为差异序列 Δ_{0i} 的最小值。

第四步,确定分辨系数ρ,通常在最少信息原理下取$\rho = 0.5$。其关联系数的计算公式为:

$$\gamma(x_0, x_i) = \frac{\Delta_{0i}(\min) + \rho\Delta_{0i}(\max))}{\Delta_{0i}(j) + \rho\Delta_{0i}(\max))}$$

最后,可以计算出灰色关联度:

$$R(x_0, x_1) = \frac{1}{4}\sum_{j=1}^{4} \gamma(x_0, x_i) \quad (0 < R < 1)$$

上式中,当 R 越接近于 1,说明比较序列与参考序列的关联度越高,即各因素对该产业集群的影响程度越高。

4 结果与讨论

4.1 结果

通过以上评价方法,得到最终评价结果如下表 4 所示。

表 4 比较序列对参考序列的关联度值(取小数点后 4 位)

序列对	关联度	序列对	关联度
$R(x_0, x_1)$	0.712 4	$R(x_0, x_7)$	0.712 6
$R(x_0, x_2)$	0.723 3	$R(x_0, x_8)$	0.617 0
$R(x_0, x_3)$	0.723 0	$R(x_0, x_9)$	0.577 7
$R(x_0, x_4)$	0.719 8	$R(x_0, x_{10})$	0.639 1
$R(x_0, x_5)$	0.782 6	$R(x_0, x_{11})$	0.751 2
$R(x_0, x_6)$	0.708 8	$R(x_0, x_{12})$	0.664 0

从上表可得,各影响因素与产业集群的区位熵系数的关联度从大到小的顺序依次为:技术推广水平、市场需求、海水资源、劳动力资源、基础设施、大型养殖生产者、沿海土地资源、政府经费支出、市场份额、水产饲料企业、养殖生产者、苗种企业。其中,技术推广水平的关

联度为0.782 6，其次是市场需求（0.751 2）和海水资源（0.723 3），最小的因素是苗种企业数量，关联度为0.577 7。

4.2　讨论

根据上述结果，本文从GEM模型的基础要素、市场要素、企业要素三个方面讨论相关因素对大菱鲆养殖产业集群的影响。

（1）基础要素

基础要素中技术推广人数对该产业集群的影响程度最高，关联度为0.782 6，说明在具备基本相同自然条件的环黄渤海区域的四个省份，大菱鲆养殖技术推广水平已成为影响大菱鲆产业集群发展的最核心因素。纵观我国大菱鲆养殖业的发展历程，产业开始集聚于山东省，该地区涌现出一批以雷霁霖院士为代表的科学工作者突破大菱鲆养殖的创新技术，结合水产养殖技术的知识外溢效应，在该地区迅速形成集聚式的技术创新，对大菱鲆养殖产业集群的发展奠定了坚实的基础[20]。时至今日，山东省依然是大菱鲆养殖产业技术创新最活跃的区域。另外，近些年在政府主导下这批科学工作者逐渐将山东省关于大菱鲆苗种繁育、养殖模式、饵料技术等方面的成果推广到辽宁省、河北省和天津市等地区，从而产业从开始集聚于山东省，渐渐向集聚于山东省、辽宁省和河北省转变，加快推进大菱鲆养殖产业集群的发展。特别是在辽宁省，根据国家鲆鲽类产业技术体系的跟踪调查，辽宁省于2013年大菱鲆养殖产量为24 820吨，其中葫芦岛占了98.1%，而葫芦岛又主要集中于兴城，兴城市工厂化设施养殖园区是大菱鲆的主要养殖区之一，占了83.4%[3]。该地区产业集群的快速发展，很大程度上归功于政府不断加大的技术推广力度和水产技术推广人员的不懈努力。

基础要素中另外一个重要因素就是海水资源禀赋，其关联度为0.723 3，说明我国大菱鲆养殖产业是以资源为依托的产业[22]，自然资源禀赋依然是形成我国大菱鲆养殖产业集群的基础性因素。现阶段我国大菱鲆养殖业基本实现了以“温室大棚+深井海水”为主的工厂化养殖模式，丰富充裕的地下海水资源是建造温室大棚及进行养殖的必要条件。另外，处于渤海周边的山东省、辽宁省、河北省和天津市，拥有合适的盐度和温度等基础条件的海水资源，且较东海周边的福建、江苏和浙江，其海水环境污染程度也相对较低。

（2）市场要素

从上述结果可知，市场要素中市场需求与大菱鲆产业集聚水平的关联度为0.751 2，仅次于技术推广水平，说明市场需求对大菱鲆养殖产业集群发展的拉动作用较为显著，家庭人均水产品类消费支出越高，反映出消费者对大菱鲆产品的市场需求总量越大，则对该地区大菱鲆养殖产业集群的影响程度越明显。此外，市场要素对产业集群发展的影响还表现为以下三个方面：第一，根据跟踪调查发现，大菱鲆养殖生产者多为家庭式作坊企业，生产的产品多为劳动密集型产品，产品加工能力不足，产品差异化较低，并且产品销售以专门化市场定期上门收购的方式为主，生产者对市场需求具有很强的依赖性。第二，根据演化经济地理理论，经济活动的空间集聚并非完全依赖于企业和消费者的理性区位决策，地方化的产业历史和文化的积淀为产业集聚提供了基本的需求起点。从我国鲆蝶类养殖业的历史看,我国环渤海地区一直有鲆蝶鱼捕捞、生产和消费的习惯,从而对该地区大菱鲆养殖产业集群的形成提供了条件。第三,随着大菱鲆产品在国内的市场知名度的提高,消费者对高蛋白低脂肪食物的青睐愈加明显,大菱鲆产品的消费方式在不断转型,家庭消费、网络营销等消费方式不断兴起,在产品价格低位运行、消费者收入增长、城市化及信息化浪潮等因素的综合影响下,传统消费市场进一步扩大,未来国内三线、四线城市及县城的市场需求越来越大,使得市场要素变得更加坚实。

(3) 企业要素

本文所考虑的企业要素主要分为企业结构和辅助行业。从结果上看,养殖大型企业数量对大菱鲆养殖产业集群的影响程度较高,关联度为 0.712 6,说明大型养殖企业数量的增加,可以提高市场集中度和资源配置效率,有利于市场品牌产品乃至区域性品牌的建设,提升企业定价权,避免行情的大幅波动,使得行业利润率不断增加,进而推动了产业集群的发展。但是辅助行业对产业集群的影响程度较低,饲料企业和苗种企业的关联度分别为 0.639 1 和 0.577 7,说明苗种企业与饲料加工企业在区域内较为分散,集群企业自身管理和与相关行业的合作还未能产生较大的优势,相关辅助行业的现状是大菱鲆养殖产业集群发展中的一个薄弱环节。从现实情况来看,大菱鲆产业的产业结构相对单一,产业集群内各成员间的关系更多地表现为一种直线式关系而不是网络关系,同时,产品的深加工与对外贸易在整个产业链中附加值较低,不利于产业集群的发展。

5 结论与建议

综述所述,本文得到以下结论:

第一,通过区位熵系数对我国 2013 年大菱鲆养殖主产区的产业集聚程度进行了测算,结果发现,山东、辽宁、河北和天津存在产业集群现象,其中辽宁省的集聚程度最高,具有明显的比较优势,其次是河北省与山东省,而江苏、浙江和福建产业集群尚不明显。

第二,利用灰色关联分析法对 GEM 模型下的各影响因素与大菱鲆养殖产业集群区位熵系数的关联度做了实证分析,结果表明技术推广水平是首要因素,市场需求是次要因素,影响程度均超过了海水资源禀赋,而以往学者对水产养殖领域的研究[18, 22],普遍认为资源禀赋是影响该产业集群最重要的因素,本文认为存在差异的可能原因是实证研究中选取的四个省份具备基本相近自然条件和海水资源,故其对产业集群的影响被弱化。

第三,从前文的讨论可知,基础要素对大菱鲆养殖产业集群的影响主要表现在技术推广水平和海水资源禀赋两个因素。根据产业的发展现状,大菱鲆养殖产业是比较容易产生技术外溢的产业,技术创新一经突破就迅速被周边地区获取,故区域内技术推广人数恰好反映了技术传播和创新的效率,对产业集群发展提供了支持。另外大菱鲆养殖产业是以资源为依赖的产业,特别是海水资源禀赋,其为养殖产业集群的形成与发展提供了基础条件。其次,近年来市场要素中市场需求对产业集群的影响愈加明显，在产品价格低位运行、消费者收入增长的情况下，大菱鲆产品满足了消费者对高蛋白低脂肪食物的需求，市场认知度提升及消费市场扩大，拉动了产业集群的发展。最后在企业要素下，大型养殖生产者数量的增加，提高区域市场集中度和资源配置效率，提升产业的规模经济水平，促使产业集群更有竞争力。

基于上述结论，结合目前产业集群发展的状况，提出以下四点建议：

第一，提高大菱鲆养殖水平，促进养殖技术推广与创新。首先政府应继续加大技术推广经费的投入，在人事编制和待遇上努力实现大菱鲆养殖产业技术人才入驻到乡镇一级的技术推广站，以科技减少日渐增加的人工成本。其次应注重水产行业人才培育，加强对年轻科技人才的培养。通过扩建水产类高等院校规模，吸引更多师生从事渔业技术学习与研究；再者水产行业协会和企业通过对本行业人员进行技术培训，提高其技术开发和应用能力。最后，加强大菱鲆产业研发费用投入，尤其是大菱鲆育种繁殖技术、饲料开发技术及病害防治技术方面的经费投入，加快饲料加工、投喂技术以及鱼病细胞工程技术的研发。

第二，政府与企业共同努力，实现资源的可持续利用。从宏观层面，政府加强政策引导，科学制定相关产业集群发展的规划、理念，保护本地区自然资源环境，同时政府、行业和科技合力指引产业确立以工厂化循环水养殖理念为指导，加快养殖生产由粗放型向集约型的转变，实现以节约资源、降低成本、提高养殖企业收入为目的的可持续发展。从微观层面，企业应将 ISO9001 质量管理体系、ISO14001 环境管理体系和 AS8000 社会责任体系等标准引入，提高产品的生产质量，树立企业的社会、环境责任，推行循环水养殖模式，合理利用自然资源。

第三，继续推进市场需求的拉动作用，扩大产品的消费市场。在市场良性发展的前提下，一方面企业采取合理的定价方略，使用取脂定价、渗透价格、满意价格等策略或根据季节和区域采取差别定价策略；同时迎合内外部市场的不同需求，制定合理的内部定位和外部定位，通过影视、网络等加大广告投入，努力打造大菱鲆优质集群品牌，提高市场影响力。另一方面，推进本地专业化市场的建设，形成核心供给力，吸引需求，优化投资环境，壮大市场主体；在充分尊重市场规律的基础上，开辟新的消费渠道，例如家庭消费、网络营销等方式。

第四，引进和扶持大型生产者的发展，并推行行业协会、合作社或股份公司的建设。政府引进和扶持大型企业的发展，发挥大型企业有效带动产业发展的作用。通过行业协会、合作社或股份公司的建设，引导企业间良性竞争和养殖生产者、苗种企业、饲料加工企业、流通企业的有效合作，充分发挥其在技术、溯源和质量标准等方面对养殖生产者的养殖规范、生产协调与管理的整合作用，努力将各环节个体集合后实现共赢。

参考文献

[1] 雷霁霖，门强，王印庚，等. 大菱鲆“温室大棚＋深井海水”工厂化养殖模式[J]. 海洋水产研究，2002，4: 1-7.

[2] 雷霁霖，刘新富，关长涛. 中国大菱鲆养殖 20 年成就和展望——庆祝大菱鲆引进中国 20 周年[J]. 渔业科学进展，2012，4: 123-130.

[3] 国家鲆鲽类产业技术研发中心. 国家鲆鲽类产业技术体系年度报告(2013)[M]. 青岛:中国海洋大学出版社,2014:9-13.
[4] 陈剑锋,唐振鹏. 国外产业集群研究综述[J]. 外国经济与管理,2002,8:22-27.
[5] 阿弗里德·马歇尔. 经济学原理[M]. 廉运杰译. 北京:华夏出版社,2005:213-243.
[6] 阿尔弗雷德·韦伯. 工业区位论[M]. 李刚剑,张英保译. 北京:商务印书馆,1997:12-13.
[7] 保罗·克鲁格曼. 空间经济学——城市、区域和国际贸易[M]. 梁琦译. 北京:中国人民大学出版社,2005.
[8] Porter M E. Clusters and the new economics of competition[J]. Harvard Business Review, 1998, 76(6): 77-90.
[9] 王辑慈. 创新的空间——企业集群与区域发展[M]. 北京:北京大学出版社,2001.
[10] 徐康宁. 当代西方产业集群理论的兴起、发展和启示[J]. 经济动态,2003(3):71-74.
[11] 盖文启. 创新网络——区域经济发展新思维[M]. 北京:北京大学出版社,2002.
[12] 王子龙. 产业集聚水平测度的实证研究[J]. 中国软科学,2006(3):109-116.
[13] 罗勇,曹丽莉. 中国制造业集聚程度变动趋势实证研究[J]. 经济研究,2005,08:106-115.
[14] 陈建军,陈国亮,黄洁. 新经济地理学视角下的生产性服务业集聚及其影响因素研究——来自中国222个城市的经验证据[J]. 管理世界,2009,04:83-95.
[15] 朱英明,杨连盛,吕慧君,等. 资源短缺、环境损害及其产业集聚效果研究——基于21世纪我国省级工业集聚的实证分析[J]. 管理世界,2012,11:28-44.
[16] Lebel L, Mungkung R, Gheewala S H, et al. Innovation cycles, niches and sustainability in the shrimp aquaculture industry in Thailand[J]. Environmental science & policy, 2010, 13:291-302.
[17] Anh P T, Dieu T T M, Mol A P J, et al. Carolien Kroeze, Simon R. Bush. Towards eco-agro industrial clusters in aquatic production: the case of shrimp processing industry in Vietnam[J]. Journal of Cleaner Production, 2011, 19(17/18): 2107-2118.
[18] Ha T T T, Bush S R, Van Dijk H. The cluster panacea: Questioning the role of cooperative shrimp aquaculture in Vietnam [J]. Aquaculture, 2013, 4: 89-98.
[19] 李艳红,徐忠. 山东省鲆鲽类养殖企业的集聚问题分析——基于SCP范式分析框架[J]. 中国渔业经济,2013,6:83-88.
[20] 韩振芳,杨正勇. 中国鲆鱼养殖的产业集聚:水平、原因及政策[J]. 中国工程科学,2014,9:93-99.
[21] 杨正勇,王春晓. 全球视野下中国鲆鲽类养殖业的发展[J]. 中国渔业经济,2009,6:115-121.
[22] 杨正勇,冷传慧,徐忠. 穿越转型的漩涡[M]. 北京:中国农业出版社,2011.
[23] 李春海,张文,彭牧青. 农业产业集群的研究现状及其导向:组织创新视角[J]. 中国农村经济,2011,3:49-58.
[24] Padmore T, Gibson H. Modeling systems of the innovation: A framework for industrial cluster analysis in region[J]. Research Policy, 1998, 26: 625-641.
[25] 张晗,吕杰. 农业产业集群影响因素研究[J]. 农业技术经济,2011,02:85-91.
[26] 吴晨. 基于GEM模型的我国省际间对虾产业集群竞争力研究[J]. 渔业信息与战略,2012,2:102-107.

[27] 邓聚龙. 灰色系统综述[J]. 世界科学，1983，7：1-5.

[28] 谭学瑞，邓聚龙. 灰色关联分析：多因素统计分析新方法[J]. 统计研究，1995，3：46-48.

[29] 白丁，卞子全，杨向东，等. 不同省域纺织产业集群竞争力的比较研究[J]. 生态经济（学术版），2011，2：162-167.

[30] 钟祥喜，肖美华，刘金香. 基于 G～2EM-CI 模型的物联网产业集群竞争力研究[J]. 统计与决策，2013，15：39-42.

[31] 黄书培，杨正勇. 不同养殖规模下大菱鲆工厂化养殖经济效益分析[J]. 广东农业科学，2011，16：113-116.

第四篇

轻简化实用技术

大菱鲆“多宝1号”优质苗种培育

1　技术名称

大菱鲆“多宝1号”优质苗种培育

2　技术要点

2.1　亲本选择

严格按照快速生长和高成活率性状遗传分析的结果选用亲鱼。选择规格：雌、雄鱼1.5龄以上，体重750 g以上。

2.2　亲本培育

亲鱼在3龄以上可进行人工生殖调控。亲鱼日常培育：利用人工配制的配合饲料、软颗粒饲料和饲料鱼等饲喂。配合饲料应符合NY 5072—2002的要求。引用SC/T 2031—2004大菱鲆配合饲料。配合饲料的日投饲量为鱼体重的1%～2%，鲜活饵料的日投喂量为1.5%～3.0%，每日投喂1～2次。

2.3　人工繁殖

人工挤压鱼体腹部法分别采集成熟的卵和精液，干法授精。按1×10^5粒卵加入1 mL～5 mL精液，快速搅拌均匀使精卵充分接触，再加入少量经沉淀沙滤的海水（符合NY 5052—2001的规定），使精液、卵子、水的体积比约为0.5∶100∶100，继续搅拌1分钟，然后静置5分钟，再加入海水，静置10分钟～15分钟，待吸水膨胀后，清水冲洗1～2次，放入2 000 mL量筒中，用海水使上浮卵和沉淀卵分离，记录上浮卵数，上浮卵经消毒后放入孵化器中孵化，消毒用药应符合NY 5071—2002的规定。

2.4　受精卵孵化

受精卵放入80 cm × 60 cm × 60 cm网箱中孵化，使之呈漂浮状态，孵化网箱露出水面10 cm左右。

2.4.1 孵化条件

水质：应符合 NY5052 的海水养殖用水水质的规定。

光照度：100 lx～2 000 lx，以 500 lx 为最好。

水温：12℃～16 ℃，以 13℃～15 ℃最好；盐度：28～35；pH 7. 8～8. 2；溶解氧 ＞6 mg/L；氨氮 <0. 1 mg/L。

2.4.2 孵化密度

受精卵按 30 万粒 / 立方米～60 万粒 / 立方米的密度，放入 0. 5 m^3～1 m^3 的孵化网箱中孵化。

2.4.3 孵化管理

充气：为保持孵化池中有充足的氧气，首先在孵化池中保持循环流水，并用若干充气石充气，使孵化池内溶解氧的水平保持在 6 mg/L，在每个孵化箱中央安置气石 1 个，以保持微流水和水流动，使受精卵在水体中呈均匀分布状态。

水流量：每天水的循环量保持在 2～3 个量程。

吸沉卵：根据沉卵量的多少及时吸出沉卵。

2.4.4 孵化周期

在 12 ℃～14 ℃条件下，一周以内孵化。在 13 ℃条件下，116 h 孵化。

2.4.5 出膜及管理

破膜后将用光滑的器皿将仔鱼移入饲育槽中，分离卵膜和仔鱼，调整水量，保证溶氧充足，同时清除死苗，保持清洁卫生。进入苗种培育程序。

2.5 苗种培育

2.5.1 培育池

分前期培育池和后期培育池。

前期培育池：圆形或方形水泥池，面积 10 m^2～20 m^2，深 0. 8 m～1. 0 m。

后期培育池：面积 20 m^2～40 m^2，水深 1 m～1. 5 m，有独立的进、排水口；池底向排水孔以一定的坡度倾斜，以利于排水。

2.5.2 培育水质

盐度：苗种培育的盐度以 20～40 为宜。

温度：水温 13 ℃～18 ℃。早期仔鱼培育期，水温应与孵化水温一致，第 2 d 开始缓缓升温，10 d 后升到 16 ℃～18 ℃，并稳定在 18 ℃。

光照强度：500 lx～2 000 lx，光线应均匀、柔和。

pH：7. 8～8. 2。

溶解氧 6 mg/L 以上。

2.5.3　培育密度

培育密度根据水温、溶解氧、氨氮等水平而定。一般情况下，初孵仔鱼密度 1×10^4 尾/平方米～2×10^4 尾/平方米。仔鱼体重 0.1 g，培育密度 2 000 尾/平方米～3 000 尾/平方米，换水量 5～6 个循环/日；仔鱼体重 0.5 g，培育密度 1 500 尾/平方米～2 000 尾/平方米，换水量 6～8 个循环/日；变态伏底稚鱼（体重 2 g）1 000 尾/平方米～2 000 尾/平方米，换水量 8～10 个循环/日。小球藻添加量：在苗种培育早期，从进水管以微流速加入，使水体中小球藻的浓度保持在 8 万个/毫升～10 万个/毫升水体，一方面用于保持水色，另一方面提高轮虫活力。苗种培育期间使用的小球藻应新鲜无污染。

2.5.4　轮虫添加量

轮虫作为开口饵料。从孵化后 3 d 投喂，连续投喂 15 d～20 d，每日投喂 2 次～4 次，轮虫每次投喂使水体达到 5 个/毫升～10 个/毫升，苗种培育期间使用的轮虫应冲洗干净，无病原。

2.5.5　卤虫无节幼虫

从 9 d～10 d 开始投喂卤虫无节幼体，连续投喂 20 d 左右；每日投喂 2 次～4 次，卤虫每次由开始的 0.1 个/毫升～0.2 个/毫升，逐步增加至 1 个/毫升～2 个/毫升。苗种培育期间使用的卤虫应与卤虫壳完全分离。

2.5.6　微粒配合饵料

第 12 d～15 d 开始投喂颗粒配合饲料直至育苗结束。配合饲料的安全卫生指标应符合 NY5072 的要求。投喂配合饲料，大菱鲆苗种 25 d 前颗粒饲料粒径 250 mm～400 mm，0.1 g～0.15 g 体重的仔稚鱼颗粒饲料粒径微 40 mm～600 mm，0.5 g 体重颗粒饲料粒径微 800 μm 左右，随着鱼体的生长，配合饲料粒径逐渐增加。配合饲料日投饲量为鱼体重的 5%～15%。饲料颗粒大小适口，投喂及时，宜少投勤投。

2.5.7　池底吸、排污

采用专用的清底工具（丁字形吸污器），一般每天清底 1～2 次。

2.5.8　水量管理

1 d～5 d 仔鱼可采用静水培育方式，日换水量可由 1/5 增至全部换水，日换水次数可由每天 1 次逐步增至每天两次。从 6 d 开始建立流水培育程序，水交换量随仔鱼的生长和密度的增大而逐步增加，可渐增至 3～4 个循环/日；仔鱼体重 0.1 g，换水量 5～6 个循环/日；仔鱼体重 0.5 g，换水量 6～8 个循环/日；变态伏底稚鱼（体重 2 g），换水量 8～10 个循环/日。

2.5.9　分苗

随着育苗的生长应定期进行分苗。孵化后 15 d～20 d 进行首次分苗，第 30 d～35 d 可以进行第二次分苗，第 60 d进行第三次分苗。第一次和第二次分苗可从密度上加以稀疏，第三次则需按大、中、小三个等级进行分拣,分类培育。

3 适宜区域

适宜在山东、河北、辽宁、天津、江苏等沿海省市人工可控的海水水体或地下井水水体中养殖。

4 注意事项

为保证亲鱼质量,只从国家级原良种场山东烟台天源水产有限公司获得亲鱼,并遵守授权生产协约。

5 技术依托单位及联系方式

技术依托单位:中国水产科学研究院黄海水产研究所
联系人:马爱军
通讯地址:山东省青岛市南京路 106 号
联系电话:0532-85835103
E-mail: maaj@ysfri. ac. cn

一种“北鲆2号”健康养殖模式

1　技术名称

一种“北鲆 2 号”健康养殖模式

2　技术要点

2.1　室内苗种培育

在牙鲆育苗室内把“北鲆 2 号”苗种培育到 2 cm～4 cm。

2.2　池塘混养

在池塘水温稳定在 18 ℃以上时，池塘投放 2 cm～4 cm 的“北鲆 2 号”苗种，与缢蛏、对虾、海蜇等混养。

2.3　捕获越冬或工厂化接力养殖

在池塘水温下降到 18 ℃以下时，用袖网（雁翅网）捕获“北鲆 2 号”大规格苗种，一部分入室内越冬，翌年接着池塘养殖，一部分销售给工厂化养殖户接力养殖。

3　适宜区域

辽宁、河北、天津、山东近海。

4　注意事项

池塘投放苗种规格不宜过大，以 2 cm～4 cm 为宜。

池塘养殖严防逃逸。

5 技术依托单位及联系方式

技术依托单位:中国水产科学研究院北戴河中心实验站
联系人:司飞
通讯地址:河北省秦皇岛市北戴河金山嘴路七号
联系电话:15032316117
E-mail:15032316117@163. com

大菱鲆弧菌病鳗弧菌基因工程活疫苗

1　技术名称

大菱鲆弧菌病鳗弧菌基因工程活疫苗

2　技术要点

通过注射或浸泡接种免疫，用于预防大菱鲆养殖过程中主要由鳗弧菌引发的弧菌病，对其它弧菌（如哈氏弧菌、副溶血弧菌等）也有一定交叉免疫保护作用。

3　适宜区域

我国鲆鲽养殖主产区域

4　注意事项

（1）只能为健康的大菱鲆进行免疫接种；

（2）本疫苗产品建议免疫鱼龄大于 60 日龄，60～120 日龄为浸泡免疫接种，120 日龄以上可进行腹腔注射免疫接种；

（3）免疫接种前应停食 24 小时，接种后 2 周内禁用任何抗生素药物。

5　技术依托单位及联系方式

技术依托单位：华东理工大学
联系人：刘晓红
通讯地址：上海市徐汇区梅陇路 130 号
联系电话：021-64252705
E-mail: liuxiaohong@ecust. edu. cn

半滑舌鳎循环水工厂化高效养殖技术

1 技术名称

半滑舌鳎循环水工厂化高效养殖技术

2 技术要点

采用新型封闭式循环海水工厂化养殖技术,养殖水温 15 ℃～26 ℃,溶氧不低于 5 mg/L,保持水质清新、稳定,水温 15 ℃以下,流换水量 4～6 个循环/日。水温 15 ℃以上,应加大流换水量,夏季高水温时,流换水量 8～12 个循环/日。投喂优质配合饲料,体长 6 cm,日投喂 3～4 次,体长 10 cm 以上,日投喂 2～3 次,体长 15 cm 以上,日投喂 2 次,日投饵 0.5%～1%。

3 适宜区域

河北沿海工厂化循环水养殖地区。

4 注意事项

半滑舌鳎雌、雄个体生长差异大,为避免饵料浪费,每月分选 1 次,逐渐淘汰雄性个体。

5 技术依托单位及联系方式

技术依托单位:河北省海洋与水产科学研究院
联系人:赵海涛
通讯地址:秦皇岛市山海关区龙海大道 151 号
联系电话:13633356373
E-mail: ninan-tao@163. com

第五篇

获奖、鉴定及验收成果汇编

基于养殖水质特性的生物过滤技术和系列生物滤器

成果名称：基于养殖水质特性的生物过滤技术和系列生物滤器
主要完成单位：中国水产科学研究院渔业机械仪器研究所
主要完成人员：吴凡、刘晃、张宇雷、张海耿、张成林、宿墨、车轩、王振华、顾川川、李月
工作起止时间：2008 年 1 月 1 日～2015 年 12 月 31 日
鉴定时间：2011 年
组织鉴定单位：农业部科教司
获奖时间：2016 年
获奖名称级别：中国水产科学研究院科技进步一等奖
内容摘要：

针对不同养殖工况的水质特性，以菌基生物膜净化理论为核心开展生物过滤技术研究，创新研发一批具有自主知识产权的养殖水处理技术和装备，并且在全国范围内进行了推广应用，为我国水产养殖业的健康发展提供了有力的技术支撑。主要内容包括：① 采用多相流交互作用提高滤器内部紊流状态，提升滤器的处理效率，解决滤器的生物堵塞现象；总结得到了滤器优化结构模型和工艺参数，建立了滤器优化设计方法。② 针对鱼类繁育水体固形物多、氨氮浓度低、生产周期短的特点，首创往复式微珠生物过滤技术，在 TSS 浓度 6 mg/L～33. 2 mg/L 条件下，氨氮去除负荷达 593g/（m^3•d）～685 g/（m^3•d）。③ 针对温水性鱼类养殖投喂量大，氨氮负荷高的特点，突破性地将生物移动床技术嫁接应用于水产养殖行业，在水力停留时间 6 min～8 min 条件下，氨氮去除率达到 70%～75%，氨氮去除负荷 560 g/（m^3•d）～700 g/（m^3•d）。④ 针对冷水性鱼类养殖水温较低，氨氮难降解等特点，自主研发出涡旋式流化床生物滤器，在 14 ℃～16 ℃水温条件下，氨氮去除负荷 264 g/（m^3•d）。⑤ 成果在全国建立推广示范点 32 个，推广新型生物滤器 416 台，带动的循环水养殖系统总投资额 8 955 万元；获得国家授权发明专利 6 项，发表学术论文 21 篇（SCI/EI 收录 5 篇），培养硕士研究生 3 名。

离岸养殖新型结构关键技术研究及应用

成果名称：离岸养殖新型结构关键技术研究及应用

主要完成单位：大连理工大学、浙江海洋学院、中国水产科学院黄海水产研究所、大连天正实业有限公司、大连海洋大学

主要完成人员：董国海、赵云鹏、关长涛、桂福坤、刘圣聪、李玉成、崔勇、许条建、郑艳娜、毕春伟、陈昌平、李娇、张涛、黄滨、刘彦

获奖时间：2016 年 2 月 26 日

获奖名称级别：科学技术进步一等奖

内容摘要：

海上离岸养殖工程装备相关研究是海岸和近海工程国家重点实验室的一个重要研究方向。该项目在离岸养殖装备工程结构相关基础研究、产业化应用方面做出了巨大贡献。项目构建了离岸养殖装备工程设计与安全评估技术理论，突破了设计建造离岸养殖装备的技术瓶颈，成果广泛应用于我国东、南、黄、渤海的深水养殖区，为我国深水养殖工程装备的发展奠定了技术基础。

离岸大型浮绳式围网创新养殖模式

成果名称：离岸大型浮绳式围网创新养殖模式

主要完成单位：中国水产科学研究院黄海水产研究所、莱州明波水产有限公司

主要完成人员：关长涛、翟介明、黄滨、崔勇

工作起止时间：2015 年 1 月～2016 年 10 月

验收时间：2016 年 11 月 18 日

验收地点：山东省莱州市

组织验收单位：中国水产科学研究院黄海水产研究所

内容摘要：

山东省海洋资源丰富，离岸海域大多未经开发，而海水养殖由近岸向离岸深远海区域拓展成为“十三五”渔业发展的重要方向。离岸大型围网作为一种新型养殖模式，可实现大水面规模化健康养殖，提高鱼类品质，可配套自动投饲、机械化采收、实时监控、在线监测、活鱼运输等装备实现精准化生产，具有广阔的推广前景，对于我省渔业转型升级、提质增效具有重要意义。项目组自主研制出离岸大型浮绳式围网，首次在莱州湾开阔海域建成离岸大型浮绳式围网 2 个，围网参数 100 m × 50 m × 12 m，单个围网养殖水体 60 000 m^3，围网的抗风浪、耐流性能优良；构建离岸浮绳式围网立体养殖模式，开展半滑舌鳎、斑石鲷、红鳍东方鲀养殖实验，养殖鱼类生长状况良好。围网的构建在设计、实施、养殖管理等方面具有创新

性，是一种可推广复制的养殖模式，适于在山东省沿海区域推广。

大菱鲆选育苗种的生产和推广

成果名称：大菱鲆选育苗种的生产和推广
主要完成单位：中国水产科学研究院黄海水产研究所、烟台开发区天源水产有限公司
主要完成人员：马爱军、王新安、黄智慧、杨志、曲江波等
工作起止时间：2015 年～2016 年
验收时间：2016 年 4 月 6 日
验收地点：烟台开发区天源水产有限公司、福建大正海珍品养殖有限公司
组织验收单位：中国水产科学研究院黄海水产研究所
内容摘要：

2016 年 4 月 6 日，中国水产科学研究院黄海水产研究所科研处邀请有关单位专家，对现代农业产业技术体系—国家鲆鲽类产业技术体系良种选育岗位（CARS-50-G01）和烟台综合试验站（CARS-50-Z06），有关大菱鲆耐高温品系选育苗种在福建霞浦推广养殖工作进行现场验收。验收结果如下：福建大正海珍品养殖有限公司从良种选育岗位基地—烟台开发区天源水产有限公司购买了选育的大菱鲆耐高温苗种 320 000 尾，平均体重 80 g，平均全长 15. 8 cm，经过 5 个月养殖，至 2016 年 4 月大菱鲆平均体重 650 g，平均全长 30. 1 cm，成活率 95%，3 月下旬已陆续上市。

牙鲆新品种“北鲆 1 号”的选育及推广应用

成果名称：牙鲆新品种“北鲆 1 号”的选育及推广应用
主要完成单位：中国水产科学研究院北戴河中心实验站、中国水产科学研究院资源与环境研究中心
主要完成人员：刘海金、于清海、薛玲玲、王玉芬、姜秀凤、杨立更、司飞、孙朝徽、王桂兴、张晓彦
工作起止时间：2001 年～2014 年
鉴定时间：2016 年
组织鉴定单位：中国水产学会
获奖时间：2016 年
获奖名称级别：中国水产学会范蠡科技奖二等奖
内容摘要：

本成果优选出了雌性化率高达 90% 以上，生长速度比普通牙鲆快 20% 以上的新品种

“北鲆1号”。取得的“低温培育—全雌苗种”和“反季节繁育—全年供卵”等成果，解决了提早产卵和环境调控等关键技术，实现了每年7个月供卵，形成了“伪雄鱼诱导—雌亲鱼配组—低温培育—全雌牙鲆”的综合配套技术体系，制定了亲鱼培育、全雌苗种培育、全雌牙鲆养成等技术规范，可在全国范围内为牙鲆养殖单位提供受精卵和苗种培育技术。该成果技术先进，成本低廉，操作简单，效果明显，是应用现代生物技术解决渔业关键技术问题的成功实例。

牙鲆抗淋巴囊肿新品种选育

成果名称：牙鲆抗淋巴囊肿新品种选育

主要完成单位：中国水产科学研究院北戴河中心实验站

主要完成人员：侯吉伦、王玉芬、王桂兴、张晓彦、于清海、姜秀凤、孙朝徽

工作起止时间：2015年～2016年

验收时间：2016年6月

验收地点：秦皇岛

组织验收单位：中国水产科学研究院北戴河中心实验站

内容摘要：

2016年6月19日，中国水产科学研究院北戴河中心实验站邀请相关专家，对承担的鲆鲽类产业技术体系全雌苗种生产岗位（CARS-50-G02）和北戴河综合试验站（CARS-50-Z03）任务中有关“牙鲆抗淋巴囊肿新品种选育”情况进行了阶段性现场验收。验收结果如下：2015年培育牙鲆抗淋巴囊肿试验鱼10个家系，截止验收时共培育出1 999尾，其中有3个家系的抗病保护率在80%以上，分别为84.49%、90.00%和97.20%。抗病保护率为90.00%家系的平均体重为405.26 g ± 89.77 g，平均全长为36.53 cm ± 2.31 cm，平均体高为12.78 cm ± 0.94 cm，均显著高于其他家系（$P < 0.05$）。

大菱鲆全雌苗种生产技术研究

成果名称：大菱鲆全雌苗种生产技术研究

主要完成单位：中国水产科学研究院黄海水产研究所、青岛通用水产养殖有限公司、烟台开发区天源水产有限公司

主要完成人员：刘新富、孟振、张和森、杨志、刘滨

工作起止时间：2011年1月～2016年10月

验收时间：2016年10月13日

验收地点：青岛通用水产养殖有限公司

组织验收单位：中国水产科学研究院黄海水产研究所

内容摘要：

（1）课题组保有3龄有丝分裂型雌核发育亲鱼31尾，其中超雌鱼15尾，平均体重1 203.9 g（563 g～1 864 g），雄鱼16尾，平均体重1 162.3 g（776 g～1 956 g）。

（2）以3尾超雌亲鱼为母本培育苗种21 030尾，每尾超雌鱼的后代各取样20尾，共计取样60尾，平均全长12.6 cm ± 0.9 cm，雌性比例为96.7%，未定性别比例为3.3%，雄性比例为0%，对照组平均全长11.7 cm ± 0.8 cm，雌性比例为52.5%，雄性比例为47.5%，未定性别0%。全雌苗种规格整齐，活力和生长良好。

大菱鲆雌雄生长差异较大，20月龄时雌鱼体重可达雄鱼的1.8倍，其全雌苗种生产技术为法国和西班牙等欧洲大菱鲆养殖国家的研究热点。本项目承担单位黄海水产研究所、合作单位青岛通用水产养殖有限公司和烟台开发区天源水产有限公司紧密合作，经过8年努力，在大菱鲆性别决定机制的明确和超雌鱼制种技术两个方面取得重要突破，建立了全雌苗种生产技术体系，为我国大菱鲆养殖生产和经营模式的多样化发展奠定了良好的基础。

第六篇

专利技术简介

工业化分层取水清水循环的养殖系统和方法

专利名称：工业化分层取水清水循环的养殖系统和方法

专利类型：国家发明专利

专利授权人（发明人或设计人）：黄滨、王秉心、梁友、关长涛、高淳仁、崔勇、李娇

专利号（授权号）：ZL201310182725.2

专利权人（单位名称）：中国水产科学研究院黄海水产研究所

专利申请日：2013年5月4日

授权公告日：2016年1月20日

授权专利内容简介：

本发明公开了一种工业化分层取水清水循环的养殖系统和方法，该系统包括依次相连的养殖池、缓冲调节池、气浮池、生物净化池和对水体进行杀菌、增氧、调温的水质优化池，最终所述水质优化池通过进水管与所述养殖池相连，形成循环系统；所述养殖池中央竖直设有可分层取水的排水管，所述排水管通过总回水管路与所述缓冲调节池相连，所述养殖池的池底中央设有颗粒物汇集排除器，本发明所公开的工业化分层取水清水循环的养殖方法采用排水管取养殖池内的中上层清水进行循环，池底的废弃物通过颗粒物汇集排除器定期排出系统之外，该工艺流程简单，投资成本低，节约能耗，也特别适合老旧车间的改造，应用前景广阔。

一种紫外照射辅助人工挑鱼刺的方法

专利名称：一种紫外照射辅助人工挑鱼刺的方法

专利类型：发明专利

专利授权人（发明人或设计人）：曹立民、王晟、林洪、年睿、隋建新

专利号（授权号）：ZL201410078430.5

专利权人（单位名称）：中国海洋大学

专利申请日：2014年3月6日

授权公告日：2016年3月2日

授权专利内容简介：

本发明公开了一种紫外照射辅助人工挑鱼刺的方法，利用鱼骨刺在紫外光照射条件下

可以发出荧光的特性，以紫外照射辅助人工挑鱼刺，提高了鱼骨刺的检出率，减少了挑刺的用时，降低了鱼肉的损失率，提高了鳕鱼无骨产品的生产效率和出成率。

一种评价大菱鲆变态反应期营养状态的方法

专利名称：一种评价大菱鲆变态反应期营养状态的方法
专利类型：国家发明专利
专利授权人(发明人或设计人)：贾玉东、孟振、牛化欣、刘新富、曲江波、杨志、高淳仁、雷霁霖
专利号(授权号)：ZL201310351318. X
专利权人(单位名称)：中国水产科学研究院黄海水产研究所
专利申请日：2013 年 8 月 13 日
授权公告日：2016 年 3 月 23 日
授权专利内容简介：

本发明提供一种判断处于变态反应期的大菱鲆营养状况良好的标准，所述的标准如下：IGF-I 含量为(105. 07 ± 13. 16)U/g prot，且 RNA/DNA 比值为 2. 92 ± 0. 77。上述的标准用于评价处于变态反应期的大菱鲆的营养状况。本发明针对仔鱼变态期核酸检测，结合变态反应后仔鱼存活率和仔鱼畸形率的验证，减少了外界环境因子(因素)的影响，客观准确的反映了仔鱼变态反应期间营养，所用设备简单，价格低廉，可节省时间和成本。

一种大菱鲆胚胎发育期原始生殖细胞(PGCs)的定位标记方法

专利名称：一种大菱鲆胚胎发育期原始生殖细胞(PGCs)的定位标记方法
专利类型：国家发明专利
专利授权人(发明人或设计人)：刘清华、李军、林帆、徐世宏
专利号(授权号)：ZL201310274379. 0
专利权人(单位名称)：中国科学院海洋研究所
专利申请日：2013 年 7 月 2 日
授权公告日：2016 年 4 月 13 日
授权专利内容简介：

本发明涉及 PGCs 的定位标记方法，具体地说是一种大菱鲆胚胎发育期原始生殖细胞(PGCs)的定位标记方法。采集大菱鲆各个阶段胚胎样品固定，固定后用 50% 甲酰胺于 −20 ℃ ～ −40 ℃保存，待用；将上述保存各个阶段胚胎样品去卵膜之后用不同浓度的 1 × PBST-甲

醇洗脱液依次进行梯度脱水处理后于 −20 ℃至 −40 ℃保存；将上述脱水处理后各个阶段胚胎样品采用梯度甲醇复水，再用 1 × PBST 缓冲液洗涤，洗脱后于 60 ℃～65 ℃预杂交 1 h～2 h；预杂交后在含有大菱鲆 RNA 探针的杂交液中于 60 ℃～65 ℃下杂交过夜；杂交后洗涤，抗体孵育显色，进而定位标记。本发明提供了一种更加便利实用的样品保存方法，解决了传统胚胎整体原位杂交经甲醇脱水保存后导致去卵膜卵黄困难的问题，同时省去蛋白酶 K 消化步骤，简化了操作步骤。

一种鲆鲽鱼类工程化池塘循环水养殖系统

专利名称：一种鲆鲽鱼类工程化池塘循环水养殖系统

专利类型：发明专利

专利授权人（发明人或设计人）：柳学周、徐永江、史宝、刘新富、孙中之、孟振

专利号（授权号）：ZL201410468716.4

专利权人（单位名称）：中国水产科学研究院黄海水产研究所

专利申请日：2014 年 9 月 15 日

授权公告日：2016 年 5 月 11 日

授权专利内容简介：

一种鲆鲽鱼类工程化池塘循环水养殖系统，属海水养殖技术领域，它包括蓄水池塘、回水处理池塘、粗滤池塘、联体组合养殖池塘、进排水系统、增氧设备和水质监测系统。系统运行过程中，自然海水进入蓄水池塘后，经由进水系统进入联体组合养殖池塘，养殖排放水由联体组合养殖池塘进入排水系统，由排水系统进入粗滤池塘，在粗滤池塘内经过滤、沉淀后溢流入回水处理池塘，养殖排放水在回水处理池塘内经处理后进入蓄水池塘，然后再进入联体组合养殖池塘循环利用。本发明系统一方面沿用传统的土池养殖模式，养殖鲆鲽鱼类在类似野生的环境下生长，提高养殖鲆鲽鱼类的品质；同时本发明对室外养殖土池进行工程化和系统化设置和布局，实现了鲆鲽鱼类池塘养殖的工厂化和信息化管理，养殖牙鲆等鲆鲽鱼类产量可达 1 000 千克 / 亩～2 000 千克 / 亩，较传统大型单体池塘养殖提高 5～10 倍，养殖效率和经济生态效益大大提高。

一种牙鲆工程化池塘高效养殖方法

专利名称：一种牙鲆工程化池塘高效养殖方法

专利类型：发明专利

专利授权人（发明人或设计人）：柳学周、徐永江、杨洪军、史宝、刘新富、孙中之

专利号（授权号）：ZL201410469240.6

专利权人(单位名称):中国水产科学研究院黄海水产研究所

专利申请日:2014 年 9 月 15 日

授权公告日:2016 年 6 月 8 日

授权专利内容简介:

一种牙鲆工程化池塘高效养殖方法,属水产养殖技术领域,它包括工程化池塘养殖系统构建、养殖池塘处理与苗种放养、饵料精准投喂技术、水环境管理与微生态调控技术和商品鱼出池收获。本方法可实现牙鲆池塘养殖的工厂化管理和高效生态养殖,养殖产量可提高 9 倍以上,养殖效益大大提升,应用推广潜力巨大。

凹形湾口和峡湾悬索式拦网设施及牧场化生态养殖方法

专利名称:凹形湾口和峡湾悬索式拦网设施及牧场化生态养殖方法

专利类型:发明专利

专利授权人(发明人或设计人):黄滨、关长涛、梁友、刘宝良、刘滨、洪磊、王薇芳、孟振、贾玉东

专利号(授权号):ZL201510687734. 6

专利权人(单位名称):中国水产科学研究院黄海水产研究所

专利申请日:2015 年 10 月 21 日

授权公告日:2016 年 7 月 6 日

授权专利内容简介:

一种凹型湾口和峡湾悬索式拦网设施及生态养殖方法,属于渔业设施与装备技术领域,所述设施,是在沿海和海岛周边凹型海湾或峡湾的湾口设置拦网设施,形成一个一面透水三面环山或两面透水两面有岛的巨大的天然生态养殖场,既可利用水体中丰富的浮游生物为饵料,还可实现虾、鱼、贝藻等多品种混合养殖,以达到充分利用水域生产潜力,提高海域生态和经济效益的目的,形成节能、无污染、低成本的生态养殖模式,在我国海水养殖业上将具有广阔的发展前景。

一种用于水产动物养殖及用药评估的循环水养殖装置

专利名称:一种用于水产动物养殖及用药评估的循环水养殖装置

专利类型:实用新型

专利授权人(发明人或设计人):张元兴、刘晓红、王蓬勃、刘琴、王启要、肖婧凡、张华、张阳

专利号(授权号):ZL2016 2 0180249. X
专利权人(单位名称):华东理工大学
专利申请日:2016 年 3 月 9 日
授权公告日:2016 年 8 月 3 日
授权专利内容简介:

本实用新型涉及一种用于水产动物养殖及用药评估的循环水养殖装置。本实用新型的循环水养殖装置综合了水产动物养殖、水处理、水质监测,可实现淡水或海水类水产动物的循环水或静水养殖、养殖水处理及排放、养殖水理化参数检测,可确保用药评估(如疫苗效力评价)过程中养殖环境优良、实验条件可控。

一种层叠货架式立体水产养殖装置

专利名称:一种层叠货架式立体水产养殖装置
专利类型:发明专利
专利授权人(发明人或设计人):张宇雷、陈翔、单建军
专利号(授权号):201410576812.0
专利权人(单位名称):中国水产科学研究院渔业机械仪器研究所
专利申请日:2014 年 10 月 24 日
授权公告日:2016 年 10 月 5 日
授权专利内容简介:

发明涉及一种层叠货架式立体水产养殖装置,属于水产养殖装置技术领域。主要应用于室内工厂化循环水养殖,利用堆叠结构有效提高纵向空间利用率,提高单位土地面积的养殖产量;本装置在鱼池一端设有用于推流的推流板,其侧边装有小型电机驱动推流板运转,加快水的流速,改善鱼池集排污效果;本装置在每个鱼池底部装有滑轮式导轨,通过外接的空气压缩机控制气缸内的进气,从而实现单层鱼池装置在导轨上沿横向移动,便于针对单层鱼池进行操作,由于鱼池移动时落水孔也进行移动,因此设置导水槽,导水槽覆盖落水孔的运动轨迹,结构紧凑、高效。

海水养殖育苗生产换水装置

专利名称:海水养殖育苗生产换水装置
专利类型:实用新型
专利授权人(发明人或设计人):司飞、王青林、任玉芹、孙朝徽、于清海、王玉芬、姜秀凤、马爱春

专利号(授权号):ZL201520995815.8
专利权人(单位名称):中国水产科学研究院北戴河中心实验站
专利申请日:2015年12月2日
授权公告日:2016年4月27日
授权专利内容简介:

本实用新型提供了一种海水养殖育苗生产换水装置,为解决现有技术中存在的换水装置的使用便捷性较低、换水效率低、育苗初期易造成苗种死亡的问题而发明。所述海水养殖育苗生产换水装置,包括挂壁:可沿竖直方向伸缩,其顶端上设有与育苗水池的边沿厚度匹配的第一挂钩;挂箱:包括网衣,以及位于网衣的外围、且使网衣成型为立体型的立体式挂架,挂架的顶端向其中心延伸有倾斜杆,倾斜杆上位于挂架中心的末端设置有钢圈,挂架的顶端上设有与挂壁匹配的第二挂钩;虹吸管:其两端分别为进水口、出水口,进水口上装设有逆水阀;使用时,挂箱挂设于挂壁上,虹吸管通过逆水阀安置在钢圈中。所述海水养殖育苗生产换水装置用于海水养殖育苗池的换水。

一种用于鲆鱼和舌鳎类亲鱼精卵采集的辅助装置

专利名称:一种用于鲆鱼和舌鳎类亲鱼精卵采集的辅助装置
专利类型:国家实用新型专利
专利授权人(发明人或设计人):张博、刘克奉、贾磊、郑德斌、肖广侠
专利号(授权号):ZL201521070700.4
专利权人(单位名称):天津渤海水产研究所
专利申请日:2015年12月17日
授权公告日:2016年5月25日
授权专利内容简介:

本发明涉及一种用于鲆鱼和舌鳎类亲鱼精卵采集的辅助装置,包括工作台板,在工作台板下部固装有四条可伸缩的支腿,在工作台板的左右及后侧边缘分别固装挡板,在工作台板的前侧边缘固装一排挂环,在工作台板的后部板面上固装一排挂钩,在任意一对挂钩与挂环之间穿装弹性绑带,在左右两侧挡板后部固装穿装轴,在穿装轴左侧紧配合安装遮光板套管,在遮光板套管上固装遮光板,在穿装轴右侧紧配合安装灯具套管,灯具套管上安装灯具曲臂,灯具曲臂顶端安装照明灯,在工作台板右前侧制有采集圆孔。本发明避免亲鱼应激状态下过分挣扎导致外伤,便于采集过程的操作。

一种用于鲟鲽活鱼血液采集的辅助装置

专利名称：一种用于鲟鲽活鱼血液采集的辅助装置
专利类型：国家实用新型专利
专利授权人（发明人或设计人）：张博、贾磊、刘克奉、郑德斌、马超、肖广侠
专利号（授权号）：ZL201521082046.9
专利权人（单位名称）：天津渤海水产研究所
专利申请日：2015 年 12 月 21 日
授权公告日：2016 年 5 月 25 日
授权专利内容简介：

本实用新型涉及一种用于鲟鲽活鱼血液采集的辅助装置，包括底座板，在底座板右端固装有圆弧形扁尺，在底座板左端绞装有搭载斜板，搭载斜板通过扁尺穿孔套在扁尺上并绕绞轴转动，搭载斜板通过卡套限位固定，在搭载斜板中部表面制有条形滑道孔，在搭载斜板上安置滑块，在滑块内部嵌装固定螺母，位于搭载斜板下方的固定螺栓穿过条形滑道孔与固定螺母旋接，旋紧后的固定螺栓与固定螺母将搭载斜板与滑块固定，在滑块上安置有采血的注射器，注射器通过绑带固定在滑块上。本实用新型可调节入针角度及入针深度，提高入针角度和深度的准确性，降低随机性操作造成的失误，提高采集成功率。

一种鲟鲽鱼类装袋辅助装置

专利名称：一种鲟鲽鱼类装袋辅助装置
专利类型：国家实用新型专利
专利授权人（发明人或设计人）：张博、贾磊、郑德斌、刘克奉、马超、肖广侠
专利号（授权号）：ZL201521117634.1
专利权人（单位名称）：天津渤海水产研究所
专利申请日：2015 年 12 月 28 日
授权公告日：2016 年 5 月 25 日
授权专利内容简介：

本实用新型涉及一种鲟鲽鱼类装袋辅助装置，包括弧底条状凹槽，在凹槽的上开口两侧一体制有凹槽侧挡板，在靠近凹槽左侧端开口的一侧挡板内侧固装纵向铰轴，在铰轴上绞装分割挡板，绕铰轴旋转的分割挡板控制凹槽左侧端开口的尺度，在铰轴左侧的侧挡板上安装位置卡块，位置卡块通过卡块顶丝固定在侧挡板上，在凹槽下方的四角处安装有四条凹槽上支管，在每条凹槽上支管内插装凹槽下支腿，在凹槽上支管内抽拉移动的凹槽下支腿通过支腿顶丝固定。本实用新型避免了鱼体损伤，可对过鱼规格及不同种的鱼进行分筛，大大提高

了鱼类的装袋效率。

一种鲆鲽类亲鱼多层保活运输专用袋

专利名称:一种鲆鲽类亲鱼多层保活运输专用袋

专利类型:国家实用新型专利

专利授权人(发明人或设计人):张博、宋文平、刘克奉、贾磊、郑德斌、肖广侠

专利号(授权号):ZL2015210703978

专利权人(单位名称):天津渤海水产研究所

专利申请日:2015 年 12 月 17 日

授权公告日:2016 年 5 月 25 日

授权专利内容简介:

本实用新型涉及一种鲆鲽类亲鱼多层保活运输专用袋,包括塑料袋桶状侧片及热压封装的塑料袋底片,自塑料袋底片向上,在桶状侧片上匀距间隔固装有多个水平的侧片圈梁,在每个侧片圈梁上安装有多个挂环,在每个侧片圈梁的内部安置有圆形塑料网片,在塑料网片的边缘对应挂环位置固装有多个挂钩,每个挂钩与对应的挂环相挂接,由此构成塑料袋内部的多个体积相同的空间。本实用新型将塑料袋桶内分层,实现同一袋内多层放置多条亲鱼,节省运输空间,提高装卸效率,同时,侧片圈梁使桶状塑料袋直立稳定,相互并排依靠,省去外面的泡沫箱,节省了空间和成本。

一种舌鳎类成鱼高效保活无水运输专用袋

专利名称:一种舌鳎类成鱼高效保活无水运输专用袋

专利类型:国家实用新型专利

专利授权人(发明人或设计人):张博、宋文平、刘克奉、贾磊、郑德斌、肖广侠

专利号(授权号):ZL201521063475.1

专利权人(单位名称):天津渤海水产研究所

专利申请日:2015 年 12 月 17 日

授权公告日:2016 年 5 月 25 日

授权专利内容简介:

本实用新型涉及一种鲆鲽类亲鱼多层保活运输专用袋,包括塑料袋桶状侧片及热压封装的塑料袋底片,自塑料袋底片向上,在桶状侧片上匀距间隔固装有多个水平的侧片圈梁,在每个侧片圈梁上安装有多个挂环,在每个侧片圈梁的内部安置有圆形塑料网片,在塑料网片的边缘对应挂环位置固装有多个挂钩,每个挂钩与对应的挂环相挂接,由此构成塑料袋内

部的多个体积相同的空间。本实用新型将塑料袋桶内分层，实现同一袋内多层放置多条亲鱼，节省运输空间，提高装卸效率，同时，侧片圈梁使桶状塑料袋直立稳定，相互并排依靠，省去外面的泡沫箱，节省了空间和成本。

一种凹形湾口和峡湾悬索式拦网设施

专利名称：一种凹形湾口和峡湾悬索式拦网设施

专利类型：国家实用新型专利

专利授权人（发明人或设计人）：黄滨、梁友、刘宝良、刘滨、洪磊、王蔚芳、孟振、贾玉东

专利号（授权号）：ZL201520819481.9

专利权人（单位名称）：中国水产科学研究院黄海水产研究所

专利申请日：2015 年 10 月 21 日

授权公告日：2016 年 9 月 14 日

授权专利内容简介：

一种凹型湾口和峡湾悬索式拦网设施，属于渔业设施与装备技术领域，是在沿海和海岛周边凹形海湾或峡湾的湾口设置拦网设施，形成一个一面透水三面环山或者两面透水两面有岛的巨大的天然生态养殖场，既可利用水体中丰富的浮游生物为饵料，还可实现虾、鱼、贝、藻等多品种混合养殖，以达到充分利用水域生产潜力，提高海域生态和经济效益的目的，形成节能、无污染、低成本的生态养殖模式，在我国海水养殖业上将具有广阔的发展前景。

循环水养殖保温棚

专利名称：循环水养殖保温棚

专利类型：国家实用新型专利

专利授权人（发明人或设计人）：庞尊方、李文升、马文辉、张淞琳、王国栋、王清滨、侯云霞、吴彦甫、肖娜、张克俭、毛东亮、李波、翟介明

专利号（授权号）：ZL201620460695.6

专利权人（单位名称）：莱州明波水产有限公司

专利申请日：2016 年 5 月 19 日

授权公告日：2016 年 10 月 5 日

授权专利内容简介：

本实用新型是一种循环水养殖保温棚，它包括圆形的养殖池，养殖池中心位置设有圆形的生物净化池，生物净化池的井壁上安装有支柱，并在支柱上端安装有圆形顶棚；养殖池的外边沿处安装有若干个固定桩；圆形顶棚的边沿处设有若干个与固定桩一一对应的连接环；

在互相对应的固定桩与连接环之间连接有钢索骨架;相邻钢索骨架之间铺设有支撑网,支撑网上铺有保温膜。本实用新型由中心向四周辐射,辐射面积大,适合在大面积养殖池塘中使用,并具有耗材少,抗风能力强的特点。

成片养殖池虹吸排污装置真空集中控制系统

专利名称:成片养殖池虹吸排污装置真空集中控制系统

专利类型:国家实用新型专利

专利授权人(发明人或设计人):马文辉、李文升、庞尊方、刘江春、孙礼娟、邵光彬、孙芳芳、林好蔚、张君华、张克俭、滕军、翟介明、李波

专利号(授权号):ZL201620460731.9

专利权人(单位名称):莱州明波水产有限公司

专利申请日:2016年5月19日

授权公告日:2016年10月12日

授权专利内容简介:

本实用新型是一种成片养殖池虹吸排污装置真空集中控制系统,它包括并列设置的至少两条支路抽气管和分别用于连接虹吸排污装置的多条单元抽气管;每一条单元抽气管上设有一个单元电磁阀,各条单元抽气管的出气端均连接在支路抽气管上;每一条支路抽气管上各自设有一个支路电磁阀;支路抽气管的出气端连接有抽气总管;还包括进口端与抽气总管出气端相连接的真空泵。通过控制开启某一支路电磁阀,然后开启该支路上的某一单元电磁阀,该单元虹吸排污装置启动排污。实现了对多个养殖池塘池虹吸排污装置的集中控制。提高了工作效率和操作准确度,并降低了人力成本。

循环水养殖池塘自动排污系统

专利名称:循环水养殖池塘自动排污系统

专利类型:国家实用新型专利

专利授权人(发明人或设计人):庞尊方、李文升、马文辉、郑维丹、孙维栋、董在英、朱吉刚、张平、任宝辉、王晓梅、滕军、毛东亮、李波、翟介明

专利号(授权号):ZL201620460656.6

专利权人(单位名称):莱州明波水产有限公司

专利申请日:2016年5月19日

授权公告日:2016年10月12日

授权专利内容简介:

本实用新型是一种循环水养殖池塘自动排污系统，包括设有水封井的排水井，排水井一侧安装有排污排水管；还包括倒U型管形式的虹吸管，虹吸下降管的端口位于水封井中；虹吸管的上端连接有真空控制阀；还包括穿过养殖池塘坝体的排污管，该排污管内端位于养殖池塘底部的集污坑处，外端与虹吸上升管的下端相连接；还包括穿过养殖池塘坝体的虹吸破坏管，该虹吸破坏管的外端与所述虹吸管相连接，内端连接有端口用于深入到养殖池塘液面下方的下倾管。既不需要水泵做动力以节约能源；又能够实现在池塘水位达到设定高度时，系统自动停止工作以确保每次排污量能够精确控制。

循环水养殖池塘紫外线消毒系统

专利名称：循环水养殖池塘紫外线消毒系统

专利类型：国家实用新型专利

专利授权人（发明人或设计人）：李文升、庞尊方、马文辉、张淞琳、王国栋、王清滨、侯云霞、吴彦甫、肖娜、姜燕、毛东亮、李波、翟介明

专利号（授权号）：ZL201620460675.9

专利权人（单位名称）：莱州明波水产有限公司

专利申请日：2016年5月19日

授权公告日：2016年10月12日

授权专利内容简介：

本实用新型是一种循环水养殖池塘紫外线消毒系统，它包括设在养殖池塘中心位置的生物净化池，在生物净化池外侧设有消毒井；生物净化池和消毒井的水处理池底板位于养殖池塘底板的上方；生物净化池和消毒井共用墙壁下端开设有消毒井进水口；还包括位于养殖池塘底板下方的预埋出水横管，该预埋出水横管的内端连接有出水内纵管，外端连接有出水外纵管；出水内纵管的上端口位于消毒井中。消毒井靠近养殖池塘中部，通过预埋出水横管将消毒处理的水输送到养殖池塘边沿部位，尽量远离循环回水。进出于消毒井的水流全靠自流，均不需要外部动力，达到了节能目的并大幅度提高了紫外线消毒处理效率。

坐落于生物净化池内的循环水分离井

专利名称：坐落于生物净化池内的循环水分离井

专利类型：国家实用新型专利

专利授权人（发明人或设计人）：李文升、庞尊方、马文辉、郑维丹、杨传军、宁欣欣、姜长海、徐长有、孙振翔、王晓梅、毛东亮、王恩东、翟介明、李波

专利号（授权号）：ZL201620460591.5

专利权人(单位名称):莱州明波水产有限公司
专利申请日:2016 年 5 月 19 日
授权公告日:2016 年 10 月 12 日
授权专利内容简介:

本实用新型是一种坐落于生物净化池内的循环水分离井,它包括设在养殖池塘中心位置的生物净化池,生物净化池的池底板与养殖池塘基础底板之间周向均布地设置有若干个混凝土支架用于支撑池底板;生物净化池的中心位置设有分离井,分离井壁的下端与池底板相连接;分离井的中心位置设有循环泵井;循环泵井中安装有提升泵;分离井底板内侧与循环泵井壁的下端相连,外侧与分离井壁的下端相连。本实用新型设于养殖池塘中部的生物净化池之内,在养殖池内完成循环水过滤,仅在过滤之前的回水提升过程需要水泵动力,其他循环过程不需要动力,从而达到降低动力消耗并提高水处理效率的目的。

一种用于即时在线水产动物营养代谢研究的装置

专利名称:一种用于即时在线水产动物营养代谢研究的装置
专利类型:国家实用新型专利
专利授权人(发明人或设计人):王蔚芳、张跃锋、黄滨、雷霁霖、杨巧莉
专利号(授权号):ZL201520863085. 6
专利权人(单位名称):中国水产科学研究院黄海水产研究所
专利申请日:2015 年 10 月 30 日
授权公告日:2016 年 11 月 30 日
授权专利内容简介:

一种用于即时在线水产动物营养代谢研究的装置,包括养殖池、粪便及饲料收集装置、氧气供应装置、在线检测系统、支架装置、外联检测装置和后台处理存储系统,养殖池为圆柱型缸体,底部为圆锥体,底部圆锥体最下方连接有粪便及饲料收集装置;氧气供应装置和在线检测系统分别位于养殖池圆柱型缸体的两侧;支架装置焊接于养殖池圆柱型缸体的两侧,并且支立于地面起到支撑固定的作用;外联检测装置通过养殖池圆柱型缸体上的外联检测设备传感器接口与养殖池相连;外联检测装置和在线检测系统均与后台处理存储系统相连。本装置可以即时连续检测水产动物摄食后各种营养素在体外的变化,也可以即时监测水产动物摄食水产饲料后各营养素在体内的保留情况。

海水工厂化养殖水质在线监测预警系统

专利名称：海水工厂化养殖水质在线监测预警系统 V1.1

专利类型：计算机软件著作权

专利申请人（发明人或设计人）：天津渤海水产研究所

专利申请号（受理号）：2016SR131956

专利权人（单位名称）：天津渤海水产研究所

专利申请日：2016 年 6 月 4 日

专利内容简介：

本系统使用户可以通过手机、PAD 等终端实时查看养殖水质环境信息，及时获取水质预警信息，并根据养殖生物的生理生化特点，设定适宜其快速生长的水质范围，当养殖水体水质低于或超出限定值的时候，移动端系统发出预警，提示管理人员迅速采取措施，使水质回归正常范围，以保障养殖生物在其适宜的水质环境中生长，减少疾病的发生，达到少用或不用渔药，保证水产品质量安全的目的。

一种鱼类形态学指标测量装置和方法

专利名称：一种鱼类形态学指标测量装置和方法

专利类型：国家发明专利

专利申请人（发明人或设计人）：吴志昊、尤锋、宋宗诚、路允良

专利申请号（受理号）：201610020325.5

专利权人（单位名称）：中国科学院海洋研究所

专利申请日：2016 年 1 月 13 日

专利内容简介：

本发明属于水产品指标测量领域，具体地说是一种鱼类形态学指标的测量装置及其测量方法，装置包括测量平台、框架、刻度尺 A、刻度尺 B、反光装置、标定尺及摄像装置，测量时框架及标定尺分别放置在测量平台上，反光装置转动安装在框架内，在框架与标定尺之间预留待测鱼类放置位置的上方设有用于拍摄鱼体图像的摄像装置，框架朝向标定尺的一侧分别安装有与测量平台垂直的刻度尺 A 及与测量平台平行的刻度尺 B；通过转动反光装置的角度，使摄像装置采集到框架上的刻度尺与标定尺读数一致，此时一次照相可采集鱼体包括体厚在内的主要外观形态学和体重指标，能快速、无损采集鱼类形态参数。

大菱鲆 *Sm*LTL 重组蛋白及其制备与应用方法

专利名称:大菱鲆 *Sm*LTL 重组蛋白及其制备与应用方法
专利类型:国家发明专利
专利申请人(发明人或设计人):马爱军 等
专利申请号(受理号):201610076867.4
专利权人(单位名称):中国水产科学研究院黄海水产研究所
专利申请日:2016 年 2 月 4 日
专利内容简介:

一种大菱鲆 *Sm*LTL 的重组蛋白及其制备与应用方法,属于分子生物学和基因工程领域,它是由 118 个氨基酸组成,相对分子质量为 13.47×10^3,等电点为 8.52,其氨基酸序列 SEQ NO:1。本发明制备了大菱鲆 *Sm*LTL 重组蛋白,且大菱鲆 *Sm*LTL 重组蛋白能够抑制纤毛虫活力,并与虫体具有结合现象,为大菱鲆抗纤毛虫药物的研发发挥积极、重要的推动作用。

一种评估海水鱼类黏液凝集素抑制盾纤毛虫的方法

专利名称:一种评估海水鱼类黏液凝集素抑制盾纤毛虫的方法
专利类型:国家发明专利
专利申请人(发明人或设计人):黄智慧 等
专利申请号(受理号):201610076868.9
专利权人(单位名称):中国水产科学研究院黄海水产研究所
专利申请日:2016 年 2 月 5 日
专利内容简介:

一种评估海水鱼类黏液凝集素抑制盾纤毛虫的方法,属于水产病害防治技术领域,它包括 Lily-type lectin (LTL)蛋白对盾纤毛虫的活力抑制评估和结合评估。本发明可以较为简便的对海水鱼类 Lily-type lectin 蛋白具有抑制盾纤毛虫活力的抑制性进行评估,并通过荧光显微镜清晰的观察到 LTL 蛋白与盾纤毛虫结合现象。

一种鱼类早期生活史耳石样品制备方法

专利名称:一种鱼类早期生活史耳石样品制备方法
专利类型:国家发明专利

专利申请人(发明人或设计人):任玉芹、周勤、王玉芬、王青林、司飞、姜秀凤、于清海、张红涛
专利申请号(受理号):201610125432.4
专利权人(单位名称):中国水产科学研究院北戴河中心实验站
专利申请日:2016年3月4日
专利内容简介:

本发明涉及生物标本制备领域,具体而言,涉及一种鱼类早期生活史耳石样品制备方法,该方法为:于镜下滴加腐蚀性溶液以暴露耳石;用纯化水清洗暴露的耳石多次,同时吸除清洗后的液体和/或组织残余;待耳石室温干燥后用封片剂将耳石包埋,待封片剂铺平后烘干;鱼类早期生活史中后期耳石封片前还需要进行磨片处理。本发明提供的方法降低了鱼类早期生活史阶段耳石获取的难度,即便样品组织不透明也可以使用本方法,且制得的标本更洁净,杂质更少,更有利于标本的长期保存。

一种与牙鲆生长性状相关的SNP位点、其筛选方法及应用

专利名称:一种与牙鲆生长性状相关的SNP位点、其筛选方法及应用
专利类型:国家发明专利
专利申请人(发明人或设计人):侯吉伦、王桂兴、张晓彦、王玉芬、刘海金、于清海、杨立更
专利申请号(受理号):201610172207.6
专利权人(单位名称):中国水产科学研究院北戴河中心实验站
专利申请日:2016年3月24日
专利内容简介:

本发明涉及本发明提供了一种与牙鲆生长性状相关的SNP位点、其筛选方法及应用。通过候选基因关联分析法,对GH基因中SNP与牙鲆生长性状进行关联分析,获得了一个与生长性状相关的SNP标记,所述SNP标记可用于牙鲆标记辅助选育,缩短育种周期。

一种与牙鲆数量性状相关的SNP位点、其筛选方法及应用

专利名称:一种与牙鲆数量性状相关的SNP位点、其筛选方法及应用
专利类型:发明专利
专利申请人(发明人或设计人):张晓彦、王桂兴、侯吉伦、王玉芬、刘海金、于清海、杨立更
专利申请号(受理号):CN201610172234.3

专利权人(单位名称):中国水产科学研究院北戴河中心实验站
专利申请日:2016 年 3 月 24 日
专利内容简介:

本发明提供了一种与牙鲆数量性状相关的 SNP 位点、其筛选方法及应用。通过候选基因关联分析法,对 MSTN 基因中 SNP 与牙鲆数量性状进行关联分析,获得了一个与数量性状相关的 SNP 标记,所述 SNP 标记可用于牙鲆标记辅助选育,缩短育种周期。

一种提高牙鲆有丝分裂雌核发育双单倍体诱导效率的方法

专利名称:一种提高牙鲆有丝分裂雌核发育双单倍体诱导效率的方法
专利类型:国家发明专利
专利申请人(发明人或设计人):王桂兴、侯吉伦、张晓彦、王玉芬、刘海金、于清海、杨立更
专利申请号(受理号):CN201610172055. X
专利权人(单位名称):中国水产科学研究院北戴河中心实验站
专利申请日:2016 年 3 月 24 日
专利内容简介:

本发明提供了一种提高牙鲆有丝分裂雌核发育双单倍体诱导效率的方法,在诱导有丝分裂雌核发育双单倍体前,先对所用母本的纯合度进行鉴定,筛选出纯合度高的亲本,再进行诱导,从而提高了诱导效率。利用本发明,可以在牙鲆上实现有丝分裂雌核发育双单倍体的批量化、规模化诱导,从而为牙鲆生物学性状的功能基因挖掘、数量性状控制位点定位以及良种的克隆化奠定坚实基础。

一种鱼类卵母细胞总 RNA 提取方法

专利名称:一种鱼类卵母细胞总 RNA 提取方法
专利类型:国家发明专利
专利申请人(发明人或设计人):史宝、柳学周、徐永江、王滨、徐涛、孙中之
专利申请号(受理号):2016102025089
专利权人(单位名称):中国水产科学研究院黄海水产研究所
专利申请日:2016 年 4 月 1 日
专利内容简介:
利用化学法和超声波破碎物理学方法相结合获得卵母细胞中含有的总 RNA,并保证

RNA 完整性，可以用于下一步基因克隆、mRNA 表达和高通量测序等研究。本发明在微量卵母细胞中加入 1 mL RNAiso Plus，可有效去除卵母细胞中核糖体等成分；分别在加入氯仿和异丙醇后，离心时间均延长了 5 min，有利于去除蛋白质等，并提高所获的 RNA 纯度与质量。

池塘循环水养殖水处理系统

专利名称：池塘循环水养殖水处理系统
专利类型：国家发明专利
专利申请人（发明人或设计人）：李文升、庞尊方、马文辉、王秉心、张克俭、郑维丹、孙维栋、董在英、朱吉刚、张平、任宝辉、王晓梅、毛东亮、李波、翟介明
专利申请号（受理号）：201610332643.5
专利权人（单位名称）：莱州明波水产有限公司
专利申请日：2016 年 5 月 19 日
专利内容简介：

本发明是一种池塘循环水养殖水处理系统，包括设在养殖池塘中心位置的生物净化池，生物净化池的中心位置设有分离井；分离井的中心位置的循环泵井中安装有提升泵；还包括设置在生物净化池壁外侧的紫外线消毒井；位于养殖池塘基础底板下方的预埋出水横管内端连接有出水内纵管，外端连接有出水外纵管；出水内纵管的上端口位于消毒井中，出水外纵管的出水口位于养殖池塘外边沿部位液面以下；养殖池塘基础底板中心位置设有集污坑，集污坑连接有排污管；分离井和集污坑之间安装有分离井排污管。在养殖池内完成循环水过滤、生物净化和紫外线消毒处理，仅在回水提升过程需要泵动力，其他循环及排污过程不需要动力，降低了动力消耗并提高了水处理效率。

循环水养殖池塘内置处理池

专利名称：循环水养殖池塘内置处理池
专利类型：国家发明专利
专利申请人（发明人或设计人）：庞尊方、肖志忠、李文升、马文辉、刘江春、孙礼娟、邵光彬、孙芳芳、林好蔚、张君华、姜燕、毛东亮、滕军、王恩东、翟介明、李波
专利申请号（受理号）：201610332599.8
专利权人（单位名称）：莱州明波水产有限公司
专利申请日：2016 年 5 月 19 日
专利内容简介：

本发明是一种循环水养殖池塘内置处理池,它包括设在养殖池塘中心位置的生物净化池,生物净化池的中心位置设有分离井;分离井的中心位置的循环泵井中安装有提升泵;还包括设置在生物净化池壁外侧的紫外线消毒井;位于养殖池塘基础底板下方的预埋出水横管内端连接有出水内纵管,外端连接有出水外纵管;出水内纵管的上端口位于消毒井中,出水外纵管的出水口位于养殖池塘外边沿部位液面以下;养殖池塘基础底板中心位置设有集污坑,集污坑连接有排污管;分离井和集污坑之间安装有分离井排污管。在养殖池内完成循环水过滤、生物净化和紫外线消毒处理,仅在回水提升过程需要泵动力,其他循环过程不需要动力,降低了动力消耗并提高了水处理效率。

一种降低大菱鲆肝脏脂肪沉积的复合添加剂

专利名称:一种降低大菱鲆肝脏脂肪沉积的复合添加剂

专利类型:国家发明专利

专利申请人(发明人或设计人):王雅慧、麦康森、艾庆辉

专利申请号(受理号):CN201610345674.4

专利权人(单位名称):中国海洋大学

专利申请日:2016年5月23日

专利内容简介:

本发明的目的是提供一种降低大菱鲆肝脏脂肪沉积的复合添加剂,包含有3.7%～4%氯化胆碱、28%～31%磷酸二氢钙、2.8%～3.1%丙酸钙、5.5%～6.8%甜菜碱、5.5%～6.8%牛磺酸、1.8%～2.2%甘氨酸、1.3%～1.6%乙氧基喹啉、43%～46%海藻酸钠、0.5%～1.8%甘草次酸和1%～1.3%姜黄素。本发明中的环保复合添加剂在大菱鲆幼鱼饲料中能够有效降低大菱鲆幼鱼鱼体、肝脏和肌肉脂肪含量以及部分血浆生化指标的含量,并且对大菱鲆幼鱼的生长没有显著影响。

一种快速筛选鲆鲽鱼类耐高温群体/家系的方法

专利名称:一种快速筛选鲆鲽鱼类耐高温群体/家系的方法

专利类型:国家发明专利

专利申请人(发明人或设计人):路允良、吴志昊、宋宗诚、尤锋

专利申请号(受理号):201610422052.7

专利权人(单位名称):中国科学院海洋研究所、威海圣航水产科技有限公司

专利申请日：2016年6月13日

专利内容简介：

本发明涉及海水鱼类养殖和育种技术领域，具体地说是一种基于动态法无损快速筛选鲆鲽鱼类耐高温群体/家系的方法。将待筛选群体/家系个体，经荧光标记后，以0.5 ℃/h～1 ℃/h的速率逐渐升高培养温度，观察鱼的游泳状态变化，记录其丧失翻身能力时的温度，即为鲆鲽鱼类可耐受的临界高温水平（Critical thermal maximum，CTM），统计分析各群体/家系的CTM值。CTM高者为耐高温群体/家系；CTM水平无统计学差异，即为相同耐高温能力的群体/家系。本发明操作过程简单、快速，检测成本低，同时有效避免了筛选期间的个体死亡，实现了鲆鲽鱼类耐高温群体/家系的无损快速筛选。

一种准确鉴别回捕褐牙鲆中放流鱼的方法

专利名称：一种准确鉴别回捕褐牙鲆中放流鱼的方法

专利类型：国家发明专利

专利申请人（发明人或设计人）：王青林、司飞、于姗姗、王桂兴、张红涛、任建功、于清海、孙朝徽、任玉芹、姜秀凤

专利申请号（受理号）：201610422623.7

专利权人（单位名称）：中国水产科学研究院北戴河中心实验站

专利申请日：2016年6月15日

专利内容简介：

一种准确鉴别回捕褐牙鲆中放流鱼的方法，属于增殖放流领域，该方法首先进行形态学上区分，然后对褐牙鲆的线粒体DNA进行PCR扩增，通过倍型分析进一步确定疑似放流褐牙鲆，最后再分有针对性地进行荧光标记微卫星标记多重PCR扩增，确定放流个体，本发明方法减少了放流个体鉴定的盲目性，减轻了工作量，提高鉴定效率。

一种向半滑舌鳎垂体细胞添加维生素E后提高促性腺激素表达水平的方法

专利名称：一种向半滑舌鳎垂体细胞添加维生素E后提高促性腺激素表达水平的方法

专利类型：国家发明专利

专利申请人（发明人或设计人）：王娜、王蔚芳、黄滨、史宝、王若青、马佳璐、陈松林

专利申请号（受理号）：201610237755.2

专利权人（单位名称）：中国水产科学研究院黄海水产研究所

专利申请日:2016 年 6 月 17 日

专利内容简介:

本发明涉及一种向半滑舌鳎垂体细胞添加维生素 E 后提高促性腺激素表达水平的方法,步骤如下。① 半滑舌鳎垂体细胞的分离与培养:首先配置 L-15 细胞培养基,然后用所配培养基进行原代细胞的培养;② 向半滑舌鳎垂体细胞中添加维生素 E:配置维生素 E 溶液,向步骤①中所培养的细胞中添加维生素 E,在添加后第四天进行样品收集;③ 向半滑舌鳎垂体细胞中添加维生素 E 后 FSH 和 LH 的定量 PCR 分析;取步骤②中的细胞沉淀进行 RNA 的提取及反转录,然后进行实时荧光定量 PCR 反应;④ 向半滑舌鳎垂体细胞中添加维生素 E 后的 FSH 和 LH 的 ELISA 分析;以步骤②中收集的上清液为材料依次进行标准品的稀释与加样、加样、温育、配液、洗涤、加酶、温育、洗涤、显色、终止、测定。

一种箱型转子碎气式气浮装置

专利名称:一种箱型转子碎气式气浮装置

专利类型:国家发明专利

专利申请人(发明人或设计人):庄保陆、单建军、顾川川、陈翔

专利申请号(受理号):201610526476. 8

专利权人(单位名称):中国水产科学研究院渔业机械仪器研究所

专利申请日:2016 年 7 月 6 日

专利内容简介:

发明提供了一种箱型转子碎气式气浮装置,包括箱体、电动机、传动轴、气泡发生器和泡沫收集罩;所述箱体的一侧设置有进水口,箱体的另一侧设置有出水口,所述箱体的顶部设置有所述的电动机,电动机下方连接所述的传动轴,传动轴的下方连接所述气泡发生器;所述箱体的上部设置有所述的泡沫收集罩;所述的传动轴为空心轴,传动轴上设置有进气口。装置具有能耗低、处理效果显著的特点,并且结构紧凑,体积小,维护方便的优点。

一种近岸鱼类养殖岩礁池塘的生态工程化设置方法

专利名称:一种近岸鱼类养殖岩礁池塘的生态工程化设置方法

专利类型:国家发明专利

专利申请人(发明人或设计人):徐永江、柳学周、史宝、张凯、王滨、蓝功岗

专利申请号(受理号):201610629886. 5

专利权人(单位名称):中国水产科学研究院黄海水产研究所
专利申请日:2016年8月4日
专利内容简介:

一种近岸鱼类养殖岩礁池塘的生态工程化设置方法,属于水产养殖技术领域,它包括池壁与塘埂构筑、水系统设置、污染物减排控制系统设置、增氧环流系统设置和水质及行为监测系统设置;本发明通过对近岸鱼类养殖岩礁池塘的池壁与塘埂、底质、进排水系统、污染物减排控制系统、水质与行为监测系统方面等进行工程化设置,大大提高岩礁池塘对自然灾害的抵御能力,减少岩礁池塘养殖废弃物对近岸海域环境的污染压力,实现对鱼类养殖岩礁池塘的工厂化管理和高产稳产,达到海水池塘养殖提质增效和转型升级的目的。

一种鱼类养殖池塘用内循环增氧装置及其使用方法

专利名称:一种鱼类养殖池塘用内循环增氧装置及其使用方法
专利类型:国家发明专利
专利申请人(发明人或设计人):徐永江、柳学周、王滨、史宝、张凯、蓝功岗
专利申请号(受理号):201610633858.0
专利权人(单位名称):中国水产科学研究院黄海水产研究所
专利申请日:2016年8月5日
专利内容简介:

一种鱼类养殖池塘用内循环增氧装置,属于水产养殖技术领域,它包括底座、潜水泵、支架、取水管、三通管、进气管、出水管、射流管、远程控制开关和电缆,潜水泵安置在底座上,潜水泵连接取水管,取水管通过三通管连接进气管和出水管,所述的三通管一水平端口与取水管连接,另一水平端口与出水管连接,与水面垂直的端口连接进气管,取水管与出水管通过三通管的两个水平端口连接在同一直线上在池塘内设置内循环增氧装置,实现上下水层交换和增氧,同时推动池塘内养殖水体形成内循环流动,便于大颗粒养殖废弃物集中清理,营造了优良的鱼类生长环境,大大降低因池塘底质腐败和水质恶化引发的病害发生几率,提高鱼类池塘养殖成活率和养殖产量,达到池塘养殖生产提质增效的目的。

鱼类黏性受精卵人工孵化设备

专利名称:鱼类黏性受精卵人工孵化设备
专利类型:国家发明专利

专利申请人(发明人或设计人):马爱军 等
专利申请号(受理号):201610809677.9
专利权人(单位名称):中国水产科学研究院黄海水产研究所
专利申请日:2016年9月8日
专利内容简介:

一种鱼类黏性受精卵人工孵化设备,属于鱼类养殖领域,它包括倒锥形外壳、圆台形网圈、网兜、圆形顶盖、主体支架和进排水系统;本发明鱼类黏性受精卵人工孵化设备,创造性的采用了倒锥形外壳内套倒锥形网的结构,倒锥形外壳与内部锥形网间留有空隙,借助于底部上升水流可有效避免卵子附壁、结块现象,锥形网亦可有效分散水流,避免水流直接冲击鱼卵,进而提高鱼卵孵化率,同时可拆卸的底部锥形网兜方便了孵化设备内部的清理。本发明采用了环状水管清理系统,可简单、有效清理孵化过程中的发霉胚胎,极大地避免了过多人为干扰对卵子孵化的不良影响。并且它制造成本低廉,操作简便,特别适合于现今本领域的大规模工厂化繁育应用。

Plasmid library comprising two random markers and use thereof in high throughput sequencing

专利名称:Plasmid library comprising two random markers and use thereof in high throughput sequencing
专利类型:美国发明专利
专利申请人(发明人或设计人):Haijin Liu, Xiao Liu, Jilun Hou, Zhichao Xu, Xiaolin Wei, Guixing Wang, Yufen Wang
专利申请号(受理号):15128557
专利权人(单位名称):中国水产科学研究院北戴河中心实验站、清华大学
专利申请日:2016年9月23日
专利内容简介:

Provided is a plasmid library comprising a DNA insertion site and two marker sequences located upstream and downstream of the site. The combinations of two marker sequences of any two plasmids selected from the library are different. Also provided is a method for high throughput two-end sequencing of an inserted DNA using the plasmid library.

一种大菱鲆精原干细胞分离方法

专利名称：一种大菱鲆精原干细胞分离方法
专利类型：国家发明专利
专利申请人（发明人或设计人）：刘清华、李军、王文琪、杨敬昆、徐世宏、王彦丰
专利申请号（受理号）：201610882409. X
专利权人（单位名称）：中国科学院海洋研究所
专利申请日：2016 年 10 月 10 日
专利内容简介：

本发明涉及精原干细胞的分离方法。采集 1～1. 5 龄大菱鲆，解剖，分离获得完整性腺，性腺经过培养液洗涤，切块剪碎、酶消化、过滤、密度梯度离心等一系列过程，获得高纯度和高密度的精原干细胞。本发明提供了一种分离海水鱼类精原干细胞的方法，该方法分离获得的精原干细胞细胞形态结构完整，纯度高，且密度大，可用于精原干细胞的保存以及干细胞移植研究。该方法的为海水鱼类种质资源的保存以及新品种的培育提供了新方法和新思路。

一种提高鱼糜品质的改良剂及其制造方法和使用方法

专利名称：一种提高鱼糜品质的改良剂及其制造方法和使用方法
专利类型：国家发明专利
专利申请人（发明人或设计人）：林洪、李钰金、周文锦、刘远平、励建荣、王锡昌、李丽迪
专利申请号（受理号）：201611041582. 3
专利权人（单位名称）：荣成泰祥食品股份有限公司、中国海洋大学
专利申请日：2016 年 11 月 22 日
专利内容简介：

本发明提供了一种提高冷冻鱼糜品质的改良剂，包括如下组分：褐藻胶、氯化钙、明胶、去离子水。其中褐藻胶、氯化钙、明胶、去离子水的质量百分比为：褐藻胶为 3%～7%，氯化钙为 1. 5%～3. 1%，明胶为 0. 5%～1. 5%，去离子水 90%～95%。本发明还提供了该改良剂的使用方法和制造方法。本发明达到的有益技术效果为：① 该改良剂，制作材料廉价易得，制造的冷冻鱼糜的凝胶强度显著增强；② 该改良剂的制造方法，对设备要求低，可操作性强，有利于工业化推广；③ 该改良剂的使用方法步骤少，操作简单，对环境要求低，使用方便，可以大量应用于冷冻鱼糜的生产。

一种冷冻分割多宝鱼的加工方法

专利名称:一种冷冻分割多宝鱼的加工方法
专利类型:国家发明专利
专利申请人(发明人或设计人):郭晓华、董浩、申照华、孙爱华
专利申请号(受理号):201611044895.4
专利权人(单位名称):山东美佳集团有限公司
专利申请日:2016年11月24日
专利内容简介:

本发明属于水产品加工技术领域,具体涉及一种冷冻分割多宝鱼的加工方法。该方法是通过以下步骤实现的:首先将多宝鱼置于冷水中养殖,并不断投放啤酒酵母,然后去除鳃丝内脏后进行浸泡消毒,消毒结束后在瓜尔豆胶水溶液中浸泡,冻结,最后进行分割后真空包装即可。本发明采用低温下浸泡联合电解水消毒,保持鱼肉鲜度的同时,杀菌效果好、生产成本低、用后无残留,对人体不产生伤害,对环境无污染;通过预处理液进行浸泡,能够去除鱼肉腥味,联合后期的处理,能够显著降低产品解冻损失率,改善鱼柳质构,改善口感。

一种循环水养殖池卷帘式隔膜装置

专利名称:一种循环水养殖池卷帘式隔膜装置
专利类型:国家发明专利
专利申请人(发明人或设计人):顾川川、王振华、张海耿、李敬超
专利申请号(受理号):201611039376.9
专利权人(单位名称):中国水产科学研究院渔业机械仪器研究所
专利申请日:2016年11月26日
专利内容简介:

发明提出一种养殖池卷帘式隔膜装置,包括圆形的鱼池、圆形的支架底、若干根支撑杆、圆形的支架顶、一对锥形卷筒;支架底固定在鱼池上端,若干根支撑杆沿支架底一周均匀布置,支撑杆下端与支架底固定,上端与支架顶固定,支架顶与支架底同轴设置;支架顶的一周设有倾斜向下的环状上滑道导向条,支架底的一周设有倾斜向上的环状下滑道导向条,上、下滑道导向条与支撑杆之间形成滑道槽;锥形卷筒可拆卸的固定在其中一根支撑杆的两侧,锥形卷筒的大端在下,小端在上,卷筒的内芯展开后嵌在滑道槽中,一对锥形卷筒分别与另一端的支撑杆的两侧连接;支架顶内固定设置顶部薄膜。本发明可根据需要更换不同内芯的锥形卷筒,十分方便。

一种大菱鲆增食欲素 Orexin 基因、重组蛋白及其制备方法

专利名称：一种大菱鲆增食欲素 Orexin 基因、重组蛋白及其制备方法
专利类型：国家发明专利
专利申请人（发明人或设计人）：刘滨、穆小生、黄滨、刘新富、迟庆宏、赵德辉
专利申请号（受理号）：201611157216. 4
专利权人（单位名称）：中国水产科学研究院黄海水产研究所
专利申请日：2016 年 12 月 15 日
专利内容简介：

本发明克隆得到了大菱鲆增食欲素的编码基因并将增食欲素 Orexin 基因与 PET-24a 表达载体相连接构建表达载体 pET-24a-orexin，转化入表达菌株进行原核表达，后经纯化复性获得大菱鲆增食欲素重组蛋白，本发明将为研究大菱鲆的食欲调控机理和开发可应用于生产的促进鱼类摄食和生长的新产品打下良好基础。

一种大菱鲆生长泌乳素（SL）基因、重组蛋白及其制备方法

专利名称：一种大菱鲆生长泌乳素（SL）基因、重组蛋白及其制备方法
专利类型：国家发明专利
专利申请人（发明人或设计人）：刘滨、穆小生、黄滨、刘新富、赵德辉、迟庆宏
专利申请号（受理号）：201611157202. 2
专利权人（单位名称）：中国水产科学研究院黄海水产研究所
专利申请日：2016 年 12 月 15 日
专利内容简介：

本发明克隆得到了大菱鲆生长催乳素的编码基因并将生长泌乳素 SL 基因与 PET-24a 表达载体相连接构建表达载体 pET-24a-SL，转化入表达菌株进行原核表达，后经纯化复性获得大菱鲆生长催乳素重组蛋白，本发明将为研究大菱鲆的生长发育调控机理和开发可应用于生产的调控大菱鲆生长发育的新产品打下良好基础。

附 录

附录 1 鲆鲽类产业技术体系 2016 年发表论文一览表

序号	作者	论文名称	发表刊物	年、卷、期、起止页
1	关长涛	国家鲆鲽类产业技术体系年度报告(2015)	中国海洋大学出版社	2016, 11
2	徐永江	中国海洋意识教育必读丛书——多彩海鱼	青岛出版社	2016, 10
3	田岳强,郭建丽,黄智慧,等	大菱鲆家系选育二代 7 种免疫因子的分析	海洋科学	2016, 40(9): 9-17
4	王广宁,马爱军,李猛,等	大菱鲆微卫星标记在亲本后代中的分离分析	海洋科学	2016, 40(4): 1-10
5	商晓梅,马爱军,王新安,等	六种鲽形目鱼类无眼侧头部皮肤形态特征与功能探讨	水产学报	2016, 40(2): 189-197
6	赵艳飞,马爱军,王新安,等	卵生型鱼类卵子发育潜能评估研究进展	海洋科学	2016, 40(8): 157-167
7	X. A. Wang, A. J. Ma	Comparison of four nonlinear growth models for effective exploration of growth characteristics of turbot *Scophthalmus maximus* fish strain	African Journal Biotechnology	2016, 15(40): 2251-2258(体系资助)
8	Z. H. Huang, A. J. Ma, D. D. Xia, et al	Immunological characterization and expression of lily-type lectin in response to environmental stress in turbot (*Scophthalmus maximus*)	Fish and Shellfish Immunology	2016, 58: 323-331(体系资助)
9	A. J. Ma, X. M. Shang	Morphological variation and distribution of free neuromasts during half-smooth tongue sole *Cynoglossus semilaevis* ontogeny	Chinese Journal of Oceanology & Limnology	2016, 35(2): 244-250(体系资助)
10	D. Y. Ma, A. J. Ma, Z. H. Huang, et al	Transcriptome analysis for identification of genes related to gonad differentiation, growth, immune response and marker discovery in the turbot	PLoS ONE	2016, 11(2): e0149414(体系资助)
11	J. L. Hou, G. X. Wang, X. Y. Zhang, et al	Cold-shock induced androgenesis without egg irradiation and subsequent production of doubled haploids and a clonal line in Japanese flounder, *Paralichthys olivaceus*	Aquaculture	2016, 464: 642-646
12	J. L. Hou, G. X. Wang, X. Y. Zhang, et al	Cytological studies on induced mitogynogenesis in Japanese flounder *Paralichthys olivaceus* (Temminck et Schlegel)	Zygote	2016, 24: 700-706

（续表）

序号	作者	论文名称	发表刊物	年、卷、期、起止页
13	Z. Meng, P. Hu, J. L. Lei, et al	Expression of insulin-like growth factors at mRNA levels during the metamorphic development of turbot (*Scophthalmus maximus*)	General and Comparative Endocrinology	2016, 235: 11-17
14	J. L. Hou, G. X. Wang, X. Y. Zhang, et al	Production and verification of a 2nd generation clonal group of Japanese flounder, *Paralichthys olivaceus*	Scientific Reports	2016, 6: 35776
15	王学颖，徐世宏，刘清华，等	渗透压、pH、葡萄糖及离子溶液对夏牙鲆精子激活及运动特征的影响	海洋科学	2016, 40(3): 40-46
16	Z. F. Fan, Z. H. Wu, L. J. Wang, et al	Characterization of embryo transcriptome of gynogenetic olive flounder *Paralichthys olivaceus*	Marine Biotechnology	2016, 18(5): 545-553
17	L. M. Peng, Y. Zheng, F. You, et al	Comparison of growth characteristics between skeletal muscle satellite cell lines from diploid and triploid olive flounder *Paralichthys olivaceus*	PeerJ	2016, 4: e1519
18	C. Y. Zhao, S. H. Xu, Y. F. Liu, et al	Gonadogenesis analysis and sex differentiation in cultured turbot (*Scophthalmus maximus*)	Fish Physiology & Biochemistry	2017, 43(1): 265-278
19	Y. L. Lu, Z. H. Wu, Z. C. Song, et al	Insight into the heat resistance of fish via blood: Effects of heat stress on metabolism, oxidative stress and antioxidant response of olive flounder *Paralichthys olivaceus* and turbot *Scophthalmus maximus*	Fish & Shellfish Immunology	2016, 58: 125-135
20	S. D. Weng, F. You, Z. F. Fan, et al	Molecular cloning and sexually dimorphic expression of wnt4 in olive flounder (*Paralichthys olivaceus*)	Fish Physiology and Biochemistry	2016, 42(4): 1167-1176
21	张宇雷，曹伟，蔡计强	基于氨氮平衡的水产养殖换水率计算方法研究	渔业现代化	2016, 43(5): 1-5
22	朱学武，徐永江，柳学周，等	半滑舌鳎黑色素聚集激素重组制备与生物活性分析	水产学报	2016, 40(10): 1595-1605
23	李斌，徐永江，柳学周，等	半滑舌鳎食欲素A体外重组表达及活性分析	水产学报	2016, 40(9): 1462-1471
24	张金勇，柳学周，史宝，等	半滑舌鳎孕酮受体膜组分1基因的克隆及组织和时空表达规律	中国水产科学	2016, 23(5): 1080-1090
25	徐永江，柳学周，史宝，等	茜素络合物对半滑舌鳎(*Cynoglossus semilaevis Günther*)苗种耳石的染色标记效果	渔业科学进展	2016, 37(6): 1-8
26	王滨，柳学周，徐永江，等	鱼类促性腺激素抑制激素及其受体的研究进展	水产学报	2016, 40(2): 278-287

(续表)

序号	作者	论文名称	发表刊物	年、卷、期、起止页
27	Y. J. Xu, X. Z. Liu, B. Shi, et al	Complete mitochondrial genome of summer flounder *Paralichthys dentatus* (Pleuronectiformes Paralichthyidae)	Mitochondrial DNA Part B	2016, 1: 889-890
28	Y. J. Xu, B. Wang, X. Z. Liu, et al	Evidences for involvement of growth hormone and insulin-like growth factor in ovarian development of starry flounder (*Platichthys stellatus*)	Fish Physiology Biochemistry	2017, 43(2): 527-537
29	B. Shi, X. Z. Liu, P. Thomas, et al	Identification and characterization of a progestin and adipoQ receptor (PAQR) structurally related to Paqr7 in the ovary of *Cynoglossus semilaevis* and its potential role in regulating oocyte maturation	General and Comparative Endocrinology	2016, 237: 109-120
30	B. Shi, X. Z. Liu, Y. J. Xu, et al	The complete mitochondrial genome of southern flounder *Paralichthys lethostigma* (Pleuronectiformes, Bothidae)	Mitochondrial DNA: Part B	2016, 1(1): 200-201
31	Y. Zhang, J. F. Xiao, Q. Y. Wang, et al	A modified quantum dot-based dot blot assay for rapid detection of fish pathogen *Vibrio anguillarum*	Journal of Microbiology and Biotechnology	2016, 26(8): 1457-1463
32	Y. Ma, Q. Y. Wang, X. T. Gao, et al	Biosynthesis and uptake of glycine betaine as cold-stress response to low temperature in fish pathogen *Vibrio anguillarum*	Journal of Microbiology	2017, 55(1): 44-55
33	D. Gu, H. Liu, Z. Yang, et al	Chromatin immunoprecipitation sequencing technology reveals global regulatory roles of low cell density quorum-sensing regulator AphA in pathogen *Vibrio alginolyticus*	Journal of Bacteriology	2016, 198(21): 2985-2999
34	K. Y. Yin, Q. Y. Wang, J. F. Xiao, et al.	Comparative proteomic analysis unravels a role for EsrB in the regulation of reactive oxygen species stress responses in *Edwardsiella piscicida*	FEMS Microbiology Letters	2017, 364(1): fnw269
35	S. L. Cui, J. F. Xiao, Q. Y. Wang, et al	H-NS binding to *evpB* and *evpC* and repressing T6SS expression in fish pathogen *Edwardsiella piscicida*	Archive of Microbiology	2016, 198(7): 653-661
36	Z. Y. Sun, J. Deng, H. Z. Wu, et al	Selecting stable reference genes for Real-time quantitative PCR analysis in *Edwardsiella tarda*	Journal of Microbiology and Biotechnology	2017, 27(1): 112-121
37	代伟伟,麦康森,徐玮,等	复合植物蛋白源替代鱼粉对半滑舌鳎生长、生理生化指标和肠组织结构的影响	中国水产科学	2016, 23(1): 125-137

（续表）

序号	作者	论文名称	发表刊物	年、卷、期、起止页
38	刘经纬，麦康森，徐玮，等	谷氨酰胺对半滑舌鳎稚鱼非特异性免疫相关酶活力和低氧应激后 HIF-1α 表达的影响	水生生物学报	2016，40(4)：736-743
39	王雅慧，王裕玉，麦康森，等	饲料中添加姜黄素对大菱鲆幼鱼生长、体组成及抗氧化酶活力的影响	水产学报	2016，40（9）：1299-1307
40	赵敏，梁萌青，郑珂珂，等	饲料中添加南极磷虾粉对半滑舌鳎（*Cynoglossus semilaevi*）雄性亲鱼繁殖性能及抗氧化功能的影响	渔业科学进展	2016，37（6）：49-55
41	H. G. Xu, L. Cao, Y. Q. Zhang, et al	Dietary arachidonic acid differentially regulates the gonadal steroidogenesis in the marine teleost, tongue sole（*Cynoglossus semilaevis*）, depending on fish gender and maturation stage	Aquaculture	2016，468：378-385
42	K. K. Zhang, K. S. Mai, W. Xu, et al	Effects of dietary arginine and glutamine on growth performance, nonspecific immunity, and disease resistance in relation to arginine catabolism in juvenile turbot（*Scophthalmus maximus* L.）	Aquaculture	2017，468：246-254
43	D. W. Liu, K. S. Mai, Y. J. Zhang, et al	Tumour necrosis factor-α inhibits hepatic lipid deposition through GSK-3β/β-catenin signaling in juvenile turbot（*Scophthalmus maximus* L.）	General and Comparative Endocrinology	2016，228：1-8
44	D. W. Liu, K. S. Mai, Y. J. Zhang, et al	Wnt/β-catenin signaling participates in the regulation of lipogenesis in the liver of juvenile turbot（*Scophthalmus maximus* L.）	Comparative Biochemistry and Physiology Part B: Biochemistry and Molecular Biology	2016，191：155-162
45	胡记东，刘远平，孙爱华，等	X 射线检测海水鱼片中鱼刺	食品科学	2016，20：151-156
46	胡记东，年睿，林洪，等	鲐鱼片中鱼骨刺 X 射线图像不同增强处理技术	中国渔业质量与标准	2016，06：36-41
47	滕瑜，赵丽静，王平，等	大菱鲆熏制预调味技术研究	食品工业	2016，1：133-135
48	滕瑜，王洪军，袁勇，等	大菱鲆鱼排加工工艺研究	中国调味品	2016，9：101-103
49	宁劲松，段元慧，尚德荣，等	前处理方法对 ICP-OES 测定水产品中无机元素的影响	中国渔业质量与标准	2016，3：41-48
50	林洪，李梦哲，米娜莎	我国水产品质量安全监管技术新动向	食品安全质量检测学报	2016，3：1018-1023

（续表）

序号	作者	论文名称	发表刊物	年、卷、期、起止页
51	M. Li, M. Z. Li, H. Lin, et al	Characterization of the novel T4-like *Salmonella enterica* bacteriophage STP4-a and its endolysin	Archives of Virology	2016, 161: 377-384
52	X. D. Wang, H. Lin, L. M. Cao, et al	Isolation, characterization, and identification of proteins interfering with enzyme-linked immunosorbent assay of antibiotics in fish matrix	Food Science and Biotechnology	2016, 25(5): 1265-1273
53	W. Y. Wang, M. Z. Li, H. Lin, et al	The *Vibrio parahaemolyticus*-infecting bacteriophage qdvp001: genome sequence and endolysin with a modular structure	Archives of Virology	2016, 161: 2645-2652
54	黄滨，马腾，刘宝良，等	不同浓度臭氧对循环水养殖系统生物膜活性及其净化效能的影响	渔业科学进展	2016, 37(3): 143-147
55	R. Jia, B. Liu, C. Han, et al	Effects of ammonia exposure on stress and immune response in juvenile turbot (*Scophthalmus maximus*)	Aquaculture Research	2017, 48(6): 3149-3162
56	B. Liu, R. Jia, C. Han, et al	Effects of stocking density on antioxidant status, metabolism and immune response in juvenile turbot (*Scophthalmus maximus*)	Comparative Biochemistry and Physiology C-toxicology & Pharmacology	2016, 190: 1-8
57	R. Jia, B. Liu, C. Han, et al	influence of stocking density on growth performance, antioxidant status, and physiological response of juvenile turbot, *Scophthalmus maximu*, Reared in land-based recirculating aquaculture system	Journal of the World Aquaculture Society	2016, 47(4): 589-599
58	L. Zeng, B. Liu, C. W. Wu, et al	Molecular characterization and expression analysis of AMPK a subunit isoform genes from *Scophthalmus maximus* responding to salinity stress	Fish Physiology and Biochemistry	2016, 42: 1595-1607
59	Y. Liu, B. Liu, J. Lei, et al	Numerical simulation of the hydrodynamics within octagonal tanks in recirculating aquaculture systems	Chinese Journal of Oceanology and Limnology	2017, 35(4): 912-920
60	R. Jia, B. Liu, W. R. Feng, et al	Stress and immune responses in skin of turbot (*Scophthalmus maximus*) under different stocking densities	Fish & Shellfish Immunology	2016, 55: 131-139
61	R. Jia, B. Liu, C. Han, et al	The physiological performance and immune response of juvenile turbot (*Scophthalmus maximus*) to nitrite exposure	Comparative Biochemistry and Physiology Part C	2016, 181/182: 40-46

（续表）

序号	作者	论文名称	发表刊物	年、卷、期、起止页
62	杨正勇，李佳莹，王春晓	鲆鲽类产业发展及市场动态简报 2016年•第1期	水产前沿	2016，5：101-102
63	王微波，杨正勇，李佳莹	鲆鲽类养殖不同产业链模式的经济效益比较研究	上海海洋大学学报	2016，25（6）：954-960
64	杨正勇，侯熙格	食品可追溯体系及其主体行为的演化博弈分析	山东社会科学	2016，4：132-137
65	杨正勇	渔业资源与环境经济学：发展反思与逻辑框架	教育教学论坛	2016，49：64-67
66	曹自强，杨正勇，李佳莹，等	中国大菱鲆养殖产业集群的影响因素分析	上海农业学报	2016，32（4）：147-153
67	侯熙格，杨正勇	中国大菱鲆养殖业集聚演变特征及成因分析	山东农业农业大学学报（自然科学版）	2016，47（2）：304-310
68	王微波，杨正勇，李佳莹	鲆鲽类养殖不同产业链模式的经济效益比较研究	上海海洋大学学报	2016，6：954-960
69	S. Li，Z. Y. Yang，D. Nadolnyak，et al	Economic impacts of climate change: profitability of freshwater aquaculture in China	Aquaculture Research	2016，47：1537-1548
70	Z. Y. Yang，S. Li，B. Chen，et al	China' s aquatic product processing industry：Policy evolution and economic performance	Trends in Food Science & Technology	2016，58：149-154
71	韩现芹，陈春秀，贾磊	汉沽养殖区海水重金属含量分布特征及潜在生态危险评价	安徽农业科学	2016，44（20）：66-68
72	肖广侠，徐文远，贾磊，等	悬浮物对褐牙鲆肌肉抗氧化酶活性及相关基因表达的影响	海洋科学进展	2016，34（4）：542-552
73	何忠伟，宫春光，殷蕊，等	水中Mn（Ⅱ）对大菱鲆幼鱼生长及碱性磷酸酶和超氧化物歧化酶活性的影响	水产科学	2016，35（4）：364-369

附录2　鲆鲽类产业技术体系2016年进行的产业技术宣传与培训

序号	培训班/现场会/调研/技术咨询等主题名称	时间	地点	培训人数				主办(参加)试验站/岗位
				培训基层技术人员	培训养殖大户	培训渔民	发放培训资料	
1	养殖大菱鲆使用药物调研及指导	2016.1.14	山东日照	30	26	—	—	苗种繁育岗位,山东、日照综合试验站
2	辽宁省东港宋利官养殖场疾病技术咨询	2016.1.16	辽宁东港	—	—	—	—	全雌苗种生产岗位,北戴河综合试验站
3	大菱鲆健康养殖+品牌文化+互联网	2016.1.18—19	辽宁兴城	200	—	—	—	良种选育岗位
4	大菱鲆腹水病疫苗生产免疫接种操作培训	2016.1.20	山东烟台	6	—	—	6	疾病防控岗位,烟台综合试验站
5	鲆鲽类产业需求调研	2016.1.25	河北秦皇岛	—	—	—	—	全雌苗种生产岗位,北戴河综合试验站
6	极寒天气情况对鲆鲽类养殖生产产生影响调研及指导	2016.1.25	山东荣成	15	9	—	—	苗种繁育岗位
7	"十三五"鲆鲽类产业技术需求调研	2016.1.26—27	天津市滨海新区大港、汉沽、塘沽	—	—	—	30	天津综合试验站,工厂养殖设施设备岗位
8	鲆鲽类产业需求调研	2016.1.26	河北省唐山市丰南区、昌黎县、黄骅市、滦南县	—	—	—	—	全雌苗种生产岗位,北戴河综合试验站 河北综合试验站
9	半滑舌鳎产业需求调研	2016.1.27	河北唐山曹妃甸	—	—	—	—	全雌苗种生产岗位,北戴河综合试验站
10	鲆鲽类产业需求调研	2016.1.28	辽宁大连	—	—	—	—	全雌苗种生产岗位,北戴河综合试验站 河北综合试验
11	日照鲆鲽类养殖产业调研	2016.1.28—29	山东日照	20	—	—	—	苗种繁育岗位,山东综合试验站
12	昆明水产品市场调查	2016.2	云南昆明	—	—	—	—	产业经济岗位

（续表）

序号	培训班/现场会/调研/技术咨询等主题名称	时间	地点	培训人数				主办（参加）试验站/岗位
				培训基层技术人员	培训养殖大户	培训渔民	发放培训资料	
13	大菱鲆和牙鲆市场价格调研	2016. 2. 24	河北昌黎	—	—	—	—	全雌苗种生产岗位，北戴河综合试验站
14	唐山曹妃甸区半滑舌鳎养殖园区调研	2016. 3. 3	河北唐山曹妃甸	—	—	—	—	全雌苗种生产岗位，北戴河综合试验站
15	鲽鱼产品的加工工艺、加工过程质量控制以及对鲆鲽鱼类文化的推广培训	2016. 4. 20	山东日照	80	—	—	90	日照综合试验站
16	劳动节前大菱鲆市场价格调研	2016. 4. 27	河北昌黎	—	—	—	—	全雌苗种生产岗位，北戴河综合试验站
17	5月份珠海、长沙、株洲等地水产品消费市场调查	2016. 5	广东、湖北	—	—	—	—	产业经济岗位
18	牙鲆与河鲀混养情况调研	2016. 5. 1	河北唐山滦南	—	—	—	—	全雌苗种生产岗位，北戴河综合试验站
19	为上海海洋大学建立4个雌核发育家系	2016. 5. 7	河北北戴河	—	—	—	—	全雌苗种生产岗位，北戴河综合试验站
20	为黄海锶标记放流牙鲆耳石课题提供受精卵	2016. 5. 11	河北北戴河	—	—	—	—	全雌苗种生产岗位，北戴河综合试验站
21	快检试剂盒现场培训	2016. 5. 11	辽宁省葫芦岛市	10	—	—	40	葫芦岛综合试验站
22	2016年水产科技活动周启动仪式	2016. 5. 13	福建厦门	96	—	—	96	池塘养殖工程岗位
23	大连、兴城面对面技术服务培训	2016. 5. 20—2016. 12. 8	辽宁大连、兴城	—	—	50	60	辽宁综合试验站，葫芦岛综合试验站
24	北京、武汉鲆鲽类产品市场调查	2016. 6	北京、武汉	—	—	—	—	产业经济岗位
25	省农业新品种选育重大专项——高产抗逆大黄鱼新品系培育	2016. 6. 2	浙江舟山	40	—	—	—	良种选育岗位

(续表)

序号	培训班/现场会/调研/技术咨询等主题名称	时间	地点	培训人数				主办(参加)试验站/岗位
				培训基层技术人员	培训养殖大户	培训渔民	发放培训资料	
26	绥中大菱鲆养殖技术培训	2016.6.8	辽宁绥中	10	2	50	80	葫芦岛综合试验站
27	鲆鲽类产业技术体系种苗生产培训与研讨会	2016.6.16	山东威海	15	5	30	100	苗种繁育岗位,青岛、山东、葫芦岛综合试验站
28	国家鲆鲽类产业技术体系新型健康养殖技术培训暨研讨会	2016.6.17	山东烟台	60	46	—	90	疾病防控岗位,专用养殖网箱岗位,烟台、山东综合试验站,苗种繁育岗位,高效养殖模式岗位
29	为河北蓝翼海洋生物科技有限公司提供牙鲆苗种和技术	2016.6.23	河北北戴河	—	—	—	—	全雌苗种生产岗位,北戴河综合试验站
30	企业技术培训	2016.6.29	福建宁德	36	—	—	36	池塘养殖工程岗位
31	山东鲆鲽类养殖生产情况调查	2016.7	山东	—	—	—	—	产业经济岗位
32	上海崇明、浙江温岭、三门、舟山等地水产品市场及养殖生产情况调查	2016.7—8	上海、浙江	—	—	—	—	产业经济岗位
33	上海、烟台、大连、昆明等地开展了“中国消费者鱼类消费意愿”问卷调查	2016.7—8	上海、山东、辽宁、云南	—	—	—	—	产业经济岗位
34	葫芦岛市大菱鲆质量安全培训	2016.7.19	辽宁兴城	—	—	60	100	葫芦岛综合试验站
35	国家鲆鲽类产业技术体系与省级水产创新团队加工与质量安全岗位建设规划研讨会	2016.7.19	山东日照	25	—	—	50	日照综合实验站
36	2016年莱州市新型渔民培训班	2016.7.21	山东莱州	60	40	—	100	莱州综合试验站
37	海水鱼类工厂化养殖病害应急防控与生物安全管理(网络授课)	2016.7.27	《水产前沿》中国水产频道	500	—	—	500	疾控岗位主办

（续表）

序号	培训班/现场会/调研/技术咨询等主题名称	时间	地点	培训人数				主办(参加)试验站/岗位
				培训基层技术人员	培训养殖大户	培训渔民	发放培训资料	
38	辽宁省兴城市鲆鲽类产品可追溯技术推广情况调查	2016. 8	辽宁兴城	—	—	—	—	产业经济岗位
39	意大利罗马水产品消费情况调查	2016. 8	意大利	—	—	—	—	产业经济岗位
40	山东半岛鲆鲽类常规苗种培育、标粗苗种生产情况调研	2016. 8. 3—5	山东威海、烟台、日照	20	50	—	40	苗种繁育岗位，山东省鱼类创新团队
41	鲆鲽类产业发展研讨会暨中国水产流通与加工协会大菱鲆分会成立大会	2016. 8. 17	山东青岛	100	—	—	—	体系各岗位和试验站
42	福建水产品消费市场调查	2016. 9	福建	—	—	—	—	产业经济岗位
43	葫芦岛、兴城鲆鲽类养殖生产情况调查	2016. 9	辽宁	—	—	—	—	产业经济岗位
44	中国水产学会渔文化分会成立大会	2016. 9. 18	浙江舟山	—	—	—	—	产业经济岗位
45	鲆鲽类产业技术培训与交流会	2016. 9. 20	辽宁兴城	60	10	—	360	葫芦岛综合试验站，辽宁综合试验站，疾病防控岗位
46	大菱鲆价格调研	2016. 9. 21	辽宁兴城	—	—	—	—	全雌苗种生产岗位，北戴河综合试验站
47	鲆鲽类产品市场营销研讨交流会	2016. 9. 24	烟台	63	15	30	65	产业经济岗位，疾病防控岗位，专用养殖网箱岗位，池塘养殖工程岗位，良种选育岗位，高效养殖模式岗位，加工与质量控制岗位，烟台、山东、葫芦岛、日照综合试验站
48	上海青浦水产品市场及养殖生产情况调查	2016. 10	上海	—	—	—	—	产业经济岗位

(续表)

序号	培训班/现场会/调研/技术咨询等主题名称	时间	地点	培训人数				主办(参加)试验站/岗位
				培训基层技术人员	培训养殖大户	培训渔民	发放培训资料	
49	东港"北鲆1号"和"北鲆2号"池塘养殖情况调研	2016.10.7	辽宁东港	—	—	—	—	全雌苗种生产岗位,北戴河综合试验站
50	中国林牧渔业经济学会渔业经济专业委员会2016年年会暨学术研讨会	2016.10.22	浙江舟山	—	80	—	—	产业经济岗位
51	水产品质量安全知识	2016.10.27	辽宁兴城	5	—	60	100	葫芦岛综合试验站
52	大连、兴城鲆鲽类养殖技术状况调查	2016.11	辽宁大连、兴城	—	—	—	—	产业经济岗位
53	新型渔民培训	2016.11.1	河北昌邑	130	—	—	130	池塘养殖工程岗位
54	东港"北鲆1号"和"北鲆2号"养殖及销售情况调研	2016.11.2	辽宁东港	—	—	—	—	全雌苗种生产岗位,北戴河综合试验站
55	中国国际渔业博览会	2016.11.2	山东青岛	—	—	—	—	产业经济岗位
56	威海市全市现代海洋与渔业专题培训班	2016.11.4	山东威海	200	—	—	—	国家环保局海洋司,威海市海洋与渔业局,产业经济岗位
57	疫苗免疫使用与操作技术现场培训	2016.11.5—12.8	辽宁大连	—	2	—	10	辽宁综合试验站,疾病防控岗位
58	"北鲆1号"和"北鲆2号"工厂化接力池塘养殖情况调研	2016.11.6	河北昌黎	—	—	—	—	全雌苗种生产岗位,北戴河综合试验站
59	"北鲆1号"和"北鲆2号"工厂化养殖情况调研	2016.11.7	河北唐山	—	—	—	—	全雌苗种生产岗位,北戴河综合试验站
60	鲆鲽类养殖技术培训班	2016.11.10	河北昌黎	64	14	18	200	全雌岗位,北戴河、河北试验站,苗种繁育岗位,疾病防控岗位
61	莱州市基层渔业技术推广体系改革与建设项目培训班	2016.11.10	山东莱州	160	—	—	160	池塘养殖工程岗位

（续表）

序号	培训班/现场会/调研/技术咨询等主题名称	时间	地点	培训人数				主办(参加)试验站/岗位
				培训基层技术人员	培训养殖大户	培训渔民	发放培训资料	
62	鲆鲽工厂化养殖疫病防控现场咨询与培训会	2016.11.11	天津汉沽	20	20	—	60	天津综合试验站，疫病防控岗位
63	中国林牧渔业经济学会第四届会员代表大会暨中国林牧渔业供给侧结构改革研讨会	2016.11.12—13	北京	—	60	—	—	产业经济岗位
64	大菱鲆养殖技术培训	2016.11.23	辽宁兴城	—	30	—	30	葫芦岛综合试验站
65	循环水养殖半滑舌鳎情况调研	2016.11.24	河北昌黎	—	—	—	—	全雌苗种生产岗位，北戴河综合试验站
66	循环水养殖技术培训班	2016.11.26	辽宁兴城	20	—	—	20	葫芦岛综合试验站
67	兴城、大连养殖技术调研	2016.11.29—30	辽宁大连、兴城	—	/3	—	/9	辽宁、葫芦岛综合试验站，产业经济岗位
68	日本水产品消费市场调查	2016.12	日本	—	—	—	—	产业经济岗位
69	美国水产品消费情况调查	2016.12	美国	—	—	—	—	产业经济岗位
70	大菱鲆疫苗生产免疫接种操作培训与示范推广	2016.12.7	天津汉沽	10	10	—	10	天津综合试验站，疫病防控岗位
71	大菱鲆疫苗生产免疫接种操作培训与示范推广	2016.12.9	辽宁瓦房店	2	2	—	5	疫病防控岗位
72	2016年日照国际海洋城新型渔民培训	2016.12.14	山东日照	100	—	—	100	池塘养殖工程岗位
73	山东省海水良种培育及高效养殖技术高级研修班	2015.12.22	山东莱州	45	60	30	105	莱州综合试验站，专用养殖网箱岗位
74	全雌牙鲆养殖情况跟踪调查	2016全年	辽宁东港、河北昌黎、河北乐亭、河北曹妃甸	—	—	—	—	全雌苗种生产岗位，北戴河综合试验站

(续表)

序号	培训班/现场会/调研/技术咨询等主题名称	时间	地点	培训人数				主办(参加)试验站/岗位
				培训基层技术人员	培训养殖大户	培训渔民	发放培训资料	
75	牙鲆养殖技术方面、病害防治、苗种质量、饲料使用厂家等技术咨询服务	2016 全年	养殖现场(辽宁东港、河北昌黎、河北乐亭、河北曹妃甸等)、通过电话、微信等通讯手段咨询	—	—	—	—	全雌苗种生产岗位,北戴河综合试验站
76	第二届"七好大菱鲆黄金养殖模式技术经济论坛"	2016. 5. 11	山东昌邑、莱州、日照、龙口、胶南、海阳、乳山、辽宁绥中、兴城以及河北省昌黎地区。	3 500	—	—	15 000	青岛七好生物科技股份有限公司
77	青岛综合试验站及黄岛区鲆鲽类养殖实地采访调研(陪同农业部科教司邀请的中央媒体记者团)	2016. 4. 7	青岛	—	—	—	—	鲆鲽类体系研发中心
78	葫芦岛、烟台主产区大菱鲆质量安全状况和产业发展现状专题调研(陪同农业部渔业渔政管理局科技与质量安全处)	2016. 4. 27—29	葫芦岛、烟台	—	—	—	—	鲆鲽类体系研发中心
79	天津天世农养殖有限公司臭氧添加技术咨询	2016. 10	天津	2				工厂化养殖设施设备岗位
80	全雌牙鲆盾纤毛虫病防治技术指导	2016. 07. 23—2016. 08. 03	昌黎县启民养殖场	—	—	—	—	河北综合试验站
81	暴雨灾害后鲆鲽类养殖技术指导及资料发放	2016. 07. 23—27	昌黎、丰南	—	—	—	—	河北综合试验站

（续表）

序号	培训班/现场会/调研/技术咨询等主题名称	时间	地点	培训人数				主办(参加)试验站/岗位
				培训基层技术人员	培训养殖大户	培训渔民	发放培训资料	
82	组织辽宁、福建、广西等省级科研院所领导20余人进行半滑舌鳎循环水工厂化养殖技术观摩	2016. 8. 17	昌黎粮丰公司	—	—	—	—	河北综合试验站
合计				5 704	484	328	17 782	